THE EUROPEAN ENVIRONMENT

STATE AND OUTLOOK 2005

Cover design: EEA
Layout: EEA

Photos © Monica Rzeszot, 2005
Pages 15, 19, 21, 23, 24, 35, 91, 111, 181,
250, 254, 267, 283, 307, 319, 411, 524, 528.

Legal notice
The contents of this publication do not necessarily reflect the official opinions of the European Commission or other institutions of the European Communities. Neither the European Environment Agency nor any person or company acting on behalf of the Agency is responsible for the use that may be made of the information contained in this report.

All rights reserved
No part of this publication may be reproduced in any form or by any means electronic or mechanical, including photocopying, recording or by any information storage retrieval system, without the permission in writing from the copyright holder. For translation or reproduction rights please contact EEA (address information below).

Citation
European Environment Agency, 2005. *The European environment — State and outlook 2005.* Copenhagen

Information about the European Union is available on the Internet. It can be accessed through the Europa server (http://europa.eu.int).

Luxembourg: Office for Official Publications of the European Communities, 2005

ISBN 92-9167-776-0

© EEA, Copenhagen 2005

Environmental production
This publication is printed according to high environmental standards.

Printed by Scanprint a/s
— Environment Certificate: ISO 14001
— Quality Certificate: ISO 9001: 2000
— EMAS registered — licence no. DK- S-000015
— Approved for printing with the Nordic Swan environmental label, licence no. 541 055

Paper
— Multicopy original — woodfree mat fine paper, TCF, ISO 9706
— The Nordic Swan label

Printed in Denmark

European Environment Agency
Kongens Nytorv 6
1050 Copenhagen K
Denmark
Tel.: +45 33 36 71 00
Fax: +45 33 36 71 99
Web: www.eea.eu.int
Enquiries: www.eea.eu.int/enquiries

THE EUROPEAN ENVIRONMENT

STATE AND OUTLOOK 2005

List of contents

Foreword	8
Introduction	10
Executive summary	16

PART A | Integrated assessment

Setting the scene
    1 Environment and quality of life 28
    2 The changing face of Europe 36
Atmospheric environment
    3 Climate change 62
    4 Air pollution and health 92
Aquatic environment
    5 Freshwaters 112
    6 Marine and coastal environment 132
Terrestrial environment
    7 Soil 168
    8 Biodiversity 182
Integration
    9 Environment and economic sectors 216
    10 Looking ahead 232

PART B | Core set of indicators

Setting the scene 255
Air pollution and ozone depletion
    01 Emissions of acidifying substances 256
    02 Emissions of ozone precursors 260
    03 Emissions of primary particles and secondary particulate precursors 264
    04 Exceedance of air quality limit values in urban areas 268
    05 Exposure of ecosystems to acidification, eutrophication and ozone 272
    06 Production and consumption of ozone depleting substances 276
Biodiversity
    07 Threatened and protected species 280
    08 Designated areas 284
    09 Species diversity 288
Climate change
    10 Greenhouse gas emissions and removals 292
    11 Projections of greenhouse gas emissions and removals 296
    12 Global and European temperature 300
    13 Atmospheric greenhouse gas concentrations 304

Terrestrial
    14 Land take.. 308
    15 Progress in management of contaminated sites... 312

Waste
    16 Municipal waste generation.. 316
    17 Generation and recycling of packaging waste.. 320

Water
    18 Use of freshwater resources... 324
    19 Oxygen consuming substances in rivers... 328
    20 Nutrients in freshwater.. 332
    21 Nutrients in transitional, coastal and marine waters...................................... 336
    22 Bathing water quality.. 340
    23 Chlorophyll in transitional, coastal and marine waters.................................. 344
    24 Urban wastewater treatment.. 348

Agriculture
    25 Gross nutrient balance... 352
    26 Area under organic farming... 356

Energy
    27 Final energy consumption by sector.. 360
    28 Total energy intensity.. 364
    29 Total energy consumption by fuel... 368
    30 Renewable energy consumption... 372
    31 Renewable electricity... 376

Fisheries
    32 Status of marine fish stocks.. 380
    33 Aquaculture production... 384
    34 Fishing fleet capacity.. 388

Transport
    35 Passenger transport demand.. 392
    36 Freight transport demand.. 396
    37 Use of cleaner and alternative fuels... 400

PART C | Country analysis

Setting the scene — main results... 408
EEA scorecard... 412
Thematic assessment
    Greenhouse gas emissions.. 414
    Total energy consumption.. 416
    Renewable electricity... 418
    Emissions of acidifying substances.. 420
    Emissions of ozone precursors.. 422
    Freight transport demand.. 424

PART C | Country analysis

Area under organic farming	426
Municipal waste generation	428
Use of freshwater resources	430
Country perspectives	**432**
Austria	434
Belgium	436
Bulgaria	438
Cyprus	440
Czech Republic	442
Denmark	444
Estonia	446
Finland	448
France	450
Germany	452
Greece	454
Hungary	456
Iceland	458
Ireland	460
Italy	462
Latvia	464
Liechtenstein	466
Lithuania	468
Luxembourg	470
Malta	472
The Netherlands	474
Norway	476
Poland	478
Portugal	480
Romania	482
Slovak Republic	484
Slovenia	486
Spain	488
Sweden	490
Switzerland	492
Turkey	494
United Kingdom	496
Methodology and main decision points	**498**
Ranking graphs	**508**

PART D | Bibliography

Introduction 529
Previous state-of-the-environment reports 530
Signals reports 531

Reports 2000–2005, by theme:
- Air pollution 532
- Climate change 534
- Terrestrial environment and biodiversity 536
- Waste and material flows 537
- Water 539
- Agriculture 542
- Energy 543
- Transport 544
- Other issues 546
- Policy measures and instruments 549
- Eionet development and information systems, by theme 552

Condensed list of reports 2000–2005, by series:
- Environmental assessment reports (2000–2003) 566
- Environmental issue reports (2000–2004) 566
- EEA reports (2004–2005) 567
- EEA briefings (2003–2005) 567
- Topic reports (2000–2003) 567
- Technical reports (2000–2003) 568
- EEA technical reports (2004–2005) 570

Foreword

This is the third state and outlook report on the European environment produced by the European Environment Agency (EEA) since 1994. Looking back, the last report, published in 1999 concluded that, despite 25 years of Community environmental policy, environmental quality in the European Union (EU) was mixed and that the unsustainable development of some key economic sectors was the major barrier to further improvements. That remains the EEA's key conclusion, despite significant progress on some issues demonstrating that environmental policy works. Were we to fast-forward to the year 2010, it would be my strong hope that in its next state and outlook report, the EEA would be able to report significant environmental improvements, not least as a result of reversing unsustainable trends in sectors such as energy, agriculture and transport.

Much has of course changed in Europe since 1999, most notably through the biggest ever enlargement of the EU. Despite this and other considerable successes, the EU now recognises that it has to do more to convey to Europeans that it understands their concerns and that it has credible strategies in place to address them. One response has been to reinvigorate the Lisbon agenda initially launched in 2000 with the target of creating new and better jobs through higher economic growth. The EU has set itself the deadline of 2010 for delivering on this agenda and against this backdrop environmental initiatives are being rigorously scrutinised. At the same time, opinion polls show that the European public believes that environment protection policies are an incentive for innovation and not an obstacle to economic performance. The evidence in this report shows that they are right.

The years since 1999 have also witnessed the adoption of the EU's sustainable development strategy. At the UN World Summit on Sustainable Development in the summer of 2002, the EU showed global leadership by setting out its broad, long-term vision for the future. But long-term objectives must not mean postponing action now. In formulating policy today, Europe has an obligation to look beyond 2010 and beyond its own borders. As this report shows, the international stakes are high. Europe cannot continue down the path of achieving its short-term objectives by impacting disproportionately on the rest of the world's environment through its ecological footprint.

Tackling clearly unsustainable trends in Europe will require real integration of environmental objectives across policy areas such as energy, transport, agriculture, industry and spatial planning. Consumers must also be given the information and incentives to change the way in which their households and lifestyles impact upon their local — and global — environments. It is not easy to bring about such shifts in behaviour but many of the environmental improvements that we need in the coming years can only be brought about through such changes. At the same time, the EU must remain vigilant to ensure that the policy measures already in place are fully implemented and properly enforced.

The European public's intuitive understanding that environmental protection and economic growth can go hand in hand is confirmed by this report. The same applies to the widely held view that prevention works better than cure. Many environmental problems stem from incorrect pricing of what we consume at the time we consume it. The challenge for European policy-makers is to ensure that the real costs of pollution and resource inefficiency are internalised into the prices of products and services, as opposed to later on at the end of the chain in the form of a pollution clean-up bill, damaged health or diminished ecosystems. The tools of the market can provide incentives to bring about some of the very necessary behavioural changes. And as the EEA has shown in some of its earlier reports, the cost of inaction can be many times the cost of sensible preventive measures.

Some phenomena are more difficult to cost in conventional market terms. As societies we have come to rely on a range of ecosystem services upon which we have not traditionally placed monetary values. Healthy ecosystems deliver an abundance of free and life-sustaining services such as productive land and soil, forest products, or other, sometimes less well recognised services, such as watershed protection and water filtration, flood protection or carbon storage. Our economies — like our societies — are

Foreword

dependent upon these often hidden services. The seas and coasts around Europe provide one example of a vital resource upon which many millions of people depend, both economically and culturally. And yet, the current deterioration of our marine ecosystems, as seen through the depletion of fish stocks, puts at risk the services they provide. Similarly, climate change is already altering ecosystems in such a way that we must ask ourselves about their capacity to adapt. While ecosystems are extremely resilient and typically only change rather slowly, there are growing signs from across the world that they can reach a critical point where they undergo 'greenlash' — or a sudden flip into a new structure.

We need to keep a strong time perspective uppermost in our minds when assessing Europe's changing environment. Changes can often be subtle and long-term and as such they may require sustained policy responses over a long period. At the same time, by taking the longer view, we may see that slow changes can be dramatic and that there is cause to sound the alarm now. For example, the Millennium Ecosystem Assessment has recently shown that humans have changed ecosystems more rapidly and extensively in the last 50 years than in any other period of human history.

For its part, the EEA is responding to the need for long-term, integrated analyses of environmental issues that take account of the global context. We are working with others on environmental accounting methods that link environmental pressures to economic sector activities in ways that are consistent with national economic accounts. In 2006 we will start to review with our partners, how we produce environmental outlooks with a view to making future modelling work more policy relevant. The EEA will also extend its work on scenarios development to provide decision-makers with new tools and insights into possible environmental futures and their associated socio-economic developments. These could provide a more robust basis for longer-term policy considerations.

By the time of the fourth state and outlook report on the European environment in 2010, the EEA intends to have played its part in helping to reverse the unsustainable trends that have been consistently identified in our work since 1994. We hope that the present report can help show governments and the public what is at stake in needing to move towards the more eco-efficient and equitable economic activities upon which Europe's future growth depends.

In conclusion, I would like to take this opportunity to thank Agency staff, members of the European environment information and observation network (Eionet), other EEA stakeholders, and external consultants who contributed to the drafting, editing, review and production of the report. In addition, I would like to thank everyone who provided data and information, especially Eionet, the European Commission services, international organisations and the secretariats of international conventions.

Professor Jacqueline McGlade
Executive Director

Introduction

1. Mandate and purpose

The European Environment Agency (EEA) together with its European Environment Information and Observation Network (Eionet) began working in 1994 to provide the EU Member States, other EEA member countries, and the European Commission with objective, reliable and comparable information on the state, pressures and sensitivities of the environment in the European Union (EU) and its surroundings. Specifically, under its founding Regulation (EEC/1210/90), the EEA is required to 'publish a report on the state of, trends in and prospects for the environment every five years, supplemented by indicator reports focusing upon specific issues'.

The present report *The European environment — State and outlook 2005* is the third such report produced since the EEA was established. The reports are intended to support strategic environmental programming in the EU and beyond. Previous reports were *Environment in the European Union 1995: report for the review of the 5th environment action programme*, published in 1995 and *Environment in the European Union at the turn of the century*, published in 1999. The latter contributed to the global assessment of the 5th environment action programme, a process that acted as a prelude to the development of the 6th environment action programme (6th EAP), adopted in 2002. This 2005 state and outlook report provides timely input to strategic policy review processes, the most important of which is the mid-term review of the 6th EAP scheduled for completion in 2006.

According to its founding regulation, the EEA's five year reports are intended to describe past, present and future perspectives for the environment. The overall purpose is to provide an integrated assessment that forms the basis for understanding what are the main challenges facing Europe's environment today and in the future, within the context of changing economic and social dynamics both in Europe and across the world.

The rapidly changing economic and demographic landscapes of Europe and the world, together with the shift in the source of environmental problems in Europe from production to consumption patterns in the last decades are two of the main socio-economic changes that underpin the environmental assessment presented in this report. In addition, the increasing relevance of international trade, as well as the growing competition for resources between developed economies such as Europe, and emerging economies such as China, India and Brazil, are two key developments that influence both European and global environments as a result of the external footprint of Europe's activities.

It is in the context of these main socio-economic changes that the report addresses the resulting environmental and human health impacts from increasing urbanisation and rural land abandonment, climate change, air pollution and other chemical exposures, depletion of natural resources such as water, soil and biodiversity and the disturbance caused to land-based and marine ecosystems.

These past and future trends for the environment and society are considered alongside the wide range of policy measures adopted at the national, EU and global levels. The report describes the range and role of policy instruments available and shows, in a more limited way, how effective these policies have been in meeting their objectives. Regulations and market measures are given particular attention in view of their roles in reducing environmental pressures through encouraging developments in environmental technologies and behavioural change.

The EEA uses its driving forces — pressures — state — impact — responses (DPSIR) assessment framework to guide its work on integrated assessment. The DPSIR framework provides a basis for analysing the interrelated factors that impact the environment, including societal driving forces such as production and consumption patterns, their pressures on the environment, the resulting changes in the state of the environment, its impacts on ecosystems and health, and the role and effectiveness of policy measures in managing these interlinked dynamics.

Data and information used in the report for assessing past trends have been gathered through Eionet, a partnership of the EEA and its member and participating countries. It consists of the EEA itself, five European Topic Centres, and a network of around 1 000 experts from 31 member countries and 6 collaborating countries, representing over 300 national environment agencies and other bodies dealing with environmental information. Other key data sources used in the report include the European Commission services, especially DG Environment, DG Eurostat and DG Joint Research Centre, and the secretariats of international conventions. The EEA has also worked in cooperation with these and other partners to produce information on outlooks for the environment using established modelling and scenarios tools to assure consistency with other processes.

2. Report structure and target audiences

This report is divided into four parts: Part A: an integrated assessment; Part B: core set of indicators; Part C: country analysis; and, Part D: bibliography of EEA publications since 2000.

Part A
Part A provides an integrated assessment at the European level across the main environmental challenges that Europe faces and how these are influenced by socio-economic activities at home and more widely across the world. It considers both past trends and future prospects. It is structured in five main sections, each containing two chapters.

The first section, called 'setting the scene', has chapters addressing environment's contribution to our quality of life and how the face of Europe has changed in recent times through its changing landscape patterns. The first chapter looks at how citizens perceive the role of the environment alongside social and economic aspects in improving their overall welfare. The chapter on landscapes combines an assessment of the potential of Europe's land area to continue providing ecological goods and services in the face of changing pressures, and an explanation of the main policy instruments at the European level that are influencing change.

The section on atmospheric environment addresses climate change, the foremost environmental challenge facing Europe and the world today, and air pollution and its impacts on people's health. In both chapters, particular attention is given to the challenges posed by pollution from households, energy supply and transport, and the sensitivities in dealing with these through policy responses aimed at behavioural change. Particular attention is also given to the future challenges and the costs of action/inaction in the face of uncertainty.

The section on aquatic environment has chapters covering freshwater, and marine and coastal environments. The freshwater chapter focuses on the main pollution sources and trends — with less focus on ecological and ecosystem aspects because of lack of data — and on how managing freshwaters will continue to be a long-term and costly endeavour for Europe. The marine and coastal chapter focuses much more on ecosystem aspects and especially on the important and fragile nature of coasts and oceans in the context of climate change.

The terrestrial environment section includes chapters on soil and biodiversity. The soil chapter reflects the present paucity of data and analysis for this area. Nevertheless, it provides some useful insights as well as a clear basis for understanding the challenges ahead for Europe on soil monitoring and modelling. The biodiversity chapter provides a comprehensive analysis of terrestrial ecosystems and species and of how Europe influences the use of natural resources globally as shown by its ecological footprint.

The final section of chapters on integration assesses the environmental impacts of main economic sectors — agriculture, transport, energy and households — in the context of improving eco-efficiency. It concludes with some reflections on where priorities could lie for future action through improved policy design and coherence, better governance structures and investments in eco-innovation that could together deliver cost-effective

Introduction

improvements to Europe's environment, economy and overall quality of life in the face of changing demographics.

The assessment for Part A is written in a more accessible style and as such is targeted to an informed political and public audience working mainly outside the environment sphere but for whom environmental concerns play a role in their day-to-day activities and decision-making. Examples include politicians and policy-makers working in the economic sectors that have most influence on the environment — agriculture, transport, industry, tourism and trade; the financial community, where environmental issues are of increasing relevance to decisions around insurance risk and investment choices; the research and development community, which would be the source of many of the innovations that could contribute to reducing future environmental pressures and impacts; and the informed public. All of these groups have an interest in how Europe's environment is developing, from the local through to the global level, and in how environmentally motivated integrated approaches can help improve European environments, economies and societies.

Part B

Part B of the report presents a first detailed assessment at the European level that uses the 37 EEA/Eionet core set of indicators which were agreed with EEA's stakeholders in March 2004. The core set was selected to provide a stable basis for indicator-based reporting by EEA and others in Europe (e.g. the European Commission) and to provide a focus for prioritising improvements in data quality.

The core set includes selected indicators relevant to the main environmental problems — air pollution and ozone depletion, climate change, biodiversity, waste, terrestrial environment and water — as well as main economic sectors — agriculture, transport, energy and fisheries. Other relevant priority areas (chemicals, noise, material flows, industry and household sectors) have not yet been included in the core set, which reflects the lack of sufficient data and methodological development. These areas will be the main focus for future work on the core set.

All of the indicators in the core set are relevant to the assessments presented in Part A of the report. The bibliography section at the end of each chapter in Part A provides a check-list of relevant indicators. While there is some repetition in the use of analysis and graphs from Part B into Part A, in the vast majority of cases every effort has been made to avoid such overlaps. The main exception concerns the soil chapter which has drawn extensively on the core set because of the relative paucity of good quality data and analysis for this issue.

Each indicator is presented in a standard four-page template that includes information on policy questions and messages, trends assessment, data quality and methodological developments. The four pages are summaries of more detailed indicator profiles that are available on the EEA website.

The assessments for each indicator are written in a more technical language and as such are targeted at a more informed audience working on, or interested in, environmental issues, or for those who want to have more in-depth information about these issues after reading relevant chapters of Part A.

Part C

Part C of the report provides a more detailed country level analysis of progress on environmental issues using a scorecard based on nine indicators from the 37 in the core set. Since the scorecard aims to give insights into progress with environmental performance, the nine indicators relate to points that policy can affect and on which policy is targeted. Thus, most of the indicators either have specific policy targets associated with them that allow analysis of progress towards these targets, or aspirational targets that allow for a similar though less definitive analysis. These indicators also have data available at the country level with trends covering a sufficient number of years to enable robust analysis of change.

The nine indicators are relevant to main policy priorities in the environment and in the economic sectors that have most impact. They are: greenhouse gas emissions and removals, total energy consumption (energy intensity), share of renewables in electricity, emissions of acidifying substances, emissions of ozone precursors, freight transport demand, share of organic farming, municipal waste generation and use of freshwater resources.

Part C goes beyond a standard indicator-by-indicator assessment by providing a composite scorecard of results across the nine indicators. The scorecard also acts as a communication tool by bringing together information in one relatively easy-to-understand format and presentation.

The methodology underpinning the scorecard distinguishes between those indicators that have targets that facilitate distance to target analysis and those that do not; and between assessing progress over a given time period (usually 10 years from the early 1990s) and the status of countries' environmental performance for the latest year for which data are available. Both measures (progress and status) are used to distinguish between countries who have had policies in place for a long time and those who are relative newcomers to the implementation of EU policies.

For the first time ever in an EEA report, Part C also includes more detailed information on the situation in each country across the nine indicators, including the different types of actions and challenges each country faces. This analysis is based on contributions provided by the countries that have been subsequently edited by the EEA. The EEA takes full responsibility for the final result. This unique feature of the report is very much a first step towards what the EEA and Eionet hope will be a long-term process of building into European reports information from countries on the state of their environments and actions which reflect their often diverse social, environmental and economic conditions.

Part D
This part of the report provides an overview of the reports that the EEA has published since the previous five-year report Environment in the European Union at the turn of the century. It includes, in the electronic version, hyperlinks to all environmental reports published in the period from January 2000 to November 2005. Corporate documents, such as annual reports and work programmes, are excluded, as are promotional brochures.

3. Supporting activities

This report represents the culmination of a process started in 2003 from which a number of other specific published products and working documents have emerged. The most important of these is a series of sub-reports that have been developed in line with the priorities of the 6th environment action programme and the EU sustainable development strategy, and other reports that have been prepared to strengthen the EEA's information in the area of policy analysis, in particular ex-post policy effectiveness analysis and economic analysis.

Sub-reports
The EEA has worked on sub-reports across eight areas of relevance to strategic policy processes, plus a ninth area, environmental outlooks, which provides an integrated assessment of future prospects for the environment.

The eight areas relevant to the main policy processes are: household consumption and the environment; an enlarged European Union; halting biodiversity loss; sustainable use and management of natural resources; environmental policy integration; climate change and a European low carbon energy system; Europe and the global environment; and, environment and health. The EEA is presently engaged in a process of publishing reports in each of these areas

Other reports
This report also draws extensively on a range of other reports produced by the EEA in recent years and especially since 2003. The most notable of these are the 2004 climate change impacts report, the 2004 EEA signals report, the 2004 transport and environment reporting mechanism (TERM) report, the 2003 Kiev report *Europe's environment: the third assessment*, and the 2003 water indicator report. More information on these and other reports published by the EEA since 2003 can be found in the Bibliography in Part D.

Introduction

More recently, the EEA has analysed extensively the use of market-based instruments in environmental policy making across Europe. These cover environmental taxes, charges, subsidies and trading permits, amongst others. The analysis has been used extensively in Part A and especially in the final chapter on how the use of market measures in environmental policy could be expanded in future years. Two reports on market-based instruments are in the publication process.

4. Policy context

Since the previous five year state and outlook report in 1999 there have been a wide range of policy developments that provide, to differing degrees, relevant contexts for the assessments presented in this report. Four developments merit particular consideration: the enlargement of the European Union to 25 Member States in May 2004, the 6th environment action programme adopted in July 2002, the EU sustainable development strategy adopted in June 2001; and the Lisbon strategy adopted in March 2000.

The enlargement of the EU has brought a unique set of new environmental assets — including rich biodiversity and landscapes and vast areas of relative wilderness — but also represents an important challenge for EU environmental policy given the capacity building and financing needs required to support implementation of the acquis communitaire. The report does not deal with the implications of enlargement as a separate issue. The main challenges are instead addressed across the chapters in Part A. The progressive adoption by the EU-10 Member States of the environment acquis has already contributed to an enhanced environment in many places and where there are opportunities for mutual learning about better policy design and implementation.

The 6th EAP sets out the EU's environmental roadmap for the 10 years to 2012. It is the main vehicle by which to achieve the environmental goals of the sustainable development strategy. It sets ambitious, long-term goals for environmental protection and in so doing provides a stable framework within which public and private sector actors in Europe and the rest of the world can take action. The programme focuses on four priority areas: climate change, nature and biodiversity, environment and health and quality of life, and natural resources and waste. It is underpinned by the preparation of seven thematic strategies covering: soil protection; protection of the marine environment; sustainable use of pesticides; air pollution; urban environment; sustainable use and management of resources; and, waste prevention and recycling.

Two of the four priority areas — climate change and nature and biodiversity — are covered by separate chapters in Part A of this report. The other priority areas are addressed in several chapters, reflecting the more cross-cutting character of these issues. Part A also includes chapters on some of the thematic strategy areas, notably air pollution, marine environment and soils. Aspects relevant to the remaining thematic strategies areas are included in other chapters.

The EU sustainable development strategy requires environmental objectives to be considered alongside their economic and social impacts (and vice-versa) so that integrated policies can be implemented for the benefit of the economy, employment and the environment. The sustainable development strategy provides a longer term perspective than either the 6th EAP or the Lisbon strategy. Part A of this report considers issues relevant to the strategy, namely sustainable production and consumption patterns in the EU and global perspectives.

The Lisbon strategy adopted in 2000 seeks to make the EU 'the most dynamic and competitive knowledge-based economy in the world capable of sustainable economic growth with more and better jobs and greater social cohesion, and respect for the environment by 2010'. The strategy was reviewed in 2004 and relaunched in 2005 with a strengthened focus on economic growth and employment and 'win-win environmental economic strategies through the development and use of eco-efficient technologies'. This new policy direction offers new opportunities to take forward the development of cleaner environmental technologies.

The final chapter in Part A considers how to contribute to these objectives, while at the same time ensuring that environmental concerns are taken into account. This can be achieved through a combination of increased integration between environmental concerns and sectoral policies, the wider use of market based measures to internalise external costs and improvements in resource productivity supported by eco-efficient technologies.

5. Environmental data: situation and prospects

The quality of this report relies heavily on the quality of the underlying data and information. Since the previous report in 1999, there have been a wide range of initiatives that have improved the quality of environmental data. The EEA has been working with Eionet on a select number of priority data flows across greenhouse gas and air pollution emissions, air quality, water quality, land cover and contaminated soils. Two particularly noteworthy developments are the update of the Corine land cover data set (used extensively in Part A) which was completed in 2005, and the processes underlying the implementation of the EU water framework directive which are driven by the data needs explicitly linked to policy support over the next 10–15 years.

Despite improvements, the quality of environmental data does not match the relatively high quality of socio-economic data. The reasons for this are several but include the fact that socio-economic data are longer established, they are collected from easier to manage administrative sources, and are often linked to explicit monetary policy decisions, so giving data providers every incentive to report accurately and on time. Environmental data on the other hand are, for the most part, only recently established, based on scientific sources that are often less stable and where funding is uncertain from year to year, and collected often for a purpose — compliance with EU legislation — which differs from that of producing the type of integrated assessment in this report.

Looking ahead there are some key developments in Europe — notably on the development of spatial data and analysis linked to the Infrastructure for Spatial Information in Europe (Inspire) and the Global Monitoring of Environment and Security (GMES) initiatives. If fully and coherently implemented these could deliver substantial improvements in the coverage and quality of environmental data and provide a link to socio-economic developments. GMES also offers the opportunity to improve European scale modelling of environmental phenomena not easily captured by conventional data collection methods.

The EEA will continue to work with Eionet and other partners to improve the quality of environmental data and their integration with socio-economic data sources with the aim of providing a better underlying basis for its next five-year state and outlook report due in 2010.

Executive summary

European improvements, local choices, global impacts

Europeans value their environment — Eurobarometer polls show that a large majority (over 70 %) want decision-makers to give equal weight to environmental, economic and social policies. As individuals, Europeans are prepared to take some environmental action, though they would do more if they had better information on environmental choices that cost little or nothing. They would also do more if they felt confident that their fellow citizens were doing the same.

Over the past 30 years, much has been done to improve Europe's environment. Lead has been eliminated from most petrol. Ozone depleting chlorofluorocarbons (CFCs) have been phased out. Nitrogen oxide emissions from road transport have been reduced by around 90 % compared to what they would have been had catalytic converters not been introduced.

Increasing treatment of urban wastewater is allowing Europe's rivers, lakes and estuaries to recover from pollution. Designation of protected natural areas in the European Union now amounts to 18 % of all the territory, helping to maintain ecosystems and preserve biological diversity. Forests are slightly increasing and in some regions are regenerating at a faster rate than before. These and many other advances translate into benefits for people's health and for their quality of life.

But major challenges remain for the future. The most pressing is climate change whose impacts are already thought to be evident in ever more frequent extreme weather events, regional water shortages and melting polar ice. Other environmental priorities are: air pollution and regulation of chemicals so as to reduce impacts on health and on the environment; the preservation of land as a productive resource and as a reservoir for biodiversity; improving the quality and quantity of freshwater; and ensuring the health of the oceans. Oceans, in particular, are key ecosystems that sustain many of the ecological goods and services on which we depend.

Answers to some of these challenges can be found in increased use of renewable energy resources such as wind and solar power to replace some of the finite non-renewable resources that both developed and emerging economies are competing to exploit.

Many of the environmental problems we currently face are rooted in the way Europe uses its land, and in its economic structure and our ways of life. These are difficult to change. Most notably, there has been a shift in environmental emphasis from production to consumption issues. Better awareness about environmental and health effects would positively impact our daily choices on what to buy, where to live and work, and where and how to travel.

Household expenditure increased by a third in the EU-15 between 1990 and 2002. It is projected to double across the EU-25 by 2030, with major differences between income groups and regions. In an increasingly globalised economy, consumer choices everywhere increasingly impact not just Europe's environment but also many other parts of the world. Better understanding of potential impacts through keener research is needed to help reverse some of the current and future downward trends.

At around five 'global hectares' per person, the 'ecological footprint' of the EU-25 — the estimated land area required to produce the resources we consume and to absorb the wastes we generate — is approximately half that of the United States, but bigger than that of Japan. It is more than double the average for countries like Brazil, China or India. Already, the total global use of natural resources is some 20 % higher than the rate of replacement each year. This has been called, 'living off the capital rather than off the interest'.

Increasing urbanisation, abandoning land

Almost three-quarters of Europe's population live in urban and suburban areas which account for some 10 percent of the total EU land area. This seems manageable, yet the intensity and conflicts over the

multiple uses of land can have repercussions on valuable portions of Europe's territory, far away from where the initial land use is actually taking place.

Recent analysis shows that more than 800 000 additional hectares of naturally productive land were converted into artificial surfaces for homes, offices, shops, factories, and roads, adding 6 % to the continent's urban areas between 1990 and 2000. This is equivalent to three times the area of Luxembourg and represents a significant shrinking of natural capital. The low price of good agricultural land compared with that of already urbanised land is among the key factors influencing this urban expansion.

Tourism, too, keeps growing rapidly, driven in part by cheap prices for air travel and by Europe's increasingly affluent and ageing population. Tourism also contributes to urban sprawl, particularly in the hinterland of coastal agglomerations, such as along the heavily developed Mediterranean coast. Poorly planned tourism development can also increase pressures on areas already experiencing water stress.

As urban areas grow, so their use of land and water from surrounding areas intensifies. This growth impacts key 'services' assumed to be provided free by nature, such as the natural filtration of groundwaters into drinking water aquifers, the preservation of wetlands and of the genetic diversity found in areas of small-scale extensive agriculture. The removal of woodland cover can radically alter rainwater run-off, provoking mudslides and other problems, while increasing the areas at risk from flooding.

Climate change is here

Climate change is underway. Average European temperatures have risen over the past 100 years by 0.95 °C, and are expected to rise by 2–6 °C in this century. In some places, agriculture is likely to benefit from longer growing seasons, but in others, severe water shortages and more severe (and less predictable) weather events will make farming more risky.

Rising sea temperatures and increased nutrients levels bring a greater probability of algal blooms — toxic phytoplankton, harmful both to marine life and humans. Zooplankton — at the bottom of the food-chain — and the fish that rely on them as a main food source tend to follow temperature trends. In fact, some species have already migrated a thousand kilometres north. Land-based animal and plant species are also on the move. For some species migration is unfortunately not an option. Alpine species living at the highest altitudes are running out of options for where to go next.

In response, EU Ministers have agreed to a target to limit the long-term global increase in average temperature to no more than 2 °C above pre-industrial levels. They have also indicated that stabilisation of CO_2 concentrations well below 550 ppm may be needed to achieve this target, requiring cuts in greenhouse gas emissions in developed countries of some 60–80 % by 2050 compared to 1990 levels. In the short term, the EU is broadly on track to meet its Kyoto targets as a result of the EU emissions trading scheme and other measures including the European climate change programme. However, its mid-term goal for 2020 — a 15 to 30 % reduction in greenhouse gas emissions from 1990 levels — will be more difficult to achieve.

EEA scenario studies conclude that the key to a low carbon emissions economy lies primarily with three measures: reducing energy consumption, increasing the share of renewable energy, and improving energy efficiency in power generation and use, notably through further energy conservation measures. The use of renewables for power generation is increasing gradually, while the possibility of increased nuclear power remains an open — and hotly debated — issue in most countries.

Slow progress on energy demand management

Since 2000, improved efficiency in energy generation and declining energy demand from industry have been

Executive summary

offset by rising energy consumption by consumers and the service sector. More electrical appliances are being used in increasing numbers of households. Studies indicate that electrical appliances left on stand-by mode, for example, now account for 3–13 % of household electricity consumption.

By 2030, the demand for energy across Europe is expected to rise by close to 20 %, a much slower rate than foreseen for gross domestic product (GDP), but nonetheless in the wrong direction vis-à-vis the requirements to combat climate change. Cost-effective measures for improving energy efficiency remain underused. More efficient combined heat and power stations could improve energy supply efficiencies. Carbon capture and storage could serve as a transition technology. Efficiency measures for buildings, vehicles and consumer goods stimulated by market-based instruments and regulations would help reduce demand.

In the medium term, sustained investment in renewable energies, energy efficiency and in hydrogen as an energy source could help reduce European dependency on fossil fuels. The latter would especially help the transport sector which is the fastest growing contributor to Europe's growing energy demand and CO_2 emissions. Worryingly, this trend is expected to continue in coming decades. Air travel in particular is expected to double its share of overall transport between 2000 and 2030.

The EU has shown committed leadership by setting ambitious targets and goals for greenhouse gas reductions. It has also accepted that inaction poses too great a risk. Shifting to low-carbon energy sources, as suggested by EEA scenarios, will entail higher energy bills for the consumer. But doing nothing also has a cost, as several studies looking at this issue are beginning to show. One suggests that the 'social costs of carbon' — the costs to global society of every tonne of carbon emitted to the atmosphere — are around EUR 60 per tonne. Other studies suggest much higher costs. Different estimates depend on how long-term impacts on climate, agriculture, air quality, pests, water supplies and diseases are given a monetary value.

These costs can be put into perspective when considering that greenhouse gas emissions in the EU-25 range from 5 tonnes to 25 tonnes of carbon per person depending on which country you live in (equivalent to EUR 300/person to EUR 1500/person in social costs). This compares with the estimated additional costs of EUR 45/person in 2030 for a low carbon economy: the latter is considerably less expensive.

We are healthier, but exposure to pollutants remain

Europe has made great strides in reducing many forms of air pollution. In particular, it has eliminated smog in many areas and reduced acid rain. However, high concentrations of fine particulates, and ground-level ozone in particular, are still causing health problems in many cities and surrounding areas. Ground-level ozone is also damaging for ecosystem health and for crops across large areas of rural Europe.

Despite reductions in emissions, concentrations of these pollutants remain high, often above existing targets. Exposure to concentrations leads to reduced life expectancy and causes premature death and widespread aggravation to health. The increasing volume of transport, 30 % for freight and 20 % for passengers over the past 10 years, has meant that significant technological improvements have not resulted in much overall reduction in emissions.

Europe loses 200 million working days a year to air pollution-related illness. Moreover, the OECD estimates that 6.4 % of deaths and illnesses in young European children are caused by outdoor pollution. This figure is disproportionately more in the new EU Member States. Analysis underpinning the Thematic Strategy for air pollution published in September 2005 has shown that substantial impacts on people's health and ecosystems will persist even with full implementation of existing legislation.

Improvements in transport technologies, from hybrids to hydrogen fuelled vehicles, all have their parts to play in reducing the exposure. So, too, does urban planning, which could offer integrated transport approaches as real alternatives to car transport in many urban areas.

Europe's citizens are also exposed to a growing cocktail of chemical pollutants generated from food and modern consumer goods, including furniture, clothing, and household products. Links between chemicals and rising trends in cancers in reproductive organs (testes, prostate and breast cancer) and in childhood leukaemia are being increasingly highlighted. Strong evidence is lacking, but the ubiquitous presence of chemical traces in people's blood samples and in the environment is an obvious cause for concern. Less use of hazardous chemicals in farming and lower residues in consumer products would help reduce the impacts of such chemical mixtures.

Pollution prevention pays off

Great efforts have been made to clean up Europe's wastewater and to reduce water polluting wastes from industry. However, there is still some way to go before the urban wastewater treatment directive is fully implemented. Progress so far has been achieved through capital investments and advanced forms of treatment.

Future trends show that further reductions in wastewater pollution will be achieved especially in the EU-10 Member States, supported by the EU structural and cohesion funds from 2007. Experience of wastewater treatment policies over the past 20 years shows that investments in treatment capacities, combined with realistic economic incentives to reduce pollution at source offers the most cost-effective way to reduce such pollution.

The EU, through such policies as the nitrates directive, has sought to reduce pollution from agriculture. Investments by the water industry continue to ensure the quality of drinking water. However, leaching into Europe's rivers and groundwater continues from the use of organic and mineral fertilisers and pesticides. While it is expected that the use of such chemicals will diminish in the EU-15, the use of mineral fertilisers is expected to rise by 35 % by 2020 in the EU-10 as agriculture intensifies.

Problems with the quality of Europe's groundwater will remain in many areas as it can take decades for pollutants entering the ground to reach our rivers,

Executive summary

lakes and water supplies. Prevention, through changing farming practices, is more cost–effective than cleaning up, especially in the longer term.

Depleting our natural resources

The state of the world's fish stocks illustrates the dangers of over-using natural resources and damaging the functions of ecosystems. Fish are the last major wild food source. The Food and Agriculture Organization of the United Nations (FAO) estimates that 75 % of the world's fish stocks are over-fished and top level predators such as tuna and sharks are becoming scarcer.

With many of Europe's stocks depleted, the European fishing fleet has moved further afield supported through bilateral agreements and subsidies. These fleets have played a role in 'fishing down the food chain' by removing significant tonnages of top-level species. This has left many commercially important species at risk and the ecosystem structure under threat.

On land, the designation of 18 % of Europe's land area as protected areas under the Natura 2000 network will contribute to securing the health and diversity of its ecosystems. Nevertheless, Europe's landscapes, which are a critical part of its cultural heritage and essential homes for biodiversity, are undergoing widespread and potentially irreversible changes. These changes impact both on species and ecosystem functioning.

The largest losses of habitats and ecosystems for biodiversity across the continent during the 1990s were in heath, scrub and tundra, and wetland mires, bogs and fens. Many of remaining wetlands have been lost to coastal development, mountain reservoirs and river engineering works. Similarly, although more of Europe is tree-covered today than in the recent past, many forests are harvested more intensively than before.

These losses are having an impact on individual species. Despite protection policies as part of the European strategy to conserve its critical wildlife habitats, many species remain threatened, including 42 % of native mammals, 15 % of birds, 45 % of butterflies, 30 % of amphibians, 45 % of reptiles and 52 % of freshwater fish.

Europe's soil is uniquely varied, with more than 300 major soil types found across the continent. Lost soil may eventually be replaced through natural processes but it can take as much as 50 years to produce just a few centimetres of new soil. Soil should be regarded as a non-renewable resource. There are many threats to soil — erosion, sealing, contamination, salinisation. These have proven difficult to tackle up to now and are expected to continue to be a challenge in line with expected future developments in Europe in urbanisation, intensive agriculture and industrialisation/deindustrialisation.

Across the continent demand for water continues to increase, particularly in the household sector. In the new Member States household water use is expected to rise by 70 % in the coming decade. More water is also being used for the irrigation of food crops, particularly in southern Europe where there are already signs of water stress. Climate change is expected to extend and intensify this problem. The long-term availability of abundant, reliable and clean water supplies will become more important in the context of future land-use planning, especially around the Mediterranean.

During the past decade, Europe has achieved a relative decoupling of economic growth from materials and energy use. Absolute resource use, however, has remained steady. There are large differences between EU countries with materials intensity varying from around 11 kg/euro of GDP to less than 1 kg/euro. These differences can in part be explained by the balance of economic activity between industry and services. Nonetheless, resource and energy productivity in western Europe is, on average, four times higher than in the new EU Member States. This provides substantial opportunities to achieve greater balance in resource productivity between the EU-15 and EU-10 through technology transfer and other measures.

Executive summary

Integration, innovation and market reform

The EU's successful environmental policies over the past 30 years have largely concentrated on easily visible point sources. These problems have been dealt with mainly by regulation and technological innovations. The challenge now is to develop and implement long-term policies for those economic sectors that contribute most to diffuse sources of pollution.

Significant progress is likely to take several decades of coherent, long-term yet flexible, policy-making that has the broad support of citizens. This means that public information-provision and awareness-raising measures will be increasingly essential for effective policy-making.

Effective policies will also need to encourage behavioural changes amongst Europe's consumers as well as to focus, in particular, the transport, energy and agricultural sectors on less environmentally damaging activities. Long-term institutional reform and financial planning that encourages greater eco-efficiency can help promote such activities. These could be supplemented by the use of market-based instruments. For example, a move away from environmentally damaging subsidies towards supporting the development and use of eco-innovations in manufacturing, energy transport and agriculture could greatly help the transition towards more sustainable economic activities.

Many EU policies already include environmental objectives and substantial budgets are being deployed to encourage actions and behaviour in line with environmental goals, for example under the Common Agricultural Policy. Nevertheless, given the broad nature of changes arising from land use, Europe could benefit from increased cooperation across sectors to tackle a balanced territorial cohesion, for example between regional urban and transport planning and the use of EU structural and cohesion funds.

The transport sector provides a good test case for underlining the benefits of more integrated approaches. For this sector, we see a myriad of inter-linked driving forces and pressures that impact the environment. On the one hand, the sector has achieved substantial reductions in emissions of air pollutants such as ozone precursors and acidifying substances. On the other hand, however, emissions of greenhouse gases continue to rise as the demands for transport (freight and passenger) outstrip improvements made in energy-related emissions through technological improvements and stricter regulations.

In line with urban development, transport infrastructure has a threefold impact on land. It contributes to the consumption of good agricultural land, the sealing of soil at increasing rates and the fragmentation of habitats across the European Union. Furthermore, it exposes an extensive part of the population to high noise-levels.

Our increased appetite for mobility by road and by air has resulted in transport issues rising to the top of the environmental/sustainability agenda from city level to world governance. This reflects the wide range of challenges surrounding transport from local concerns (urban planning and design) to global ones (greenhouse gases and climate change).

More integrated long-term actions have delivered substantial benefits. Taxation on petrol illustrates the effectiveness of long-term shifts in economic incentives via market-based instruments. American and European vehicle technologies are basically the same. Nevertheless, European fuel taxes of around 50 % have stimulated changes in consumer behaviour.

Together with political pressure to use technologies, these factors have made new European cars almost twice as fuel-efficient in recent decades as their American counterparts, where fuel taxation is much lower. Studies suggest that considerable savings in energy intensity could be made by similar approaches to energy pricing.

The European environment | State and outlook 2005

Executive summary

What we can do

Tax reform can contribute to a more sustainable, healthy environment. A gradual shift of the tax base away from taxing 'good resources' such as investment and labour, towards taxing 'bad resources' such as pollution and inefficient use, would also help to internalise environmental costs into service and product prices. This would in turn create more realistic market price signals.

Policy-makers could also design flanking measures to ensure that environmental taxes are not socially inequitable. Poorer members of society generally spend a greater proportion of their income on basic needs such as food, water and energy. Studies have found that particularly taxes on electricity fall on the poor, while transport taxes are relatively benign for the poor who have less access to private transport. Pollution taxes are generally neutral in their impact across social groups.

Policies that derive more revenues from consumption, and less from labour, can also provide a wider and expanding tax-base as a response to both the declining workforce and an ageing society.

The seven thematic strategies being developed under the 6th environment action programme, along with sector integration policies, and the EU's sustainable development strategy, all encourage long-term planning.

Long-term coherent policies can encourage the re-structuring of incentives from financial instruments such as market prices and taxes that will be necessary to reduce the rising and increasingly evident costs of using the planet's natural resources. The resulting gains in eco-efficiency could also help improve the competitiveness of the European economy. Better energy and resource productivity in Europe could also help partly offset other competitive advantages enjoyed by the emerging economies of Asia and South America.

Nevertheless, there are substantial barriers to effective and efficient implementation of policies at all levels of governance in the EU. EEA studies indicate that the institutional set-up can be as important as the design of the policy itself.

Public support for the environmental gains achieved over the past decades is reflected in the findings of the Eurobarometer 2005 survey, which also suggests that European citizens are prepared to do more. This report demonstrates that more does indeed need to be done by both governments and citizens in order to bring economic development into line with the Earth's carrying capacities.

Europe is well placed to lead the way by creating a smarter, cleaner, more competitive and more secure European society. Such advances would encourage improvements in global eco-efficiency and equity that ultimately secure Europe's quality of life.

Executive summary

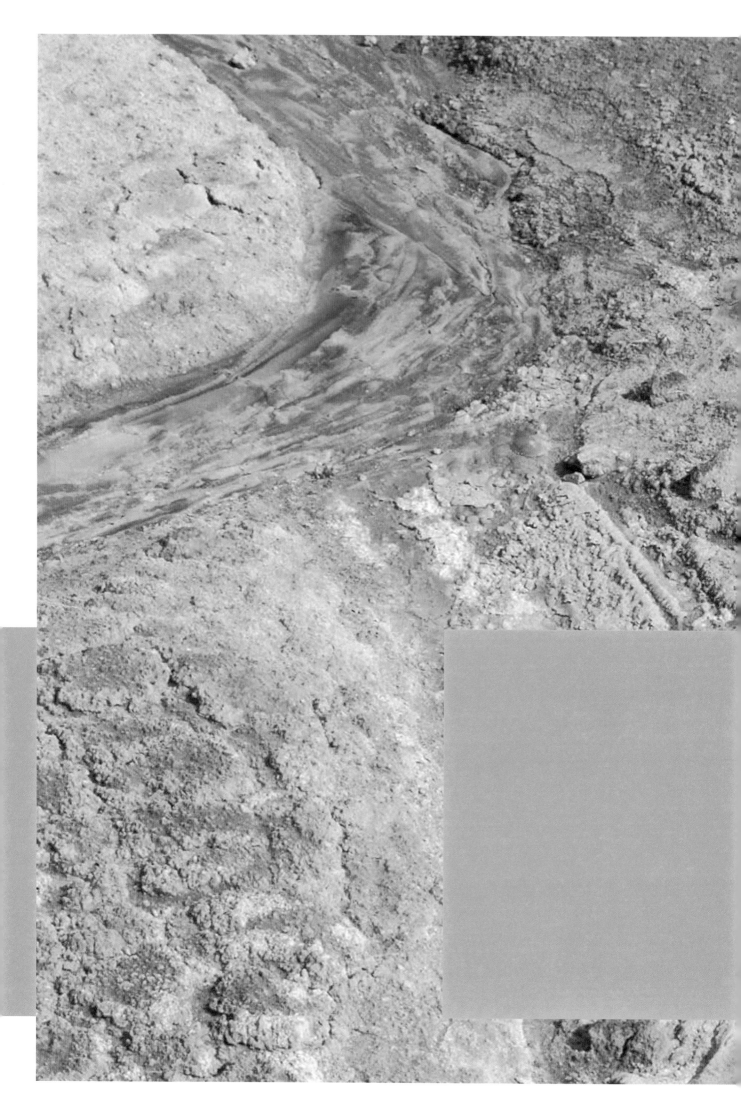

A Integrated assessment

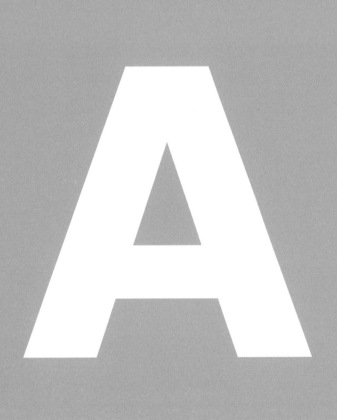

Integrated assessment

Setting the scene
   1 Environment and quality of life 28
   2 The changing face of Europe 36

Atmospheric environment
   3 Climate change 62
   4 Air pollution and health 92

Aquatic environment
   5 Freshwaters 112
   6 Marine and coastal environment 132

Terrestrial environment
   7 Soil 168
   8 Biodiversity 182

Integration
   9 Environment and economic sectors 216
  10 Looking ahead 232

PART A

1 Environment and quality of life

1.1 Europe's environment — rich and diverse but under pressure

Europe has a rich and diverse environment. With its beautiful landscapes, historical cities and cultural treasures, it remains one of the world's most desirable and healthy places in which to live, and to invest, and is one of the most frequently visited global travel destinations.

Extending from the Arctic Circle to the Mediterranean and from the Caucasus to the Azores, Europe is home to an array of natural and semi-natural habitats and ecosystems, featuring a wide range of species and genes. This biodiversity, though limited in comparison with other continents, is our environment's 'insurance', as it ensures the environment's ability to adapt to change and renew itself.

In Europe, as everywhere, humankind depends on the Earth's ecosystems for the services they provide — for resources such as food, water, timber, fibre and fuel; for functions such as climate regulation, the absorption of wastes and the detoxification of pollution; and for protection as afforded by the atmospheric ozone layer. Over the last 50 years we have changed these ecosystems more rapidly than ever before to improve human well-being and sustain economic development. At the same time, the full ecological and economic costs associated with these gains are only now becoming apparent.

The alteration or loss of natural resources, together with changing climate conditions, are making us ever more vulnerable to the forces of nature. In 2004, weather-related disasters across the world caused economic losses of more than EUR 86 billion (USD 105 billion), almost twice the total in 2003. Some 12 000 weather-related disasters since 1980 have caused more than 600 000 deaths and cost just over EUR 1 trillion (USD 1.3 trillion).

Europe is one of the most urbanised continents. Today some 75 % of Europe's population lives on just 10 % of its land area. Urbanisation is beneficial to the environment, inasmuch as resource consumption and soil sealing tend to be lower per person, and the provision of environmental services such as waste management and wastewater treatment are cheaper per person, than for more dispersed populations. Nonetheless, in recent decades, the increasing trend towards the dispersal and sprawl of urban settlements is leading to an increasing fragmentation and loss of valuable landscape amenities.

Europeans are now living in a part of the world where rapid changes are shaping landscapes as never before, and bringing with them a different quality to our surroundings. Wetlands are being drained to make way for urban development; the use of mountains and upland areas are changing as farms make way for skiing and other kinds of recreation. Forest management has also had to adapt in the face of developments in the timber trade brought on by increased competition in the global economy.

The European environment remains under pressure, but now, to sustain our standards of living, we are exporting this pressure by importing more and more resources from elsewhere in the world to meet our European needs. We have become disproportionately more responsible for the consumption of global resources than almost any other region. At around 5 'global hectares' per person, the ecological footprint of the 25 Member States of the European Union (EU-25) — the estimated area required to produce the resources we consume and absorb the wastes we generate — is approximately half that of USA, but is still bigger than other large economies, including Japan.

The average European's footprint is also more than double his or her counterpart in Brazil, China or India, as well as the global average. Already the total global use of ecological resources is some 20 % higher than what the planet's natural systems can renew each year. Hence, unless Europe, and other developed nations, reduce their ecological footprint through using fewer resources and through efficiency measures, and make ecological space available for emerging economies, then more severe ecosystem damage, more material shortages and greater pressures on the global climate are likely.

Growing appreciation of the links between economic performance and the environment are encouraging much greater 'eco-efficiency' in our use of energy and resources. Such 'eco-innovation' has the double benefit of optimising the use of scarce resources — both renewable and non-renewable — and helping Europe compete in the global economy.

The operation of the global marketplace and the liberalisation of trade are expected to continue changing Europe's ecological footprint. Food, clothing and electronic goods now routinely come from the other side of the planet and these trends are expected to continue. Since the price of few products properly reflects the environmental damage caused by the production process and resource depletion, Europe will often be buying foreign environmental assets at a discount.

In the second half of the 20th century, global trade in raw materials grew by a factor of six to eight, and in manufactured goods as much as 40-fold. Thus, Europe is not alone in its increasing dependence on ecological creditors abroad. However, with the planet's resources expected to come under increased pressure as the demands from other parts of the world grow, this dependence is expected to become less sustainable for both the EU and the rest of the world.

1.2 Connecting to Europe's citizens

The task for environmental authorities and other actors is to address these new challenges adequately whilst maintaining the support of voters and other stakeholders. Such support, at least from results of opinion polls, appears considerable.

According to Eurobarometer polls, a large majority of EU-25 citizens want policy-makers to consider the environment equally alongside economic and social

Figure 1.1 European opinions on the environment's influence on the quality of life and the perception of the environment's importance in the policy-making process

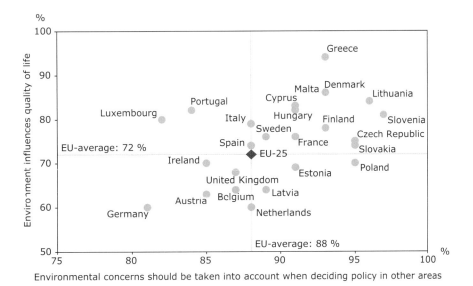

Source: Eurobarometer 217, 2005.

policies (Figure 1.1). Furthermore, they consider environmental protection policies an incentive for innovation (67 %) and not an obstacle to economic performance (80 %).

In the same survey almost two-thirds of respondents gave priority to protecting the environment over economic competitiveness. Additionally, they felt that the EU was the most suitable level at which to address environmental issues, given the transboundary nature of many problems and their wish for more harmonised approaches to relevant policy-making. An endorsement for the EU, which has stimulated up to 80 % of environmental policy measures at Member State level over the last 25 years.

People's main environmental concerns are, however, related to their local day-to-day living conditions, such as the state of their water, air pollution and the perceived threats from chemicals. Even anxieties about such global issues as climate change were expressed in local terms. Over 70 % of Europeans therefore see the environment as having a significant influence on their quality of life and want to see the environment taken into account when policies are decided in other areas. They understand the interconnections between their environment and the activities in economic sectors such as transport, energy and agriculture, and perceive the benefits of more integrated approaches.

Our well-being and quality of life depend on the state of the environment and the services, such as climate regulation, provided by natural ecosystems. Improving human well-being and human development in the coming decades will thus largely depend on our ability to ensure the sustainable use of the environment — a task made more complex by the changing nature of human activities that impact it most.

1.3 Europe's changing environmental problems

Within Europe, progress in dealing with environmental pressures has been evident in several areas, and these are to a large extent consistent with people's day-to-day concerns. There have been substantial reductions in acidifying air emissions, and consequent improvements in some aspects of air quality, in substances that deplete the ozone layer, and in point source emissions to water. Much of this has been achieved through the application of abatement technologies and through resource substitution, both of which have been encouraged by EU and Member State environmental regulations.

The protection of biodiversity, through the designation and protection of habitats, has gone some way in improving the maintenance of ecosystem productivity and landscape amenity. At the same time, action on waste management has not led to an overall reduction in waste volumes, which reflects the fact that progress here is more intimately related to general economic and social development.

Many changes in climate and their impacts on ecosystems and human health are already visible in Europe, particularly in southern Europe where water shortages, fires and droughts are increasingly apparent, along with more unpredictable weather patterns. Meanwhile, the scientific evidence of climate change is getting firmer, with the manifestation of more robust indicators suggesting a much faster rate of change than previously thought.

There is also an increasing threat to human health from exposure to new forms of invisible, time-delayed and more systemic pollution and chemicals. Rising rates of cancer, asthma and neurodevelopmental diseases, particularly in children, are damaging the current and future health, and therefore wealth, of our societies.

Many of today's most challenging environmental pressures are proving more difficult to tackle than those which have registered greatest progress in recent decades. The sources of treatable pressures were then easily identifiable — industrial plants or car exhausts — and so could be dealt with adequately through regulatory standards and the application of abatement technologies.

Five sectors — transport, energy, agriculture, industry and households — contribute most to current problems and are expected to continue doing so in the future. In these sectors, many pollution sources are much more diffuse, numerous and varied, thus proving more difficult to control. Even where new technologies have been introduced, their effectiveness has often been overwhelmed by increased demand.

It is becoming clear that a mix of instruments, which encourage societal shifts to less damaging forms of behaviour and promote increased technical and economic efficiency are needed. Such integrated approaches, if well designed and fully implemented, can be cost-effective by addressing environmental and economic considerations together, and tackling cross-sectoral problems. Making progress on such approaches takes time, as can be seen when looking at the development of environmental policy over the past three decades.

1.4 Solutions to deal with change

Environmental policy measures at international level and in Europe are relatively new compared with economic and social policies. Nonetheless, over the past 30 years or so, significant progress has been made in establishing a comprehensive system of environmental regulation in the EU. Action began in Stockholm in 1972 when the United Nations conference on the human environment first drew global attention to environmental issues. At the European level six successive European environment action programmes followed, based on a combination of thematic and sectoral approaches to ecological problems.

The first environment action programme, adopted in 1973, established the principles of the polluter paying, of prevention at source and of the appropriateness of action at the European level: principles which later became EU Treaty obligations. The fifth environment action programme (1992–2000) focused on reducing pollution levels, implementing legislation that would benefit EU citizens and integrating the environmental dimension into all areas of the Commission's policies, especially its main sectors — transport, energy, agriculture and industry.

The sixth environment action programme (6EAP), which runs to 2012, gives a new sense of purpose and direction to the Community's environmental policy. The programme puts forward a series of actions to tackle persistent environmental problems in four priority areas: climate change; nature and biodiversity; environment, health and quality of life; and natural resources and waste.

The strategic approach of the 6EAP is underpinned by five major objectives: to improve the implementation of existing environmental legislation at national and regional levels; to integrate environmental concerns into other policy areas; to work closely with business and consumers in a more market-driven approach to identifying solutions; to ensure better and more accessible information on the environment for citizens; and to develop a more environment-conscious attitude towards land use planning.

Thematic strategies are one component of the actions foreseen within the 6EAP. This concept was introduced as a specific way of tackling key environmental issues which require a holistic approach because of their complexity, the diversity of actors involved and the need to find multiple and innovative solutions. Seven such thematic strategies will be developed according to a common approach — soil protection; protection and conservation of the marine environment; sustainable use of pesticides; air pollution; urban environment; sustainable use and management of resources; and waste recycling.

Policy-making in the 1970s and early 1980s focused on the local point sources of pollution which were managed, in general, by directives and regulation. The past 20 years have seen a shift in focus towards regional and global problems, caused mostly by diffuse sources of pollution. For example, it was in the late 1980s that global issues like the 'ozone hole' emerged as a serious and urgent problem which required global and regional measures to be implemented if environmental policy was to be successful.

Such issues demanded economic incentives and the provision of better information to both companies and citizens, as supplements, or sometimes substitutes, to regulation. Most Europeans would like to have more information on environmental problems, and particularly on their solutions (Figure 1.2). Indeed, people also feel that enforcing existing regulations, making them stricter, increasing fines for those who break them, and increasing public awareness are the most effective tools for solving environmental problems.

Another important shift came in the early 1990s when the emissions and environmental supply-side measures of the 1970s and 1980s were supplemented by more upstream sectoral integration and demand-side management policies in the 1992 fifth environment action programme and the Maastricht Treaty. Further, the 1998 'Cardiff process' focused on integrating environmental considerations into the thinking of those economic sectors causing the problems, such as agriculture and transport.

The 1990s also saw the emergence, for the first time, of global companies looking seriously, and in concert, at the emerging environmental agenda, as illustrated in the 1992 World Business Council for Sustainable Development report 'Changing course: A global business perspective on development and environment'. This report, by 46 major companies, also introduced the concept of eco-efficiency which the companies felt was essential in the communication of sustainable development. Ten years after, a counterpoint book 'Walking the talk: the business case for sustainable development' demonstrated results achieved by several companies and recognised that the business of business had changed.

The greater scientific complexity and uncertainty surrounding current environmental hazards such as climate change, ecosystem integrity and health hazards from chemical and other pollutants mean that more sophisticated policy-making has come into

Figure 1.2   Comparison between environmental concerns and lack of information Europeans have

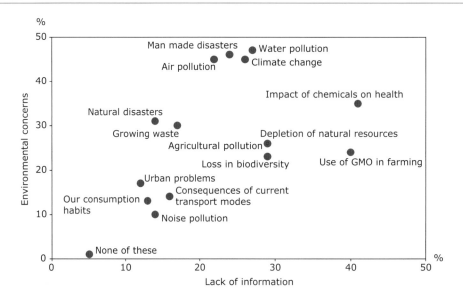

Source: Eurobarometer 217, 2005.

play. This involves a greater use of long-term tools, including scenarios and expert approaches, such as the precautionary principle which was incorporated into the EU Treaty in 1996.

The process of devising policy measures to better reflect an interconnected reality has also brought efficiencies from 'cost spreading'. For example, policies on acid rain and climate change, originally handled separately, have produced large cost-effectiveness improvements after being handled in a more integrated fashion.

Nonetheless, more integrated policy approaches bring their own transaction costs in that they are much more difficult to implement. They involve many actors from across the main economic sectors, such as transport, energy and agriculture, as well as consumers. Additionally, their increased flexibility can often mean greater implementation and enforcement difficulties at regional, national and European levels.

The lessons to be learnt from the past decades are, however, clear: environmental policies, when properly developed and implemented, have led to significant and cost-effective improvements in several fields, whilst stimulating innovation in the development of environmental technologies and services. Currently, the global market for such technologies and services is worth about EUR 425 billion (USD 515 billion) a year and is forecast to expand at a rate of about 3 % a year.

Overall, such progress has been brought about through 'traditional' measures, regulating products and production processes, and protecting important natural sites. These policy areas are covered by well-established EU legislation. However, more integrated policies, including further market-based instruments designed across environmental problems, sectors and scales, and over time, remain a challenge.

1.5 Looking ahead

This chapter began with a description of what is special about Europe's environment and how it contributes to the quality of our daily lives, then considered how Europe's citizens want to see its character maintained in the face of changing, and increasingly global, socio-economic challenges, and discussed how policy measures have been developing in response.

What is clear is that with rapidly changing economic developments within Europe and across the world, now and in the coming decades, it will become increasingly difficult to balance these various considerations. Bearing this in mind, subsequent chapters consider the environmental challenges Europe faces now and in the future, as well as how it could respond through further policy development.

Chapters 2–8 look in more detail at the changing face of Europe's land, as one of the key underlying resources needed to sustain our well-being; and the state of the continent's environment, including prospects for the future, across the main environmental priorities that underpin the 6EAP — climate change, biodiversity, use of natural resources and health issues. These chapters also explore, to differing degrees, the ways in which the benefits of our ecological resources and services are being eroded, at considerable actual and future costs to people's health, Europe's economy and the well-being of the rest of the world.

Chapter 9 summarises the main findings from previous chapters and then considers the past performance and future prospects of four economic sectors — transport, agriculture, energy and households — in creating environmental pressures and taking measures to deal with them.

The concluding Chapter 10 then analyses how these pressures and impacts on the environment can be dealt with in the future through more integrated action that focuses on three areas: the institutional structures needed to implement more coherent and integrated action; the internalisation of the costs of environmental damage into prices, through the use of market-based instruments such as emissions trading, financial incentives and taxes; and prospects for the eco-innovations needed to substantially reduce environmental pressures and to improve ecological resource productivity.

The chapter ends by considering how such measures can help Europe adapt to the challenges of ensuring continued prosperity in the face of global competition and expected demographic changes.

References and further reading

Europe's environment — rich and diverse but under pressure

European Environment Agency, 2005. Ecological Footprint database update to 2002.

Millennium Ecosystem Assessments, 2005. *Ecosystems and human well-being synthesis* (www.millenniumassessment.org/en/Products.Synthesis.aspx — accessed 10/10/2005).

European Environment Agency, 2004. *Mapping the Impacts of recent natural disasters and technological accidents in Europe*, EEA Issue Report No 35, Copenhagen.

IFRC, 2004. *World disasters report*, International Federation of Red cross and Red Crescent Societies.

IFRC, 2005. *World disasters report*, International Federation of Red cross and Red Crescent Societies.

Munich Re, 2005. *Topics Geo — Annual review: Natural catastrophes 2004*. (www.munichre.com/ — accessed 10/10/2005).

Connecting to Europe's citizens

European Commission, 2005. *Lisbon, growth and jobs — working together for Europe's future*, Special Eurobarometer 215. (www.europa.eu.int/comm/public_opinion/index_en.htm — accessed 10/10/2005).

European Commission, 2005. *The attitudes of European citizens towards environment*, Special Eurobarometer 217. (www.europa.eu.int/comm/public_opinion/index_en.htm — accessed 10/10/2005).

Europe's changing environmental problems

European Environment Agency, 1999. *Environment in the European Union at the turn of the century*, Environmental Assessment Report No 2, EEA, Copenhagen.

European Environment Agency, 2005. *Climate change and a European low-carbon energy system*, EEA Report No 1/2005, Copenhagen.

European Environment Agency, 2005. *Environment and health*, EEA, Copenhagen (in print).

European Environment Agency, 2005. *European environmental outlook*, EEA Report No 4/2005, Copenhagen.

European Environment Agency, 2005. *Sustainable use and management of resources* (in print).

WWF, 2005. *Living planet report*. (www.panda.org/news_facts/publications/general/livingplanet/index.cfm — accessed 10/10/2005).

Solutions to deal with change

European Commission, 1998. *Towards sustainability — fifth environment action programme (1992–2000)*, Decision 2179/98, 10.10.1998 OJ L275/1, Brussels.

European Commission, 2001. *Environment 2010: Our future, our choice — sixth environment action programme*, COM(2001)31 OJ L242, Brussels.

European Environment Agency, 2001. *Late lessons from early warnings: The precautionary principle 1896–2000*, Environmental Issues Report 22, EEA, Copenhagen.

European Environment Agency, 2005. *Environmental policy integration in Europe — Administrative culture and practices*, Technical Report No 5/2005, EEA, Copenhagen.

European Environment Agency, 2005. *Environmental policy integration in Europe — State of play and an evaluation framework*, Technical Report No 2/2005, EEA, Copenhagen.

Schmidheiny, S. *et al.*, with the Business Council for Sustainable Development, 1992. *Changing course: A global business perspective on development and environment.*

Schmidheiny, S., with the Business Council for Sustainable Development, 2002. *Walking the talk: the business case for sustainable development.*

Treaty on European Union — Maastricht Treaty (1992), Official Journal C 191, 29 July 1992.

United Nations Environment Programme, 1972. United Nations conference on the human environment, Stockholm. (www.unep.org/Documents.multilingual/Default.asp?DocumentID=97&ArticleID= — accessed 10/10/2005).

2 The changing face of Europe

2.1 The face of Europe: a mosaic of changing landscapes

The history of human culture suggests that 'landscape' is one of the earliest and most obvious concepts for perceiving and describing our environment. However, there is not just one idea of landscape — a landscape can be perceived from a variety of observations and viewpoints — but, in contrast to the notion of 'wilderness', the term landscape is frequently associated with human interference or influence. It is at the landscape level that changes in terms of land use, naturalness, culture or character become meaningful and recognisable for human interpretation.

Landscape is as much vision as it is reality. The way we perceive landscapes, the attraction we feel to some of them, and our feelings when conflicts arise over the use of land, are all matters of extreme importance for conservation and future human welfare. A landscape is essentially a photograph of what is going on; it reveals, in short, who we are. At the same time landscapes are also dynamic expressions of continually changing natural processes (climatic, physical, biological) and changes caused by human activity.

It is clear that landscape analysis requires the consideration of different factors not equally easy to apply. The spatial dimension must be considered, as must the temporal component. It is especially important to know both where and when change is happening, given the uneven distribution and value of ecological goods and services across Europe, the vast range of activities that impact on them and the changing character and intensity of these impacts over time.

One strategy for preserving landscapes has been the establishment of protected areas. Early protection measures focused on preservation of landscape scenery, but during recent decades nature reserves have been designated mainly to minimise the probabilities of extinction and maximise the conservation of species. However, we now know that many species require a range of habitat types during their lives, and different species use the environment on different scales. Scientists are therefore embracing the idea that biodiversity should be dealt with not only at the habitat or species level but also at the scale of landscapes.

2.2 Landscapes: photographs of human uses of land

Human decisions have a strong influence on the shape of landscapes and the social, economic and political conditions needed to let such landscapes — or environments — develop. International, national and regional policies (on agriculture or the environment, for example), demographic trends (such as the migration of populations between countries and regions, from cities to the countryside or vice versa, and population growth), together with ecological factors are all interlinked.

Scientists, planners and policy-makers are increasingly aware that adequate decisions cannot be made solely at the site level. This is especially important in a European context where landscapes are dominated by human influence. Most human activities, particularly industrial activity, urban development and transport, have an impact on the landscape, but these impacts are relatively localised compared with the wide role played by agriculture in shaping our surroundings. Land use patterns have undergone revolutionary changes in the past; today, though less drastic and visible, changes keep on altering our environment leaving large, often irreversible, land use footprints. Patterns of changing land use across Europe show that tensions are rising almost everywhere between the needs of society for resources and space, and the capacity of the land to support and absorb these needs.

There is increasing confirmation that the drivers of many environmental problems affecting European land originate outside the actual territory where the changes are observed. A global market economy, the measures of the common agricultural policy (CAP), trans-European traffic networks, large-scale demographic and socio-economic changes, cross-boundary (e.g. airborne) pollution, as well as differences in land-planning mechanisms at the national, regional and local level, are

The changing face of Europe   PART A

Map 2.1 Availability of Corine land cover data

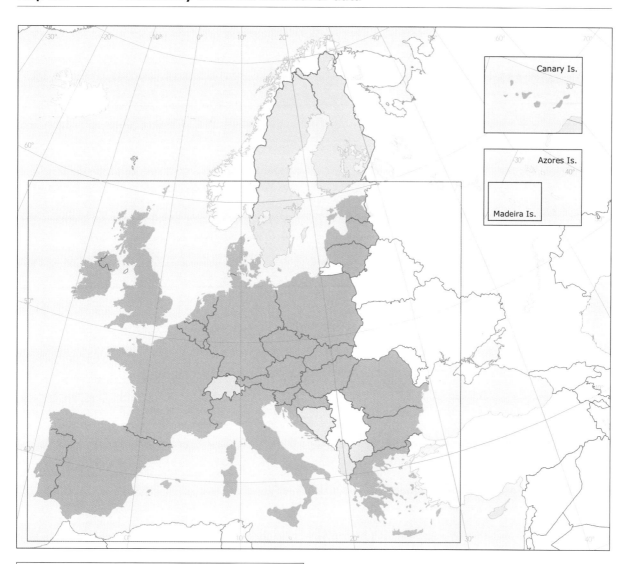

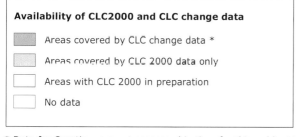

Availability of CLC2000 and CLC change data

- Areas covered by CLC change data *
- Areas covered by CLC 2000 data only
- Areas with CLC 2000 in preparation
- No data

* Data for Croatia were not processed in time for this publication

Note: The large box in the map indicates the geographical coverage for Maps 2.3., 2.4 and 2.5 that follow later in this chapter.

the main drivers of change and environmental pressure. There is now a rising awareness of the added benefits of considering the territory as a unit of analysis and as a basis for stimulating better coordination of policies.

Europe is debating a stronger and more balanced territorial focus for its policies. This debate has been developed by the Member States and the European Commission within the 1999 European spatial development perspective (ESDP). This process has led to commonly agreed policy orientations around better territorial balance and cohesion, improved regional competitiveness, access to markets and knowledge, as well as to wiser management of natural and cultural resources.

The policy orientations reflect the ongoing geographical concentration of many parts of European society in highly urbanised areas. The long-term aim is to see a European territory with many prospering regions and areas, geographically well spread, all playing an important economic role for Europe and providing a good quality of life for their citizens.

Polycentric spatial development is the main concept related to the aim of territorial cohesion. The concept can be described as a bridging mechanism between economic growth and balanced development. Thus, polycentric development can bridge the different interests of the Member States by encouraging more balanced and coordinated competitiveness. Interest in polycentric development is also fuelled by the hypotheses put forward in the ESDP that polycentric urban systems are more efficient, more sustainable and more equitable than either monocentric urban systems or dispersed small settlements.

2.3 Maintaining landscapes in future

While territorial cohesion is the subject of continuing debate, the links between territorial cohesion and economic and social cohesion — two fundamental aims of the European Union (Article 16 of the Treaty) — remain to be further clarified. There is thus a need to have a broader vision of cohesion that encompasses many dimensions of the development of territories and of their inter-relationships.

In this respect, a territorial dimension has been proposed for the conception of structural policies after 2007. The Commission has also proposed European territorial cooperation as an objective for Structural Funds interventions for 2007–2013 in support of territorial cohesion within the EU.

At the same time, although the Lisbon strategy has no explicit territorial dimension, one of its three main priorities calls for Europe to be made an attractive area in which to invest and work. This priority includes considerations relating to access to markets and the provision of services of general interest as well as to factors relating to the creation of a healthy environment for enterprises and families.

The implementation of the Lisbon strategy and of future structural policies will take place in regions, in national territories and at European level. Therefore, a key question for policy-makers at different levels is to explore, identify, understand and select potential areas for development within their own territory in order to contribute effectively to this overall European strategy.

The rest of this chapter analyses and discusses changes in Europe's territory (land cover) from both a spatial (landscape) and temporal (statistical change) perspective. In the context of the factors mentioned earlier, this allows us to understand what is happening and where it is happening, and to put these in the context of the particular policies that most influence change.

2.4 Dominant landscape types and changes in land cover

Wherever we live in Europe, and whether surveying our surroundings or taking in the view from an aeroplane, landscapes powerfully characterise our sense of place. Their slowly shifting patterns both reflect and support Europe's many cultures, societies, economies and environments. We see many different images when looking across Europe, but the EEA has categorised these landscapes into seven dominant types (Map 2.2), which reflect prominent functions of the land. These seven landscape types in turn give an indication of where the highest potential remains for preserving the amenities and services that land provides, and hence where changes in land cover (and land use) can have the most impact on nature.

The diversity and distribution of the landscape types in 2000 indicate where the continent's main reservoirs of 'naturalness' are located: in the Mediterranean and northern European regions, and in many coastal zones and major mountain ranges such as the Alps and the Carpathians. Forested land dominates in the Baltic states, Germany, Scandinavia and Slovenia. Agricultural landscapes are widespread across the continent, with broad patterns of arable land seen, for example, in Denmark and the United Kingdom (England), while pasture and mosaics, which make room for more symbiosis with nature, can be observed in Alpine and other regions. Urban settlements represent an important share of the total territory in terms both of the space they consume and of their much higher impact on natural habitats. The famous north-western urban 'pentagon' is visible from the dominant landscape map, as well as concentrations in other areas, including along coastlines and river corridors.

The image of seven dominant *landscape types* for the year 2000 was preceded by decades of rapid change in land cover and land use across Europe. The changes for the decade from 1990 are presented for the eight aggregate *land cover types* in Table 2.1 below (a total of 23 countries have included an assessment of change in their CLC2000 programme: Austria, Belgium, Bulgaria, Czech Republic, Denmark, Estonia, France, Germany, Greece, Hungary, Ireland, Italy, Latvia, Lithuania, Luxembourg, the Netherlands, Poland, Portugal, Romania, Slovenia, Slovakia, Spain, United Kingdom).

Table 2.1   Land cover 1990, 2000 and change — sum of 23 EEA member countries

	Artificial areas	Arable land and permanent crops	Pastures and mosaics	Forested land	Semi-natural vegetation	Open spaces/ bare soils	Wetlands	Water bodies	Total km^2
Land cover 1990	160 785	1 171 098	798 607	1 003 905	257 503	515 60	45 283	125 334	3 614 073
Consumption of initial land cover	1 821	24 456	17 400	39 119	8 929	2 284	1 357	198	95 563
Formation of new land cover	10 493	18 096	15 066	44 602	4 087	1 772	181	1 267	95 563
Net Formation of land cover (formation-consumption)	**8 658**	**– 6 400**	**– 2335**	**5 474**	**– 4 816**	**– 454**	**– 1 043**	**916**	**0**
Net formation as % of initial year	5.4	– 0.5	-0.3	0.5	-1.9	-0.9	-2.3	0.7	
Net formation as % of total land cover	0.24	– 0.18	– 0.06	0.15	– 0.13	– 0.01	– 0.03	0.03	
Total turnover of land cover (consumption and formation)	**12 313**	**42 552**	**32 466**	**83 721**	**13 016**	**4 056**	**1 538**	**1 464**	**191 127**
Total turnover as % of initial year	7.7	3.6	4.1	8.3	5.1	7.9	3.4	1.2	5.3
Total turnover as % of total land cover	0.34	1.18	0.90	2.32	0.36	0.11	0.04	0.04	5.3
No land cover change	**158 964**	**1 146 642**	**781 206**	**964 786**	**248 574**	**49 276**	**43 926**	**12 5136**	**3 518 510**
No land cover change as % of initial year	98.9	97.9	97.8	96.1	96.5	95.6	97.0	99.8	97.4
Land cover 2000	**169 443**	**1 164 698**	**796 271**	**1 009 379**	**252 687**	**51 106**	**44 240**	**126 250**	**3 614 073**

PART A | Integrated assessment | Setting the scene

Map 2.2　　Dominant landscape types of Europe, based on Corine land cover 2000

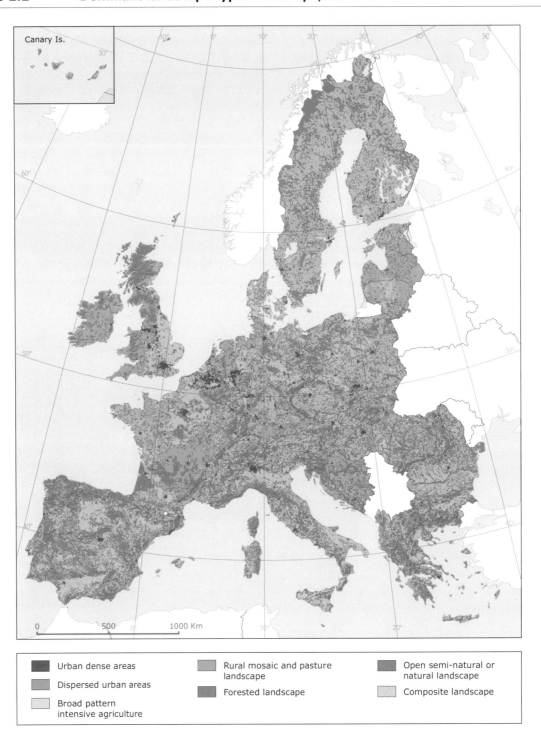

Land cover change is important both in terms of the total amount or net change in types of cover, and the actual locations where these changes occur. To understand the potential impacts on nature, both change information and spatial information are needed.

Starting with Europe as a whole, the net change in land cover between 1990 and 2000 highlights the increases in urban and other artificial land development and forest area, and the decrease in agricultural and natural area (Figures 2.1–2.3). The net change in artificial land area is a good indicator of urban sprawl, which is mostly an irreversible one-way process. The trends for total turnover confirm that urban sprawl was a key process in Europe in the 1990s, driven by economic growth and increasing consumption, suburbanisation and the implementation of the internal market (including transport infrastructure).

This sprawl is partly at the expense of natural land, and this development has important consequences for the long-term potential of the land to continue to provide ecological services and amenities.

In addition to demographic trends in rural areas, which in many places took the form of depopulation, the changes in agriculture and forestry can be ascribed mainly to the extension of the common agricultural policy, combined in some countries with rapid economic growth fostered by their accession to the EU and access to the internal market.

In the following sections, the three major components of the overall change in land cover are analysed in more detail, both at the European level and for some selected regions where the observed patterns and dynamics illustrate interesting policy perspectives. The three major components are:

- development of urban and other artificial land;
- decreases in the agricultural area resulting from a range of changes in use; and
- increases in the area of forest and decreases in the area of natural land.

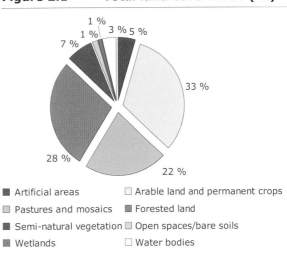

Figure 2.1 Total land cover 2000 (%)

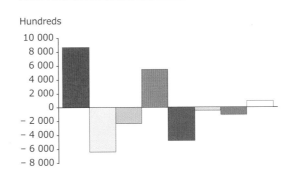

Figure 2.2 Net change in land cover 1990–2000 — EEA-23 (ha)

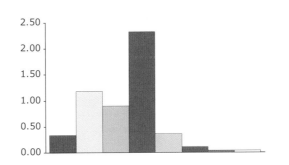

Figure 2.3 Total land cover turnover 1990–2000 as % of total territory for EEA-23

2.5 Urban sprawl and other artificial land development

Spatial perspective

Urban areas and infrastructure increased by more than 800 000 ha between 1990 and 2000, a 5.4 % increase and equivalent to the consumption of 0.25 % of the combined area of agriculture, forest and natural land. These percentages may seem small, but urban sprawl is concentrated in particular areas which tend to be where the rate of urban growth was already high during the 1970s and 1980s, and running alongside the emerging problems of rural depopulation. On a straight extrapolation, a 0.6 % annual increase, although apparently small, would lead to a doubling of the amount of urban area in little over a century. This needs careful consideration as we look ahead to the type of Europe we would like to see in the next 50–100 years, in the context of possible climate change and the many impacts and adaptation challenges it would pose.

A closer look reveals that sprawl around large agglomerations is continuing, but new development patterns can also be observed (Map 2.3). Urban development often takes place at a distance from large cities, around smaller towns or in the countryside. Further analysis shows that this is more visible for residential sprawl and the development of economic activities, in turn linked to the development of transport networks. Together, these factors contribute to the sealing of soil and the fragmentation of the natural landscape. This is largely a consequence of increasing passenger and freight transport demand, as well as increases in the price of urban land. The attractiveness of living in cities has fallen, while the quality of life associated with more rural areas, being closer to nature, has increased. This represents a planning challenge for small municipalities attempting to maintain their populations and attract small and medium-sized enterprises.

The extremely low price of agricultural land (in most cases good agricultural land) compared to already urbanised land (e.g. brownfield sites) or former industrial sites, is also an important factor underlying urban sprawl. In many development projects, the cost of agricultural land acquisition is relatively low and enables better profits to be realised than for already urban land or the use of former industrial waste land, even if no remediation is needed (non polluted sites). This factor is particularly important in the economic heart of Europe (also known as the Pentagon zone). The trend of good agricultural land being deliberately and artificially maintained at a low value is reinforced by the broad use of expropriation tools. A direct side effect of these combined tools — low value, future use not taken into account, and expropriation — is clearly demonstrated by the development of villages near cities, for residential or business purposes.

Urban sprawl is particularly important in coastal areas, and not just in the hinterland of urban coastal agglomerations. One of the world's 34 biodiversity hot spots, the Mediterranean area is particularly affected by these changes, although the level of artificiality of the coastline was already high before 1990. In the long run, this calls into question the sustainability of economic development based on tourism. Consequences in the immediate hinterland include the knock-on need for road infrastructure to accommodate the inland spread of individual housing.

Other areas with visible impacts of urban sprawl are in countries or regions with high population density and economic activity (Belgium, the Netherlands, southern and western Germany, northern Italy, the Paris region) and/or rapid economic growth (Ireland, Portugal, eastern Germany, the Madrid region), particularly where countries or regions have benefited from EU regional policies. New Member States, where little urban sprawl is detected, may follow the same path of urban development, and the accompanying environmental impacts will be all the higher because the very areas that are poised for change still host large amounts of natural landscape.

Drivers and impacts of artificial land development

On the European scale, the main drivers of urban development are housing (including related services), recreation, and industrial and commercial sites outside the urban fabric (Figure 2.4).

In several western countries, residential sprawl is accompanied by growth in recreational facilities, dominated by golf courses (Austria, Denmark, Ireland, Luxembourg, Spain, Portugal and the United Kingdom). Most of the development of these areas has been at the expense of agricultural land, mostly arable, but the picture varies from country to country. As much as 15 % of the land used for construction has been forest or semi-natural land, even higher than this in some particular regions.

Some 59 000 ha previously used by agriculture and 23 000 ha of forested and natural land on the 10-km strip of Mediterranean coast (five countries) were developed for housing, transport infrastructure and other needs between 1990 and 2000 (Figure 2.5). During the same period, 24 000 ha of natural land were converted to agriculture. This situation is typical of coastal zones on which agricultural land is scarce.

Figure 2.5 Origin of artificial land uptake 1990–2000, EEA-23 (%)

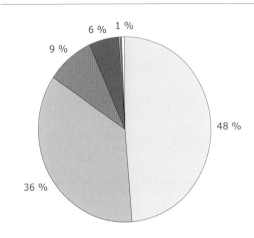

- Open spaces with little or no vegetation
- Natural grassland, heathland, sclerophyllous vegetation
- Forests and transitional woodland shrub
- Pastures and mixed farmland
- Wetlands
- Water bodies
- Arable land and permanent crops

Figure 2.4 Drivers of artificial land development

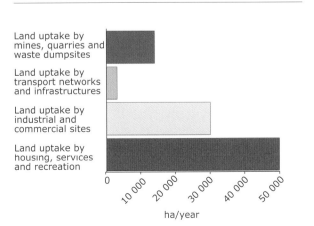

Country comparisons

At the country level, urban sprawl and associated developments during 1990–2000 were most intense in the densely populated the Netherlands and in Ireland, which was until recently particularly rural. Looking at the overall annual increase in urban/artificial land cover during 1990–2000, Ireland has the highest rank due to its very low initial level of urbanisation and strong economic development, closely followed by Portugal and Spain (Figure 2.6). All these countries were recipients of substantial transfers of funds under the EU cohesion policy. Germany, Greece and Luxembourg are among a group of countries close to the European average. The lowest values are generally found in the new Member States but also in Belgium and the United Kingdom.

Map 2.3 Sprawl of urban and other artificial land development, 1990–2000

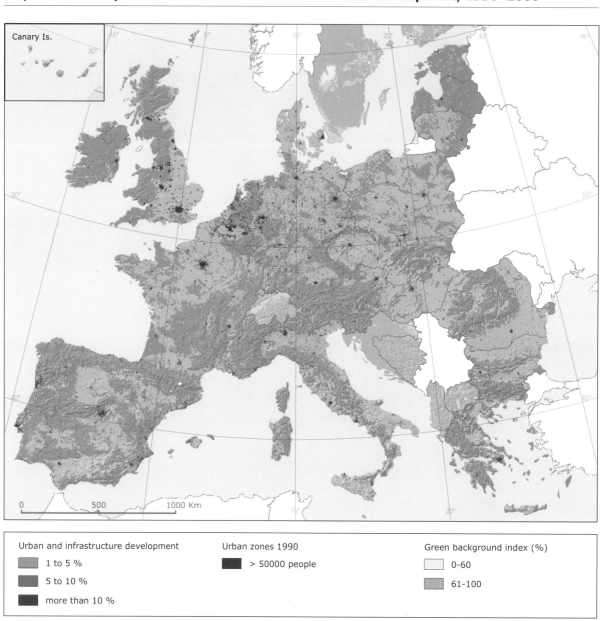

Urban and infrastructure development
- 1 to 5 %
- 5 to 10 %
- more than 10 %

Urban zones 1990
- > 50000 people

Green background index (%)
- 0-60
- 61-100

Typical change patterns

Sprawl in the countryside is observed in most countries or regions. Northern Italy, Ireland, the United Kingdom, and several regions of France, Germany and Spain can be given as examples. The contrast is marked between sprawl in the EU-15 and what we see in the other European countries. It is linked mainly to spatial planning developments for commerce and housing, which drive land price increases and conversion from agriculture, as well as the growing dependence on cars for commuting. This kind of diffuse urban sprawl meets the desires of people for more space, but also creates higher pressures on surrounding natural habitats. The type of discontinuous urban fabric that covers most of Belgium and the Netherlands is a good example of this phenomenon.

Map 2.3a

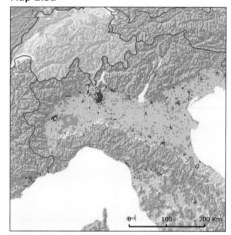

Sprawl along transport axes and the coastline: In large countries, transport networks — especially roads — often follow river corridors towards the sea. The so-called 'inverse T' of urban sprawl along the Rhône river down to the Mediterranean coast can be observed. The coasts themselves attract urban development for a range of reasons related to their attractiveness to tourists, and to urbanites looking for a higher quality of life by buying second homes. As a result, 1990–2000 was a period of marked change for the Mediterranean.

Map 2.3b

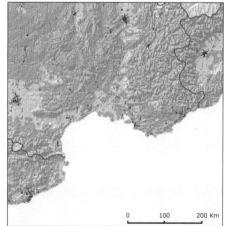

Time lags and uneven development. The 1990–2000 window is too early to capture many of the developments in the new EU Member States and the accession countries. Economic development in many of these countries is now accelerating, partly through their own dynamism and partly through their greater access to EU markets and the cohesion and structural funding that accompany membership. Comparisons between eastern Germany and Poland for the period 1990–2000 may provide insights for the future. Eastern Germany has benefited from large monetary transfers from western Germany since 1990, making it one of the fastest changing regions in Europe. Further east, in Poland, where EU membership is more recent, there has been less change during 1990–2000 and the contrast with Germany is still marked. This contrast is accentuated because of the region's history.

Map 2.3c

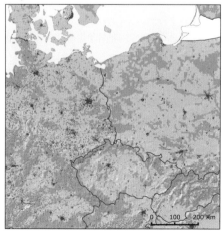

Do these numbers matter?

From cross-checking Corine land cover results for land uptake for artificial surfaces with other statistical surveys, it is most likely that there is an underestimation in the CLC results. This results, in particular, from the resolution of CLC, which cannot monitor small villages (< 25 hectares) and most roads and railways (narrower than 100 metres). Hence the overall picture of extension of artificial surfaces and their impact on landscapes and nature is probably more extensive than CLC reveals. Further information on data quality and methodological issues can be found in the two boxes at the end of the chapter.

Even though the annual increases in land uptake for most countries seem small, extrapolation into the future merits consideration. In order to see how the future might appear under certain assumptions, the 'rule of 70' — according to which an annual increase of 1 % in land uptake for artificial areas sees a doubling of urban development in 70 years — can be applied, as shown in the following table:

Annual rate of increase	1 %	2 %	3 %	4 %	5 %	7 %	10 %
Number of years for doubling	70	35	23	18	14	10	7

Source: Levy, Michel Louis, Comprendre les Statistiques, Seuil, Paris, 1979

We can conclude that if countries follow the Irish urban development rate of more than 3 % per year, they will double their artificial areas in just over 20 years; the Spanish rate brings a doubling in 40 years, the Dutch rate in 50 years, and so on. It is also possible, using this perspective, to ascertain the future(s) of the new Member States and accession and candidate countries which are just starting new development of their urban and transport infrastructures. This may be especially relevant in the context of how European cohesion funds are allocated and spent in the 2007–2013 period.

Figure 2.6   Mean annual urban and infrastructures land take as % of artificial land cover 1990

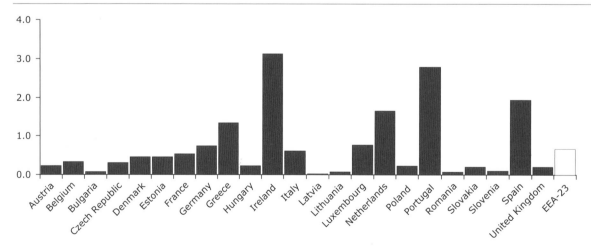

It is also interesting to consider the contribution of different countries to total urban land uptake in Europe (Figure 2.7). On this measure, Germany (21 %), France (14 %) and Spain (13 %) are the main contributors, due to their large surface area, followed by Italy (9 %) and the Netherlands (6 %). While the contributions of Portugal and Ireland are both below 5 %, these nevertheless represent large areas given the size of these countries.

The share of urban land uptake can be compared with total land cover turnover in the period 1990–2000 (Figure 2.8). This indicator needs to be interpreted carefully. For example, Ireland, Portugal and Spain have very low values because of the size and dynamism of their agricultural and forest sectors. Urban sprawl accounts for more than 50 % of total land cover change in the Netherlands, highlighting the competition for land between agriculture and urban development. Luxembourg, where agriculture is not as important, has a similar value to those of Austria, Belgium, Denmark and Germany.

2.6. The differentiation of European rural landscapes

Agriculture is the most dominant land use in Europe, covering twice as much land as forestry and more than 10 times as much as urban areas. European agriculture

Figure 2.7 Mean annual urban and infrastructures land take as % of total EEA-23 urban land take

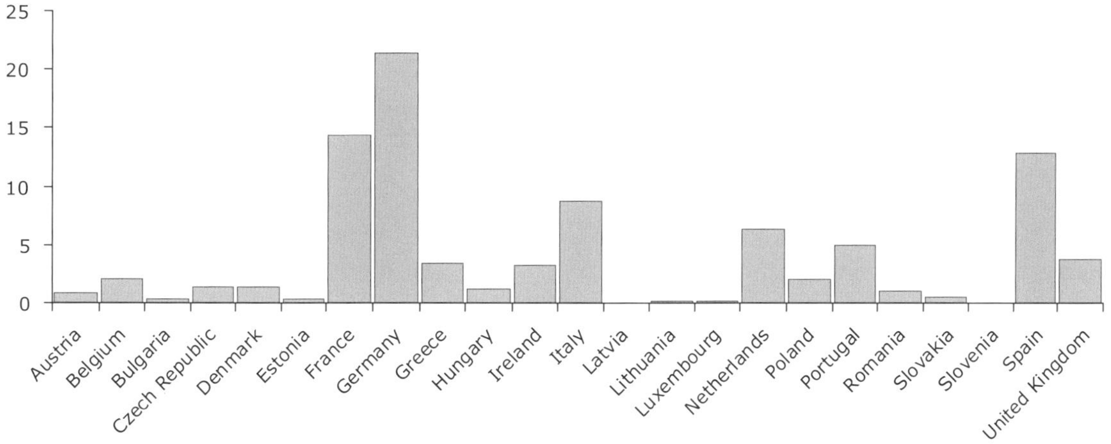

Figure 2.8 Mean annual urban and infrastructures land take as % of total land cover change 1990–2000

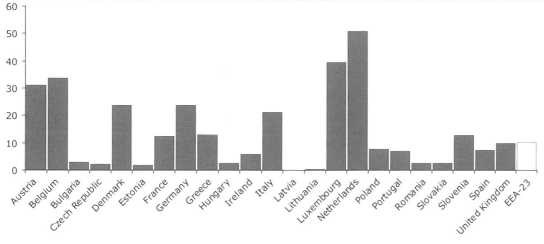

comprises a diverse mosaic of farming systems. The second half of the 20th century saw the transformation of many areas of traditional rural landscape into modernised, more intensive agriculture in response to the post-war drive for food security in Europe. This objective was initially at the heart of the common agricultural policy and has largely been achieved. The CAP has now been reorientated towards a wider rural policy perspective, more explicitly embracing environmental considerations and rural development issues. The accession of the new European countries, where western agricultural productivity levels have not yet been reached, has opened a new debate on reconciling development needs with the protection of semi-natural areas, in particular the dry grasslands that are such a characteristic element of the landscapes of Europe.

Spatial perspective

Due to the multiple drivers at work during the decade, land cover change in agriculture shows highly contrasting trends. Farmland abandonment coexists with intensification in the same countries, sometimes even in the same regions (Map 2.4).

The patterns that have emerged are largely a result of farmers' responses to changing economic and market conditions. Important contrasts have appeared between the more dynamic and productive areas, and the more stable areas that are prone to abandonment. Withdrawal of farming is often associated with knock-on conversions between pasture and crops elsewhere.

Conversion of new marginal land to agriculture appears to have taken place in Portugal and Spain and to a smaller extent in the south-west of France, eastern Germany and Hungary. This process is due in part to the scarcity of good land in some countries, where farmland is being used for other purposes, especially urban development.

Conversions between pasture and cropland are shown, with extensification — a possible prelude to farmland abandonment — sometimes occurring in the same region as intensification. Trends in eastern Germany and Hungary are very typical of these diverging trends, and can be linked to economic reforms in agriculture. The protection of pastureland in the Czech Republic is clearly evident, as well as the conversion from pasture to crops in south-east Ireland and other regions, often driven by more intensive livestock farming and the resulting demand for animal feed. Farmland abandonment has taken place in some mountain regions in southern Europe, in some parts of Germany and in new Member States such as Hungary and Slovakia. Abandonment and conversion of marginal land to agriculture are seen to coexist in some regions. Both trends are potentially detrimental to biodiversity.

Drivers and impacts

The main trend in Europe has been towards a conversion of arable land and permanent crops to pasture, set-aside and fallow land (Figure 2.9). There are three major aspects to consider: the conversion of agricultural land to urban sprawl (described in the previous section); conversion and rotation from pasture to arable land and vice versa within agriculture; withdrawal of farming with or without forest creation and conversion of forested and natural land to agriculture.

Long-term conversion between pasture and arable land is often linked to a switch between intensive arable agriculture and extensive livestock grazing. However, this is rarely the full story: for example, some pastures are intensively managed and cannot be considered as an extensive, low-input use of land. Country differences are important, with the Czech Republic and Germany accounting for more than half of the total extension of set-aside, fallow land and pasture.

At the European level, conversion of forest and natural land to agriculture is balanced by withdrawal of farming with or without woodland creation (Figure 2.10). National variations are important, and the maps show that opposite processes can happen in neighbouring regions and even in the same region.

The above conversions, even within the same region, seem to be either market orientated, with a clear link to scarcity of land in some places, or a purely individual choice linked to farmers deciding to retire for example.

Where conversions are not the desired ones, some tailor-made policies would be useful. Clearly, extensive practices might not be economically viable in their own right.

Country comparisons

Internal rotations within agriculture and conversions to and from agriculture represent more than half of the total turnover of land cover (2.8 % out of a total of 5.3 % land turnover as a percentage of initial year).

In most countries, the agricultural area has decreased at the expense of cropland or pasture/mosaic land (Figures 2.11 and 2.12). These changes are moderate in net terms except, as previously mentioned, in Ireland, where there has been an increase in crop production for animal feed, and in the Czech Republic, where farmland abandonment has been mitigated by an incentive policy to farmers for keeping or expanding pastures. Also notable is a small extension of arable land cover in the Baltic countries.

These overall net changes mask a range of changes and conversions that have taken place within countries. While no trend can be detected at the national level in most countries, major regional and local conversions can be identified.

Withdrawal of farming with and without woodland creation, and conversion of forest and other semi-natural land to agriculture vary between countries (Figure 2.13). High turnovers are observed in Hungary and Slovakia, where withdrawal of farming is the main component; in Spain, where conversions to agriculture are the main change; and in Portugal, where both processes take place.

Figure 2.10 Main annual conversions between agriculture and forest/semi-natural land cover in ha per year 1990–2000, EEA-23

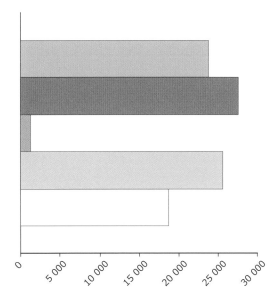

☐ Withdrawal of farming without significant woodland creation
■ Withdrawal of farming with woodland creation
■ Conversion from wetlands to agriculture
☐ Conversion from dry semi-natural and natural land to agriculture
☐ Conversion from forest to agriculture

Figure 2.9 Main annual flows of agricultural conversions in ha per year 1990–2000, EEA-23

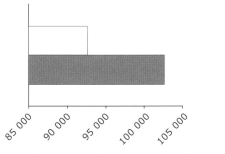

☐ Conversion from pasture to arable and permanent crops
■ Extension of set aside fallow land and pasture

Part A | Integrated assessment | Setting the scene

Map 2.4   Internal and external conversions of agriculture 1990–2000

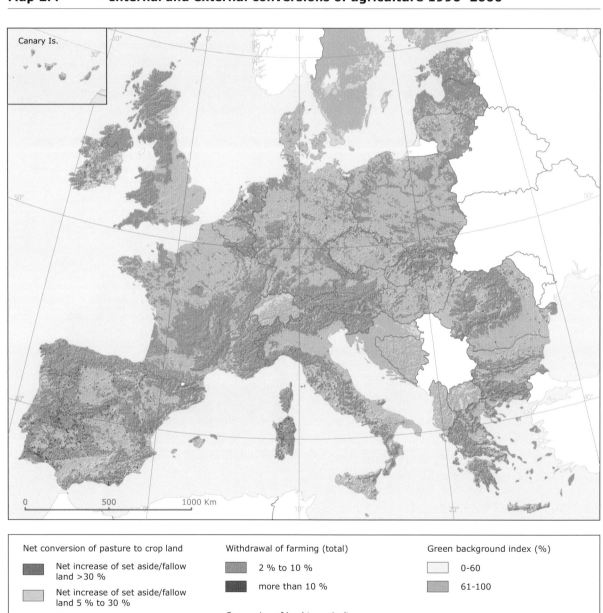

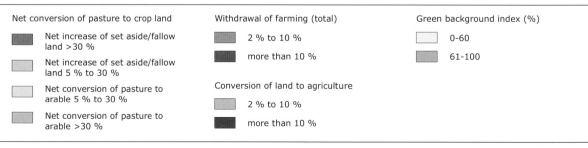

Typical change patterns: differentiation of agricultural landscape

Conversion of arable land to pasture or forest: To mitigate the effects of transition to a market economy, the Czech Republic created incentives for farmers to keep farmland managed as pasture wherever possible. This policy has been a huge success, resulting in a wide extension in pastures (bright green) during the period. A different approach was adopted in Slovakia where land was returned to its previous owners, who may not necessarily have been interested in farming. As a result, some withdrawal of farming with woodland creation has taken place. These two situations coexist in many parts of Europe.

Map 2.4a

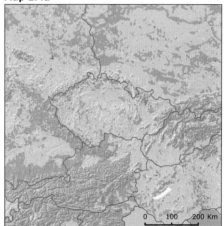

Withdrawal of farming and conversion of marginal land to crops: In the Iberian Peninsula, withdrawal of farming accompanied by woodland creation can coexist with new cultivation of open natural land. Part of the process is due to multiannual rotations between forested land (including transitional woodland and shrubs) and agro-forestry, with alternations of clearings and natural recolonisation. The rest results from reforestation policies, development of tree plantations and agricultural subsidies for crops such as olives. If not managed carefully, such changes can lead to the loss of valuable extensively managed habitats.

Map 2.4b

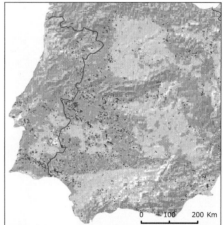

Conversion of arable land to pasture and withdrawal of farming: In overall terms between 1990 and 2000, France showed a slight decrease in agricultural area. This small overall change masks some regional contrasts, however. Areas south of Paris (dark blue) show withdrawal of farming, but conversions from pasture to arable land are visible (pink and yellow) in the wider *Bassin Parisien*.

Map 2.4c

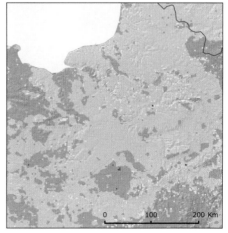

Figure 2.11   Net formation of agricultural land 1990–2000, as % of initial year, EEA-23

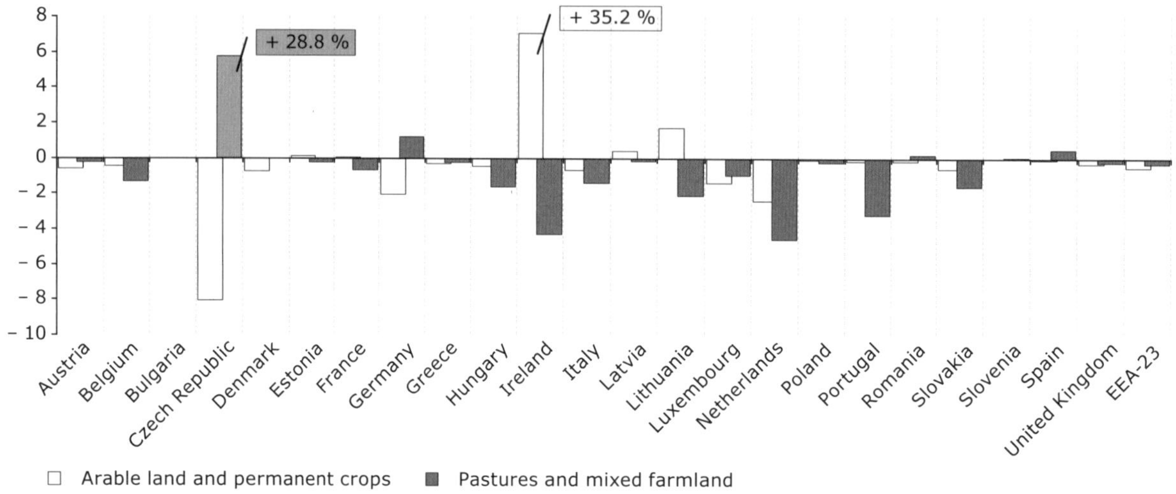

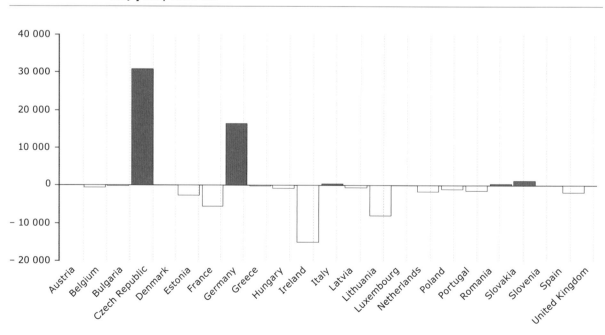

Figure 2.12   Net conversion from pasture (+) to arable land and permanent crops (−) ha/year, EEA-23

The changing face of Europe PART A

Figure 2.13 Conversions between agriculture, forest and natural land, as % of country area 1990–2000

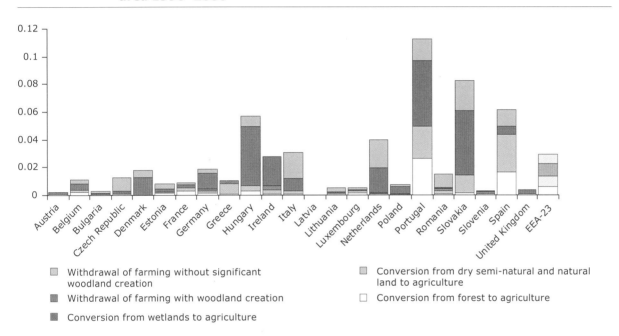

- Withdrawal of farming without significant woodland creation
- Withdrawal of farming with woodland creation
- Conversion from wetlands to agriculture
- Conversion from dry semi-natural and natural land to agriculture
- Conversion from forest to agriculture

2.7 The extension of forested land in peripheral regions

The total forested area of Europe has increased by 0.5 % in 10 years. During the decade, however, the forest territory has experienced significant rotations, up to 8 %, mainly as a result of felling and replanting. Of 1 million ha of new forested land, a quarter is the result of the withdrawal of farming (Map 2.5).

Spatial perspective

There has been major afforestation in Ireland, Portugal, Spain and the United Kingdom (Scotland). Afforestation of farmland is often an alternative source of income for farmers in regions where agriculture faces difficulties and has been subsidised by the CAP. For instance, Regulation (EEC) N° 1257/1999 provides for an aid scheme to promote afforestation as an alternative use of agricultural land and to develop forestry activities on farms.

Part A | Integrated assessment | Setting the scene

Map 2.5   **Afforestation in Europe, 1990–2000**

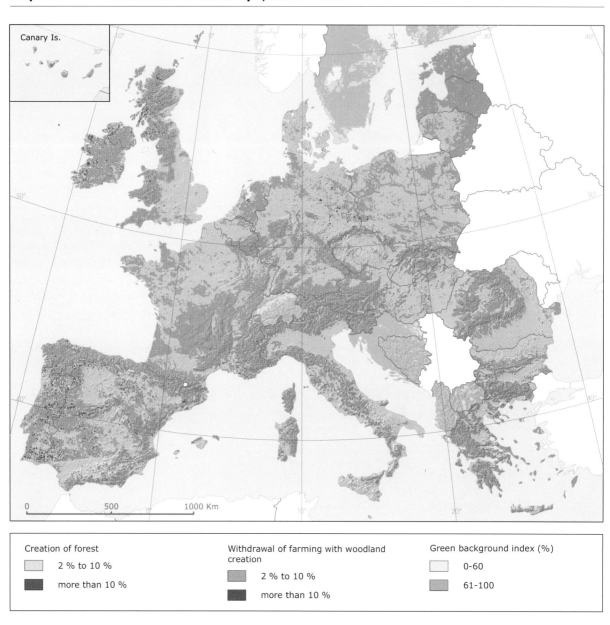

Typical change patterns: afforestation of semi-natural land

In Scotland, efforts to protect and plant native woodland (especially birch and oak) have continued; however, most new plantations are coniferous, which accounted for around 20 % of forested land in 2000. Ireland's forest cover has increased to approximately 10 % of the total land area, towards a target of 17 % by 2030. A limiting factor has been the scarcity of suitable and affordable land, with planting on blanket bog in the past because of its low agricultural value. Since the mid-1990s, policies have been aimed at switching away from planting on upland blanket bogs to wet mineral soils — of marginal value for agriculture, but very productive for forestry.

Map 2.5a

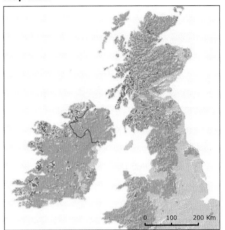

The total forest area of Spain increased during the 1990s, a sign of the success of afforestation plans. Policies have also contributed to maintaining the most valuable forests. New forest areas consisting of broad-leaved and mixed — rather than coniferous — trees, have mainly replaced transitional woodlands or dry semi-natural areas. In Portugal, forest creation was the main land cover change recorded. Continuing land abandonment, coupled with the withdrawal of management by burning, cutting and grazing, has allowed scrub encroachment and tree establishment in many areas throughout the country.

Map 2.5b

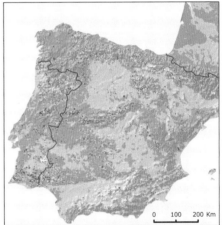

In Italy, withdrawal of farming and afforestation in the Alps and Apennine mountains has resulted from the abandonment of pastures and the decline of arable farming on terraces. This has been supported by common agricultural policy reform measures, notably EC Regulation 2080/92 on afforestation of agricultural land. In the French Mediterranean, forest creation is largely a result of the reforestation of semi-natural degraded land that had been damaged by fire.

Map 2.5c

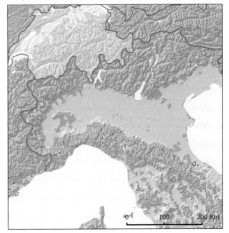

Drivers and impacts

Due to their role in maintaining the equilibrium of landscapes across Europe, changes in forest cover and types are important. There are specific ecological factors at work: for example, the rapid development of production forests in southern Europe not only creates poor ecosystems, but may also contribute to making these forests more prone to recurrent fires. Afforestation can lead to adverse effects as well: some natural dry land or wetlands used for plantations may have a high conservation value that is destroyed by afforestation.

Between 1990 and 2000 some deforestation took place for urban/infrastructure and agricultural uses (Figure 2.14). The deforested areas were on average small, but these changes can in some cases impact on the regional ecosystem. Forest creation on previously agricultural land, together with afforestation of open natural land, is a significant development in some countries (e.g. Ireland, the Netherlands, Spain, and the United Kingdom).

Woodland creation is also observed in peripheral countries or regions of the Atlantic and in some new Member States, as well as to a more limited extent in the Mediterranean mountain areas.

The other two categories of land cover change for forests are conversion from transitional woodland to forest and recent felling (Figure 2.14). The data for these two Corine land cover classes are not as accurate as forest inventories in each country, but the observed patterns are similar. The main advantage of the Corine approach is that it allows users to track the spatial distribution of forest trends in a consistent way across Europe.

Country comparisons

In general, the area of forest land across Europe has increased only slightly with the exception of Ireland, which had been the country with the least forest in Europe but where major afforestation has been undertaken (Figure 2.15). However, the area of open semi-natural and natural land (wetlands, dry grassland, heathland, sand and bare rocks, and glaciers in Austria and Italy) has generally decreased.

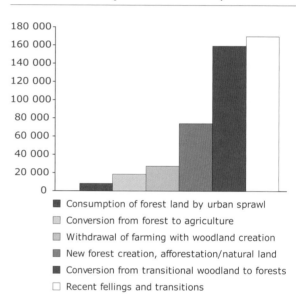

Figure 2.14 Main trends in woodland and forest formation, ha per year 1990–2000, EEA-23

- Consumption of forest land by urban sprawl
- Conversion from forest to agriculture
- Withdrawal of farming with woodland creation
- New forest creation, afforestation/natural land
- Conversion from transitional woodland to forests
- Recent fellings and transitions

The net formation of forest and natural land masks the much larger internal rotations that are taking place. These are important because they are a significant factor in determining forest age and ecological quality.

Careful management is a critical determinant of the ecological health of a forest. Extensive felling can degrade ecological quality, which is restored only when trees are allowed to come to maturity. If changes in the internal rotations of forests seem to balance out overall in Europe, significant rotations take place at the country level, including in countries where land cover change during the period has been slow such as Denmark, Latvia, Lithuania, and Luxembourg (Figure 2.16).

Afforestation on open natural land and woodland creation resulting from withdrawal of farming, has been an important shift in countries such as Hungary, Portugal and Slovakia. In terms of relative increase of forest area, Ireland is followed by Portugal, Slovakia, Spain, Hungary and the United Kingdom (Figure 2.17). When expressed as a share of total European forest and woodland creation, Spain and Portugal are the largest contributors, followed by Ireland and the United Kingdom.

Figure 2.15 Net formation of forest and natural land 1990–2000 as %, EEA-23

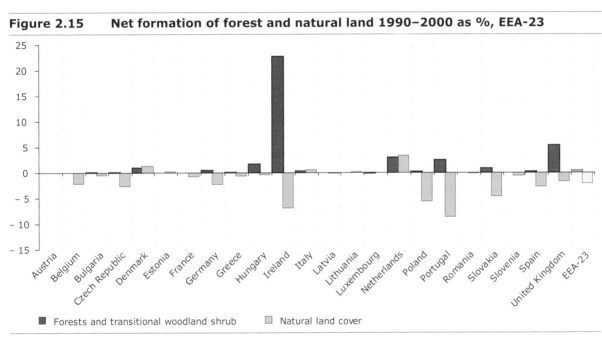

Figure 2.16 Internal rotations of forests ha per year as % of forest territory 1990, EEA-23

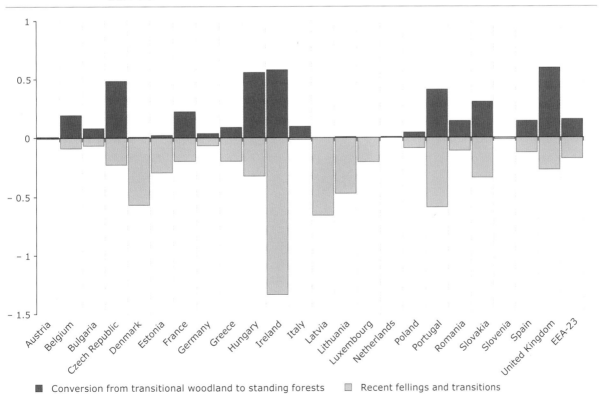

Figure 2.17   Contributions to total European forest and woodland creation (%)

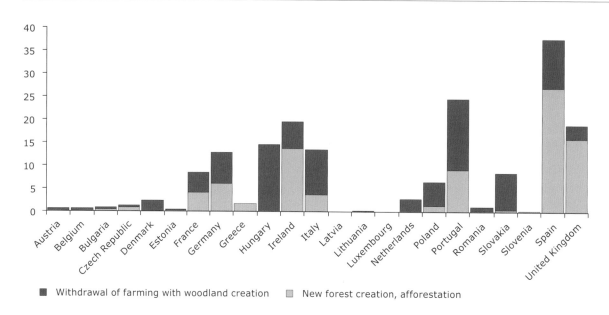

■ Withdrawal of farming with woodland creation   ■ New forest creation, afforestation

Figure 2.18   Change in the composition of European forests in ha 1990–2000, EEA-23

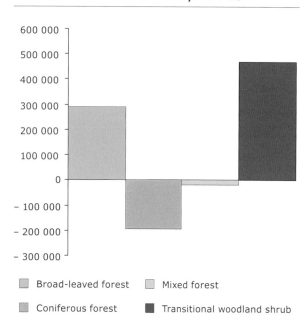

■ Broad-leaved forest   ■ Mixed forest
■ Coniferous forest   ■ Transitional woodland shrub

Analysis of the composition of forests shows the importance of internal rotations linked to the forest cycle of felling and replantations as well as a slight decrease in coniferous forest, and an increase in broad-leaved forest (Figure 2.18).

2.8   Summary and conclusions

The way in which we perceive landscapes and are attracted to some of them, and our feelings when conflicts arise over the use of land, are all matters of extreme importance for conservation and future human welfare. Landscapes change as a result of natural processes and human influence. It is as important to know where change is happening as to know when. This is especially so given the uneven distribution of ecological goods and services across Europe, the vast range of activities that impact on them and the changing character and intensity of these impacts over time.

Land use patterns across Europe show that tensions are arising almost everywhere between our need for resources and space and the capacity of the land to support and absorb this need. Globalisation, agriculture, transport networks, demographic changes and land-planning mechanisms at the national level are the main sources of environmental pressure. There is now rising awareness of the value of considering the territory as a unit of analysis and as a basis for stimulating better coordination of sector policies.

In the 1990s in Europe, changes in land cover were mainly characterised by increases in urban and other artificial land development and forest area at the expense of agricultural and natural areas. Urban areas and infrastructure increased by 6 %; on a straight extrapolation this would lead to a doubling of the amount of urban area in Europe in little over a century. Urban sprawl is concentrated in particular areas, which tend to be those where the rate of urban growth was already high during the 1970s and 1980s. Urban sprawl is also significant in the coastal zones. In the context of possible climate change and the many impacts and adaptation challenges we will face as a result, these perspectives deserve careful consideration.

Some 1 million ha of new forested land were created in Europe in the 1990s, with about a quarter the result of the withdrawal of farming. There has been major afforestation in Ireland, Portugal, Spain and the United Kingdom (Scotland). Afforestation of farmland has been subsidised by the CAP and is often an alternative source of income for farmers where agriculture is difficult.

Agriculture is the most dominant land use in Europe and comprises a diverse mosaic of farming systems. The accession of the new European countries, where western agricultural productivity levels have not yet been reached, has opened new debates on reconciling development needs with the protection of semi-natural areas, in particular dry grasslands. In the 1990s, land cover change in agriculture showed highly contrasting trends with farmland abandonment coexisting with intensification in the same countries and sometimes even in the same regions.

These diverging trends can be linked to economic reforms in agriculture. Conversion from pasture to crops is often driven by more intensive livestock farming and the resulting demand for animal feed. Farmland abandonment has taken place in some mountain regions in southern Europe and in some new Member States. Abandonment and conversion are both potentially detrimental to biodiversity. Future CAP reforms could help mitigate such impacts.

In policy terms, Europe is debating a stronger and more balanced territorial focus for its policies through the European spatial development perspective. The long-term goal is to see a European territory with many prospering regions and areas each playing an important economic role for Europe and providing a good quality of life for its citizens.

References and further reading

ESPON, 2005. *Synthesis report II, In search of territorial potentials — Mid-term results by spring 2005.* (See www.espon.lu/online/documentation/programme/publications/index.html — accessed on 18/10/2005).

European Environment Agency, 2002. Towards an assessment of European landscapes —methodological developments. Unpublished working document.

European Environment Agency, 2004. *Corine Land Cover 2000, Mapping a decade of change*. Brochure, EEA, Copenhagen.

Data source and quality

Corine land cover (CLC) is a globally unique, independent inventory: it is built on a single European classification of land cover types which makes it an invaluable tool for Europe-wide assessments and for comparisons between countries, regions and other zones of interest.

The first Corine land cover map was finalised in the early 1990s. The updated Corine land cover 2000 (CLC2000) is based on the results of IMAGE2000, a satellite imaging programme undertaken by the Joint Research Centre of the European Commission, together with the EEA. Today 29 countries and more than 100 organisations are involved in the production and dissemination of the CLC2000 data. This updated Corine used the same methodological rules and comprised an independent mapping of the change in land cover and a revision of the 1990 database.

The strength of CLC is in its use with other spatial environmental databases. Across the European territory, 44 different land cover types are distinguished and mapped using photo-interpretation of satellite images by national teams in the participating countries. These national land cover inventories are then integrated into a seamless land cover map of Europe. The resulting European database uses a standard methodology and nomenclature, resulting in a powerful tool for use both within and between participating countries. Given the huge amounts of satellite data and other information used, processing and validation across the 29 participating countries takes several years to complete. This is why the use of data from the 2000 inventory only began in earnest in 2005.

However, as with any data set, CLC has limitations related to the observation tool and methodology used. CLC is an analysis and mapping of landscape units on the basis of their physiognomy and radiometric characteristics. It is not, however, a classification of pixels, nor a survey of hectares of a given homogeneous type (as monitored by farm surveys or area-sampling surveys). Rather it is an appropriate background reference for analysing potential conflicts in the use of land and the impacts of land use pressures on biodiversity, and for organising and integrating other sources of information accordingly.

The smallest mapped and classified unit in CLC is 25 ha. Thus, more or less all CLC classes, monitored from satellite imagery, may include significant heterogeneous micro-areas of less than 25 hectares. CLC therefore cannot deliver a very accurate assessment of surfaces (e.g. as needed for agriculture statistics used for calculating crops and related subsidies). As a result of the 25-ha limitation, the Corine classification also includes mixed classes ('discontinuous urban fabric' and 'land principally occupied by agriculture with significant areas of natural vegetation'). These classes have a high interest from an ecological perspective.

CLC land units will disappear or pop up when they come just below or just above the 25-ha threshold. This is consistent with monitoring landscape systems. Considering the mapping of changes in CLC2000, the smallest mapped change is 5 ha. It may therefore happen (but very rarely) that a 5–24 ha change results in the creation or deletion of a small zone. In order to avoid any misinterpretation, the user will have access to and can compare three datasets: CLC1990-revised, CLC change 1990–2000 and CLC2000. These will be available on the EEA website from early 2006.

CLC2000 has been prepared and its quality controlled by the EEA. CLC1990, an experimental programme using images from 1986 to 1994, did not meet the same standards but, after 10 years of extensive use, can now also be considered as of good quality. Moreover, the original CLC1990 was revised during the CLC2000 process in order to fix possible errors and eliminate geometric discrepancies that could generate false change. However, problems still remain, especially for some of the countries which were pioneers of applying the Corine methodology in the 1980s and because of the different time periods in countries between the production of CLC1990 data and the CLC2000 updates. These are being resolved through using the data and improving it in consultation with national experts.

Using accounting methods to analyse spatial changes

The land and ecosystem accounts method (LEAC), as developed by EEA, provides a framework for analysing spatial land cover changes. Considering Corine's 44 land cover classes, there are approximately 1 900 possible one-to-one changes from one Corine class to another. LEAC essentially presents a typology of these changes, classifying the changes into types of flows. The flows are classed as: 'urban land management', 'urban residential sprawl', 'sprawl of economic sites and infrastructure', 'agriculture internal conversions', 'conversion from forest and natural land to agriculture', 'withdrawal of farming', 'forest creation and management', 'water-body creation and management' and 'changes due to natural and multiple causes'. Flows are then combined with the 1990 and 2000 stocks to assess the relative importance of the various processes. Making full use of CLC, land cover accounts are computed at the most detailed level, and tables and indicators can be produced and mapped for any kind of geographic zone, from countries or river basins to regions or small areas. The EEA's full land and ecosystem accounts compendium and associated statistics are available at www.eea.eionet.eu.int/Public/irc/eionet-circle/leac/library?l=/leac_stat&vm=detailed&sb=Title — accessed on 18/10/2005.

In addition to indicating land cover, land accounts are designed as a framework into which other data and statistics can progressively be introduced. Some of these data will relate to changes in the structure, patterns, productivity, species composition and quality (health) of the land cover units considered as images of ecosystems. Other statistics will specifically address the land use issue. Land use relates to the many economic and social functions of land: housing, food production, industrial activities, services, transport, recreation and nature protection. There can be many uses of land on the same land cover unit, and their various functions need to be described using socio-economic statistics. Because of the common infrastructure provided by land cover accounts (based on CLC), ecosystem accounts and land use accounts are bridged in a system that facilitates analysis of the interactions between the economy and the environment.

Land cover change in terms of numbers of total changes or a net balance of surfaces is not particularly useful in terms of interpreting environmental impacts. The actual locations where the changes take place matter more, particularly when looking at the potential impacts of land use on nature. These impacts result from the sealing of soil and fragmentation by the development of artificial surfaces and linear infrastructures, which lead to the quasi-irreversible destruction or degradation of natural ecosystems, and from noise and pollution, generated by transport and other intensive land uses. Other degradation may result from the conversion of forests and natural land to agriculture and, in particular contexts, from the use of natural land (including wetlands) for productive afforestation. In addition to the direct and irreversible loss of land occupied by natural habitats, these various intensive uses contribute to creating barriers that risk fragmenting the ecological network. Background landscape maps have proved to be efficient for analysing and presenting land cover changes in their proper context. These 'dominant landscape types' and 'green background' maps are presented and discussed in this chapter.

3 Climate change

3.1 What is climate change?

Weather is something we experience every day. It relates to whether the sun is shining or whether it is raining, what the temperature is, and the direction and force of the wind. Climate is the average weather over a long time period.

The climate is not static: it has changed in the past, over centuries, millennia and even longer periods of time. The natural causes of this include fractional changes in solar radiation, volcanic eruptions that can shroud the Earth in dust, and natural fluctuations in the climate system itself, such as, for example, the North Atlantic Oscillation.

Recent research into past climate — involving detailed analysis of tree rings, ice cores, ocean sediments, and coral and plant remains — reveals a period of about 8 000 years of overall stability, with global average temperatures moving only by small fractions of a degree Celsius. Over the last millennium, the first 900 years saw only small fluctuations in average global temperatures in the northern hemisphere of less than 1 °C, followed by rapidly rising temperatures in the last 50 years or so (Figure 3.1).

Average global temperatures are now about 0.7 °C above pre-industrial levels, and are currently rising faster than at any time in modern human society. Nine of the 10 warmest years in a detailed thermometer record extending back for 150 years have occurred in the past decade, with the four hottest years globally in 1998, 2002, 2003 and 2004. Projections for the next 100 years show a continuation in this trend, estimates of the global increase ranging from 1.4 °C to 5.8 °C.

In Europe, temperature rise has been even greater than the global average during the 20th century, namely 0.95 °C. The greatest warming has been in the Iberian Peninsula, north-west Russia and parts of the European Arctic. In Europe, the eight warmest years on record have all been since 1990, with the hottest in 2000. The European average temperature is projected to further rise by 2.0 °C to 6.3 °C in the next 100 years.

Initial scientific concern that this global warming might to a large extent be due to emissions of greenhouse gases caused by human activities has now grown to near certainty. The UN Intergovernmental Panel on Climate Change (IPCC), a global organisation of scientists, was set up by the World Meteorological Organization and the United Nations Environment Programme in 1988 to review the evidence. It concluded in 2001 that, while many of the fluctuations in temperature until the mid-20th century could be due to natural events such as volcanic eruptions and variations in solar activity, 'there is new and stronger evidence that most of the warming observed over the last 50 years is attributable to human activities, in particular to the emission of greenhouse gases'.

The important factor is the large rise in concentrations of greenhouse gases in the atmosphere. These gases trap heat that is radiated from the surface of the Earth and prevent it escaping to space. The effect has been known for more than a century, and is now directly measurable in the atmosphere. The prime culprit is carbon dioxide (CO_2), a gas emitted when (fossil) fuels are burnt. The main fossil fuels are coal, oil and natural gas. These are made up of plant and animal matter millions of years old. Another cause of the increase of CO_2 in the atmosphere is the large-scale cutting of forests (deforestation).

Human activity is currently sending around 25 billion tonnes of CO_2, the most relevant greenhouse gas, into the atmosphere each year. The gas typically persists in the atmosphere for around a century before being absorbed by the oceans and ecosystems on land. Because of its long atmospheric lifetime, these CO_2 emissions have caused a steady rise in concentration of the gas in the atmosphere: the current rate is between one and two parts per million each year. A pre-industrial atmospheric concentration of the gas of between 250 and 280 parts per million (ppm) has risen to around 375 ppm today — higher than at any time in the past 500 000 years.

Man-made emissions of other greenhouse gases such as methane, nitrous oxide and fluorocarbons have raised concentrations of these gases in the atmosphere too.

These increases have been sufficient to have the same warming impact as a further 50 ppm of CO_2. The IPCC scientists have concluded that, taken together, these accumulations of greenhouse gases are the prime cause of recent climate change — and the likely cause of future warming.

3.2 Indications of climate change

Indications of climate change are already visible across the world. Most obviously, warming is leading to most of the world's mountain glaciers and the Greenland ice sheet melting. In general, warming is highest in polar regions. There, melting ice means that more of the solar energy reaching the Earth's surface is absorbed, and less is reflected back into space. Rises in Arctic winter temperatures have reached 5 °C in some places already, seven times the global average rise.

There are other indications that weather patterns are shifting around the world, due to extra heat energy in the climate system caused by rising temperatures. In the Pacific Ocean, the periodic fluctuations known as

Figure 3.1 Reconstructed and measured temperature over the last 1 000 years (northern hemisphere) and projected temperature rise in the next 100 years

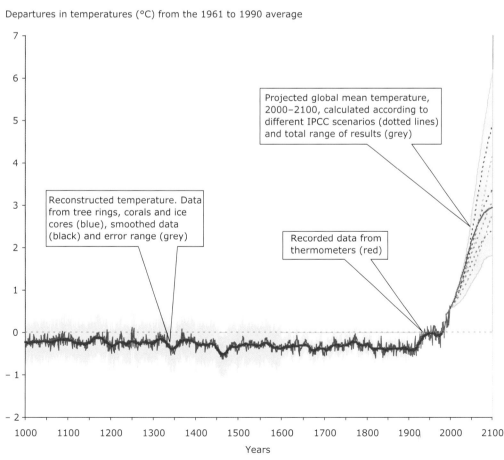

Source: Mann et al., 1999 (last 1 000 years); IPCC, 2001 a (projections for the next 100 years).

El Niño events appear to be becoming more frequent and intense. Tropical storms are afflicting new areas. In the Southern Ocean, weather systems that once brought rain to south-west Australia now often do not make landfall. Other weather systems are hitting the Antarctic Peninsula where once they were unknown.

The greater energy in the atmosphere is also causing a rise in extreme conditions of all sorts, including drought, heavy rain, heatwaves and sometimes even intense cold. Europe has seen an increase in floods in recent years — there were 238 flood events between 1975 and 2001, and 15 major floods in 2002 alone — and in heatwaves and forest fires. As crops fail and floods make some areas increasingly inhabitable, these events are starting to have a negative effect, especially on vulnerable societies and economies. Rises in temperatures in the Arctic and the loss of sea ice are damaging ecosystems and the indigenous cultures that depend on them.

Two of the most visible impacts of higher temperatures in Europe are melting ice and reduced snowfall. Eight out of nine glaciated regions in Europe show significant retreat of glaciers in the past century. In the Alps, glaciers lost a third of their area and half of their mass between 1850 and 1980. This retreat has gathered pace since 1980, in line with accelerating climate change. A further quarter of the Alpine glaciers had gone by 2003, with 10 % disappearing during the hot summer of 2003 alone. Studies into past

Figure 3.2   Number of reported deaths and minimum and maximum temperatures in Paris during the heatwave in summer 2003

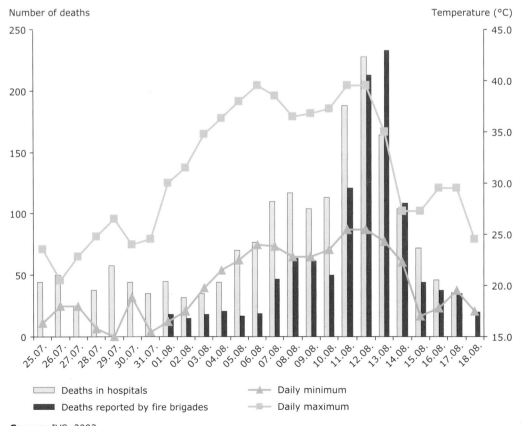

Source: IVS, 2003.

climates suggest the region has not seen a change of this scale for at least 5 000 years.

Across Europe there is less snowfall and more rain. As a result, winter snow-cover has decreased significantly across Europe since the 1960s.

In the Arctic north of Europe, warmer air and water have caused melting of sea ice. Recent measurements point to the lowest recorded area of sea ice since 1978 when satellite records became available. The current rate of shrinkage is estimated at 8 % per decade; if this continues, there may be no ice at all during the summer of 2060. Meanwhile, ice thickness has also decreased by an average of around 40 %, while the period of summer melt across the Arctic has increased by more than five days since 1979.

Chapter 8 examines the effects of climate change on biodiversity. At a landscape level, it is worth noting that the average annual growing season for plants has lengthened by 10 days across Europe since 1960, and plant productivity has risen by 12 % in the same period. Taken together, these two factors have increased the continent's 'greenness', although the picture is variable.

Growing water shortages and excessive temperatures in southern Europe are starting to shut off this trend, and climate models suggest that much of the continent may start to 'yellow' in future, as deserts advance.

It is often difficult to disentangle the effects of climate change from other factors such as changing land use. Across Europe, however, climate change already appears to be impacting many sectors of society. Higher temperatures and more intense droughts are producing a rising trend in the number and severity of forest fires in the Mediterranean. These threaten forestry, farming, tourism and the suitability of the land for habitation. Meanwhile, disappearing glaciers are damaging winter tourism in the Alps. Changes in rainfall and flows from glaciers are altering river flows, sometimes causing floods or emptying reservoirs. Higher summer temperatures are intensifying photochemical smogs, raising ozone concentrations to levels that increase the likelihood of damage to health.

It is impossible to say whether the heatwave across Europe in 2003 was caused directly by climate change. Extreme events usually have many causes, but undoubtedly, by raising average temperatures,

Figure 3.3   Number of flood events

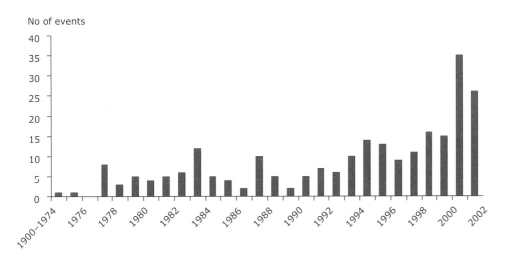

Source: WHO-ECEH, 2003.

climate change makes such extreme heatwaves more likely. Climate models suggest that the likelihood of the occurrence of heatwaves has doubled in recent years and that they are likely to become increasingly frequent in future.

High temperatures are a threat to human health. The 2003 heatwave saw 20 000 more people die in Europe than in the same period in other years, some 14 000 of them in France. Most people died from heat stroke, and heart and respiratory ailments, as daily maximum temperatures rose to 40 °C and, perhaps of equal importance, night-time minimum temperatures stayed above 25 °C on the warmest nights (Figure 3.2).

The World Health Organization (WHO) is concerned that the annual death toll from heatwaves might multiply by mid-century as a result of climate change. Personal efforts to stave off the worst effects of the heat are likely to result in a big increase in the use of air conditioning across much of Europe. This will, of course, have a knock-on effect on energy generation and consumption and of subsequent greenhouse gas emissions.

According to WHO, higher temperatures also increase the incidence of a wide range of diseases, from allergies such as hay fever through asthma attacks triggered by ozone smogs to food poisoning, which has a well-established association with temperature, and even the spread of tick-borne ailments such as Lyme disease. Potential malaria risk areas may increase, and a doubling of epidemic potential over Europe is predicted by UNEP-Grid/Arendal.

Figure 3.4 Economic and insured losses caused by weather and climate related disasters in Europe

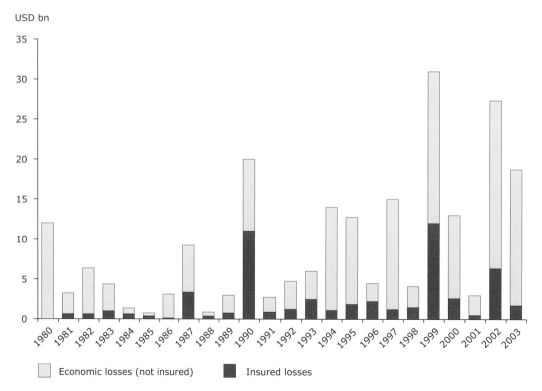

Source: NetCat Service, Munich Re, 2004.

Climate extremes are creating ever greater risks of catastrophic events of all sorts. Flood events in particular have soared in Europe and, although better warning and rescue systems have prevented a commensurate rise in deaths, the loss to property has been substantial (Figures 3.3 and 3.4). The severe flooding in Austria, the Czech Republic, Germany, Hungary and Slovakia in August 2002 caused economic losses of about EUR 25 billion. Flooding occurred again in eastern Europe in 2005.

3.3 Possible future impacts

Rising temperatures and changes in precipitation

The IPCC has stated that if the world continues on its current economic and technological trajectory, without any specific climate change policy being introduced, the projected increase in temperatures worldwide is expected to be between 1.4 °C and 5.8 °C by 2100.

The future temperature will depend on how sensitive the climate is to the 'forcing' effect of greenhouse gases, and on the pace and type of global development. Recent studies during preparation for the next assessment from the IPCC, due in 2007, suggest the temperature may be towards the upper end of this range.

According to model calculations, on current trends Europe can expect a rather bigger increase in temperatures than the global average over the coming century — between 2.0 °C and 6.3 °C — but the change will not be uniform across the continent. Within Europe, temperature rise is expected to be marginally greater in Greece, Italy and Spain, as well as in the north-east of the continent, but it is likely to be less along the Atlantic coastline, where the moderating influence of ocean temperatures will continue to be felt. By 2080, on current trends, nearly every summer in many parts of Europe will be warmer than the current hottest summers.

Meanwhile, precipitation rates are also changing. There are strong differences in regional and local trends, of course, but during the 1990s rainfall in northern Europe was 10–40 % greater than long-term averages, while southern Europe was 20 % drier. Such changes may be exceptional, due in part to natural climate cycles like the North Atlantic Oscillation, but climate models suggest this continent-wide trend of a wetter north and a drier south will persist and strengthen. Furthermore the current trend towards more drought and intense rainfall events in different parts of Europe will probably continue.

People will try to adapt to these changes. For example, in farming, more crops may be able to be grown as the growing season extends, particularly in northern Europe. In some places new farming areas might open up or new crops be grown. However, it is expected that such changes will be offset by adverse effects on agriculture in many parts of Europe.

In the droughts and higher temperatures in southern Europe, there will probably be lower yields and abandonment of farmland. High temperatures will mean that the effective period when some plants grow may actually shorten. Farmers will need more irrigation water (and to use it more efficiently) to survive in southern Europe. The expected decline in rainfall will often leave rivers running dry, and the impact of fewer water resources could be even more damaging for farmers than higher temperatures. Meanwhile, crops may be at greater risk from pests and diseases, including invaders against which the plants have no defence.

Adaptation will not only be necessary for agricultural activities. As climate zones shift, the flora and fauna associated with them will show different distribution trends as well. Some species will be able to adapt within a certain range, some will expand into new territories, whilst others, including many of those in mountain ecosystems, will be left with little habitat available. Studies suggest that in the Alps a 1 °C warming could lead to the loss of 40 % of the area's endemic plants and a 3 °C warming up to 90 %, while a 5 °C warming could bring about a 97 % loss. An evaluation of the coherence and adaptability of networks of protected areas is urgently needed in order to identify ways of reducing this risk.

Melting of ice and snow can be expected to continue apace. By 2050, three-quarters of today's glaciers in parts of the Alps are expected to have disappeared. Melting in the Arctic will be even greater if, as expected, warming continues at more than double the rate in lower latitudes. The area of the Arctic Ocean covered by sea ice is projected to shrink by 80 % by 2050.

The disappearance of ice could open up sea lanes in the Arctic, increasing the potential for trade, industry and the exploitation of resources such as oil and natural gas. Warming will cause permafrost to melt, damaging infrastructure such as roads, buildings and pipelines. As shore ice melts, low-lying coastal areas will be exposed to flooding during storms at sea. Indigenous lifestyles around hunting fish and polar bears and herding reindeer are already being affected as changes in the ice alter migration patterns. Such lifestyles could be extinguished if the changes continue.

Sea level rise and the impact on the marine environment

Sea levels are already rising globally. This is a result of both the thermal expansion of ocean water as it warms and the melting of ice on land. The rise in sea levels around the shores of Europe during the 20th century has been between 0.8 centimetres a decade in the western approaches of Brittany in France and Cornwall in the United Kingdom, and up to 3 centimetres a decade on the Atlantic coast of Norway. The variable trend is caused by local conditions and movements in the height of the land surface. Although the changes in sea level may not seem substantial, in low-lying areas even small changes can flood large areas of land.

This trend of rising sea levels is expected to double or even quadruple in the 21st century. There will be even more to come, because of the very long time lags in transmitting rising air temperatures into the ocean depths and through large masses of ice: it takes decades or centuries for the heat to penetrate.

Warming in the open ocean has so far been restricted to the topmost 200 or 300 metres, but ultimately it will work its way down to the ocean floor. As the warming penetrates, thermal expansion will continue. Even if air temperatures were stabilised today, a combination of thermal expansion of the oceans and melting ice pouring more water into the oceans would continue to raise sea levels.

Rises in sea levels, when coupled with increased risk of extreme storms, will often require big increases in investment in sea walls along Europe's long shoreline. Some governments, for example in the United Kingdom, have endorsed the concept of 'managed retreat', where in some low-lying rural areas the sea is allowed to have its way.

Rising sea temperatures are also having direct effects on Europe's coastal ecosystems. Warming has so far been greatest in isolated basins like the Baltic and in the western Mediterranean. Blooms of phytoplankton grow in the warmer waters, especially when fertilised by flows of nutrients from the land.

These blooms reduce oxygen levels and are sometimes toxic for fish and other wildlife, and even to humans. Meanwhile, zooplankton and the fish that feed on them have followed temperature trends and migrated up to 1 000 kilometres north.

The threat of abrupt climate change

There is growing scientific concern that climate change may prove to be more rapid and pronounced than current projections from the IPCC suggest. The next report of the IPCC is expected to reflect this. In particular, there are fears that the climate system contains the potential for abrupt change, that is, change that, once triggered by warming, can not be reversed even by the subsequent lowering of greenhouse gas concentrations or global temperatures.

The IPCC scientists are still uncertain, but there are theories that much of the climate system may be programmed to operate in a series of relatively steady states, but that, under stress, it may jump from one state to another in just a few years.

One such change could be the rampant melting of the large ice sheets on Greenland and West Antarctica.

These two huge masses of ice have the potential to raise sea levels worldwide by 13 metres. Once begun, say some glaciologists, melting of the Greenland ice cap would be hard to stop because melting itself would raise local temperatures. It would do this in two ways: first by reducing the ice cover that reflects solar radiation back into space, leaving more to be absorbed; and, second, by lowering the level of the ice surface, where it would experience higher air temperatures.

The irreversible melting of the Greenland ice cap could be triggered by local warming of less than 3 °C, according to recent research. The accelerated warming of Arctic regions so far suggests that local warming of 3 °C might be triggered by a global warming of just 1.5 °C, so we are already more than halfway to this point as a result of past emissions.

Another abrupt impact of climate change with potentially large consequences for western Europe in particular is the collapse of the ocean thermohaline circulation. This is a global ocean circulation, part of which includes the North Atlantic Current that brings warm water north from the tropical Atlantic. It largely prevents Europe experiencing temperatures more typical of its latitude — such as those of a Siberian winter.

The thermohaline circulation appears to be either switched on or off, with little in between. It may have switched off thousands of years ago, plunging Europe into much cooler temperatures. This thermohaline switch may have been one of the triggers that pushed the world into and out of ice ages.

The circulation itself is driven by salinity differences in the ocean, particularly within European territory in the far North Atlantic. Circulation might be switched off within a few decades, if the water in that region of the ocean became less saline. This could happen as a result of greater melting of ice in Greenland or greater precipitation in the Arctic region generally, both of which would cause large amounts of freshwater to flow into the critical region, so reducing seawater salinity. Both are possible consequences of climate change.

The effects of a collapse of the North Atlantic circulation on European climate remain uncertain. It might simply moderate the effects of global warming in western Europe, but at the other extreme it could push temperatures further downwards, creating what some have termed a 'new ice age' in Europe. Given our limited state of knowledge on the ocean climate at present, it is not possible to predict whether or when this might happen.

Other potentially catastrophic events could include:

a. The release of large amounts of the greenhouse gas methane from frozen tundra and continent shelves, where it is known to be trapped in frozen lattices known as hydrates. This could send global temperatures soaring even faster than current models suggest.

b. A change in how terrestrial ecosystems exchange CO_2 with the atmosphere. Currently, they act as a net sink for atmospheric CO_2, absorbing some of the emissions from burning fossil fuels. Some models suggest that as temperatures rise and ecosystems such as the Amazon rainforest die, they could by 2050 be converted into net sources of CO_2 releases into the atmosphere. This could again accelerate climate change.

3.4 International efforts to halt climate change

In 1992, at the Earth Summit in Rio de Janeiro, Brazil, most of the world's governments signed the UN Framework Convention on Climate Change (UNFCCC). It set as its long-term objective 'the stabilisation of greenhouse gas concentrations in the atmosphere at a level that would prevent dangerous anthropogenic interference with the climate system. Such a level should be achieved within a time-frame sufficient to allow ecosystems to adapt naturally to climate change, to ensure that food production is not threatened and to enable economic development to proceed in a sustainable manner'. More than

175 countries have ratified the climate convention, including all large industrialised countries.

The first legally binding outcome of this declaration was the agreement in 1997 of a supplement to the climate convention, called the Kyoto Protocol. After prolonged negotiations over its rule-book, and a long period to get sufficient industrialised countries to ratify, the Kyoto Protocol finally came into force in February 2005. The protocol agreed targets for emissions of six key greenhouse gases: carbon dioxide, methane, nitrous oxide, and three groups of fluorinated gases. For the time being, these targets apply to 35 industrialised countries and cover the period 2008 to 2012, known as the protocol's first compliance period. USA and Australia decided not to ratify the protocol, although they remain committed to the climate convention declaration to prevent dangerous climate change.

The commitment of industrialised countries, as a whole, in the Kyoto Protocol was to reduce their emissions of a basket of six greenhouse gases to 5.2 % below their levels in a given base year (1990 in most cases) by the period 2008–2012. Since not all those countries have ratified the protocol, the total reduction target of those that did ratify is about 2.8 % below their 1990 emissions.

Countries are meant to meet their targets by cutting domestic emissions but are entitled to also use the protocol's 'flexible mechanisms'. These include direct trade in emissions permits (called assigned amount units, or AAUs) between countries with targets, and investment in projects in other developed or developing countries, known respectively as Joint Implementation and the Clean Development Mechanism, that cut emissions which would otherwise be made. Countries are also allowed to use increasing carbon uptake by forests and other ecosystem sinks.

Figure 3.5   Kyoto burden-sharing targets for EU-15 countries

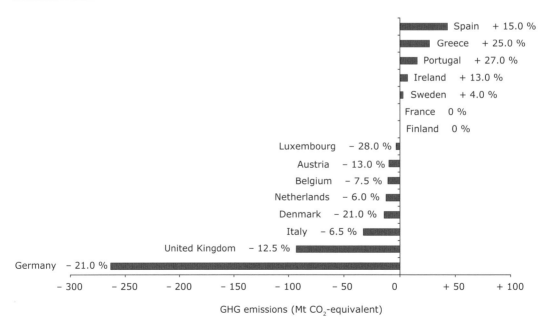

Source: EEA, 2004.

The then 15 Member States of the European Union (EU-15) accepted an 8 % reduction target at Kyoto, and subsequently have agreed to a burden-sharing agreement among themselves (Figure 3.5). Thus each of the 15 was given a national target. Eight countries received reductions targets, two had targets for emissions on parity with 1990 levels, while five countries were allowed increases.

Since the burden-sharing targets were negotiated, 10 more countries have joined the EU. With the exception of Cyprus and Malta, these countries all have their own targets under the protocol, with cuts in emissions ranging from 6 to 8 per cent.

As part of its effort to reach its Kyoto target, the EU has introduced an emissions trading system. At the heart of the scheme is the common trading 'currency' of emission allowances. One allowance represents the right to emit one tonne of CO_2. Member States have drawn up national allocation plans for 2005–2007 which give each installation in the scheme permission to emit an amount of CO_2 that corresponds to the number of allowances received. Allowances that are not needed can be traded between companies either directly or through exchanges or sold to any person within the EU.

The aim is to stimulate innovation and give reductions in emissions a market value. This ensures that emissions are reduced in the most cost-effective way. The emissions trading scheme is linked to Kyoto's Joint Implementation and Clean Development Mechanism, which will allow European companies to earn carbon credits by investing in climate-friendly technologies in other countries. A formal market for the first trading period (2005–2007) opened for business in March 2005.

3.5 Reaching the Kyoto targets

Although EU-15 emissions in 2003 were 1.7 % below their 1990 level, it seems that policy measures already decided upon in Member States will not be sufficient to allow them to reach their collective target under the Kyoto Protocol through domestic action. While emissions fell during the 1990s, they have grown overall since 2000, driven by ever-increasing transport demand and small increases in the use of coal and lignite in power generation, which had previously decreased substantially during the 1990s.

Since 1990, reductions in emissions have been mostly from waste (largely methane) and industrial processes. There have also been more modest reductions in the energy sector and in agriculture, but emissions from transport have increased by more than a fifth. Within the transport sector, emissions from aviation and shipping rose the most. Among the EU-15, emissions from domestic transport are projected to increase by 31 % between 1990 and 2010, with increased mileage more than offsetting improvements in the energy efficiency of new vehicles.

The most recent estimate is that emissions within the EU-15 in the first compliance period from 2008 to 2012 will be 2.5 % below the 1990 level, compared with a targeted 8 % reduction. Nevertheless, if all the planned domestic measures and use of Kyoto mechanisms that Member States have so far stated they intend to implement are introduced, then emissions are expected to be reduced by more than the target (9.4 %).

The prospects for the eight new Member States meeting their Kyoto commitments — Cyprus and Malta have no targets — are rather better. Many of them are still recovering from the economic breakdown and restructuring of the 1990s, which caused emissions to drop sharply. As a group, they are expected to have emissions in the first Kyoto compliance period that are about 19 % below 1990 levels, substantially below their national targets.

3.6 Strategy for the future

Setting future targets

As the Kyoto Protocol came into force, countries began to discuss what should follow it, taking into account the commitment in the UNFCCC to prevent 'dangerous' climate change. The UNFCCC has not defined this term, so what it means is inevitably a political as much as a scientific judgment. In March 2005, the EU Council of Environment Ministers concluded that — based on the scientific evidence of likely consequences, including the risk of abrupt irreversible changes to the climate system — the world should strive to avoid exceeding an average of 2 °C warming above pre-industrial temperatures. In addition, scientists have proposed that to help natural systems and human society adapt to inevitable change, the world should act to prevent warming ever proceeding faster than 0.2 °C a decade. (The current rate is 0.18 °C a decade).

The European Council meeting in March 2005 reaffirmed this position, saying that the European Council confirmed that 'with a view to achieving the ultimate objective of the UN Framework Convention on Climate Change, the global annual mean surface temperature increase should not exceed 2 °C above pre-industrial levels'.

What does such a target imply? Temperature rise around the world is only a third of the way to a 2 °C warming, but on current trends a 2 °C warming is likely to be exceeded between 2040 and 2070. Time lags in the natural system of two or three decades mean that, in practice, there is little time left to head off such a rise.

Preventing a rise of 2 °C will require stabilising concentrations of greenhouse gases in the atmosphere at some level. Although in practice it will involve an aggregation of a number of greenhouse gases, this level is usually expressed as being equivalent to a certain concentration of the prime gas of concern, CO_2.

Unfortunately, it is as yet unclear precisely what concentration of greenhouse gases can ensure that the world does not exceed a 2 °C average warming. This is because of continuing scientific uncertainty about how sensitive the climate system is to the 'forcing' of greenhouse gases. The EU Council of Environment Ministers in 2004 suggested that, taking an estimate of 'medium climate sensitivity', the world could sustain a rise to the equivalent of around 550 ppm of CO_2. Allowing for expected changes in other greenhouse gases, this figure roughly corresponds to a concentration of CO_2 itself of around 450 ppm.

That compares with baseline scenarios that suggest concentrations would rise to 935 ppm CO_2-equivalent by 2100, or 675 ppm for CO_2 alone.

Since the Council took its decision in 2004, the situation has begun to look even more difficult. New estimates have suggested that 550 ppm CO_2-equivalent may be too high to prevent a 2 °C warming. On new estimates of climate sensitivity, it could still leave a 70 % chance of temperatures exceeding the 2 °C threshold, and, in order to minimise the risk of an overshoot, it may eventually be necessary to bring concentrations back down to 450 ppm CO_2-equivalent, or less than 400 ppm of CO_2 itself.

With current levels less than 25 ppm short of these concentrations, that would be exceedingly hard to do. A concentration of 450 ppm CO_2-equivalent could be reached in little more than a decade on current trends.

To meet the 2 °C temperature target, the EU Council of Ministers proposed in December 2004 that global greenhouse gas emissions should peak around 2020 and then fall by anything between 15 and 50 % below 1990 levels by 2050. The precise figure would depend on future scientific judgments about the sensitivity of the climate system and the greenhouse gas concentration targets chosen.

Whatever the appropriate target, it is clear that if the world is to reach any sensible, stable level of greenhouse gas concentrations in the atmosphere, major cuts in emissions will have to be made. These cuts will have to come first from the industrialised countries, which currently have the highest per capita emissions, but ultimately will have to involve almost every nation.

Discussions at an international level on how to continue after the Kyoto commitment period were initiated at the UNFCCC conference in Buenos Aires in 2004 and will continue at forthcoming UNFCCC conferences, starting in Montreal, Canada, in November/December 2005.

The G8 Summit in Gleneagles in June 2005 affirmed the commitment of the leaders of the world's richest nations. The longer-term perspective — to consider what actions are necessary after 2012, the end of the first commitment period of the Kyoto Protocol — taken by the G8 is another step on general political action to adapt to and remediate global climate change.

Ensuring globally fair shares

Once the international community has decided on appropriate maximum allowable global greenhouse gas emissions, it will have to address the question of how those emissions are shared out between countries.

A number of different models have been proposed. One is the per capita approach, often known as 'contraction and convergence', in which emissions permits are given to countries strictly on the basis of their populations. Another approach is a system based on 'carbon intensity' targets that allocate emissions

Figure 3.6   GHG emissions for baseline and climate action scenarios

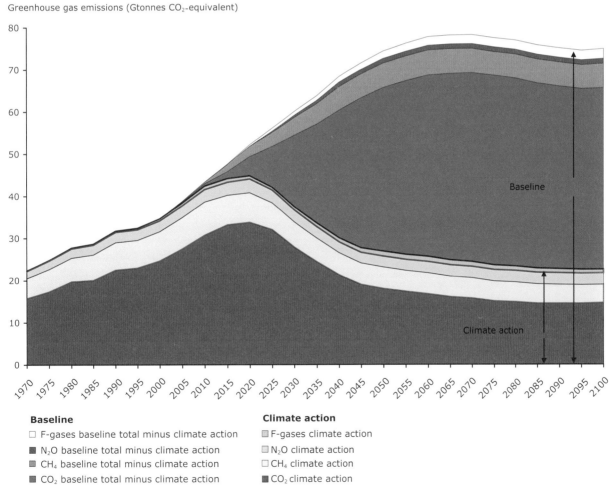

Source: EEA, 2005.

permits according to the amount of gross domestic product (GDP) countries generate for each tonne of carbon emitted. Likely formulas could combine these approaches. These and other options are expected to be discussed at UNFCCC conferences in years to come.

The EU Environment Council in March 2005 mentioned that to provide emissions 'space' for developing countries to raise their emissions sufficiently to develop their economies, it will be necessary for industrialised countries to reduce their emissions by the order of 15–30 % by 2020 and 60–80 % by 2050. In the light of those calculations the EU has been trying to chart paths to such a sustainable 'low-emissions' future.

3.7 Pathways to a low-emissions future

Among other institutions, the EEA has worked through a number of scenarios aimed at assessing what changes would be needed to ensure a low-emissions future (Figure 3.6). They all use existing technology and rely on a market in carbon emissions to make investments cost effective. This section does not aim at reviewing them all but at sketching some of the conclusions reached and constraints addressed.

The central assumption in the EEA scenarios is that EU greenhouse gas emissions should fall to 20 % below 1990 levels by 2020, 40 % below by 2030 and as much as 65 % by 2050. In the early years, the EU would rely considerably on using Kyoto Protocol flexible mechanisms in order to meet these targets. The reliance on such mechanisms would fall in subsequent years, when it is hoped that internal EU and national low-emissions policies will have become fully effective.

As already stated, CO_2 emissions in the EU-15 have been rising since 2000. On present policies — and despite continued reductions in the energy intensity of the European economy, through improvements in energy efficiency and structural changes such as the reduced importance of high-energy manufacturing — this rise will continue after 2010. The EEA baseline scenario projects for the EU-25 an overall 14 % rise above 1990 levels by 2030 (Figure 3.7).

EEA studies conclude that the key to switching from that trajectory to a low-emissions development pathway will ultimately lie primarily in reducing energy consumption and improving energy efficiency, and changing the way Europe generates and uses energy for all purposes, including transport. There are a number of ways to do this, and most will have to be used.

The low-carbon energy pathway (LCEP) scenario analyses how the European energy system would change if a CO_2 permits price rise were introduced, one which reaches EUR 65 per tonne of CO_2 in 2030. This would, the scenario suggests, lead to energy-related CO_2 emissions being 11 % lower in 2030 than in 1990 (Figure 3.7). A higher penetration of renewables could increase that to a possible 21 % emissions decline from 1990; a nuclear phase-out could reduce it to 8 %. This range represents a reduction on the 2030 baseline emissions of between 17 and 31 %.

Energy efficiency
Many cost-effective strategies for improving energy efficiency remain heavily underused. This occurs on both the energy supply side, where more efficient power stations could be employed (those, for example, that use heat that would otherwise be wasted), and on the demand side, where many homes and workplaces use energy wastefully. More goods, including computers, stereo systems, mobile phones, household appliances and air-conditioning systems, are being purchased, and households are generating more waste and using more water and energy. Although new equipment is sometimes less wasteful of resources, this is not always the case. For example, many electronic goods run on stand-by mode when not in use, and so use substantially more electricity than their predecessors.

Supply-side improvements in efficiency will rely considerably on market mechanisms, but those on the demand side will probably be more dependent on awareness-raising among end-use consumers and regulations on technical standards. However, improved energy efficiency does not necessarily mean that absolute cuts in energy consumption will follow, because the baseline is on a rising trend. Since 2000, gains from improved efficiency in energy generation and declining energy demand from industry have been wiped out by rising energy consumption by consumers/households and in the service sector.

A proposed EU directive on energy efficiency on the demand side sets a target for Member States to save 1 % of energy put into supply each year between 2006 and 2012, compared with the baseline scenario. If this progress on energy efficiency was extended beyond 2012 along the lines of the EU energy efficiency action plan, it could cut energy consumption by almost a fifth from the baseline between 2000 and 2030. The recent green paper on energy efficiency states that as much as 20 % of energy savings could be realised in a cost-effective way by 2020, according to available studies. However, this would require both the implementation of adopted legislation as well as additional policies and measures. The EEA scenarios suggest that improved efficiency and reduced consumption could account for almost half of emission reductions by 2010, decreasing to a one-third contribution beyond 2012.

Passenger cars, in addition to freight transport, have been the largest element of rising consumer demand. There have also been significant increases in energy use from domestic electrical appliances and for heating

Figure 3.7   Total greenhouse gas emissions in EU-25 (baseline and LCEP scenarios)

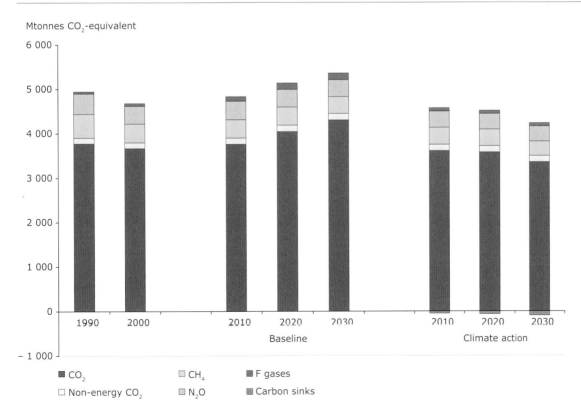

Source: EEA, 2005.

and air conditioning. Europeans are demanding ever greater energy services in their homes and workplaces. There is a great deal of potential to counter this trend in the household and service sectors, for example by adopting cost-effective improvements to energy efficiency in electrical appliances, and through better thermal insulation of buildings. Stemming the rising demand for energy for transport, however, is likely to be a bigger challenge, with the aviation sector in particular needing attention.

Fuels switch and renewables

If the EU is to make the progress it desires towards a low-emissions economy, a change in the fuel mix, especially for electricity generation, appears inescapable. Indeed, CO_2 emissions from public power plants (EU-15) between 1990 and 2002 remained almost stable despite a substantial increase in electricity production, through a combination of efficiency improvements and fuel switching, which produces a one-off benefit (Figure 3.8). However, as a result of increased power generation overall, the increased uptake of coal in the power generation mix and the loss of the once-off fuel switch benefit, CO_2 emissions in this sector are currently increasing again.

There is no blueprint for the right mix of low- and no-carbon energy technologies. Much will depend on technological developments, markets and political developments. The EEA scenarios suggest that further changes in the methods of power generation would account for more than 70 % of the emissions cuts likely

Air travel: a growing issue

Air travel is increasing, and rapidly. Globally, air passenger traffic has risen by an average of 9 % every year for the past 45 years — more than twice as fast as GDP. In large part, this increase has been driven by falling prices. The real cost per passenger-kilometre of air travel has fallen by 80 per cent since 1960, and halved since the late 1980s. The trend is forecast to continue, and the world aircraft fleet is expected to double by 2020.

Emissions have risen accordingly. CO_2 emissions from international aviation rose by 73 % between 1990 and 2003. They now amount to 12 % of national emissions from transport.

For frequent air travellers, emissions from the aircraft in which they travel are likely to be their biggest personal contribution to climate change. A return flight for two passengers across the Atlantic produces as much CO_2 as an average European passenger car does in a whole year.

This is only part of the climate impact of aircraft. Aircraft also emit nitrogen oxides and water vapour, both of which contribute directly or indirectly to climate change. They also create condensation trails, which may affect cirrus cloud coverage and thus add to global warming. The IPCC estimates that the total impact of aviation on climate is between twice and four times the effect of its CO_2 emissions alone.

Greenhouse gas emissions from international flights are, however, not accounted for under the Kyoto Protocol because there has been no agreement on how the emissions should be allocated. Moreover, international treaties on civil aviation prevent national or EU initiatives to impose taxes on kerosene or other restrictions without the approval of the International Civil Aviation Organization.

Aircraft, and the road transport they generate around airports, create other environmental problems. There is growing concern about aircraft noise near airports, particularly at night, and about emissions on the ground from both aircraft and other traffic. Nitrogen oxide emissions from major airports can also threaten local air quality targets.

There is a growing concern about implementing policy instruments that aim to reduce the environmental effects of international aviation by giving aircraft manufacturers incentives to improve fuel economy and reduce emissions of nitrogen oxides, or incentives to airlines to operate in an environmentally better way. Including the aviation sector in the emissions trading scheme is one option being considered within the EU as recently proposed by the European Commission in a communication on reducing the climate change impact of aviation (COM (2005) 459 final).

The rise of low-cost airlines is a double-edged trend. The operators are carrying more people on fewer flights than conventional airlines, but their low prices stimulate more journeys. Overall, air travel is expected to double its share of passenger transport between 2000 and 2030, from 5.6 % to 10.5 %, representing almost a tripling of air passenger-kilometres.

by 2030. For instance, under the LCEP scenario, the share of electricity generated by burning fossil fuels would be substantially lower (13 %) by 2030 compared with the baseline development. Renewable sources and perhaps nuclear power would take up a greater share. Within the fossil fuels sector, natural gas, which contains about 40 % less carbon than coal or oil per unit of energy, would increase its share from 18 % in 2002 to 42 % in 2030 at the expense of solid fuels. Additionally, natural gas power plants are more efficient than existing power plants and new coal-fired power plants. The size of the fossil-fuel burning industry today is such that even modest improvements in the thermal efficiency of its power plants could have major impacts on Europe's CO_2 emissions.

Fuel switching could be substantially stimulated by market-based emissions trading. Carbon dioxide permit pricing would improve efficiency in both the supply and use of energy by, for example, stimulating the spread of more efficient fossil-fuel technologies such as combined cycle turbines and combined heat and power (CHP). It would also stimulate the further substitution of low-carbon fuels like natural gas for coal and it could promote investment in zero-carbon renewable energy sources, even though additional measures will be needed to substantially increase their share.

There would be significant additional benefits from the spread of renewables as a replacement for fossil-fuel burning. Besides reducing CO_2 emissions, renewables would improve the diversity, security and self-sufficiency of European energy supplies. A vibrant renewables industry would also generate jobs and exports. The EU has already charted renewables as an appropriate way forward and has set 'indicative' targets to generate 12 % of total energy consumption from renewables by 2010 in the EU-15 and to generate 21 % of electricity from renewables in the EU-25 by the same year. However, so far since 1990, the share of renewable electricity in gross inland electricity consumption has risen only marginally, from 12.2 % to 12.7 % in 2002. The share of renewables in total energy consumption increased from 4.3 % to 5.7 % over that period. Significant further efforts will have to be made if the 2010 targets are to be met (Figure 3.9).

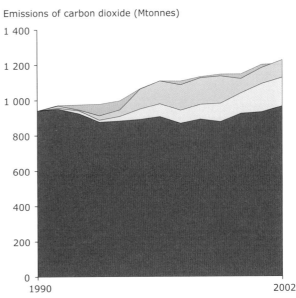

Figure 3.8 CO_2 **reductions in EU-15 for electricity and heat production, 1990–2002**

- Change due to share of nuclear and all renewables
- Change due to efficiency improvement
- Change due to fossil fuel switching
- Actual CO_2 emissions

Notes:
1. Emissions data for Luxembourg is not available and so this country is not included in the calculation for the European Union.

2. The chart shows the contributions of the various factors that have affected CO_2 emissions from electricity and heat production. The top line represents the development of CO_2 emissions that would have occurred due to increasing electricity production between 1990 and 2002, if the structure of electricity and heat production had remained unchanged from 1990 (i.e. if the shares of input fuels used to produce electricity and heat had remained constant and the efficiency of electricity and heat production also stayed the same). However, there were a number of changes to the structure of electricity and heat production that tended to reduce CO_2 emissions and the contributions of each of these changes to reducing emissions are shown above. The cumulative effect of all these changes was that CO_2 emissions from electricity and heat production actually followed the trend shown by the red area at the bottom of the graph.

Source: EEA and Eurostat, 2005.

Today biomass and hydropower account for around 90 % of total energy and electricity produced from renewables. Due to environmental constraints and lack of suitable sites, large hydropower is not expected to increase substantially in the EU-25 as a whole, while wind and biomass are expected to continue to grow rapidly. Wind is already a significant energy source in several countries, including Denmark, Germany, Spain and the United Kingdom.

In 2007, the EU is to set formal European targets for the use of renewable fuels in the period after 2010. Currently a 20 % target for renewable energy in 2020 has been proposed as an EU-25 target, building on the EU-15 target of 12 % by 2010. Such targets should provide long-term signals for industry, investors and researchers. Yet energy research and development in Europe has been in decline since 1990, despite the growing public acceptance of the need for innovation in the sector. So what is the longer-term potential?

Of renewable sources of electricity generation, the LCEP scenario suggests that wind and biomass are the most promising. At least until 2030, solar and geothermal power will make only modest contributions to energy production. The study sees renewables generating 28 % of EU electricity in 2030, roughly double the present contribution. There could also be a substantial expansion in the burning of biomass fuel in CHP plants. If additional incentives for the deployment of renewables were introduced, the share of renewables electricity would increase to almost 40 % in 2030, and account for 22 % of total energy consumption (Figure 3.10). Such a variant of the LCEP scenario indicates that this would further significantly lower CO_2 emissions to 21 % below 1990 levels.

Figure 3.9   Share of renewable electricity in gross electricity consumption in the EU-25 in 2002

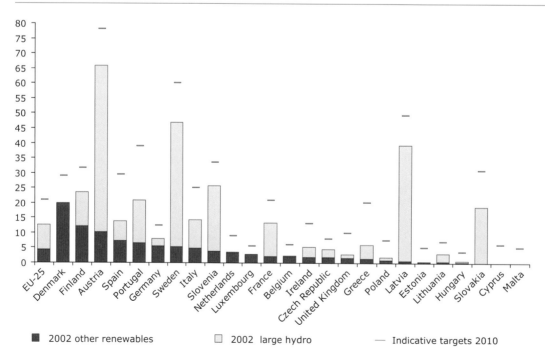

Source: EEA, 2005.

There is potential for biofuels in the transport sector in the next two decades. Due to competing demands for the land needed to grow the bioenergy crops, attention has to be paid to nature conservation requirements and other environmental objectives such as less intensive agriculture.

While a CO_2 permits price is expected to stimulate renewables development in coming decades, this alone will not be enough. Other instruments are likely to be necessary. They will include the removal of harmful subsidies for other fuels and government intervention to ensure that fuel prices reflect environmental externalities, such as the effects of acid deposition on ecosystems and the effects of particulates and ozone on human health. Energy subsidies within the EU-15 amounted to almost EUR 30 billion in 2001, with more than 73 % oriented towards the support of fossil fuels.

One conclusion of the LCEP scenario is that, all things being equal, increasing the share of renewables would further substantially decrease European CO_2 emissions. If nuclear energy were phased out, that would increase CO_2 emissions, while a higher share of nuclear energy could contribute to further reductions (Figure 3.10). Increasing nuclear energy, however, would have to take into account other considerations, including cost, public concerns, waste disposal and the global politics of nuclear proliferation.

Figure 3.10 Development of gross inland energy consumption and energy related CO_2 emissions according to different scenarios — EU-25

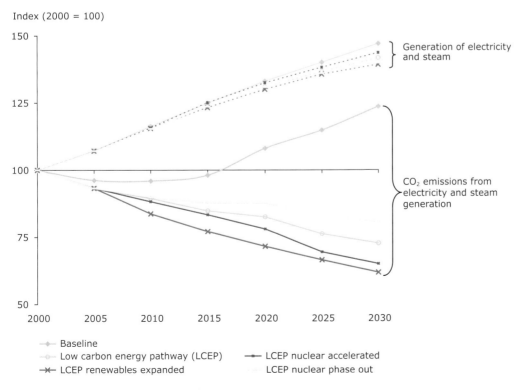

Source: EEA, 2005.

Carbon capture and storage

An emerging new option, not considered in LCEP scenarios, is the capture and storage of CO_2 from power stations and industrial stacks. The technology could potentially contribute significantly to the mix of measures that is required to meet the tough long-term targets on cutting emissions.

The International Energy Agency suggests that by 2030 substantial amounts of CO_2 could be captured in Europe. The gas would be sent by pipeline or tanker for burial in geological formations that are impermeable to CO_2, and so kept out of the atmosphere for a long period of time. These stores might include emptied oil and gas wells, unmineable coal seams and saline aquifers. There are, however, still some questions regarding storage in saline aquifers.

Some of the technology's promoters argue that carbon capture and storage offers the potential for continued use of fossil fuels, while dramatically reducing CO_2 emissions. Others see it as a transitional technology, as 21st century economies move to lower-carbon energy systems.

The technology works best at large stationary sources such as power stations, oil refineries and coal gasification plants, where there are economies of scale in extracting and moving the gas. Some of these facilities could be attached to hydrogen production facilities in any future hydrogen economy (see Section 3.10), provided they use a pre-combustion technology.

One possibility of carbon capture involves passing flue-gas emissions through chemical scrubbers containing amines that react with and trap CO_2. Similar technology is already used in some places to remove CO_2 from natural gas to boost the proportion of hydrogen. Separation of CO_2 prevents the release to the atmosphere of 85 % or more of CO_2 emissions, but requires energy and reduces the efficiency of the power station or production plant.

After capture, the CO_2 is best compressed and transported by pipeline for injection underground. This technology, too, has been developed, largely in USA, where CO_2 is pumped into oil wells to help the removal of remaining hydrocarbons. Similarly, injection of CO_2 into coal mines could recover methane, another valuable fuel. An EU trial on this is under way in Poland.

Europe's largest potential CO_2 storage arena is likely to be deep saline aquifers and depleted oil and gas fields in the North Sea, mainly in the Netherlands, Norway and the United Kingdom. It is, however, not yet clear to what extent deep saline aquifers provide a long-term safe storage opportunity. The Norwegian state oil company Statoil already strips 1 million tonnes of CO_2 each year from natural gas at its Sleipner gas field and buries it in a saline aquifer beneath the ocean floor, without ever bringing it to land.

Non-CO_2 emissions cuts

Significant cuts in greenhouse gas emissions can be achieved by tackling gases other than CO_2. Several of these gases have large projected emissions increases under the baseline scenario, and the first target would be to moderate those increases. Up to 2030, it may be cost effective to achieve about a quarter of the overall reductions in greenhouse gas emissions in this way.

Methane is the most important man-made greenhouse gas after CO_2. Methane emissions have more than doubled since pre-industrial times. Contributions have come from a wide range of activities, from agriculture to the exploitation of fossil fuels and waste disposal. Molecule for molecule, methane is a much more potent greenhouse gas than CO_2. However, its relatively short lifetime in the atmosphere means that emissions have a strong warming effect for about a decade. So cuts in emissions would have a substantial short-term effect in reducing the greenhouse gas burden in the atmosphere.

Methane is produced in large quantities as organic waste biodegrades. Gases seeping out of landfill sites are a major source. The EU's 1999 landfill directive aims to cut these emissions by requiring alternative disposal routes for biodegradable waste, such as incineration, composting and recycling. The directive also requires the recovery of methane emissions from new landfill sites from the start of operations and from existing sites from 2009. The aim is to cut emissions from waste by at least 50 % by 2030. Further cuts could be made by capping old landfill sites to prevent methane venting.

Methane is also emitted from farm manure and directly from the guts of ruminant farm animals. Anticipated declines in animal numbers in the EU should reduce these emissions by 25 % by 2030. Further cuts may be possible by changing animal diets.

Other potential ways of cutting Europe's methane emissions include reducing the emission from coal mines, natural gas pipelines and other parts of the hydrocarbon supply chain. Low-cost measures to seal leaking pipes and tap gas as it leaves mines could cut mining emissions by 60 % and gas industry emissions by about a third by 2030.

Nitrous oxide is another significant greenhouse gas with a variety of sources. Major steps have already been taken to reduce industrial emissions. These include preventing releases from plants manufacturing adipic acid, which is used in nylon production. The chemicals industry overall has cut emissions by some 60 % since 1990. Another source to be tackled is soils treated with nitrogen fertilisers. Anticipated reductions in fertiliser application on Europe's farms are expected to reduce these emissions by 8 % by 2030, matching a similar decline since 1990.

Fluorinated gases such as the hydrofluorocarbons (HFCs), which are used in refrigeration and air conditioning, currently amount to about 1 % of overall EU greenhouse gas emissions. Baseline scenarios see emissions continuing to grow substantially, particularly in new EU countries. However, low-cost measures to reduce leakage and adopt alternatives should be able to cut projected emissions by 50 % in 2030, but that would still represent an increase of around 60 % on 1990 levels.

There is an additional factor to be considered in the overall picture: as ozone depleting substances are phased out of use as refrigerants and also for certain other uses, in accordance with the Montreal Protocol and Regulation (EC) No. 2037/2000, they are to a large extent replaced by substances that are greenhouse gases, such as HFCs.

3.8 Adaptation measures required

The Kyoto Protocol also includes provisions for limiting the impacts of climate change. Considerable climate change is now inevitable because of time lags, partly in climate systems and partly in economic, political and technological systems. Considerable adaptation will be necessary to cope with changing climate zones, with the growing risk of extreme events and with the continuing rise in sea levels. The EU Environment Council has recognised the challenge and the need for actions to adapt both in developed and developing countries.

Adaptation measures will range from improved flood defences and holding back rising sea levels to changed farming systems and climate-proofing infrastructure, and better public health systems to fight new diseases. The circumstances and thus the priorities for action will differ among EU Member States but common methodologies can be used within the EU for assessing vulnerability. Equally, it will be vital to integrate these assessments with other strategies on biodiversity, water, agriculture and other areas to ensure maximum effectiveness.

As ever it is the poorest, least developed countries around the world that are among the most vulnerable to the effects of climate change, having the least financial and technical capacity to adapt to droughts, floods and other climatic disasters. The EU is taking up its responsibility to help the developing world meet the challenge of climate change through aid programmes.

3.9 Carbon sinks

The EU has not included expansion of natural carbon sinks substantially — through, for example, extending forests or changing agricultural practices — in its portfolio of actions to meet the Kyoto targets, even though there is provision to do this in the protocol. Efforts to meet future long-term targets would likely involve an expansion of Europe's carbon sinks, and they have been included in the low-emissions scenarios.

Most of the world's forests are currently absorbing more CO_2 than they release because of the fertilising effect of rising CO_2 levels in the atmosphere and because of changes in forest management, for example when remote areas cannot be harvested economically. In 2010, the EEA estimates that the forests and other natural carbon sinks of the 25 EU countries will be absorbing around 50 million tonnes of CO_2 annually. This is equivalent to around 1 % of emissions from fossil-fuel burning.

Scientists warn that by mid-century forests could begin to release some of this absorbed CO_2 as temperatures increase further. So there is a danger that these carbon sinks could one day switch from being part of the solution to being part of the problem.

3.10 A possible hydrogen economy

The transport sector is one of the most difficult areas in which to reduce CO_2 emissions. With continued fast rises in demand for transportation, the transport sector shows increased CO_2 emissions. A baseline scenario sees a 31 % rise in transport emissions above 2000 levels by 2030. Four-fifths of these expected emissions will come from road transport.

There are possible technical improvements for road transport to reduce emissions of individual vehicles. They include: better internal combustion engines, hybrid vehicles that combine the internal combustion engine with an electric motor, and replacing hydrocarbon fuels with biofuels such as alcohol made from starch crops and diesel from oil seeds (Table 3.1).

Governments could encourage all these developments through research and development, regulation, market-based mechanisms or consumer information to promote efficient vehicles and their more efficient use. However, the LCEP scenario still sees a rise in transport emissions of 20 % above 2000 levels by 2030.

Technical changes should thus be complemented by government strategies to improve average loading factors of vehicles, to shift transport from energy-intensive to more efficient transport modes and to provide mobility of people and goods with less transportation, e.g. through reduced travel distances. This could be achieved through transport charges that better reflect environmental costs, investment in more energy-efficient transport modes and through improvements in urban planning to reduce distances and streamline itineraries.

In the longer run, hydrogen may be the basic energy carrier for a low-carbon society. It can be used for electricity production as well as for fuel in transport systems.

Hydrogen production is most commonly carried out by steam reformation of natural gas and electrolysis. The problem is that this process itself requires large amounts of energy. In terms of the contribution of hydrogen fuel to reducing climate change, everything therefore depends on the original source of energy to manufacture the hydrogen.

If the hydrogen is manufactured using electricity generated by burning fossil fuels, the gains are small or even negative. However, if renewable sources are used — or if the chance is taken to capture and store CO_2 emissions from a hydrogen manufacturing plant — then the benefits could be considerable. In the medium term, renewable electricity will in many cases contribute more to CO_2 reduction if it substitutes fossil fuels directly rather than if it is used to produce hydrogen. Some suggest that places with profuse geothermal, hydroelectric or wind power could become global centres for the clean manufacture of hydrogen. Policy-makers in Iceland, for instance, have discussed the possibility of the country becoming the hydrogen-economy equivalent of an oil state.

Although hydrogen, when used in combustion, is relatively pollution free, it can be transported into the stratosphere very rapidly, where, through its reaction with ozone, it can increase the amount of stratospheric water. This could, in turn, rapidly intensify stratospheric ozone depletion. A prerequisite of any hydrogen-based energy or transport system, therefore, should be the strict control of losses of hydrogen.

Although the basic vehicle technology already exists for hydrogen-fuelled transport, further development is needed to achieve mass production at a reasonable price. There would also be substantial costs in developing the global infrastructure to deliver fuel to a whole new generation of hydrogen filling stations. Hence, widespread hydrogen use will take at least 20 years to happen.

3.11 Costs and benefits

Converting Europe to a low-carbon energy system will not be a cost-free enterprise. However, many early initiatives, particularly in improving energy efficiency in the household and service sectors, may have low or even negative costs, and there is still potential for reducing fossil fuel consumption at low or even no cost. The costs of a transition to a global and European low-carbon energy system can be minimised by implementing policies and measures in all sectors, by the participation of all major emitting countries in an international effort to address climate change, by the optimal use of the Kyoto flexible mechanisms (and internal EU emissions trading); by international cooperation on technology research and development, and by the removal of potentially environmentally harmful subsidies.

Globally, the costs of a low-carbon energy system increase the lower the assumed level at which greenhouse gas concentrations should stabilise. Assuming a stabilisation of 550 ppm CO_2-equivalent (or about 450 ppm CO_2) would lead to costs of about 1 to

Table 3.1 Attributes of alternative engine and fuel technologies

Attribute	Engines			Fuels	
	Advanced ICE	Hybrid	Fuel cell electric	Biofuels	Hydrogen
Vehicle emissions	Reduces CO_2 and regulated pollutants	Reduces CO_2 and regulated pollutants	Virtually no tailpipe emissions, may be upstream emissions	Tailpipe emissions reduced; fuel cycle CO_2 emissions reduced, but may be some N_2O and PM increase	Tailpipe emissions reduced or eliminated; fuel cycle emissions vary greatly according to production method
Speed and drivability	Probably improved	Probably improved	Probably improved	Some types may adversely affect performance of conventional engines	Engine-dependent
Refueling infrastructure	Uses existing infrastructure	Uses existing infrastructure	Probably requires major new infrastructure	Significant new infrastructure	Major new infrastructure
Cost of motoring	Potentially higher, but lower fuel consumption	Potentially higher, but lower fuel consumption	Uncertain	Probably increased costs	Probably increased costs
Timescale for widespread deployment	Short (from 2005)	Short and medium (2005–2030)	Long (post 2030)	Short and medium (2005–2030)	Long (post 2030)

Source: Adapted from Kroger et al., 2003.

4 % of GDP by 2050, depending on the IPCC scenario used. The EEA scenario work showed costs of about 1 % of GDP by 2040, in line with IPCC lower estimates.

Current estimates in EEA scenarios indicate that for the EU-25 the additional annual costs of low-emissions scenarios would represent about 0.6 % of EU GDP in 2030, or some EUR 100 billion. Average electricity generation costs in 2030 in a low-emissions scenario are likely to be 25 % higher than in the baseline scenario. The additional energy bill for households would be EUR 110–120 per household per year compared with the baseline, which already projects an average increase in household energy costs in the EU-25 of about EUR 2 300 per year by 2030. Scenarios that include stronger emphasis on renewables — which offer the greatest long-term potential for emissions reductions — would add another EUR 10–20 to household bills.

There are great uncertainties in such calculations, especially in the longer term beyond 2030. Many economic models with high estimates of the cost of cutting emissions assume a close relationship between carbon emissions and GDP, which would be very costly to break. They see a baseline future in which cheap carbon fuels remain the main source of energy. Models with lower estimates, however, assume that even without efforts to halt climate change the world is slowly moving towards lower use of carbon fuels. Such a transition would be faster and cost less with appropriate policies and measures, as mentioned above.

A second important element that distinguishes the models is their handling of the nature of technological change. Many treat technological change as largely independent of economics, as something that just happens. Others take a more sophisticated view in which innovation is mainly driven by need, economic incentives and the day-to-day process of 'learning by doing'. It is, in the jargon, 'induced technological change'.

The two approaches have important implications for policy. Traditional models suggest that there is an advantage in delaying the adoption of new technology because it will become cheaper with time. However, if most technological change is induced, then early adoption is vital to encourage further innovation and drive down costs. Models that include induced technological change also predict much lower eventual costs for meeting stabilisation targets.

Investment in diversifying away from fossil fuels would bring substantial ancillary benefits. They range from improved energy security and self-sufficiency to reduced urban pollution from fossil-fuel emissions, with resulting better health and ecological recovery. There would also be gains in jobs and exports as similar technologies are adopted around the world, especially if alternatives to fossil fuels are more labour intensive.

There are other reasons to believe that costs to combat climate change may not be too high for society. Predictions of high economic costs often assume that energy costs are a major element in the global economy. In fact, in recent decades energy costs have represented 3–4 % of world GDP. The dire predictions also tend to disregard the fact that funds put towards combating climate change will only delay the continued growth that economists regard as almost inevitable. Thus even a 4 % cut in global GDP by 2050 — among the highest cost estimates to achieve a CO_2 stabilisation level of 450 ppm according to the IPCC — would only delay a given level of global production by two or three years.

Reducing greenhouse gas emissions generates benefits in the form of avoided damages from climate change. The potential benefits depend to a large extent on the availability and costs of adaptation technologies and policies, and the sensitivity of the climate to rising concentrations of greenhouse gases in the atmosphere. It is especially relevant to analyse the global damage cost if the EU target of a global average temperature increase of 2 °C were not achieved. However, only a few studies are available on the costs of inaction. A recent study has found that the 'social cost of carbon', i.e. the cost to society of the emissions of every tonne of CO_2 into the atmosphere, was around EUR 60, with a range from EUR 30–120. Other studies have produced cost estimates per tonne as far apart as more than EUR 1 000 and virtually zero.

There are a number of reasons for the wide range. An important difference between studies is the extent to which different types of impacts are included in the analysis. For example, many studies do not properly address agriculture, ecosystem changes, biodiversity loss, loss of wetlands and impacts on water resources. Another difference is in the way that economists give a cash value to the lives and welfare of the poor. National accounting would largely ignore such lives, but most models adjust this cold calculation with an equity weighting. The value of the weighting varies. Models that give a value to the lives of the poor closer to those of the rich will give a high social cost for carbon emissions.

Some experts believe that very long-term impacts, such as the rising of sea levels due to the melting of the Greenland ice cap over thousands of years, should be discounted to zero. Most just ignore the potentially enormous cost of such catastrophic irreversible changes. Others argue that this is immoral, given that we have no alternative planet to inhabit.

The economic consequences of climate change can already be seen today. In Europe, over the past 20 years, the insurance sector has seen more than a doubling of economic losses (measured in real terms), partly resulting from weather and climate-related events, though other factors such as increasing pressure on coastal areas and floodplains, and more widespread insurance coverage, have also contributed to this increase. As will be seen in later chapters, considerable impacts on different economic sectors in different regions can be expected in future, although not all regions and locations, and not all economic sectors, will be equally affected.

3.12 Summary and conclusions

Global temperatures are rising faster than ever before and Europe exceeds the global average. Increasing precipitation, melting glaciers and ice sheets, increased frequency of extreme weather events, rising sea levels and increasing stress on terrestrial and marine ecosystems and species are among the most visible impacts on the environment. Moreover, more extreme weather is becoming a real threat to human health and our economic well-being, causing deaths and economic disruption from excessive heat, forest fires and flooding.

Burning fossil fuels remains the number one source of greenhouse gas emissions, and neither renewable energy nor nuclear energy is being developed fast enough to replace fossil fuels. In addition, increasing transport demands (road, aviation and shipping) now pose a serious threat. While emissions fell during the 1990s, they have grown overall since 2000. The EU's short term (Kyoto) targets for greenhouse gas emissions reductions are expected to be met only if all actual and planned additional policies and measures are fully implemented.

Air travel shows similar trends to other transport modes in terms of rising contributions to emissions, but to an exaggerated degree. Globally, air passenger transport has risen by an average of 9 % per year for the past 45 years, driven in large part by falling prices. Emissions have risen accordingly. International flights are currently excluded from the Kyoto Protocol targets because there is no agreement on how the emissions should be allocated. Moreover, international treaties prevent EU initiatives to impose taxes on kerosene or other restrictions without approval of the International Civil Aviation Authority. One option is for airlines to agree to participate in the EU carbon trading regime. The European Commission has recently made a proposal in this regard.

Longer-term EU targets for emissions (2020) and temperature reductions (2050) are not expected to be met. However, there is potential for a massive reduction (up to 40 % by 2020) in EU greenhouse gas emissions. This is technically feasible but requires a major shift in the EU energy system towards alternative energy sources (including nuclear) and unprecedented efficiency improvements through the increased uptake of environment-friendly technologies, especially by households.

Further emission reductions could be achieved in parallel through implementation of Kyoto flexible mechanisms in co-operation with developing countries. However, to ensure fair shares, emissions 'space' for developing countries need to be built in so that these countries can raise emissions and develop their economies. To do this, industrialised countries will have to reduce their emissions by 15–30 % by 2020 and 60–80 % by 2050. This reinforces the case for substantial reductions.

The EU has enjoyed some success with its policies, for example its emissions trading scheme. Many cost-effective strategies for improving energy efficiency remain heavily underused, such as better running of power stations and awareness-raising in households. However, efficiency measures alone will not be sufficient. Faster development of nuclear energy and renewable energy is urgent. Changes in fuel mix are now inescapable, and hydrogen needs to become the ultimate fuel. The implementation of new ideas, such as carbon capture, is also critical.

Climate change is inevitable now, and even if the correct measures are taken today, there will still be a time lag of two to three decades. The cost of inaction to society could be enormous. Some estimates put the cost at between EUR 30–120 per tonne for CO_2 emitted into the atmosphere. On the other hand, converting Europe to a low-carbon energy system will not be a cost-free enterprise either. Current estimates put the average electricity generation costs at being EUR 110–120 per household per year higher than at present.

References and further reading

The core set of indicators found in Part B of this report that are relevant to this chapter are: CSI 10, CSI 11, CSI 12, CSI 13, CSI 27, CSI 28, CSI 29, CSI 30, CSI 31, CSI 35, CSI 36 and CSI 37.

What is climate change?
Climatic Research Unit, 2005. Global average temperature change 1856–2004. See www.cru.uea.ac.uk/cru/data/temperature/.

European Environment Agency, 2004. *Impacts of Europe's changing climate. An indicator-based assessment*, EEA Report No 2/2004, Copenhagen.

Intergovernmental Panel on Climate Change, 2001. *Climate change 2001*, Synthesis report, CUP, 2001.

Mann, M.E., et al., 1999. 'Northern hemisphere temperature during the past millennium: interferences, uncertainties and limitations', *Geophysical Research Letters*, 26, pp. 759–762.

Indications of climate change
Arctic Climate Impact Assessment, 2004. *Impacts of a warming Arctic*, Final Report, Cambridge University Press, Cambridge, the United Kingdom, 146 pp. (See: www.acia.uaf.edu/ — accessed 12/10/2005)

European Environment Agency, 2004. *Impacts of Europe's changing climate. An indicator-based assessment*, EEA Report No 2/2004, Copenhagen.

European Environment Agency, 2004. *Mapping the impacts of recent natural disasters and technological accidents in Europe*, EEA Issue Report No 35, Copenhagen.

IVS, 2003. *Impact sanitaire de la vague de chaleur en France survenue en août 2003*, Rapport d'étape, 29 August 2003, Saint-Maurice, Institut de Veille Sanitaire.

Klein-Tank, Albert, 2004. *Changing temperature and precipitation extremes in Europe's climate of the 20th century*, Thesis, University of Utrecht, 124 pp.

Munich Re, 2000. *Topics-annual Review of Natural Disasters 1999*, Munich Reinsurance Group, Munich, Germany.

UNEP Grid/Arendal. www.grida.no/climate (accessed 15/9/2005).

WHO-ECEH, 2003. *Climate change and human health risks and responses*, Geneva, Switzerland.

World Health Organization, 2004. Heat-waves: risks and responses. (See: www.euro.who.int/eprise/main/WHO/Progs/CASH/HeatCold/20040331_1 — accessed 12/10/2005).

World Health Organization, 2005. Extreme weather events and public health responses (see: www.euro.who.int/eprise/main/WHO/Progs/GCH/Topics/20050809_1 — accessed 12/10/2005).

WWF International, 2005. *Europe feels the heat — extreme weather and the power sector*.

Possible future impacts
Broecker, W., 1997. *Science*, vol. 278, pp. 1582–8.

European Climate Forum, 2004. 'What is dangerous climate change?' Initial results of a symposium on key vulnerable regions, climate change and Article 2 of the UNFCCC, 27–30 October 2004, Beijing.

Hadley Centre, 2005. Stabilising climate to avoid dangerous climate change — a summary of relevant research at the Hadley Centre, Met Office, Exeter, the United Kingdom. (See: www.met-office.gov.uk/research/hadleycentre/pubs/brochures/ — accessed 12/10/2005).

Hadley Centre, 2005. International symposium on the stabilisation of greenhouse gases, 1–3 February 2005, Met Office, Exeter, the United Kingdom. (See: www.stabilisation2005.com/ — accessed 12/10/2005).

Hare, W., 2003. Assessment of knowledge on impacts of climate change — contribution to the specification of Article 2 of the UNFCCC, Background report to the WBGU Special Report No 94.

Intergovernmental Panel on Climate Change, 2001. *Climate change 2001*, Synthesis report, CUP, 2001.

Jones, C.D., *et al.*, 2003. *Geophysical Research Letters*, vol. 30, pp. 1479–82.

Parry, M.L. (ed.), 2000. *Assessment of potential effects and adaptation for climate change in Europe: The Europe Acacia Project*, Jackson Environment Institute, University of East Anglia, Norwich, United Kingdom. 320 pp.

Rial, J., *et al.*, 2004. *Climate Change*, vol. 65, pp. 11–38.

Stainforth *et al.*, 2005. *Nature,* Vol. 433, pp. 403–406.

International efforts to halt climate change
Eickhout, B., Den Elzen, M.G.J. and Vuuren, D.P. van, 2003. *Multi-gas emission profiles for stabilising greenhouse gas concentrations: emission implications of limiting global temperature increase to 2 °C*, RIVM Report 728001026, the Netherlands.

European Commission, 2005. *Communication of the Commission, Winning the battle against global climate change*, Commission staff working paper, 9 February 2005.

European Council, 2002. Council Decision 358/2002/EC, concerning the approval, on behalf of the European Community, of the Kyoto Protocol to the United Nations Framework Convention on Climate Change and the joint fulfilment of commitments thereunder (OJ L 130 of 15.5.2002, p. 1, comprising the protocol and its annexes).

European Council, 2004. *Environment Council conclusions on climate change*, 21 December 2004, Brussels.

European Council, 2005. *Environment Council conclusions on climate change*, 10 March 2005, Brussels.

European Environment Agency, 2004. *Exploring the ancillary benefits of the Kyoto Protocol for air pollution in Europe*, Technical Report No 93. Copenhagen.

Kyoto Protocol, UN Framework Convention on Climate Change (See: http://unfccc.int/resource/docs/convkp/kpeng.html — accessed 12/10/2005).

Reaching the Kyoto targets
Berk, M. and den Elzen, M., 2001. 'Options for differentiation of future commitments in climate policy:

how to realise timely participation to meet stringent climate goals?' *Climate Policy* 1(4): 465–480.

den Elzen, M.G.J. and Meinshausen, M., 2005. *Global and regional emission implications needed to meet the EU two degree target with more certainty*, RIVM report 728001031 (in print), Bilthoven, the Netherlands.

den Elzen, M.G.J. and Meinshausen, M., 2005. 'Emission implications of long-term climate targets', Scientific Symposium 'Avoiding Dangerous Climate Change', Met Office, Exeter, the United Kingdom.

European Environment Agency, 2004. *Ten key transport and environment issues for policy makers*, EEA Report No 3/2004, Copenhagen.

European Environment Agency, 2005. *European environmental outlook*, EEA Report No 4/2005, Copenhagen.

European Environment Agency, 2005. *Greenhouse gas emission trends and projections in Europe 2005*, Copenhagen.

Strategy for the future
Bartsch, U. and Müller, B., 2000. *Fossil fuels in a changing climate: impacts of the Kyoto Protocol and developing country participation*, Oxford University Press, Oxford.

European Environment Agency, 2005. *Climate change and a European low-carbon energy system*, EEA Report No 1/2005, Copenhagen.

Meinshausen, M., 2005. 'On the risk of overshooting 2 degrees C', presentation to Stabilisation 2005 conference, Met Office, the United Kingdom. www.stabilisation2005.com.

Meyer, A., 2000. *Contraction & convergence: The global solution to climate change*. Green books, London.

United Nations Framework Convention on Climate Change, 1992. United Nations General Assembly, United Nations Framework Convention on Climate Change, www.unfccc.int/resources, United Nations, New York.

United Nations Framework Convention on Climate Change, 1997. Note on the time-dependent relationship between emissions of GHG and climate change, FCCC/AGBM/1997/MISC.1/Add.3.

United Nations Framework Convention on Climate Change, 2002. Report of the Conference of the Parties on its 7th session, held at Marrakesh from 29 October to 10 November 2001. Addendum. Part Two: action taken by the Conference of the Parties. The Marrakesh Accords and Marrakesh Declaration. FCCC/CP/2001/13/Add.1.

United Nations Framework Convention on Climate Change, 2004. UNFCCC, 10th Conference of the Parties, Buenos Aires. December 2004. (See: http://unfccc.int/meetings/cop_10/items/2944.php — accessed 12/10/2005).

United Nations Framework Convention on Climate Change, 2005. Kyoto Protocol. Status of ratification. December 2004. (See: http://unfccc.int/resources/kpstats.pdf — accessed 12/10/2005).

van Vuuren, D.P., den Elzen, M.G.J., Berk,M.M., Lucas, P., Eickhout, B., Eerens H., and Oostenrijk R., 2003. *Regional costs and benefits of alternative post Kyoto climate regimes*. RIVM report 728001025/2003, National Institute of Public Health and the Environment, Bilthoven.

WBGU (German Advisory Council on Global Change), 2003. *Climate protection strategies for the 21st century: Kyoto and beyond*, Special Report 2003, Berlin.

Pathways to a low-emissions future
Bates, J., Adams, M., Gardiner, A., *et al.*, 2004. *Greenhouse gas emission projections and costs 1990–2030*, EEA-ETC/ACC Technical Paper 2004/1 in support of SOER 2005.

Criqui, P., Kitous, A., Berk, M., den Elzen, M., 2003. *Greenhouse gases reduction pathways in the UNFCCC process up to 2025*, Technical Report, European Commission, Environment DG, Brussels.

Department of Trade and Industry, 2003. *Review of the feasibility of carbon capture and storage in the UK, Cleaner Fossil Fuels programme*, London.

Department of Trade and Industry, 2003. *Our energy future — creating a low carbon economy*, Energy White Paper, London.

European Commission, 2003. Proposal for a directive of the European Parliament and of the Council on energy end-use efficiency and energy services, COM(2003) 739 final, Commission of the European Communities, Brussels.

European Commission, 2005. *Doing more with less*, Green paper on energy efficiency, COM(2005) 265 final.

European Council, 1999. Directive 99/31/EC of 26 April 1999 on the landfill of waste.

European Council, 2003. Directive 2003/30/EC of the European Parliament and of the Council on the promotion of the use of biofuels or other renewable fuels for transport. Brussels, 8 May 2003.

European Environment Agency, 2001. *Renewable energy success stories*, Environmental Issue Report No 27, Copenhagen.

European Environment Agency, 2002. *Energy and environment in the European Union*, Executive summary 2002, Environmental Issue Report No 31, Copenhagen.

European Environment Agency, 2003. *Analysis of greenhouse gas emissions trends and projections in Europe 2003*, Technical Report No 4/2004, Copenhagen.

European Environment Agency, 2004. *Energy subsidies in the European Union: A brief overview*, Technical report No 1/2004, Copenhagen.

European Environment Agency, 2005. *Climate change and a European low-carbon energy system*, EEA Report No 1/2005, Copenhagen.

European Environment Agency, 2005. *Household consumption and the environment*, EEA Report, Copenhagen (in print).

European Renewable Energy Council, 2004. *Renewable energy target for Europe — 20 % by 2020*.

Gibbins, J., *et al.*, 2005. 'Scope for future CO_2 emission reductions through carbon capture and storage', presentation to Stabilisation 2005 conference, Met Office, the United Kingdom. (See: www.stabilisation2005.com — accessed 12/10/2005).

Hadley Centre, 2005. International symposium on the stabilisation of greenhouse gases, 1–3 February 2005, Report of the Steering Committee, Met Office, Exeter, the United Kingdom.

Hadley Centre, 2005. *Stabilising climate to avoid dangerous climate change*, a summary of relevant research at the Hadley Centre, Met Office, Exeter, the United Kingdom.

International Energy Agency, 2002. *Beyond Kyoto — Energy dynamics and climate stabilisation*, IEA, Paris.

International Energy Agency, 2003. *Energy to 2050. Scenarios for a sustainable future*. IEA, Paris.

International Energy Agency, 2003. *World Energy Investment Outlook, 2003 insights*, IEA, Paris.

International Energy Agency, 2004. *World Energy Outlook 2004*, IEA, Paris.

International Energy Agency, 2004. *Prospects for CO_2 capture and storage*, OECD/IEA.

International Energy Agency, 2004. *Hydrogen and Fuel Cells*, Review of National Research and Development Programs.

Intergovernmental Panel on Climate Change, 2002. Workshop on carbon dioxide capture and storage, *Proceedings*, Regina, Canada, 18–21 November 2002, Published by ECN.

Kroger, K., Fergusson, M. and Skinner, I., 2003. *Critical issues in decarbonising transport: The role of technologies*, Tyndall Centre Working Paper 36.

Adaptation measures required
Berlin European Conference for Renewable Energy 'Intelligent Policy Options', 2004. Conclusions of session 3: Looking forward — Horizon 2020.

Gupta, J., 1998. *Encouraging developing country participation in the climate change regime*, Institute for Environmental Studies, Vrije Universiteit, Amsterdam.

Philibert, C., 2000. 'How could emissions trading benefit developing countries', *Energy Policy*, 28:947–956.

Carbon sinks
British Geological Survey, 1996. Joule II Project No CT92-0031, *The underground disposal of carbon dioxide*.

Jones, C.D., *et al.*, 2003. *Geophysical Research Letters*, vol. 30, pp. 1479–82.

A possible hydrogen economy
Akansu, S.O., Dulger, Z., Kahraman, N. and Veziroglu, T.N., 2004. 'Internal combustion engines fueled by natural gas — hydrogen mixtures', *International Journal of Hydrogen Energy* 29(14): 1527–1539.

Blok, K., Williams, R.H., Katofky, R.E and Hendriks, C.A., 1997. 'Hydrogen production from natural gas, sequestration of recovered CO_2 in depleted gas wells and enhanced natural gas recovery', *Energy* 22(2/3): 161–168.

European Commission, 2003. *Hydrogen energy and fuel cells, A vision for our future*, High Level Group for Hydrogen and Fuel Cells: 16, Brussels.

European Hydrogen and Fuel Cell Technology Platform, 2004. Steering Panel — Deployment Strategy, draft report to the Advisory Council, 6 December 2004.

Pearce, F., 2000. Kicking the habit, *New Scientist*, 25 November 2000.

Costs and benefits
Barker, T., 2005. 'Induced technological change in the stabilisation of CO_2 concentrations', presentation to Stabilisation 2005 conference, Met Office, the United Kingdom. www.stabilisation2005.com.

Bates, J., Adams, M., Gardiner, A., *et al.*, 2004. *Greenhouse gas emission projections and costs 1990–2030*, EEA-ETC/ACC Technical Paper 2004/1 in support of SOER 2005.

den Elzen, M.G.J., Lucas, P. and van Vuuren, D.P., 2005. 'Abatement costs of post-Kyoto climate regimes', *Energy Policy* 33(16), pp. 2138–2151.

Department for Environment, Food and Rural Affairs, 2003. *The social cost of carbon: a review*, report July 2003, London.

Met Office, the United Kingdom, 2005. Presentations at Stabilisation 2005 Conference: www.stabilisation2005.com.

Schneider, S., 2005. 'Overview of dangerous climate change', presentation to Stabilisation 2005 conference, Met Office, the United Kingdom. www.stabilisation2005.com.

Umweltbundesamt, 2005. *Klimaschutz in Deutschland bis 2030-Politikzenarien III*. UBAFB Nr: 000752.

Climate change | PART A

4 Air pollution and health

4.1 Introduction

Air pollution moves across both natural and political boundaries. Acidifying gases can disperse for thousands of kilometres before being deposited as acid rain on some distant habitat. Even urban smogs can spread far and wide in the calm air of a hot summer. Thus, the control of air pollution in Europe is necessarily an activity best addressed by countries in cooperation with each other. One of the early defining activities of European environmental regulation was action on the sulphur emissions that contribute to acid rain and damage human health.

Europe has made great strides in reducing many forms of air pollution in order to protect human health and ecosystems. A range of limit and target values have been set to ensure protection (Table 4.1).

In particular Europe has eliminated winter smoke smogs and reduced the threat from acid rain. However, high concentrations of fine particulates and ground-level ozone, in particular, are still causing human health problems in many cities and surrounding areas, and also problems for ecosystem health and crops across large areas of rural Europe. Despite reductions in emissions, concentrations of these pollutants remain high — often above existing targets — exposing populations to concentrations that reduce life expectancy, cause premature death and widespread aggravation to health.

Recent estimates suggest that every day people across Europe have some difficulty breathing because of air pollution. The most common effects are coughs and other respiratory problems such as bronchitis, but asthmas and allergies may also occur. Cardiovascular

Table 4.1   **EU ambient air quality limit (LV) and target (T) values for the protection of human health and ecosystems (1999/30/EC, 2002/3/EC, 2001/81/EC)**

Pollutant	Value (average time)	Number of exceedances allowed/ minimum exceedance area	To be met in
Human health			
Ozone (T)	120 µg/m³ (8h average)	< 76 days/3 year	2010
PM_{10} (LV)	50 µg/m³ (24h average)	< 36 days/year	2005
PM_{10} (LV)	40 µg/m³ (annual mean)	None	2005
SO_2 (LV)	350 µg/m³ (1h average)	< 25 hours/year	2005
SO_2 (LV)	125 µg/m³ (24h average)	< 4 days/year	2005
NO_2 (LV)	200 µg/m³ (1h average)	< 19 hours/year	2010
NO_2 (LV)	40 µg/m³ (annual mean)	None	2010
Ecosystem protection			
Ozone (T)	AOT40c of 18 (mg/m³).h (5 year average)	Daylight hours May–July	2010
Ozone	AOT40c of 6 (mg/m³).h (5 year average over 22 500 km²)	Reduction > 33 % compared to 1990	2010
Acidification	Critical load exceedances (year, average over 22 500 km²)	Reduction > 50 % compared to 1990	2010
NO_x (LV)	30 µg/m³ (annual mean)	> 1 000 km²	2001
SO_2 (LV)	20 µg/m³ (annual mean)	> 1 000 km²	2001
SO_2 (LV)	20 µg/m³ (winter average)	> 1 000 km²	2001

function, too, can be affected both by pollution-induced inflammation and even impacts on brain to heart stimulation.

There are wide variations in the susceptibility of people to air pollution. The largest effects are generally seen in people already with cardiovascular and respiratory diseases. Children, the elderly and those taking in large amounts of air while exercising outdoors in polluted conditions also appear to be vulnerable. However, the threshold below which no effects occur either do not exist or are yet to be properly identified for some air pollutants.

To meet the objectives of the sixth environment action programme (6EAP), air pollution targets need to be progressively tightened. The 6EAP called for the development of a thematic strategy on air pollution with the objective of attaining 'levels of air quality that do not give rise to significant negative impacts on, and risks to human health and the environment'. Following its communication in 2001 on the EU's Clean Air for Europe programme (CAFE), the scientific and technical underpinning of the thematic strategy, the European Commission has examined whether current legislation is sufficient to achieve the 6EAP objectives by 2020. This analysis showed that significant negative impacts will persist even with effective implementation of current legislation.

The thematic strategy on air pollution therefore, through further actions aims at cutting, by 2020, the life-years lost from particulate matter by almost half and acute mortality from ozone by 10 % compared with the 2000 levels. It also aims at reducing substantially the area of forests and other ecosystems suffering damage from airborne pollutants (acidification, eutrophication and ground-level ozone).

It is estimated that the new strategy will deliver health benefits worth at least EUR 42 billion per year through fewer premature deaths, less sickness, fewer hospital admissions, improved labour productivity, etc. This is more than five times the cost of actually implementing the strategy, which is estimated at around EUR 7.1 billion per annum or about 0.05 % of EU-25 gross domestic product (GDP) in 2020.

It is impossible to estimate the true cost to Europe's population and economy of air pollution over past years. One estimate puts the annual cost of health damage caused by air pollution at between EUR 305 billion and EUR 875 billion. From another angle, it has been estimated that in the absence of past emissions reductions brought about by regulations and technological developments, Europeans would have had to reduce their driving by 90 % in order to maintain the levels of air quality we experience today. The positive effects of past actions on Europe's social cohesion and economic competitiveness are evident.

4.2 Acid rain and ecosystem health

Removing the worst of acid rain has been a major success story for collaborative European environment policy. Acid rain is caused by fallout from emissions of sulphur dioxide, nitrogen oxides and ammonia. Sulphur dioxide comes mostly from burning coal and oil ships, power plants and industrial boilers. Nitrogen oxides also come partly from power plants and boilers, but in largest measure from ship and vehicle emissions. The main source of ammonia is from evaporation from slurry in animal stockyards and manure application on farms.

In 2002, 40 % of acid emissions came from sulphur dioxide, 32 % from nitrogen oxides and 28 % from ammonia. Of land-based sources, energy industries made up 32 % of emissions, agriculture 25 %, transport 13 % and industry 11 %. The greatest contribution to the reduction in emissions since 1990 came from the energy industry (52 %), followed by other industry (16 %) and transport (13 %). Over the same period ship emissions of SO_2 and NO_X have continued to grow, such that they are set to exceed all land-based sources combined.

Ever since the industrial revolution, these gases have posed problems. They erode buildings and statues, prevent trees growing near major industrial areas and contribute to widespread lung and heart disease. This last effect was most obvious during major smog episodes seen in European cities up to the 1960s.

Scientific verification of the extent and ecological significance of the spread of this pollution in rain clouds became apparent much more recently.

The first firm evidence of extensive ecological damage from long-range acid deposition came from the acidification of Scandinavian lakes and rivers in the 1960s and 1970s, resulting in thousands of lakes becoming too acidic for many fish species to survive. It gradually became clear that the acidification was largely caused by run-off from soils that had been chemically altered by acid rain. Later, during the 1980s, it appeared that large areas of forests in central Europe were also succumbing to acid rain, partly through the direct effects on foliage and partly through acidification of the forests' soils.

Europe began a programme to reduce acid emissions after the Stockholm environment conference in 1972. The 1979 United Nations Economic Commission for Europe (UNECE) Convention on Long-range Transboundary Air Pollution (CLRTAP) started with a protocol aimed at reducing sulphur emissions by at least 30 % and continued with protocols further cutting sulphur emissions and limiting those of nitrogen oxides. By the late 1980s, Europe had adopted an integrated approach, addressing the problems of acidification, eutrophication and ground-level ozone. The 1988 large combustion plant directive, revised in 2001, the 1999 protocol to abate acidification, eutrophication and ground-level ozone, and the 2001 national emissions ceiling directive (NEC directive) all addressed these problems taking a 'critical loads' approach, capping emissions of sulphur dioxide, nitrogen oxides, ammonia and non-methane volatile organic compounds.

The scientific study of acid emissions and their impacts has improved substantially since the first discovery of dead lakes in Scandinavia. It has become clear that acid deposition is often greater in southern and eastern Europe, even though the ecological damage has been greater further north. This is partly because the cumulative load of acid fallout onto soils over past decades has been higher in the north, and also because the soils in the north have less capacity to neutralise the acid than those further south.

Nitrogen emitted as nitrogen oxides or ammonia can cause acidification and eutrophication in freshwater and terrestrial ecosystems as well as eutrophication of marine ecosystems. Eutrophication is the consequence of an excess input of nutrients that disturb ecosystems. A common outcome is excessive blooms of algae in surface waters.

Advances in scientific knowledge have prompted a change in policy-makers' approach to reducing emissions. They have decided to target emissions reductions in those areas that cause acid deposition over the most vulnerable ecosystems. Many ecosystems now have an assessment of the 'critical load' of acid deposition that they can absorb without significant harmful long-term effects — with these limits deliberately erring on the safe side. Critical loads in regions with thin soils or those vulnerable to eutrophication are often many times lower than in areas with better-buffered soils.

Today the emission targets set by the European Union are somewhat stricter than those of the CLRTAP. In response to various legislation, many large fossil-fuel-burning power stations in Europe — the predominant source of sulphur dioxide — installed flue-gas desulphurisation equipment to remove sulphur dioxide from stack emissions. Others have reduced their emissions by burning coal or oil with lower sulphur content or by converting to natural gas.

Largely as a result of these changes, sulphur dioxide emissions peaked in the EU in the late 1970s and have fallen by two-thirds since 1980. Emissions from public electricity and heat production have been achieved as a result of efficiency improvements, fuel switching, and the use of flue gas desulphurisation technologies (Figure 4.1). Some countries have made substantially greater reductions: emissions have fallen by more than

90 % in Austria, Denmark, Germany and the United Kingdom.

The cuts in sulphur dioxide emissions, however, have not been universal. Some Mediterranean countries have seen small increases. Additionally, one important economic activity has remained largely outside the controls on sulphur dioxide emissions. That is shipping which, as a result of continued burning of high-sulphur fuel and the extensive clean-up elsewhere, now contributes 39 % of sulphur dioxide emissions among the EU-15 nations. Until recently, shipping emissions were on track to exceed all land-based emissions within 20 to 30 years; latest estimates suggest even sooner. As a result, EU environment ministers have now agreed to reduce the maximum allowed sulphur content of

Figure 4.1   Reductions in SO$_2$ emissions from public electricity and heat production in the EU-15

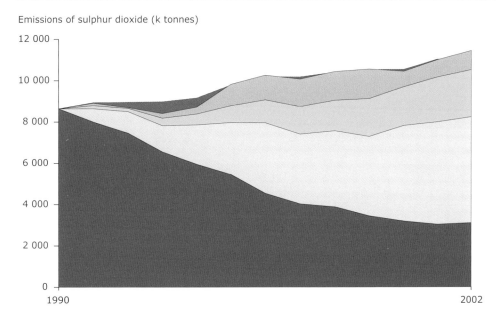

Emissions of sulphur dioxide (k tonnes)

■ Change due to share of nuclear and non-combustible renewables
▨ Change due to efficiency improvement
▨ Change due to fossil fuel switching
□ Change due to abatement
■ Actual SO$_2$ emissions

Notes:
1. Emissions data for Luxembourg are not available and so this country is not included in the calculation for the European Union.
2. The chart shows the contributions of the various factors that have affected SO$_2$ emissions from electricity and heat production. The top line represents the development of SO$_2$ emissions that would have occurred due to increasing electricity production between 1990 and 2002, if the structure of electricity and heat production had remained unchanged from 1990 (i.e. if the shares of input fuels used to produce electricity and heat had remained constant, the efficiency of electricity and heat production also stayed the same and no additional abatement technologies had been introduced). However, there were a number of changes to the structure of electricity and heat production that tended to reduce SO$_2$ emissions and the contributions of each of these changes to reducing emissions are shown by the first four coloured areas. The cumulative effect of all these changes was that SO$_2$ emissions from electricity and heat production actually followed the trend shown by the red area at the bottom of the graph.

Source: EEA and Eurostat, 2005.

marine fuel from 5 % to 1.5 % from 2006. This should have some effect in reducing emissions. The current average content of sulphur is 2.7 %.

Reductions in nitrogen oxides, which come mainly from road transport, have been less than those for sulphur dioxide. They are down by more than a quarter on 1990 levels in the EU-15. The reduction has been mainly due to the introduction across Europe of catalytic converters attached to the exhaust pipes of most cars. These remove most of the nitrogen oxide emissions, as well as other pollutants, but the effectiveness of this technological innovation has been undermined by increased road traffic. Again, shipping has been exempt from EU regulation on NO_X, and as mentioned earlier, emissions from ships in EU seas are set to exceed all land-based emissions within 15–20 years. It is harder for the EU to regulate NO_X rather than SO_2 emissions from ships, since the UN Convention on the Law of the Sea limits the ability of coastal states to regulate the construction and design of vessels flagged outside the EU. These vessels are responsible for over 50 % of the ship movements in EU seas. The International Maritime Organization (IMO) is therefore the preferred forum to tackle this problem, and the IMO is now working to develop tighter standards for NO_X emissions from ships by 2007.

Ammonia emissions from agriculture are difficult to calculate and harder to control. They are believed to have largely stabilised, along with livestock numbers on European farms. As a result of the reductions in

Map 4.1   Excess of nitrogen deposition in 2000 and 2030

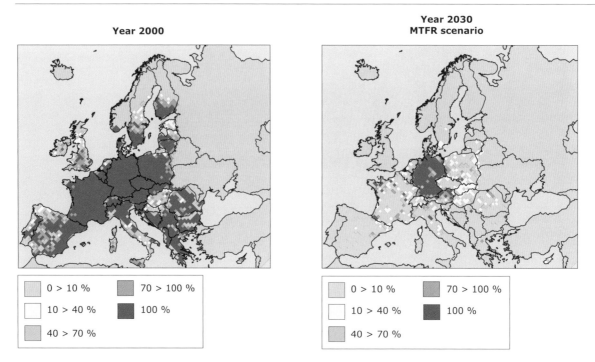

Note: Percentage of total ecosystems area receiving nitrogen deposition above the critical loads (database of 2004). Data reported for the EEA countries except from Iceland and Turkey, although the maps show areas without data as belonging to the '0 > 10 %' class. MTFR is the maximum technically feasible reduction scenario.

Source: EEA, 2005.

emissions of other acid emissions, their contribution to overall acid deposition has grown dramatically, however. They now represent 25 % of all acidifying emissions.

Overall, emissions of acidifying gases have decreased across Europe by more than 40 % in the EU-15 and almost 60 % in the EU-10, and by more than half in industry and power generation.

These emission reduction measures have resulted in more ecosystems throughout Europe receiving depositions of acidifying compounds lower than their critical load. Nonetheless, roughly 10 % of European ecosystems still received acid deposition above their critical loads in 2004. This includes 18 % of forests in the EU-15 and 35 % of forests in the EU-10.

Even when loads below the assessed critical level are attained, some ecosystems will not recover because of historic damage. Today, some 14 000 Swedish lakes remain affected by acidification, with 7 000 being limed regularly to prevent further acidification. It could be decades or even centuries before many recover.

The health of Europe's forests deteriorated until the mid-1990s. Since then a period of recovery has been followed by some further deterioration. Over a fifth of the forests are still classified as 'damaged'. The causes of these trends are not completely clear, and may not all be the result air pollution. Droughts and climate change could also be playing a role.

Europe clearly still has some way to go before it recovers from the legacy of past decades of acid deposition. So what is the prognosis and how much more can be done?

Acid deposition is expected to continue declining, thanks to the implementation of the national emissions

Figure 4.2   Emissions of air pollutants based on different scenarios — EU-25

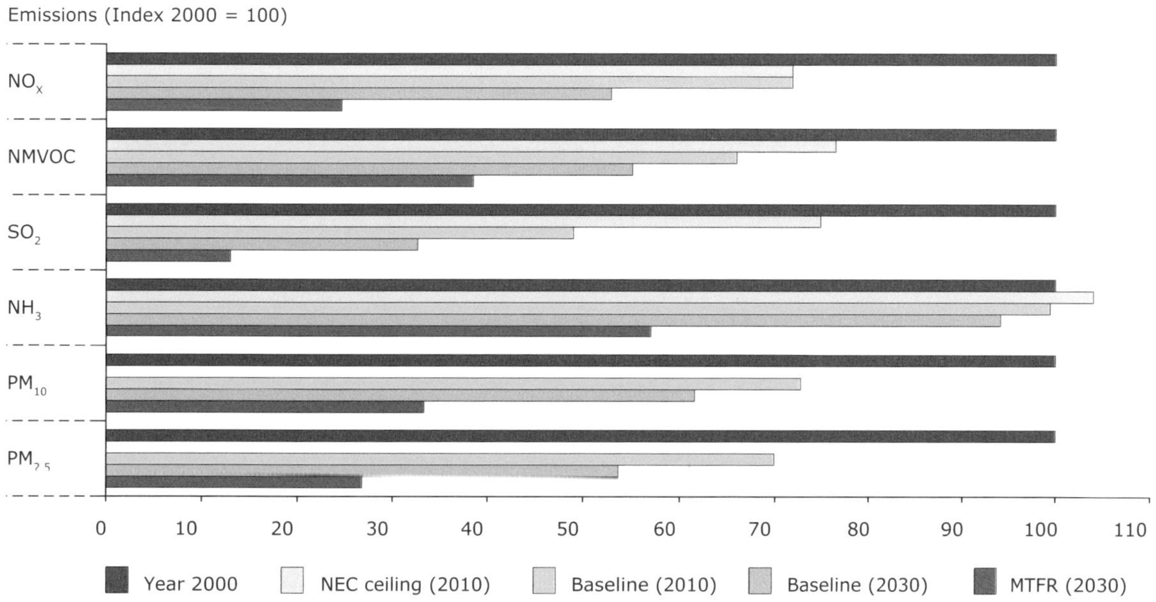

Note: MTFR is the maximum technically feasible reduction scenario.
Source: EEA, 2005.

PART A | Integrated assessment | Atmospheric environment

ceilings directive and corresponding protocols under the CLRTAP. Sulphur dioxide emissions in the EU-25, for instance, will, on current projections, drop by 51 % from 2000 by 2010, by which time they will be lower than at any time since about 1900. By 2030, under low-emissions scenarios, they will have been reduced by nearly two-thirds from a 2000 baseline (Figure 4.2).

Existing measures will see a decline of EU-25 emissions of oxides of nitrogen by 47 % between 2000 and 2030, with further reductions technically feasible. In contrast, ammonia emissions are projected to decrease only slightly by 6 % up to 2030 (Figure 4.2).

The expectation is that, overall, planned measures to reduce acid deposition will reduce forest areas at risk by more than 50 %. If the maximum feasible reductions in emissions were attained, then fallout over all Europe's forests, bar a handful in Benelux and Germany, could be brought below their critical loads. Similarly, the percentage of EU ecosystems at risk of eutrophication could be brought down from 55 % in 2000 to 10 % in 2030 (Map 4.1).

4.3 Particulates and human health

Particulate pollution is a recurring problem. Before acid rain became a concern in the 1970s, the number one air pollution problem in Europe was coal-based urban winter smogs. After a series of major disasters, many European countries took action to ban coal-burning in urban areas. The smoke problem appeared to have been solved. Illness and death from lung diseases such as emphysema and pneumonia fell as a result.

We now know, however, that smaller, largely invisible, particles continue to be hazardous to Europeans' health. These particles are generally categorised by their size. Those less than 10 millionths of a metre in diameter, known as PM_{10}s, are the most frequently measured. However, there is growing concern that a subset of these, $PM_{2.5}$s or fine particles with a diameter of less than 2.5 millionths of a metre, may be the most dangerous as they penetrate deeper into the lungs.

The primary source of most of these particulates, especially $PM_{2.5}$s, is fuel-burning in power stations, industrial plants and vehicle engines, most notably diesel engines. Some fine particulates are also produced during chemical reactions in the atmosphere, particularly during smog episodes.

Most studies conclude that particulates are the main pollutants causing deaths in Europe today. Recently the CAFE programme has put the number of premature deaths due to exposure to anthropogenic $PM_{2.5}$ particulates at 348 000 for the year 2000. Geographically, CAFE studies suggest that the greatest damage to health occurs in the Benelux area, in northern Italy and in parts of Poland and Hungary. In these areas, the average loss of life expectancy from particulates may be up to two years.

European policy-makers have reacted to this mounting evidence. Clean-up measures have substantially reduced particulate emissions since 1990 (Figure 4.3); for example, PM_{10} emissions in Germany and the United Kingdom have been cut by more than 50 %. Further cuts should follow as vehicle technology improves, particularly with the introduction of filters on diesel emissions.

Baseline scenarios, which assume that current and planned policy measures will be fully implemented, estimate that a cut could be achieved in emissions of PM_{10} between 2000 and 2030 of 38 %, and a cut in $PM_{2.5}$ emissions of 46 % (Figure 4.2). On the face of it, such cuts should be reflected in declining air concentrations of particulates. If so, they would be enough to cut the annual number of life-years lost from particulates by around a third from the current 4 million, and to reduce serious hospital admissions by a similar proportion from the current 110 000 a year.

Unfortunately, this is not certain. There is growing concern that recent declines in emissions are not being reflected in falling concentrations in the air we breathe, though we do not have long enough time series for concentrations of PM_{10} to establish clear trends. Concentrations are strongly influenced by meteorological conditions, linked to changing production of secondary particulates within smogs. There is also a concern that emissions from transport are not falling as quickly as expected due to test cycles not reflecting real-world driving conditions, the chip-tuning of diesel cars and other non-combustion emission sources (brakes, car tyres) that increase in line with traffic growth and congestion. As for SO_2 and NO_x, shipping is a major source of particle emissions which has not yet been addressed; modelling and measurement suggest ships may be contributing 20–50 % of secondary particles in port and coastal areas.

In any event, it remains likely that for some decades to come, many urban areas in the EU-25 will continue to have unsafe concentrations of particulates, largely because of the continued growth in road transport but also due to contribution from other activities such as small combustion. Passenger transport volumes in the EU-25 have risen by 20 % in the past decade, and freight transport has risen by 30 %, the rise almost exactly tracking that in GDP.

End-of-pipe technological innovation, such as the installation of particulate traps in diesel cars, is not enough to keep up with this growth in demand. Moreover, such innovations generally entail a slight increase in fuel consumption, thus potentially increasing emissions of carbon dioxide (CO_2).

Figure 4.3   Change in emissions of primary and secondary fine particles (EFTA-3 and EU-15), 1990–2002

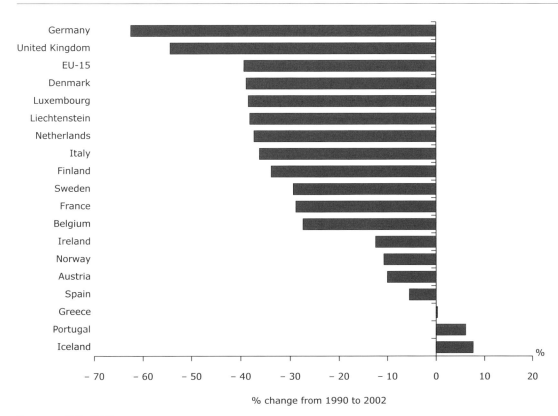

Source: EEA, 2005.

Clearly there is a need for changes in the way transport is used. In recognition of this, regulators, in addition to encouraging further technological development, are increasingly looking at the possibility of influencing the behaviour of motorists, through incentives to buy the cleanest vehicles, road pricing, the promotion of more environment-friendly modes of transport, and environmental zoning.

4.4 Ozone impacts on people and ecosystems

Ozone occurs naturally in the atmosphere, especially in the stratosphere, where it forms a chemical shield that protects life on the planet's surface from too much harmful ultraviolet radiation from the sun. That is why the world has acted to eliminate the manufacture and use of substances that have been damaging the ozone layer. Human activities also lead to ozone accumulation at ground level, where it can be a health hazard. In places, ozone levels are sometimes above what are deemed safe limits, largely because of the considerable year-to-year fluctuations caused in major part by weather conditions.

Ozone is not directly emitted into the atmosphere. It forms as a result of photochemical reactions, more intensively in summer, involving nitrogen oxides and volatile organic compounds (VOCs). A part of VOCs with high ozone formation potential, known as non-methane volatile organic compounds (NMVOCs), are produced from vehicle exhausts as well as nitrogen oxides. Nitrogen oxides are also emitted from power stations and industrial boilers, and NMVOCs also evaporate from solvents in paint, glue and printing.

Catalytic converters were introduced on petrol-fuelled passenger cars in Europe at the beginning of the 1990s. They effectively reduce emissions of carbon monoxide, nitrogen oxides and NMVOCs (Figure 4.4). Without this kind of technology, emissions would by now be

Figure 4.4   Contribution to change in ozone precursors emissions for each sector and pollutant 1990–2002

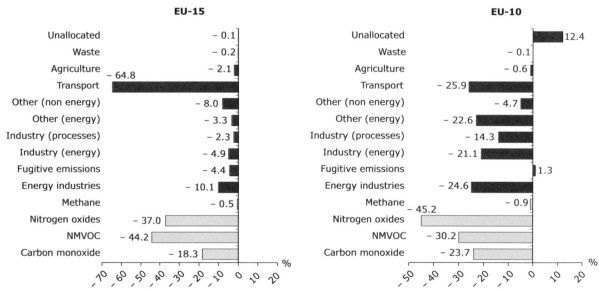

Source: EEA, 2005.

far above the levels of the early 1980s, and air quality would be falling fast.

Ozone concentrations are highest during episodes of photochemical smog, itself a complex chemical cocktail. Besides ozone and its chemical precursors and products, chemical smog can also contain other pollutants like sulphur dioxide. Fine particulates, too, are an important product of photochemical smog. Once formed, smog can persist for days and travel long distances from the urban areas in which it usually forms. Along the way, it can change its chemical composition, sometimes becoming more toxic by the time it reaches rural areas. Indeed, some of the highest concentrations of ozone eventually occur in these rural areas away from the sources of the compounds that cause the smog.

Ozone is a health hazard to humans because it inflames airways and damages lungs. It causes coughing, can trigger asthma attacks and aggravate breathing difficulties, and, ultimately, can cause death from respiratory and heart diseases. While it is hard to distinguish the health effects of ozone from other air pollutants, such as particulates, ozone is thought to hasten the deaths of up to 20 000 people in the EU each year. Further, it is responsible for people vulnerable to its effects taking medication for respiratory conditions for a total of 30 million person-days a year.

Figure 4.5 Average occurrence of exceedances for stations, which reported at least one exceedance, by EU region

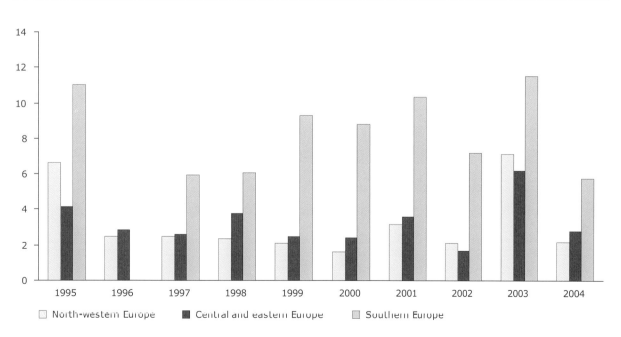

Note: North-western Europe: United Kingdom, Ireland, the Netherlands, Belgium, Luxembourg and France north of latitude 45 °.
Central and eastern Europe: Germany, Poland, the Czech Republic, Slovakia, Hungary, Austria and Switzerland.
Southern Europe: France south of latitude 45 °, Portugal, Spain, Italy, Slovenia, Greece, Cyprus and Malta.
Northern Europe has not been included in this figure because of the low number of exceedances.

Source: EEA, 2005.

Most of the damage appears to be done in the intense smog episodes that sometimes form in still summer air, when there is no rainfall or wind to remove the pollutants and slow the reactions that create them.

Public health authorities in Europe now regularly issue warnings during smog episodes so that vulnerable people can stay indoors and avoid heavy exercise.

Map 4.2   Maximum one-hour ozone concentrations observed during the summer period 2004 (April–September)

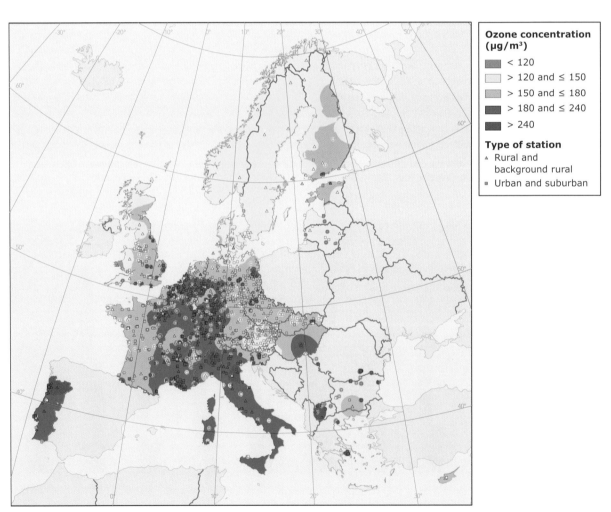

Source: EEA, 2005.

Air pollution and health

To counteract these problems, legislation has been introduced that has brought about a decline in emissions of the precursors of ozone — nitrogen oxides and NMVOCs — by about a third since 1990 (Figure 4.4). This has been mainly due to the widespread introduction of catalytic converters for cars and the EU solvents directive, which controls emissions of industrial solvents. The largest reductions have been in Germany, at 53 %, and the United Kingdom, at 46 %. However, emissions have increased in Greece, Portugal and Spain, and it is in these countries that ozone levels are highest. High NO_x and VOC emissions from shipping activities in the Mediterranean are also contributing to the southern European ozone problem.

Reductions in precursor emissions have been even greater among the 10 new EU nations, where the closure of old, heavily polluting industrial plant has helped. The Czech Republic, Estonia, Latvia, Lithuania and Slovakia have all seen reductions of more than 40 % since 1990.

Most countries should meet EU emissions ceilings for ozone precursors set to come into force in 2010. However, in the complex chemical environment of urban smog, reduced emissions of these 'precursor' pollutants will not necessarily produce equal reductions in concentrations of ozone and fine particulates in smogs. Their production depends on non-linear chemical processes, as well as on temperatures and sunlight. It is probably for this reason that the past decade has seen declining emissions of precursors accompanied by a slight increase in annual average ozone concentrations, especially in city centres.

Specifically, the EU target for ozone requires that every year the 26th worst smog (averaged over three years and measured as a daily maximum of eight-hour average ozone concentrations) should not have an ozone concentration greater than 120 micrograms in a cubic metre of air. Despite declining emissions of the ozone precursors, the average occurrence of exceedances of the EU target for ozone increased between 1997 and 2003, particularly in southern Europe. Occurrences of exceedances fell back substantially in 2004 (Figure 4.5). The highest maximum one-hour concentrations in the summer of 2004 were observed in northern Portugal, northern Italy, Albania, Macedonia, and some of the Greek islands (Map 4.2).

The toxicity of ozone smog is aggravated by other toxic compounds in the chemical cocktail. Some, such as benzene, particles and poly-aromatic hydrocarbons, are direct emissions from vehicle exhausts; others, such as nitrogen dioxide and some particulates of sulphate, are formed inside the smog itself.

Nitrogen dioxide, for instance, is created by the oxidation of nitric oxide from car exhausts. Like ozone, concentrations of nitrogen dioxide (NO_2) have stabilised in recent years, whereas before 2000

Asthma

Some of the worst and most distressing respiratory problems triggered by air pollution are among children. Asthma is now the most common respiratory disease among western European children, afflicting 7 % of children aged 4 to 10 — though there is large variation between countries.

The explanation for soaring asthma rates remains uncertain. There is a clear association between epidemics of asthma attacks in the community and local peaks in air pollution. Ozone levels in smogs may be most critical during these acute episodes, but there is much less evidence to support the thesis that long-term trends in ozone levels can explain the growing number of children who suffer asthma attacks. Nor is there much evidence that parts of Europe with more air pollution have more asthma. In fact, asthma is generally less frequent in parts of central and eastern Europe, despite air pollution levels that are higher than in western Europe.

Most researchers conclude that asthma has a range of linked causes. Air pollution seems most likely to trigger attacks among children who are already susceptible to asthma, but other factors may create that susceptibility. These could include genetic predisposition, diet and even, it has been suggested, excess hygiene in the home.

a downward tendency of NO_2 concentration was recorded. Many parts of urban Europe regularly record levels of nitrogen dioxide in the air that are above target levels. Recordings of 15–30 % above targets are typical, but some stations record levels more than twice the target level.

Ozone smogs in the lower atmosphere have ecological as well as health effects. Ozone in the air stunts crop growth and damages the foliage of trees. As long-term exposure to ozone in the lower levels of the atmosphere does the most damage to plant life, Europe has established separate targets on average ozone concentrations to reflect this. Part of Europe already complies with these limits, but a large part of southern and central Europe, from Spain to Poland, does not. The year 2003 was particularly bad for such pollution, and it is thought that high ozone levels could have been as important as high temperatures and drought conditions in that year's poor crop yields in southern Europe.

4.5 Other airborne pollution issues affecting health

Carcinogens

Little is known about the root causes of many cancers. There are genetic factors, of course, but, for some cancers at least, the environment may play a crucial role. In general, children may be at greater risk from environmental carcinogens than adults. A small but significant increase in childhood cancers has been noted since the mid-1980s, some of which may be attributable to environmental exposures. Several studies show a positive association between local traffic density and childhood leukaemia.

However, the evidence is that most cancers in children are initiated before birth, sometimes because of foetal exposure to carcinogens. Such exposure is especially dangerous because the rate of cell division in the foetus is extremely high. Thus, the chance of mutations arising from exposure to a carcinogen is that much greater.

Known carcinogens in the environment include polycyclic aromatic hydrocarbons (PAHs), a group of chemicals created by the incomplete burning of

---

The ozone conundrum

While the chemicals that form ozone smogs are most emitted in urban areas, the highest concentrations of ozone in the air are often recorded in rural areas. This is because the 'cocktails' of pollutants in smogs have a complicated life. In the lower parts of the atmosphere, under solar radiation, ozone is formed by the photolytic reaction of nitrogen dioxide (NO_2), itself an oxidation product of nitric oxide (NO). Nitric oxide is released from vehicle exhausts and other sources of emissions and is oxidised in the air to form NO_2. The molecules of NO_2 then take part in photochemical reactions with volatile organic compounds (VOC), also mainly from vehicle exhausts, to create ozone (O_3).

The prevailing way of oxidation of NO to NO_2 is through reactions with ozone. During those reactions, the ozone molecule is destroyed. Hence ozone concentrations decline in the presence of higher concentrations of NO such as in urban areas.

Actual ozone concentrations within the smog can vary greatly. Close to the sources of NO emissions — such as near dense urban traffic, major highways and industrial sources — ozone levels will be lower because significant amounts of it are being destroyed. Reciprocally, away from these areas, in the suburbs and rural areas round cities, the air still contains plenty of NO_2 and non-methane volatile organic compounds (NMVOCs) to create ozone, but little NO to destroy it. It is in these places that ozone levels are usually highest.

These complications can have important implications for efforts to reduce ozone levels. A reduction in emissions of the precursor gases will reduce the rate of formation of ozone, but they will also reduce the rate of destruction, especially in city centres. Under some circumstances, a reduction in emissions might lead to higher rather than lower ozone levels in city centres.

anything from coal to garbage. PAHs form part of vehicle emissions but they may also reach the air from incinerators, landfill sites, some factories and even fast-food restaurants. Some studies suggest that men working with PAH may pass on an increased risk of brain cancer to their children.

One omnipresent airborne cancer threat is ultraviolet (UV) radiation from the sun. This is the main cause of skin cancer, accounting for approximately 80–90 % of all cases. Rates of skin cancer are rising in Europe as Europeans sunbathe more and take more holidays in places nearer the equator, where UV levels are higher. However, rising levels of UV radiation, caused by a thinning ozone layer, may also play their part. Many sunscreens do not effectively protect against UV-A radiation, which is receiving more attention because of its potential contribution to one of the more lethal skin cancers, malignant melanoma.

Another possible threat is electromagnetic fields, including the low-frequency fields from power lines and the higher frequency fields from mobile phones and radio transmitters. There is no strong evidence of any link at typical environmental levels, but government-sponsored assessments have pointed out that studies involving mobile phone use, for instance, have not yet had time to reach firm conclusions about long-term effects. Recent studies, when considered together, indicate a correlation between low frequency electromagnetic fields and childhood leukaemia, although the evidence is not conclusive.

Many potential carcinogens are found at highest concentrations inside buildings. Indoor pollutants of concern include furnishings and paints, household cleaners and other chemicals, as well as building materials and the by-products of human activities such as cooking and smoking. Significantly, Europe's children spend 90 % of their time indoors rather than outdoors.

Concentrations of many of these pollutants have risen in many homes, particularly in northern Europe, because of better insulation and other efforts to avoid wasting heat. Any reduction in ventilation may also raise humidity in the home which can stimulate the growth of mites, moulds and bacteria, and often increase the release of toxins from construction materials, such as formaldehyde and benzene.

Another source of concern is the naturally occurring radioactive gas radon, a decay product of uranium that seeps out of some rocks and soils and can accumulate in buildings. There is a strong relationship between domestic exposure to radon and the development of lung cancer. Recent estimates suggest that radon is responsible for up to 30 000 deaths from lung cancer in Europe every year.

While scientists and health professionals are aware of this blend of problems, much less is known about the private indoor environment of Europeans than about the public outdoor one. While there are several successful European directives regulating outdoor air quality, there are none yet to control indoor air quality.

Neurotoxins and endocrine disrupters
Some toxins disrupt neurological development in children and damage their behaviour, memory and ability to learn. Symptoms can range from dyslexia to autism. The prevalence of autism and attention-deficit hyperactivity disorder (ADHD) seems to be increasing across Europe, and there is concern among health professionals that environmental factors may be involved. Finding the mechanisms and causes, however, has so far proved elusive.

Lead is mostly closely linked to neurological damage in children. Even low doses have been implicated in reduced IQ and behaviour and learning disorders in children. Since lead accumulates in bones, from where it may be released in later life, it also poses a potential hazard to the elderly. The largest source of exposure used to be lead in car exhausts, as lead was once a universal additive in petrol. Europe has been in the forefront of removing lead from petrol over the past 20 years, resulting in lead levels in the blood of most European children falling dramatically.

It nevertheless took many years to convert warnings over the neurological effects on children of lead in petrol into action. When taken, it was as much because petrol containing lead additives poisoned catalytic converters as because of health concerns.

Mercury, released in significant quantities from coal-burning power plants, is another heavy metal implicated in neurodevelopment damage. In the environment, mercury is often converted into an organic form, methyl mercury, which is toxic and easily crosses from blood to brain and, via the placenta, into the foetus. Humans mostly encounter methyl mercury when eating fish. In early 2005, Europe adopted a new tougher strategy for reducing exposure to mercury.

Also considered dangerous is a range of chemicals known as persistent organic pollutants (POPs), many of which contain chlorine or bromine. POPs tend to accumulate both in ecosystems and in the bodies of animals and humans. Many are also known to be toxic, interfering with basic body functions such as the hormonal system and neurological development. For instance, several appear to interfere with the function of thyroxin, the hormone which regulates a number of genes responsible for brain development.

Many POPs have been banned in Europe for some years. This has caused major reductions in their concentrations in the European environment and in the bodies of Europeans. Pentachlorophenol levels in the blood of Germans, for instance, have fallen more than 90 % since the chemical was banned in the late 1980s (Figure 4.6). POPs are now being phased out globally following the Stockholm Convention from 2001.

Figure 4.6   Pentachlorophenol (PCP) in German human plasma

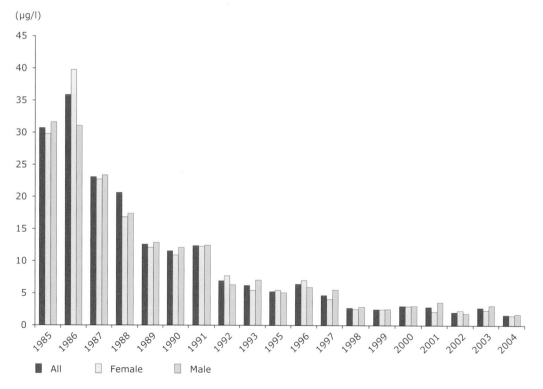

Source: German Environmental Specimen Bank, 2005.

Air pollution and health PART A

POPs in the Arctic

Some persistent pollutants in the atmosphere defy breakdown and can travel long distances, before eventually reaching the Arctic. There, the cold air can no longer hold them, they condense on to the ice or into the ocean, and enter the food chain. Many then become concentrated in the large amounts of body fat on animals such as whales, seals and polar bears living in the cold terrain.

Mercury is known to be accumulating in the European Arctic, along with metals like platinum, palladium and rhodium, which are manufactured today for use in the catalytic converters fitted to cars. Now Norwegian polar bears have been found with sufficient POPs in their systems to cause marked feminisation.

The Inuit people in Greenland and Canada face high exposure to polychlorinated biphenyls (PCBs) and mercury through eating meat from fish, whales and seals in their traditional diets (Map 4.3). Dietary intake of mercury and PCBs exceeds guidelines, and in some communities with traditional diets researchers have reported evidence of neurobehavioural effects among their children. Despite the passing of international treaties banning the use of POPs, the continued presence of these chemicals in the global environment makes it likely that their concentrations will continue to rise in parts of the Arctic.

Map 4.3   PCB levels found in human blood samples from Arctic peoples

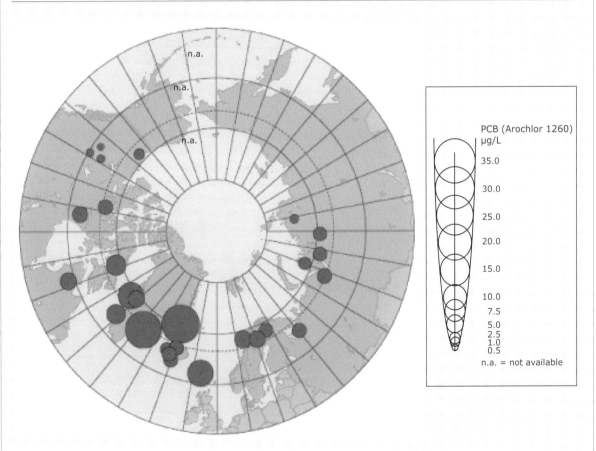

Note: PCB (as Arochlor 1260) concentrations in blood of mothers and women of child-bearing age.

Source: AMAP, 2003.

The problem, however, is not over. POPs can last for decades before degrading, and can travel long distances during that time. Many evaporate into the air and travel on the winds. Some appear to accumulate in Arctic environments where they condense out of the cold air. Thus, the far north of Europe could become a final resting place for some of them.

Some POPs form part of a wider group of chemicals commonly found in the environment — endocrine disrupters; other chemicals in this group include phthalates, which are found in many plastics. These disrupt the orderly release of hormones through the body — the endocrine system that controls almost every function of the body, from sexual differentiation before birth to digestion and the functioning of the heart. The science in this field remains uncertain, but endocrine disrupters have been linked to a worldwide decline in sperm counts in the past 50 years, while fathers exposed to a range of environmental pollutants from the air and other pathways seem to produce fewer boys.

Anti-pollution measures over the past half century have dramatically reduced the presence of many known toxins in the environment, especially those emitted into the atmosphere. However, the number of chemical additives in consumer products, pharmaceuticals and the wider environment has grown. Exposure to individual chemicals may be small but the timing of exposures, along with combined exposures from multiple sources — the 'cocktail effect' — suggest that more preventive action would be useful to take account of inherent complexities and uncertainties.

Nobody is immune. Results from biomonitoring of chemicals in our bodies clearly show an increased burden of some persistent and bio-accumulative chemical substances. When WWF, the international conservation agency, tested the blood of 14 EU environment ministers, it found that all contained traces of PCBs, pesticide residues, brominated flame-retardants and phthalates.

4.6 Summary and conclusions

Reducing acid rain has been a major success story for collaborative European environment policy. If maximum feasible reductions in emissions are attained, then fallout over Europe could be brought below critical loads and thus protecting forests and soil from further deterioration.

Particulate pollution continues to take a heavy toll on Europe's health, and represents the biggest air pollution killer in Europe today being responsible for 348 000 premature deaths in year 2000. Clean-up measures have substantially reduced particulate emissions since 1990. Further cuts should follow, particularly with the introduction of filters in diesel cars. Nevertheless, it remains likely that, for some decades to come, many urban areas in the EU-25 will continue to have unsafe concentrations of particulates resulting from road transport but also from other sources such as small combustion.

Ozone smogs are thought to hasten the deaths of 20 000 people in the EU each year. Emissions of the precursors of ozone have declined by a third since 1990 and most countries should meet EU emissions ceilings set to come into force in 2010. Unfortunately the complex chemical environment of urban smog means that, despite declining emissions of ozone precursors, annual ozone concentrations have increased slightly.

Transport is the major cause of the most intractable air pollution problems Europe faces today. The dramatic improvements made by technologies such as catalytic converters in cars are being overwhelmed by increases in demand. Without such converters, however, some emissions would be 10 times the level they are now. While our air is generally cleaner, the trends are not good enough to meet air quality targets for 2010. End-of-pipe technological innovation is not enough. Current social trends, ranging from growing suburbanisation and the declining availability and rising cost of public transport and growing demand for imported consumer goods increasing the volume of shipping in EU seas, emphasise the many dimensions of action required. Options for action include incentives to buy the cleanest vehicles, road pricing, environmental zoning and changes in spatial planning to minimise urban sprawl and port charging which reflects the external costs of shipping.

There are several other chemicals that are present in the air, including benzene and polycyclic aromatic hydrocarbons, which are carcinogenic. In general, children are at a greater health risk from exposure to them. Several studies show a positive association between local traffic density and childhood leukaemia. These chemicals are also found in high concentrations inside buildings where Europe's children spend 90 % of their time.

Lead is another pollutant most closely linked to damage to children. The largest source of exposure used to be lead in car exhausts, but Europe has been in the forefront of removing lead from petrol in the past 20 years. As a result, lead levels in the blood of most European children have fallen dramatically.

Persistent organic pollutants (POPs) such as polychlorinated biphenyls (PCBs), are produced during waste incineration and are known to be toxic. In Europe a great number of POPs have been banned for some years. They form part of a wider group of chemicals found in the environment known as endocrine disrupters. They disrupt the orderly release of hormones in the body. Endocrine disrupters have been linked to a reported 50 % decline in sperm counts in the last 60 years.

It is impossible to estimate the true cost of such a wide variety of threats from air pollution. One estimate puts the annual economic costs of health damage in Europe caused by air pollution at between EUR 305 billion and EUR 875 billion. What is clear is that there is an emerging history of threats to human health and the environment that were well understood but largely ignored. The cost of that delay was measured both in lives lost and in damaged ecosystems that ultimately cost far more to clean up than it would have cost to avoid the problem in the first place. The lesson is that even when scientific uncertainties remain, and when the cost-benefit analyses of action are hard to assemble, it is often advisable to adopt a precautionary stance.

References and further reading

The core set of indicators found in Part B of this report that are relevant to this chapter are: CSI 01, CSI 02, CSI 03, CSI 04, CSI 05 and CSI 06.

Introduction

European Environment Agency, 2004. *Air pollution and climate change policies in Europe: exploring linkages and the added value of an integrated approach*. Technical report No 5/2004.

European Environment Agency, 2003. *Air pollution in Europe 1990–2000*. Topic report No 4/2003.

European Environment Agency, 2004. *EEA Signals 2004*.

European Commission, 2001. *Environment 2010. Our future, Our choice* — The Sixth Environment Action Programme, 2001. COM(2001)31; OJ L242.

European Commission, 2005. Communication from the Commission to the Council and the European Parliament on Thematic Strategy on air pollution. COM (2005) 446 final.

EU Clean Air for Europe. CAFÉ — COM (2001) 245 final (See www.europa.eu.int/comm/environment/air/cafe/index.htm — accessed 13/10/2005).

International Institute for Applied Systems Analysis, 2004. *CAFE Scenario Analysis Report No 1. Baseline Scenarios for the Clean Air for Europe (CAFE) Programme*. Final Report. (See www.iiasa.ac.at/rains/cafe.html — accessed 13/10/2005).

SCALE Baseline report on Respiratory Health. (European Commission, DG Environment, 2004) www.europa.eu.int/comm/environment/health/finalreports_en.htm — accessed 13/10/2005).

McConnell, R., Berhane, K., Gilliland, F.D., London, S.J., Islam, T., Gauderman, W.J., Avol, E., Margolis H.G. and Peters, J.M., 2002. Asthma in Exercising Children Exposed to Ozone. *The Lancet*, Vol. 359, 386–391.

Acid rain and ecosystem health

European Environment Agency, 2001. *Air Emissions — Annual topic update 2000.* Topic report No 5/2001.

European Environment Agency, 2002. *Air pollution by ozone in Europe: Overview of exceedances of EC ozone threshold values during the summer season April–August 2002.* Topic report No 6/2002.

European Environment Agency, 2004. *Annual European Community CLRTAP emission inventory 1990–2002.* Technical report No 6/2004.

European Environment Agency, 2004. *EMEP/ CORINAIR Emission Inventory Guidebook — 2004.* Technical Report No 30.

European Environment Agency, 2002. *Emissions of atmospheric pollutants in Europe, 1990–1999.* Topic report No 5/2002.

European Environment Agency, 2004. *Exploring the ancillary benefits of the Kyoto Protocol for air pollution in Europe.* Technical report No 93.

European Environment Agency, 2005. *European environment outlook.* EEA Report No 4/2005.

European Environment Agency, 2001. *The ShAIR scenario.* Topic report No 12/2001.

Particulates and human health

EU Clean Air for Europe. www.europa.eu.int/comm/ environment/air/cafe/index.htm. (Accessed April 2005).

European Commission, 2004. SCALE Baseline report on Respiratory Health. (See www.europa.eu.int/comm/ environment/health/finalreports_en.htm — accessed 13/10/2005).

International Institute for Applied Systems Analysis, 2004. *CAFE Scenario Analysis Report No 1. Baseline Scenarios for the Clean Air for Europe (CAFE) Programme.* Final Report. (See www.iiasa.ac.at/rains/cafe.html — accessed 13/10/2005).

McConnell, R., Berhane, K., Gilliland, F. D., London, S.J., Islam, T., Gauderman, W. J., Avol, E., Margolis H.G. and Peters, J.M., 2002. Asthma in Exercising Children Exposed to Ozone. *The Lancet*, Vol. 359, 386–391.

Ozone impacts on people and ecosystems

European Environment Agency, 2001. *Air pollution by ozone in Europe in summer 2001.* Topic report No 13/2001.

European Environment Agency, 2003. *Air pollution by ozone in Europe in summer 2003 — Overview of exceedances of EC ozone threshold values during the summer season April–August 2003 and comparisons with previous years.* Topic report No 3/2003.

European Environment Agency, 2005. *Air pollution by ozone in Europe in summer 2004.* Technical report No 3/2005.

European Environment Agency, 2003. *Europe's Environment: the third assessment.* Environmental assessment report No 10.

EU COM(2004) 416 Final. The European Environment and Health Action Plan 2004–2010.

OECD Environmental Outlook 2001: *Human Health and Environment.* OECD Publications ISBN 92-64-18615-8- No 51591, 2001.

Valent, Francesca *et al.*, 2004. Burden of disease attributable to selected environmental factors and injury among children and adolescents in Europe. *The Lancet*, Vol 363, pp 2032–2039.

WHO Health report 2002. *Global estimates of burden of disease caused by the environmental and occupational risks.* (See www.who.int/quantifying_ehimpacts/global/en/ — accessed 13/10/2005).

Other airborne pollution issues affecting health

AMAP, 2003. *AMAP Assessment 2002: Human health in the Arctic.* Arctic Monitoring and Assessment Programme (AMAP), Oslo, Norway. Xiv 137 pp.

European Commission, 2004. SCALE *Baseline report on biomonitoring*. (See www.europa.eu.int/comm/environment/health/finalreports_en.htm — accessed 13/10/2005).

German Environmental Specimen Bank, 2005. (See www.umweltprobenbank.de — accessed 13/10/2005).

Meironyté Guvenius D., 2002. *Organohalogen contaminants in humans with emphasis on polybrominated diphenyl ethers*. Akademisk avhandling, Karolinska Institutet.

Norén K. and Meironyté D., 2000. *Certain organochlorine and organobromine contaminants in Swedish human milk in perspective of past 20–30 years*. Chemosphere; 40:1111–1123.

Socialstyrelsen, 2005. *Miljö och Hälsorapporten*, Sweden.

Umweltbundesamt, German Environmental Survey, 2003. (See www.umweltbundesamt.de/survey-e/index.htm — accessed 13/10/2005).

US Environmental Protection Agency, 2003. *Americas Children and the Environment — measures of contaminants, body burdens and illnesses*.

5 Freshwaters

5.1 Introduction

Water is both a vital ecological and economic resource and an essential feature of the natural landscape. It is also a renewable resource. Water abstracted from rivers and underground reserves returns to the natural environment, finding its way to the sea, and from there it evaporates and falls onto the land again as rain. Human activity is an important element in the water cycle. We need water but can do great damage to the natural aquatic environment if we abstract too much or pollute it. That damage will also impact our own ability to maximise the benefits of water.

Managing the water cycle is thus a case study in sustainable use of a key natural resource. Since 2000 the water framework directive (WFD) has been in place as the main European legislation to protect our water resources. With its two main principles focusing on the 'good status' of all water bodies, and assessing them in relation to activities in the river basin, the WFD follows an integrated approach to water resource management.

5.2 Supply and demand

European countries meet their freshwater needs from surface water such as rivers, lakes and reservoirs, and from groundwater. The share of each source varies among countries and according to regional characteristics. Countries such as Norway, Spain and the United Kingdom, for example, use more surface water, while Austria, Denmark and Germany use more groundwater. In southern Europe, there is growing use

---

Water framework directive

In 2000, Europe adopted the water framework directive to bring together and integrate work on water resource management.

The basis for the directive's work is the river basin. Most water, once it falls to the ground in precipitation, remains within a single river basin, flowing by gravity either to the sea or into groundwater reserves. Human management of the water cycle almost invariably follows this pattern. Water is sometimes moved between river basins, and this may be required more in dry climates in the future. Such bulk transfers usually involve pumping against the forces of gravity and are very expensive — cripplingly so for many uses, including agricultural irrigation.

The directive's second principle is to restore every river, lake, groundwater, wetland and other water body across the Community to a 'good status' by 2015. This includes a good ecological and chemical status for surface waters and a good chemical and quantitative status for groundwater. It requires managing the river basin so that the quality and quantity of water does not affect the ecological services of any specific water body. Thus, any abstraction has to maintain ecologically sustainable flows in rivers and preserve groundwater reserves. Discharges and land-based activities have to be restricted to a level of pollution that does not affect the expected biology of the water. In particular, the directive means that new measures will have to be taken to control the agricultural sector so as to manage both its diffuse pollution sources and its abstractions of water for irrigation.

The WFD will repeal several older pieces of legislation, such as the surface water directive, the freshwater fish and shellfish directives and the groundwater directive. In future, the objectives of these directives will be covered in a more coherent and integrated way by the WFD and daughter directives. Only four water-related directives will stay in place: the urban waste water treatment directive, the bathing water directive, the nitrates directive and the drinking water directive. Measures and objectives to combat extreme floods and droughts beyond securing a good quantity of groundwater are not covered by the WFD but will be dealt with by an action programme and a directive which are currently under development.

Europe has also recognised that, to achieve the aims of the water framework directive, 'the role of citizens and citizen groups will be crucial'. The implementation of the directive will require careful balancing of the interests of a wide range of stakeholders. The greater the transparency in the establishment of objectives, the imposition of measures and the reporting of standards, the greater the care Member States will take in implementing the legislation in good faith, and the greater the power of citizens to influence the direction of environmental protection. Caring for Europe's waters requires more involvement of citizens, interested parties and non-governmental organisations, especially at the local and regional levels. Thus the framework directive has established a network for the exchange of information and experience to ensure that implementation will not be left unexamined until it is already behind schedule or out of compliance.

of desalinated sea water, notably on Mediterranean islands where there is heavy seasonal water demand from tourists. Furthermore, several countries, including Spain, plan to greatly increase their desalination capacity as an alternative to bulk transfer of water between river basins.

Total rainfall over Europe is around 3 500 cubic kilometres a year, rather more than 10 times the 300 cubic kilometres of water withdrawn every year from the natural environment for all human activities. Although, on face value, there appears to be enough water, many of the major centres of population are in the drier parts of the continent, whereas the water is mostly in the thinly populated north. Regional demand and regional availability often do not match.

Precipitation is greatest in the west, where the winds carry moisture off the Atlantic Ocean, and in the mountains, where rising air squeezes out the last of that moisture. In western Norway, rainfall is around 2 000 millimetres a year. Downwind, inland and in the lee of mountains rainfall is much less — around 500 millimetres a year in much of eastern Europe and around 250 millimetres in southern and central Spain.

Much of Europe's water never reaches water bodies where it can be tapped for human use, particularly in hotter areas. Annual potential evaporation around the Mediterranean reaches nearly 2 000 millimetres a year, eight times the rainfall. In parts of Spain only a tenth of rainfall reaches rivers. Evaporation is also a major drain on water storage in reservoirs in the region.

Figure 5.1   Annual water availability per capita by country, 2001

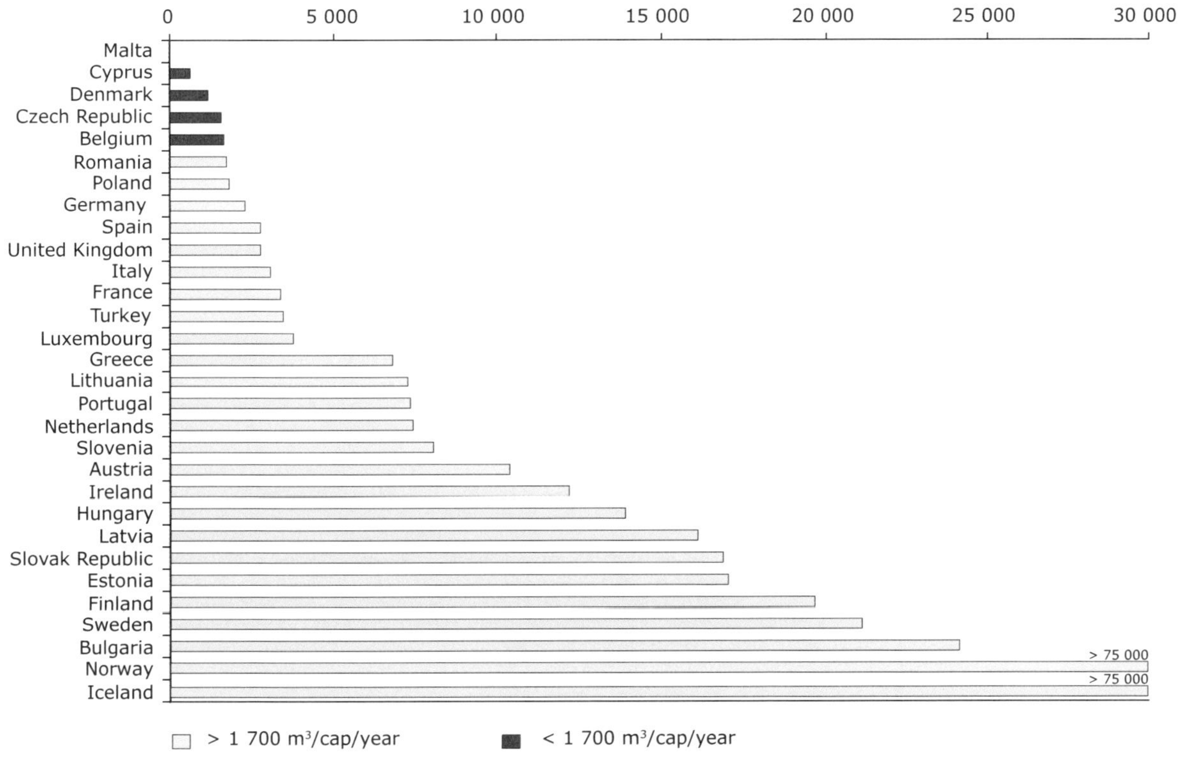

☐ > 1 700 m³/cap/year   ■ < 1 700 m³/cap/year

Source: EEA, 2003.

For such reasons, the continent's water abundance is more theoretical than real. Annual freshwater availability per head varies from less than 1 000 cubic metres in Cyprus and Malta, through around 3 000 cubic metres in France, Italy, Spain and the United Kingdom, to more than 10 000 cubic metres in mountainous countries such as Austria and Slovenia, and more than 75 000 in Norway and Iceland (Figure 5.1).

Though few Europeans suffer devastating water shortages, this imbalance of supply and demand has already created hydrological 'hot spots', where local water abstraction far exceeds supply, with knock-on effects for the functioning and long-term viability of the ecosystems. Shortages are most notable around some large cities, on small islands and in some Mediterranean coastal tourist areas. Moreover, significant fluctuations in water supply, both from month to month and year to year, can cause shortages. This is particularly the case in southern Europe where demand, especially from agriculture, is usually greatest when supply is least.

Countries where withdrawals are greater than 20 % of total available supplies are generally regarded as water stressed. Four countries — Cyprus, Italy, Malta and Spain — already fall in that category. Others are likely to join them as climate change is expected to influence both the supply and demand for water. More details on the relationship between water abstraction and renewable freshwater resources are described by the water exploitation index.

5.3 Water use

Roughly a third of the water abstracted in Europe for human use is intended to irrigate crops. Just under another third is for use in power station cooling towers. A quarter is for household use such as taps and toilets. The remainder, about 13 %, is consumed in manufacturing (Figure 5.2).

This sectoral share-out varies quite widely across the continent, however. In Belgium and Germany, for instance, more than two-thirds of water is abstracted for cooling towers at power stations. Irrigation, meanwhile, currently accounts for less than 10 % of water abstractions in most of the temperate countries of northern Europe, but in southern Europe, in countries such as Cyprus, Greece and Malta and parts of Italy, Portugal, Spain and Turkey, irrigation accounts for more than 60 % of water use. In the EU-15, 85 % of the irrigated land is in the Mediterranean countries. Among candidate countries, Romania and Turkey have the biggest share of land under irrigation.

Abstraction statistics need to be treated with care, however. They are often taken as a measure both of water use and of the potential impact of water abstraction on the aquatic environment. Some withdrawals are indeed 'consumptive', with the water being incorporated into products such as crops or manufactured goods and not returned to the river basin, but others are not. Much of the water abstracted from rivers is eventually returned in a polluted or partially cleaned form, after use in manufacture or in houses and offices. Considerable amounts are returned quickly and little changed — notably when abstracted to supply cooling towers.

Overall in Europe, 80 % of the water used in agriculture is either absorbed by crops or evaporates from fields. In manufacturing and households, 80 % is returned to the local environment, albeit often polluted and at a different location or catchment. In electricity generation, 95 % of the abstracted water is returned, a little warmer than it left but otherwise generally unchanged. Warmer water can, however, negatively impact on local ecosystem structures.

These contrasting fates of abstracted water need to be taken into account when considering recent trends and future projections for water in Europe. For instance, gross water abstraction has been in decline since the early 1990s, a trend which is expected to continue, with a predicted further reduction of about 11 % in abstractions between 2000 and 2030, to around 275 cubic kilometres a year (Figure 5.2). This, however, does not necessarily mean that there is more water in Europe's rivers.

In most places, that reduction has and will be a result of the introduction in the power sector of cooling towers that use far less water than existing cooling systems. They are expected to allow around a two-thirds reduction of water abstractions for cooling across Europe, even if current projections of a doubling in thermal electricity production come true (Figure 5.2). However, as most water abstracted for cooling is returned to the river — and since actual water losses through evaporation in these new systems are higher than for conventional cooling systems — the apparent reduction in abstractions is unlikely to result in commensurate increases in water in rivers.

Meanwhile demographic and economic trends are likely to raise water use in other sectors. Domestic use, currently around 25 % of the European total, can be expected to rise with wealth and with diminishing household size, a function, among others, of Europe's ageing population. The increase in second homes and mass tourism, including water-intensive activities such as watering golf courses, also raises per capita water use. It is possible, however, that trends to increase domestic water use could be moderated by regulations or economic incentives to encourage people to switch to more water-efficient lavatories and household appliances.

Water use in manufacturing is likely to be dependent on the future of the heavy industries that currently use around 80 % of the water in this sector (such as iron and steel, chemicals, metals and minerals, paper and pulp, food processing, engineering and textiles). Increases are expected to be greatest among the industrialising candidate EU countries, but use may decline elsewhere as heavy industry declines or adopts more water-efficient industrial technologies.

Geographically, water demand has shown different trends in different parts of Europe, and this is likely to continue. Northern Europe is likely to see substantial reductions in water withdrawals, as power plants change to modern cooling systems. With other uses probably stable until 2030, however, there may be little

Figure 5.2   Water abstraction in Europe (EEA-31 without data for Iceland)

Sector	Withdrawals in 2000
Agriculture	99.6 km³ (32 %)
Electricity	95.0 km³ (31 %)
Manufacturing	39.8 km³ (13 %)
Domestic	73.2 km³ (24 %)

Changes in annual water withdrawals (2000 to 2030): Agriculture + 11 %, Electricity − 68 %, Manufacturing + 43 %, Domestic + 3 %.

Annual water withdrawals (million m³ per year), 2000–2030, by sector: Agriculture, Electricity, Manufacturing, Domestic.

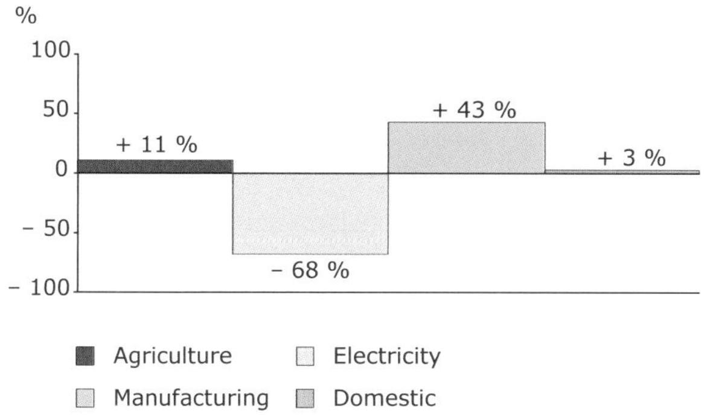

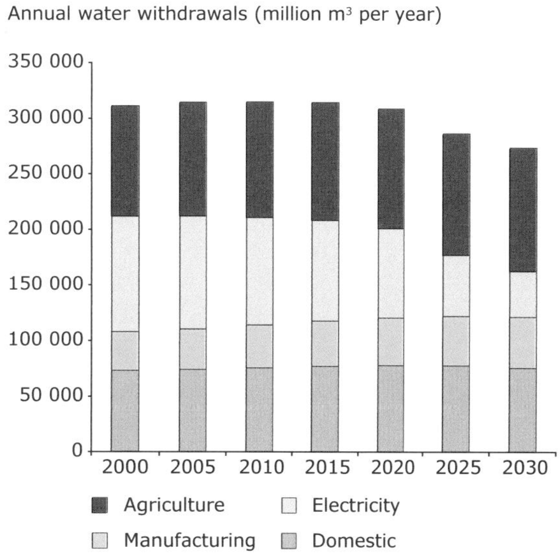

Source: EEA, 2005.

change in overall consumptive use of water. Use could, in fact, rise if climate change causes increased use of irrigation in agriculture in this region.

Higher temperatures are likely to have an even bigger impact on water demand in southern Europe, where the need for irrigation of crops will undoubtedly increase. Baseline assumptions foresee a 20 % increase in the area of southern Europe under irrigation by 2030. In many places, there is simply not the water to meet this demand, so there will be strong pressure for significant improvements in the efficiency of irrigation systems (Map 5.1).

Even allowing for such improvements, current projections see a rise of 11 % in water demand for agriculture. The question remains whether this water will be available in practice, and how countries will meet the competing needs of agriculture and the ecological protection of aquatic ecosystems. This will raise further questions about the sustainability of certain patterns of agriculture, particularly in southern Europe, in the light of projected changes in climate in already water-short areas.

Among new EU Member States, domestic water use declined during the 1990s. The collapse of some heavy

Hydroelectricity

Hydroelectricity represents 1.5 % of overall energy consumption in Europe. Countries such as Austria, Portugal, Slovakia, Slovenia and Sweden rely for significant amounts of their power on hydroelectricity generated at dams that trap river flows. Water use for hydropower does not involve abstraction of water, but is nonetheless economically and ecologically vital. River ecosystems rely on river flow, of course, as do commercial river fisheries.

Most appropriate sites for large hydroelectric dams are already occupied. Concerns about the ecological effects may constrain further development. These concerns range from altered flow and temperature regimes that destroy fish-spawning areas, handicap fish migration, kill fish in turbines and dry out wetlands, to the capture of sediment and nutrients behind dams, which can reduce the fertility of the waters downstream and may also increase erosion of river banks. For instance on the River Rhône, dams have reduced the sediment carried into Lake Geneva by some 50 %.

Climate change could make many hydroelectric power plants less reliable in future. While some plants in northern Europe could generate more power, studies suggest that output from hydroelectric dams in Bulgaria, Portugal, Spain, Turkey and Ukraine could fall by 20–50 % because of declining rainfall.

Map 5.1   Current water availability and changes expected by 2030

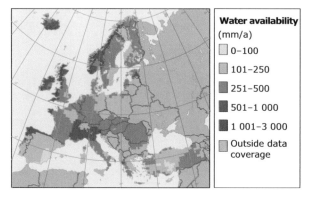
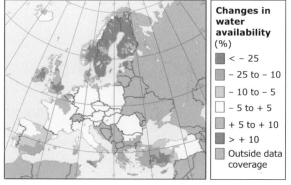

Current water availability in European river basins

Changes in average annual water availability under the LREM-E scenario by 2030

Source: EEA, 2005.

industries cut industrial water use in parts of central and eastern Europe by up to two-thirds during the decade. A crisis in farming also led to declines in abstraction for irrigation — as many irrigation districts went unwatered. Abstractions for public water supply also declined, typically by 30 %, both because of disruption of supply and because of the market effect of the introduction of water meters and more realistic water charges.

Among new members, current domestic water use is around 40 cubic metres per person per year, compared

Groundwater

Groundwater flows through the subsurface, both into and out of natural underground reservoirs, also known as aquifers, usually in the pores of porous rocks. In many areas of Europe, groundwater is the dominant source of freshwater. In a number of places water is being pumped from beneath the ground faster than it is being replenished through rainfall (Map 5.2). The result is sinking water tables, empty wells, higher pumping costs and, in coastal areas, the intrusion of saltwater from the sea, which degrades the groundwater. Saline intrusion is widespread along the Mediterranean coastlines of Italy, Spain and Turkey, where the demands of tourist resorts are the major cause of over-abstraction. In Malta, most groundwaters can no longer be used for domestic consumption or irrigation because of saline intrusion, and the country has resorted to desalination. Intrusion of saline water due to excessive extraction of water is also a problem in northern countries, for example in Sweden.

Sinking water tables can also make rivers less reliable, since many river flows are maintained in the dry season by springs that dry up when water tables fall. Groundwaters also help sustain surface reservoirs of water such as lakes and wetlands that are often highly productive ecosystems, and resources for tourism and leisure activities. These, too, are threatened by over-abstraction of groundwater.

Map 5.2   Groundwater overexploitation

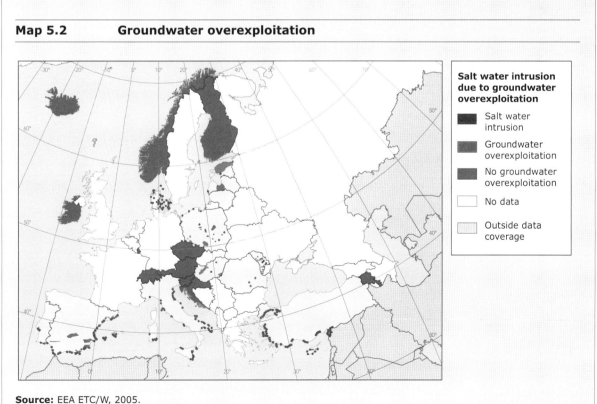

Source: EEA ETC/W, 2005.

with an EU average of 125 cubic metres. It is expected to rise substantially towards the EU average as living standards improve, though by how much is uncertain. The biggest increases in water use in the coming years, however, will probably be in the EU candidate countries, especially Turkey, where growing wealth, industrialisation and increased demand for irrigation will be compounded by continuing population growth.

Not all these expected increases need occur. The potential for greater efficiency in water use may be much greater than currently anticipated. Such improvements may be unlocked by more realistic water pricing, which would make investment in efficiency more attractive, especially in agriculture. Domestic water use could be cut through tougher water-efficiency standards for household appliances such as washing machines, dishwashers and lavatories.

Perhaps the greatest potential for water saving lies in reducing leakage rates in water distribution systems, particularly for domestic use. In some older cities in Europe, losses exceed a third. In some places this leakage is not strictly 'lost', since it recharges groundwaters, from where it can be pumped to the surface again. However, in many places this is impossible because the groundwaters beneath cities are too contaminated to be used.

Map 5.3   Water stress in 2030

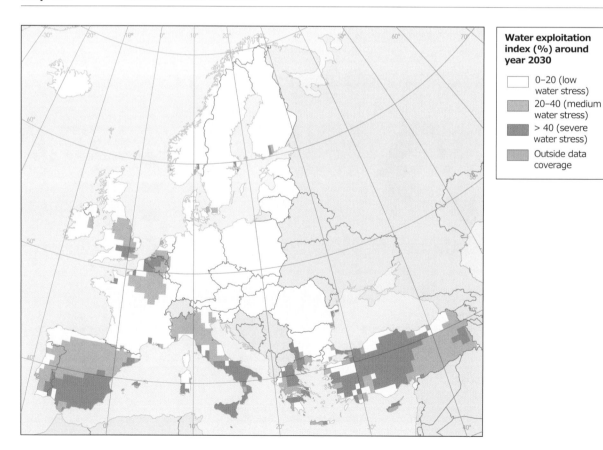

Source: EEA, 2005.

5.4 Climate change and water stress

Substantial changes in precipitation patterns, possibly linked to climate change, are already apparent in Europe. In some northern countries there has been a marked increase in precipitation in recent decades, particularly in winter, while declining rainfall is a recent feature of southern and central Europe, especially in summer. These trends are expected to continue, causing serious water stress in parts of southern Europe in particular (Map 5.3).

In parts of the north, additional rainfall will increase river flow. Water availability may increase by 10 % or more in much of Scandinavia and parts of the United Kingdom by 2030. In southern Europe a combination of reduced rainfall and increased evaporation will cause a reduction of 10 % or more in the run-off in many river basins in Greece, southern Italy and Spain, and parts of Turkey, by the same date. Most of this change is already on course to happen from emissions of greenhouse gases that have already occurred; future emissions will likely accelerate these changes.

In southern Europe, this reduced supply will be made worse by sharply rising demand, particularly from farmers needing more water to irrigate their crops. Water stress in many river basins in this part of Europe can be expected to increase (Figure 5.3). Prominent examples will include the Guadalquivir and Guadiana rivers in Spain (and the latter also in Portugal) and the

Figure 5.3 Water stress in river basins in 2000 and 2030

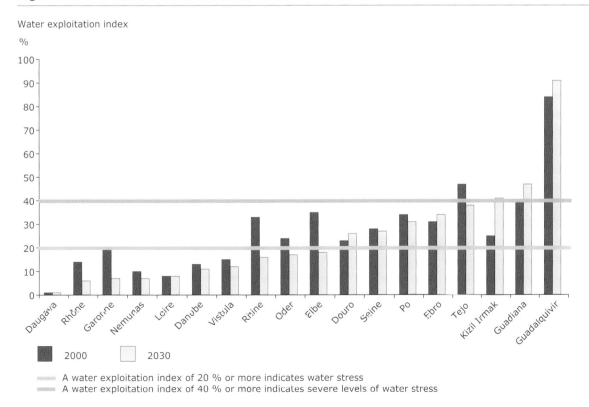

Source: EEA, 2005.

Kizil Irmak in Turkey. The Guadalquivir is expected to see more than 90 % of its flow abstracted by 2030. Spain is already responding to future anticipated shortages with plans for a large network of desalination works in the country and a push to more efficient irrigation systems. The drought conditions already apparent on the Iberian Peninsula in spring/summer 2005 underline the urgency of such measures. Where rivers cross national boundaries the demands of shared extraction add to the complex situation — for example in 2005 the flow of some rivers into Portugal were highly reduced, impacting the generation of hydropower, water available for irrigation and even water for human consumption.

In general, northern Europe is likely to become more flood prone and southern Europe more drought prone as the extra energy in the climate system increases the probability of extremes — not just droughts but also severe storms and floods, such as those in central Europe in recent years.

5.5 Water quality

River water quality across Europe is generally improving (Figure 5.4). Like water use, water quality can be a complicated concept underlined by the influence of various pressures and multi-cause/multi-effect relationships. A pristine pollution-free river flowing through an unaltered landscape may be easy to recognise, but the manner in which human activity has altered and degraded pristine rivers has taken many forms, and assessing the extent of the damage and progress towards recovery is no easy task.

Conventionally, water quality is defined by biological and chemical parameters. For instance, biochemical oxygen demand (BOD) is an index widely used to assess the amount of organic oxygen-consuming pollution in a river. BOD results for six EU Member States show markedly different distributions for water quality in rivers (Figure 5.5). However, simple statistical parameters can be misleading because the baseline natural conditions of rivers can be very different. Hence efforts are being made to conduct wider assessments of biological and ecological health. The water framework directive aims at achieving good ecological and chemical status for all water bodies in Europe by 2015.

Pollution can take many forms. Faecal contamination from sewage makes water aesthetically unpleasant and unsafe for recreational activities such as swimming, boating or fishing. Many organic pollutants, including sewage effluent, and farm and food-processing wastes consume oxygen, suffocating fish and other aquatic life. Nutrients such as nitrates and phosphates,

Figure 5.4 **Average pollution concentrations for European rivers**

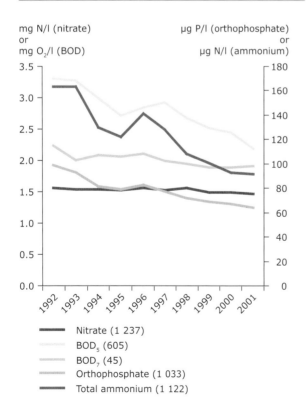

Nitrate (1 237)
BOD_5 (605)
BOD_7 (45)
Orthophosphate (1 033)
Total ammonium (1 122)

Note: Numbers in brackets refer to the number of rivers used to calculate average concentrations for each pollutant.

Source: EEA-ETC/W, 2004.

from everything from farm fertilisers to household detergents, can 'overfertilise' the water causing the growth of large mats of algae, some of which are directly toxic. When the algae die, they sink to the water bottom, decomposing, consuming oxygen and damaging ecosystems.

Pesticides and veterinary medicines from farmland, and chemical contaminants including heavy metals and some industrial chemicals can threaten wildlife and human health. Some of these damage the hormonal systems of fish, causing feminisation, even at very low concentrations. Sediment run-off from the land can make water muddy, blocking sunlight and, as a result, killing wildlife. Irrigation, especially when used improperly, can bring flows of salts, nutrients and other pollutants from soils into water. All these pollutants can also make the water unsuitable for abstraction for drinking water, without expensive treatment.

Water quality is also influenced by the physical management of rivers and the wider hydrological environment of a river basin. Canalisation, dam-building, river bank management and other changes to the hydrological flow can disrupt natural habitats such as bank side vegetation and destroy pebble riffles where salmon and other fish spawn. They also change seasonal flow patterns that are vital to many species, as well as the connectivity between habitats, a very important factor for the functioning of aquatic ecosystems and for the development of the different life stages of aquatic organisms. In urban agglomerations, storm water carrying contamination from streets and roofs can contribute to water pollution if it is not collected into the sewage system and delivered to treatment plants, but discharged directly into water bodies.

Figure 5.5 Percentage of rivers from six EU countries distributed on water quality classes for BOD (mg/O$_2$/l) in 2001 (1997 for the Netherlands)

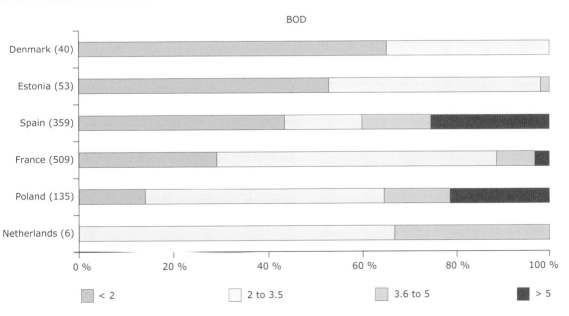

Legend: < 2 | 2 to 3.5 | 3.6 to 5 | > 5

Note: River classification based on annual average concentrations from representative subset of river monitoring stations. Numbers refer to number of river stations.

Source: EEA-ETC/W, 2005.

Integrated assessment | Aquatic environment

Most European rivers have been modified. For instance, some 90 % of Danish rivers are canalised, culverted or regulated. In Germany only 10 % of rivers are considered to be largely natural, while in France river engineering has degraded 64 (out of 76) wetlands of national importance, covering more than 11 000 square kilometres.

Groundwaters, too, suffer from the consequences of intensive agriculture and the use of nitrogen fertilisers and pesticides. Nitrates contamination is widespread across Europe, where the EU drinking standard for nitrate is exceeded in many of the groundwater bodies. Other sources of groundwater contamination are heavy metals, oil products and chlorinated hydrocarbons, mainly introduced from point sources of pollution, such as landfills.

Overall, nitrate contamination is the most commonly sited issue. It is often a particular problem in rural water supplies, which are not necessarily well monitored since they often serve only small populations and are not covered by the monitoring requirements of the drinking water directive. However, nitrate contamination should be reduced with the implementation of the nitrates directive (91/676/EEC).

5.6 Water pollution control developments

Today, around 90 % of the population in north-west Europe is connected to sewer and treatment systems. The figure is generally between 50 and 80 % among southern European members of the EU-15, but averages less than 60 % among the 10 new Member States. Most industries also have their effluent discharges connected to sewerage systems or have their own treatment plants. Some large cities, however, including the cities of Bucharest and Milan, still discharge their wastewater almost untreated into rivers.

Treatment of urban wastewater is typically divided into three categories. Primary treatment involves the filtering and physical removal of detritus; secondary treatment is biological, removing or neutralising microbiological contamination and oxygen-consuming organic material. The most advanced, tertiary, treatment involves chemical methods to remove more intractable pollutants, particularly nutrients. More than 70 % of wastewater in Austria, Denmark, Finland, Germany, the Netherlands and Sweden undergoes tertiary treatment, while in southern Europe just around 10 % of discharges receive this treatment.

Under the 1991 urban waste water treatment (UWWT) directive, the standards for collection, treatment and disposal of wastewater required in each location depend on the size of the urban area and whether the receiving waters are classified as sensitive or non-sensitive. For discharges to sensitive waters, the directive required all urban areas with more than 10 000 people to provide primary, secondary and tertiary treatment for their waste by 1998. Meanwhile, for discharges to non-sensitive waters, urban areas with populations greater than 15 000 people should provide primary and secondary treatment for their waste by 2000. For both categories, these rules will apply to all urban areas down to a population of 2 000 inhabitants from the end of 2005. The dates are extended, generally to 2010, for the 10 new Member States.

Many EU-15 countries have not yet complied fully with the directive. Several have failed to monitor water courses and assess their ecological status so they can be designated sensitive areas where appropriate. Many have not yet installed the sewage treatment capacity that the directive required by 1998 and 2000. Others are seeking postponements to requirements to extend sewage treatment to smaller urban areas by 2005.

Countries that have shown that a successful implementation of the UWWT directive is possible, leading to significant improvement in water quality, include Austria, Denmark, Germany and the Netherlands. Those further behind include France, where only 40 % of sewage discharges to sensitive areas meet the required standard. In Spain, supported by substantial subsidies from the EU Cohesion Funds, 55 % of the population is so far connected to public sewage treatment plants.

Some new EU Member States are further advanced than others. In Estonia, 70 % of the population is served by wastewater treatment plants, while in Poland 55 % of the population is connected to sewage plants.

Despite the gaps in compliance, the directive is cutting point sources of pollution to rivers substantially. In both Denmark and the Netherlands, point-source discharges to surface waters have decreased by 90 %. Estonia has also achieved a 90 % reduction in such discharges in a decade.

Assessing the outcomes of investment in the quality of river water is difficult because there is no simple measure. No two rivers are alike; and no single indicator captures all the factors. Also, the quality of water in rivers in some countries is a response to pollution control measures in upstream countries as well as the host. In places, the deposition of pollution onto water from the air may also play a role.

Nonetheless, most rivers have improved across Europe, generally the greatest in formerly badly polluted urban and industrial areas, where point sources of pollution predominated, and where clean-up investment has been concentrated. It has been less good — and in some cases there has been clear deterioration — in rural areas which until recently were near-pristine, where diffuse agricultural sources of pollution predominate, and which have been largely outside the requirements of the UWWT directive.

Most of these are smaller rivers, but there are larger rivers, too, that have not seen improvement in all parameters. These include the Duero in Spain, where BOD and phosphate levels have deteriorated in the past 25 years, and the Wisla in Poland, where ammonium concentrations rose in the 1980s.

Discharges of a wide range of trace amounts of hazardous substances into the aquatic environment — such as heavy metals including cadmium and mercury, as well as pesticides and dioxins — have been in decline in recent years, thanks to a range of EU environmental measures, some related to water and some more general in scope. For instance, loads

History of water pollution control

After the industrial revolution, most of Europe's rivers were treated not so much as natural ecosystems but more as convenient routes for transporting liquid wastes to the sea from thousands of factories and sewerage networks. The discharges often received minimal or even no treatment to reduce their toxicity or aesthetic unpleasantness. Thousands of kilometres of waterways became toxic, devoid of oxygen and often entirely lifeless. Cities turned their backs on them; some were covered over and became little more than large sewer pipes.

In recent decades, largely since the launch of the EU environmental policy at the Paris Summit in 1972, strenuous efforts have been made to clean up discharges of sewage and industrial waste, and turn these rivers into amenities for leisure and corridors for wildlife. In financial terms, it has been Europe's largest environmental endeavour.

Initially, efforts were concentrated on removing gross and offensive pollutants and oxygen-consuming organic wastes, including raw sewage, through filtration and biological treatment. Investment was made first on rivers used for drinking water, and later moved to protecting estuaries and coastal waters, to meet standards set by the bathing waters directive.

Microbiological contamination and oxygen deficiency is now largely under control in many places. During the 1990s, BOD levels in rivers improved by 20–30 %. Efforts have moved to controls on chemical pollutants such as pesticides. Here, there has been considerable success in removing such pollution from point sources, such as industrial discharges and effluent from urban sewage systems.

Phosphate concentrations in European rivers have been reduced by a third and more — with the biggest reductions in countries which had the largest point-source pollution. Eutrophication of lakes and coastal waters has been reduced as a result, but hot spots remain. The number of monitored lakes with phosphorus concentrations below 25 micrograms per litre has increased from 75 % to 82 % in the past 20 years.

There is, however, growing recognition that, in increasing numbers of water bodies, point sources are no longer the main pollution threat. As pipe discharges have been cleaned up, an increasing, and often dominant, source of pollution has been diffuse sources percolating from the land, through soils, in numerous rivulets and trickles from land drains.

of hazardous substances reaching the Baltic Sea have fallen by at least 50 % since the late 1980s. However, not all substances are monitored, and, for many, their toxicity is not clear.

5.7 Costs and benefits of water pollution control

Water pollution control has undoubtedly proved costly for many countries. Several Member States spend around 0.8 % of gross domestic product (GDP) on this, and it has used up more than 50 % of environmental investment across Europe in recent decades. This raises questions about whether it has crowded out action on other, perhaps more immediately important, problems. Nevertheless, lessons can be learned about how to do the job most effectively.

Governance problems often lie behind difficulties with meeting the objectives of the UWWT directive. In particular, sewage treatment is often the responsibility of municipal authorities which lack the financial resources and administrative competence to complete expensive treatment works in good time and to the greatest benefit of a river system. In some countries, for example France and Spain, overlaps of institutional responsibilities, together with bottlenecks in financing, appear to be important reasons for not fully implementing the directive on time.

Comparisons also show that efforts to reduce pollution at source, before it enters the sewerage system, are often cheaper than building new treatment plants. Realistic charging for effluent treatment, for instance, made it easier for the Netherlands to meet its directive requirements (and more cheaply because measures were taken by industry to prevent pollution) than in other countries where governments had to invest heavily in treatment plants.

Across Europe, direct legislative action to reduce certain widely used pollutants in consumer products has also been shown to be highly cost effective. The most dramatic change has been the reduction, by more than 50 % in many countries, of phosphorus in household detergents. Phosphorus discharges per person have typically fallen from 1.5 kilograms per person per year to less than 1 kilogram.

The main reason for delays in implementing the UWWT directive is the costs involved, so eco-efficient approaches that minimise investment deserve more attention. Greater emphasis on eco-efficiency, and economic incentives that promote wastewater reduction at source, are likely to be the keys to more timely and cost-effective implementation of the UWWT directive in Member States.

Under the EU cohesion policy, countries are eligible for considerable EU subsidies, up to 75–85 % of investments. If there are no economic instruments in place to provide industries with incentives, there appears to be a considerable risk that EU subsidies will lead to excess investment in sewage treatment plant capacity. Finding the right balance between incentives to promote eco-efficiency and prevent pollution at source, and adequate sewage treatment capacity, would help, as sewage treatment is one of the most capital-intensive environmental measures.

It is expected that the cohesion policy, through the Cohesion and Structural Funds which are designed to bring about closer economic and social integration by encouraging growth in those regions of the EU most in need, will continue to support sewage treatment plants from its proposed EUR 336 billion budget for 2007–2013 for EU-10. Support is greatly needed as current investments in, for example, Estonia and Poland are at the level of EUR 5–10 per capita (not PPP — purchasing-power parity — adjusted), and will need to be increased to a level of about EUR 40–50 per capita to comply with the agreed deadlines.

These findings suggest that EU funding for pollution control plants — e.g. through the Cohesion Fund — should be spent carefully to avoid over-reliance on large capital projects. Often, the use of economic instruments such as taxing and charges, alongside capital investments, would be more cost-effective.

5.8 Tackling diffuse sources of pollution

While the UWWT directive will continue to reduce the discharge of nutrients from point sources, the new focus for EU activity to protect water bodies from pollution is likely to be diffuse sources, which make up an increasing proportion of emissions to rivers. Where traditional point discharges may all come through a handful of large pipes, diffuse discharges trickle from soils and thousands of field drains over hundreds of square kilometres. The challenge of controlling and policing them will thus be great in both technical and logistical terms.

Recent laws, such as the nitrates directive and the water framework directive, provide the in-country basis for establishing the further regulations, new institutional frameworks and additional monitoring systems that are seen as needed to tackle diffuse pollution and manage water bodies so that their ecological functions and resources are maintained.

The main source of diffuse pollution to water is from the largest land use across most of Europe — agriculture. A particular focus of concern is nutrients, primarily nitrates and phosphates. Nitrates are generally the greatest problem. More than half of the nutrient discharges in Europe now come from diffuse sources. Most nitrate pollution in particular arises from farm fertiliser and manure. Nutrients contribute to eutrophication in lakes, coastal waters and the marine environment. They pollute rivers and groundwaters and contaminate drinking water.

During the past half century, the rising use of commercial inorganic mineral fertilisers and increased concentrations of livestock, with their resulting manure, have resulted in a sharp increase in the application of nutrients to the land over Europe. In the past decade or so, nutrient use on farms in the EU-15 has been stable at around 70 kilograms per hectare per year (surface balance), and is expected to remain stable in the coming decades.

In eastern Europe, agricultural sector activity has dropped substantially as a result of political and economic changes during the 1990s, leading to a sharp decline in fertiliser use, which typically halved from around 70 kilograms per hectare at the start of the 1990s and remained low throughout the decade. As these countries join the EU, fertiliser use is resuming its upward trend. In EU-10, increases of 35–50 % in use of phosphates and nitrates are likely.

While much of the nutrients in fertilisers is, of course, absorbed by crops — the purpose for which they are applied — much is not. Wherever fertiliser and farm manure is not absorbed, nitrates will migrate through soils. Most European soils contain a large nitrogen surplus from constant applications. These are typically around 50–100 kilograms per hectare of farmland. Most of this surplus will eventually find its way to water.

As a result of these changes, coupled with the controls on point sources, agricultural emissions are now the dominant source of pollution in many river basins. In the river catchments that drain into the North Sea, the total nitrogen load averages 14 kilograms per hectare of land per year, of which 65 % comes from diffuse sources associated with human activities, mainly farming. The equivalent figures for phosphorus are 0.9 kilograms and 45 %.

Away from the North Sea, most other catchments, with the exception of the Po basin in northern Italy, have lower absolute levels of nitrate loading, although the proportion from agriculture remains high, above 60 % in all cases. The picture for phosphorus is more varied, because of the continued importance of point sources for this nutrient, which are being dealt with largely through implementation of the UWWT directive.

5.9 Nitrates

Fertiliser application for arable farming is the main source of nitrates. In rivers where arable land covers more than half the upstream catchment, nitrate levels are three times higher than in rivers where the upstream arable land cover is less than 10 %. Across the EU, nitrate pollution of rivers is generally lower in the Nordic countries and central Europe, where arable farm intensity is less (Figure 5.6).

In 2000, 14 European countries had rivers whose waters exceeded the nitrate value set in the EU drinking water directive which is designed to keep water in public supplies safe for drinking. Five countries had rivers that exceeded maximum allowable concentrations under the directive.

The situation is even worse for groundwater reserves. In many groundwater bodies in Europe for which data are available, measurements of nitrate concentrations have found levels that exceed the specified values of the drinking water directive.

Figure 5.6   Trends in nitrate concentrations in rivers in European countries

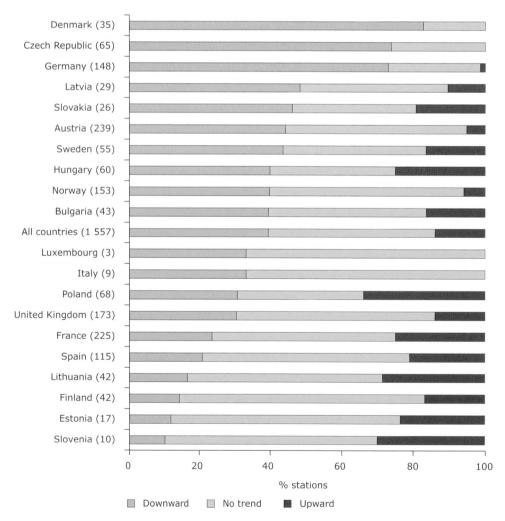

Source: EEA, 2005.

In some parts of Europe, these problems may be expected to get worse before they get better, particularly for groundwater. It can take years or even decades for nitrates to reach zones from which drinking water is taken. As the average age of groundwater used in drinking water is 40 years, much of the surplus nitrogen applied to farms in recent decades has yet to reach the water that it will eventually pollute. Indeed, a nitrate legacy may exist beneath many of Europe's fields that future generations will have to pay to clean up.

Removing nitrates from water to make it fit for drinking is expensive. To make it fit for public supply, water contaminated with nitrates is often diluted with cleaner water from other river or groundwater sources. Denitrification of UK drinking water already costs some EUR 30 million a year, and capital spending in the next two decades to meet European standards could cost the country 10 times as much.

It is generally much cheaper to prevent nitrates from reaching the water in the first place. A review of the possible costs to farmers makes an initial estimate of EUR 50–150 per hectare per year to alter farming methods to comply with nutrient management standards under the EU nitrates directive. This is considerably cheaper than the cost of removing nitrates from polluted water. Moreover, changing farming practices puts the responsibility on the farmers who caused the pollution, rather than on the consumer.

In 1991, the EU introduced a nitrates directive, aimed at stemming the flow of nitrates into the natural environment and drinking water. Member States are required to designate nitrate-vulnerable zones where the risks are highest and to impose strict controls on nitrate use in those areas.

Implementation of the nitrates directive across Europe has been generally poor. However, the synthesis of Member States' reports for 2000 concludes that 'Member States have in the last two years shown a real willingness to improve implementation. They realise that costs induced by drinking water treatment for nitrates excess, or by eutrophication damages in dams or coastal waters will still increase, and that the investments dedicated to urban wastewater treatment will be inefficient regarding nutrients if a parallel effort is not devoted to an effective reduction of agricultural nutrients losses'.

Nitrate pollution can be tackled at source. In Denmark, for instance, a national nitrate management plan began in the 1980s, before the directive came into force. It offered advice to farmers on making efficient use of fertilisers and imposed annual nitrogen 'budgets' on farms. It has substantially stemmed the leakage of nitrate from Danish farming systems.

The patchy implementation of the nitrates directive has been reflected in a patchy pattern of trends in nitrate pollution across Europe. Average nitrate concentrations in European rivers are falling. However, while 25 % of monitoring stations have shown a decline since 1992, 15 % have shown an increase. The most marked reductions have been noted in Denmark, Germany and Latvia, with additional success stories noted in regions, including the Algarve and the east of France, where intense field controls, including soil analysis, have accompanied the dissemination of good-practice advice.

The story for coastal waters is more complex, often because of complicated interactions between the riverine and marine environments. There has been a decrease in measured concentrations of both phosphorus and nitrogen in Dutch coastal waters since 1991, in line with reduced loads in the river Rhine. In Denmark, where reductions in discharges began earliest, there has been a 40 % reduction in the marine nitrogen load around Danish shores since 1989.

5.10 Summary and conclusions

The quality of river water across Europe has improved thanks to a range of EU environmental directives since the 1970s. Water abstractions have also been in decline. However, pressures from agriculture, urbanisation, tourism and climate change suggest that guaranteeing water quality will continue to be a costly issue.

Future demographic and economic trends are likely to increase water consumption by households and for tourism-related use. Northern Europe is likely to see substantial reductions in water withdrawals as power stations switch to new technologies. However, overall use could rise if climate change leads to greater demand for water for irrigation.

In southern Europe, higher temperatures are likely to increase the need for irrigation of crops, so there is a strong case for substantial improvements in the efficiency of irrigation systems. Among new EU Member States and candidate countries, water use is expected to increase, especially in households as living standards improve, suggesting scope for using technologies and market measures to manage demand.

Water quality is most severely affected by pollution from households, industry and agriculture. For the past 15 years, the main focus has been on point sources of water pollution, such as households and factories, with good results. Today, approximately 90 per cent of the population in north-west Europe is connected to sewerage and treatment plants. Nevertheless, many EU-15 countries have not yet complied fully with the UWWT directive and the new EU countries still have many years of effort ahead of them.

Treating wastewater is expensive: the EU-15 has spent around 0.8 per cent of GDP on it. Approaches which combine avoiding pollution at source, through charges, with targeted construction of treatment plants offer a cost-effective solution to implementation. Under the EU cohesion policy, new EU countries are eligible for considerable subsidies over the next decade or so to assist with wastewater treatment. Guidelines would help to steer new member countries towards a policy of charges paid for by polluters linked to EU funding for treatment plants.

As point sources of pollution show a marked improvement in terms of impact on water quality, diffuse sources of water pollution, particularly from agriculture, will dominate future water policy. Diffuse sources of water pollution are by their nature less obvious and harder to police than point sources, and this will have an impact on the success of the required legislation.

Fertiliser application for arable farming is the main source of diffuse pollution to water, with nitrates the greatest problem. Nitrate pollution is higher in the EU-15 than in the new Member States. In some parts of Europe these problems are expected to get worse before they get better, especially in groundwater where it can take decades for nitrates to reach drinking water zones. Cleaning up nitrate pollution is estimated to be around 10 times more expensive than preventing pollution in the first place through changes in farming methods.

Sustainable management will continue to be the dominant theme regarding freshwater resources. Across Europe, rivers have been canalised, culverted or regulated. Wetlands of national importance have been altered by river engineering. In other words, much of Europe's waterways have been 'managed' in ways that are damaging to the long-term condition of the environment.

The water framework directive, launched in October 2000, aims at achieving good ecological status for all water bodies in Europe by 2015, based on wider ecological principles. There is potential for much greater efficiency in water use through the introduction of market-based instruments (such as water charges and pollution taxes) and new technologies, as well as tougher standards to reduce leakage in water distribution systems.

References and further reading

The core set of indicators found in Part B of this report that are relevant to this chapter are: CSI 18, CSI 19, CSI 20, CSI 24 and CSI 25.

Introduction
European Environment Agency, 2000. *Sustainable use of Europe's water? State, prospects and issues*, Environmental Assessment Report No 7, EEA, Copenhagen.

European Environment Agency, 2004. *EEA signals 2004*, EEA, Copenhagen.

European Parliament and Council, 2000. Directive 2000/60/EC establishing a framework for Community action in the field of water policy also known as the water framework directive (WFD).

Supply and demand
European Environment Agency (1999). *Sustainable water use in Europe — Part 1: Sectoral use of water*, Environmental Assessment Report No 1, EEA, Copenhagen.

European Environment Agency, 2000. *Groundwater quality and quantity in Europe*, Environmental Assessment Report No 3, EEA, Copenhagen.

European Environment Agency, 2001. *Sustainable water use in Europe — Part 2: Demand management*, Environmental Issue Report No 19, EEA, Copenhagen.

European Environment Agency, 2003. *Europe's environment: the third assessment* — 'Chapter 8 — Water, Environmental Assessment', Report No 10, EEA, Copenhagen.

European Environment Agency, 2003. *Status of Europe's water*, Briefing No 1/2003, EEA, Copenhagen.

European Environment Agency, 2004. *EEA signals 2004*, EEA, Copenhagen.

Water use
European Environment Agency, 2003. *Europe's water: An indicator based assessment*, Topic Report No 1/2003, EEA, Copenhagen.

European Environment Agency, 2005. *European environmental outlook*, Report No 4/2005, EEA, Copenhagen.

Climate change and water stress
European Environment Agency, 2001. *Sustainable water use in Europe — Part 3: Extreme hydrological events: floods and droughts*, Environmental Issue Report No 21, EEA, Copenhagen.

European Environment Agency, 2003. *Europe's water: An indicator based assessment*, Topic Report No 1/2003, EEA, Copenhagen.

European Environment Agency, 2005. *Climate change and river flooding in Europe*, Briefing 1/2005, EEA, Copenhagen.

Water quality
European Environment Agency, 2000. *Sustainable use of Europe's water? State, prospects and issues*, Environmental Assessment Report No 7, EEA, Copenhagen.

European Environment Agency, 2003. *Europe's water: An indicator based assessment*, Topic Report No 1/2003, EEA, Copenhagen.

Water pollution control developments
European Commission, 2004. *A new partnership for cohesion: convergence, competitiveness, cooperation*, Third report on economic and social cohesion. (See www.europa.eu.int/comm/regional_policy/sources/docoffic/official/reports/pdf/cohesion3/cohesion3_cover_en.pdf — accessed 22/10/2005).

European Environment Agency, 2003. *Europe's water: An indicator based assessment*, Topic Report No 1/2003, EEA, Copenhagen.

European Environment Agency, 2005. *Effectiveness of urban wastewater treatment policies in selected countries: An EEA pilot study*, EEA 2/2005, Copenhagen.

Costs and benefits of water pollution control
European Commission, 2004. *A new partnership for cohesion: Convergence, competitiveness, cooperation*, Third report on economic and social cohesion.

European Council, 1976. Directive 76/160/EEC concerning the quality of bathing water.

European Council, 1991. Directive 91/271/EEC on urban waste water treatment.

Tackling diffuse sources of pollution
European Council, 1991. Directive 91/271/EEC on urban waste water treatment.

European Environment Agency, 2000. *Nutrients in European ecosystems*, Environmental Assessment Report No 4, EEA, Copenhagen.

Nitrates
European Council, 1976. Directive 76/160/EEC concerning the quality on bathing water.

European Council, 1991. Directive 91/676/EEC on nitrates from agricultural sources; EU nitrates directive.

European Environment Agency, 2000. *Groundwater quality and quantity in Europe*, Environmental Assessment Report No 3, EEA, Copenhagen.

European Environment Agency, 2001. *Late lessons from early warnings: The precautionary principle 1896–2000*, Environmental Issue Report 22, EEA, Copenhagen.

European Environment Agency, 2004. *Agriculture and the environment in the EU accession countries*, EEA Environmental Issue Report 37, Copenhagen.

European Environment Agency, 2004. *EEA signals 2004*, EEA, Copenhagen.

European Environment Agency, 2005. *European environmental outlook*, Report No 4/2005, EEA, Copenhagen.

European Environment Agency, 2005. *Source apportionment of nitrogen and phosphorus inputs to the aquatic environment*, draft report, EEA, Copenhagen.

European Environment Agency, 2005. *Sustainable use and management of resources*, EEA, Copenhagen (in print).

European Parliament and Council, 2000. Directive 2000/60/EC establishing a framework for Community action in the field of water policy also known as the water framework directive (WFD).

6 Marine and coastal environment

6.1 Introduction

The seas around Europe have been a vital resource over millennia. They provide a wide range of employment and environmental services, including fisheries, shipping and port development, tourism, wastewater cleansing, oil and gas production, aggregate extraction, wind, wave and tidal energy production, and much more. In many coastal regions, fish and marine mammals have been the dominant source of food and their capture the main employment activity. Balanced management of marine and coastal resources can contribute to the objectives of the Lisbon agenda and the longer term aspirations of the EU sustainable development strategy.

Recent results from European scientific programmes such as ELOISE, and from the EEA, have identified a number of key pressures, drivers and impacts affecting Europe's marine environment (Table 6.1). They derive from a variety of land and marine-based activities and the two key global processes of climate change and ocean dynamics.

The pressures arising from these global processes include elevated air and sea surface temperatures, rising sea levels and changing weather conditions. They occur at a pan-European scale but have regionally different outcomes.

The pressures arising from land-based socio-economic activities are of a more regional and local nature. The sources of such pressures include changing farming and forestry practices that alter the contents of run-off water to estuarine and coastal waters. Urbanisation and infrastructure development are changing the natural dynamics of coastal ecosystems as well as increasing pollution from effluents and storm waters. Industrial discharges, mass tourism and maritime trade are other contributors. The extraction of large volumes of aggregates also has a significant effect on coastal systems.

Table 6.1 Major impacts related to main drivers and pressures in the coastal and marine environment

Pressures/drivers	Impacts
Climate change	Erosion, biodiversity loss, increased/changed flood-risk, altered species composition
Agriculture and forestry change	Eutrophication, contamination, biodiversity/habitat loss, subsidence, salinisation, altered sediment/water supply
Urbanisation and infrastructure change	Coastal squeeze, eutrophication, contamination, habitat loss/fragmentation/human disturbance, subsidence, altered sedimentation, increased flood-risk, salinisation, altered hydrology
Tourism development	Seasonal/local impacts, beach 'management', habitat disruption, loss of species, increased water demand, altered longshore sediment transport, loss of local cultural values
Industry and trade expansion	Contamination, exotic species invasion, dredging, sediment supply/erosion
Fisheries/aquaculture expansion	Species loss/overexploitation of fish stocks, impact on migratory species, habitat loss, species introduction/genetic pollution, contamination, eutrophication
Energy exploitation and distribution	Habitat alteration, altered water temperature, changed landscape/amenity, subsidence, contamination, accident risk, noise/light disturbance

Source: ELOISE, 2004.

Marine and coastal environment PART A

Pressures from offshore marine and other coastal activities are equally apparent. Overfishing and aquaculture and the increasing demand for energy are the most pertinent, with relatively new techniques and practices, in particular, threatening marine wildlife populations on an unprecedented scale.

One of the main reasons for the build-up of pressures on the marine and coastal environments stems from

Map 6.1 Pan-European marine ecosystems

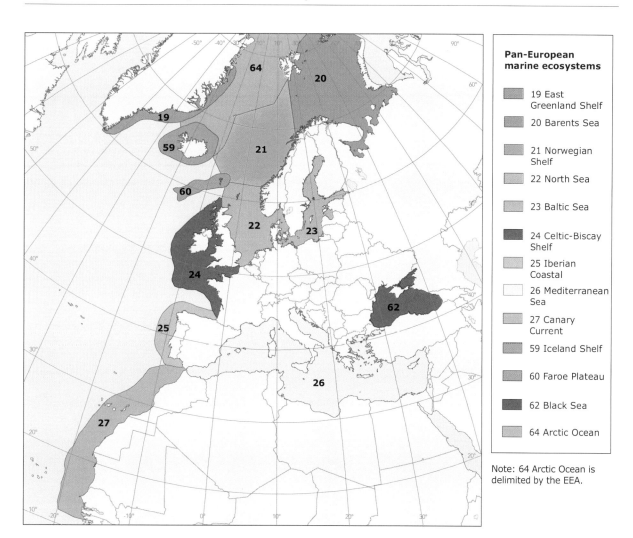

Note: 64 Arctic Ocean is delimited by the EEA.

Note: The large marine ecosystems (LMEs) project was created in support of the global objectives of Chapter 17 of Agenda 21, as a follow-up to the 1992 United Nations Conference on Environment and Development (UNCED). Out of the 64 LME defined worldwide, 13 are pertinent to the European environment. Numbering used in the map follows the one used in the LME project.

Source: UN (See www.oceansatlas.org — accessed 12/10/2005).

The European environment | State and outlook 2005

the fragmented approach to strategic development and management. Without doubt the future health of the marine environment and its living resources now depends upon Europe taking an integrated approach to conservation, management and spatial planning — with the ecosystem at its core (Map 6.1).

6.2 Regional perspectives on the state of the marine environment

The relative strength of the drivers, pressures and impacts varies regionally. This is partly because of the hydrography of Europe's marine ecosystems and surrounding coastal landscapes, and partly because of the socio-economics of the coastal states associated with them.

From both a biophysical and a political perspective, the fact that Europe's ecosystems are so different means that extra efforts must be made to achieve comparable assessments of trends in environmental conditions and policy effectiveness. In particular, existing data and monitoring schemes need to be analysed consistently, in order to be able to detect changes in trends from the various long time series that exist. In this respect the ecosystem approach, proposed in the EU's marine strategy, is vital.

Results are available from a range of analyses of environmental conditions undertaken and variously published by intergovernmental, European, regional and scientific bodies and the EEA. They are presented here in summary for key marine regions: the Baltic Sea, the Barents Sea, the Black Sea, the Celtic-Biscay Shelf Sea, the Iberian Coast Sea, the Mediterranean and the North Seas. Boxes giving further background information for each region can be found across the chapter.

Over the last decade different regions have witnessed significant alterations in coastal morphology, increases in coastal flooding, loss of ice cover, reduced water quality, and declines in biodiversity, living resources and cultural landscapes, as a result of climate change and socio-economic conditions in the coastal zone. There are early signals that Europe's marine and coastal ecosystems are also undergoing structural changes to the food chain, evidenced by the loss of key species, the occurrence of large concentrations of key planktonic species in place of others, and by the spread of invasive species, and caused by widespread human activities.

In the **Baltic Sea** there are continuing problems with eutrophication, anoxic conditions and toxic blooms of algae, overexploitation of both freshwater and marine fisheries, and with alien and accidental introductions of species. To the north, in the **Barents Sea**, ecosystem-wide disturbances have been documented; these have been caused by the decline in capelin due to overfishing and periodic booms in the herring population, and pollution levels from shipping, military activities and oil extraction. Future challenges will arise from the disposal of nuclear submarines and ecosystem changes related to reductions in ice cover and the melting of the permafrost due to global warming.

In the **North Sea** the concerns are those of damage to the food web, threatening globally important populations of seabirds and some commercially important fish species, and the wide range of discharges of pollutants such as nitrogen to water and air from the heavily populated coastal zone and larger rivers. The **Celtic-Biscay Shelf Sea** has extensive fisheries using trawls, gill-nets and long-lines; these together with oil drilling have damaged the rich cold-water coral reefs. The rough sea conditions also mean that coastal ecosystems have been seriously impacted by a series of oil and other discharges, and also make shipwrecks more likely. The **Iberian coastal sea** is heavily influenced by oceanic conditions. As a result, global warming and any alterations to ocean circulation due to climate change will have an effect in the future on the structure of the ecosystem.

The challenges facing the **Mediterranean** are associated with coastal erosion, eutrophication hot spots and toxic algal blooms, low nutrient levels leading to low productivity in the south-east, fisheries by-catches of marine wildlife and invasion of alien species. To the east, the structure of the **Black Sea** ecosystem has been

Baltic Sea

The Baltic is, in essence, a giant brackish fjord 1 500 kilometres long, where freshwater and pollutants from rivers accumulate at the surface, making the waters increasingly anoxic, until they are 'flushed out' every few years by oxygen-rich water from the North Sea.

The Baltic Sea is bordered by Denmark, Estonia, Finland, Germany, Latvia, Lithuania, Poland, Russia and Sweden. Cities on its shores include Gdansk, Helsinki, St Petersburg and Stockholm. The primary human impacts on the sea are overfishing; pollution from the land, including heavy metals, persistent organic pollutants and, particularly, nutrient discharges, arising from agriculture, forestry, urbanisation and industrial developments; changes to the aesthetic landscape and seascape from industrial and energy developments such as wind farms; coastal squeeze and coastal erosion.

The Baltic is particularly vulnerable to eutrophication, partly because it is semi-enclosed and partly because it drains an area of land four times larger than the sea itself. Eutrophication has caused large-scale replacement of coastal sea grasses, important for fish nurseries, with large beds of algae, particularly along the more densely populated shores of the southern Baltic. The associated toxic blooms of algae have caused major losses of fish and disturbed recreational activities.

The anoxic conditions on the bottom of the sea seem to be worsening. This is partly due to eutrophication, and partly because of natural variability in weather conditions.

Because of its changing salinity, the Baltic is home to stocks of both freshwater and marine fish. Catches grew through the 1990s, but stocks are now generally overexploited. Most of the catches are small herrings, but there are also significant cod and other marine stocks near the exit to the North Sea and freshwater fish such as salmon in the fresher regions of the Gulf of Bothnia in the north (Figure 6.1).

The ecosystem has been disturbed by hunting for marine mammals, which, along with pollution, has reduced seal populations to low levels. This has left the cod as the main predator in the sea. The cod in turn is now threatened both by overfishing and by episodic events. As predators progressively disappear, other fish species, such as sprat, have grown in importance.

Another problem in the Baltic is invasion by alien species which, together with accidental introductions, is having a direct effect on the viability of native species found only in the Baltic.

Figure 6.1   Landings of main commercial species in the Baltic Sea

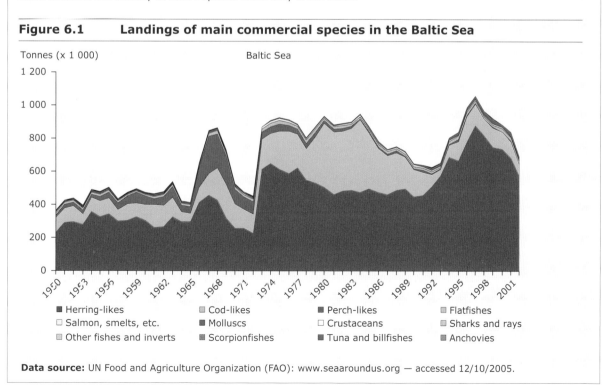

Data source: UN Food and Agriculture Organization (FAO): www.seaaroundus.org — accessed 12/10/2005.

Barents Sea

The Barents Sea is a shallow shelf area located between the northern shore of Russia, the southern edge of the Arctic Ocean and the northern tip of the Atlantic Ocean. It includes Svalbard on the far northern edge of the Atlantic and Novaya Zemlya, north of the Urals. The sea receives water from the Pechora and other Russian rivers, and it is strongly influenced by major currents that exchange water between the two oceans. Depending on the season, ice covers between a third and two-thirds of the sea.

The Barents Sea is a highly productive area with strong upwelling and a ready food supply for many commercial species. The food web is dominated by a handful of species: diatoms, krill, capelin, herring and cod. The relationship between these species is highly dynamic. At up to 8 million tonnes, the capelin stock, which feeds on the sea's abundant plankton, is potentially the largest in the world and in the past has supported extensive fishing operations.

The capelin has been in serious decline, partly because of overfishing and partly due to periodic booms in the population of young herring, which eat capelin larvae. The size of capelin and herring populations go up and down like a seesaw. Capelin numbers rocketed after herring stocks collapsed in the late 1960s, but then declined as herring recovered.

The periodic collapses in the capelin population cause food shortages for other species, including fish such as cod, mammals such as harp seals and birds such as guillemots. When the capelin last disappeared, cod switched to feeding on krill and other species. Seals left the ice and invaded the Norwegian coast looking for food. The birds mostly died.

These switchback changes are a natural phenomenon, probably partly driven by changing inflows of water from the surrounding oceans. They are further affected by fishing, mainly by Norwegian and Russian fleets. Fishing, for instance, caused a crash in the herring population in the 1970s, which saw an overall decline in the sea's fish landings of some 95 % between the late 1970s and mid-1980s. Catches have since partially recovered (Figure 6.2).

In this marine ecosystem and in the deeper waters off east Greenland, Iceland, around the Faeroes, off Norway and Svalbard, large sponge fields with very rich associated fauna exist. So far no detailed records about the impact of fisheries on the benthic community in these areas have been collected, but it is thought very likely that because of their slow growth they take many years to recover after even partial damage.

Pollution levels are not high in the Barents Sea, but there are significant sources present, including onshore oil extraction, shipping and radioactive fallout from nuclear tests and the Chernobyl accident. There is also a lot of military activity, highlighted by the loss of the Kursk nuclear submarine in the eastern Barents Sea in 2000. An expected major increase in oil and gas production in the region is likely to add to pollution risks.

Figure 6.2   Landings of main commercial species in the Barents Sea

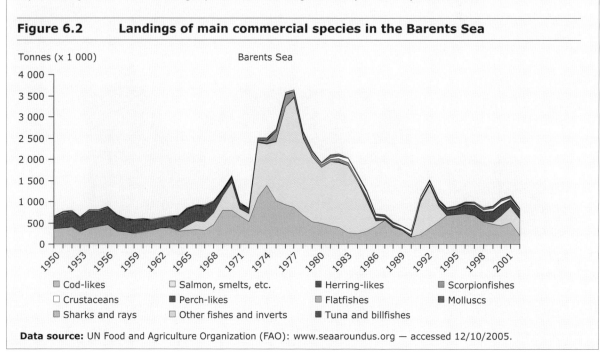

Data source: UN Food and Agriculture Organization (FAO): www.seaaroundus.org — accessed 12/10/2005.

Marine and coastal environment

North Sea

The North Sea covers some 750 000 square kilometres and, at an average depth of 90 metres, is shallow. From the results of the EU research programme Eurosion, it has been estimated that some 17 million people in nine countries live in the coastal zone under the influence of erosion. The coastline is one of the most diverse in the world, with towering fjords, wide estuaries and deltas, mudflats and marshes, rocky cliffs and sandbanks.

The sea is intensely exploited by European countries for a wide range of resources. These include fish, marine sands and gravels, and the hydrocarbons beneath the seabed, which provide half of the EU's energy needs. It is also a major shipping route, serving world ports like Hamburg and Rotterdam, and oil and gas terminals linked to offshore rigs by pipelines. It provides access to the Baltic Sea, and its narrow southerly exit through the Straits of Dover is one of the most heavily used sea routes in the world.

The ecology of the North Sea has been substantially altered by heavy levels of fishing. Landings are currently at around 2.3 million tonnes a year and include herring, sardine, anchovy, cod, mackerel and haddock for human consumption, along with shellfish and sand eels, which are used as feed for farm animals and aquaculture (Figure 6.3).

Most fish stocks are overexploited and some are in danger of crashing. As a result of low levels of spawning stocks of North Sea cod, recruitment has fallen from 390 million fish a year in the 1960s and 1970s to less than 250 million in the 1990s. The current low stocks mean that fish are caught younger and smaller, which is economically as well as ecologically bad news. Allowing stock recovery would go hand in hand with bigger profits.

Overexploitation has happened despite progressively tougher restrictions on catches and fishing technology introduced through the common fisheries policy. Overfishing is also damaging the marine food web, reducing its resilience, with sometimes unpredictable consequences for other species.

Those species under threat from the damaged food web include globally important populations of seabirds. A recent crash in stocks of sand eels in the Shetland Islands and elsewhere, principally caused by overfishing, has deprived coastal breeding populations of puffins and other species of a main source of food. More unexpectedly, recent enforced reductions in fishing are also lowering numbers of opportunistic seabirds. This is because some seabird colonies, such as some species of gulls or skua, have grown large by feeding extensively on discards and process waste from fishing vessels. The North Sea's population of skua, for example, has risen 200-fold in the past century.

The sea is a major sink for a wide range of discharges to water and air from the surrounding countries. Pollution to the sea comes from direct discharges from coastal communities and via rivers, drainage from farmland and, to a significant extent, fallout of air pollution. Eutrophication from nitrogen sources in water and air is a major threat. Wildlife also suffers from such pollution as well as from oily wastes and industrial discharges.

Figure 6.3   Landings of main commercial species in the North Sea

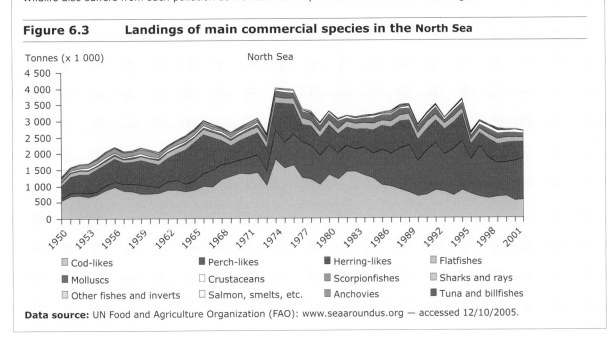

Data source: UN Food and Agriculture Organization (FAO): www.seaaroundus.org — accessed 12/10/2005.

Integrated assessment | Aquatic environment

Celtic-Biscay Shelf Sea

The Celtic-Biscay Shelf occupies the north-east Atlantic west of Scotland, Ireland, England and France. It includes the Irish Sea, the English Channel and the shallower inshore reaches of the Bay of Biscay off France. It is strongly influenced by currents in the Atlantic itself, including the Atlantic Drift in the north and the Azores Current in the south.

Its conditions are highly seasonal, and it responds strongly to periodic flips in the natural climate system known as the North Atlantic Oscillation. The North Atlantic Oscillation influences sea temperatures, currents and the numbers and distribution of many species of fish, including bluefin tuna and albacore. All this gives the Celtic-Biscay Shelf a high and dynamic biodiversity, much of which is now being actively exploited or has been in the past. Major harvests have included seaweed, whales, molluscs, herring, sand eels and mackerel. Landings of main commercial species have remained fairly constant in recent decades (Figure 6.4).

The shelf includes a number of large sea mounts on which sit rich reefs of cold coral, such as *Lophelia pertusa*. Globally important, cold corals are long lived, slow growing and provide a habitat for other marine species, including commercially valuable fish. The reefs form a chain along the edge of the continental shelf from western France, through high concentrations west of Ireland to a scattering off Scotland.

The waters near the reefs are known to contain unusually large concentrations of flatfish, resulting in their being targeted by fishing boats, often with damaging and counter-productive results. Some reefs have been badly damaged by trawling, as well as by gill-nets and bottom long-lines. The reefs are also at risk from oil drilling.

Pollution is not a great threat out on the shelf, where waves and a strong tide flush away any accidental discharges from ships. Local coastal ecosystems such as estuaries, coastal lagoons and sandy shores, however, can be damaged, and the rough waters make shipwrecks more likely. The shelf has seen a series of oil tanker disasters, including the Torrey Canyon, which ran aground off Cornwall in the United Kingdom in 1967; the Amoco Cadiz, which was shipwrecked off Brittany, France, in 1978; the Sea Empress off Wales in 1992; and the Erika, again off Brittany, in 1999. In each case, the winds and waves brought the oil ashore, and some remains of each of these ecological disasters can still be seen.

Figure 6.4   Landings of main commercial species in the Celtic-Biscay Shelf

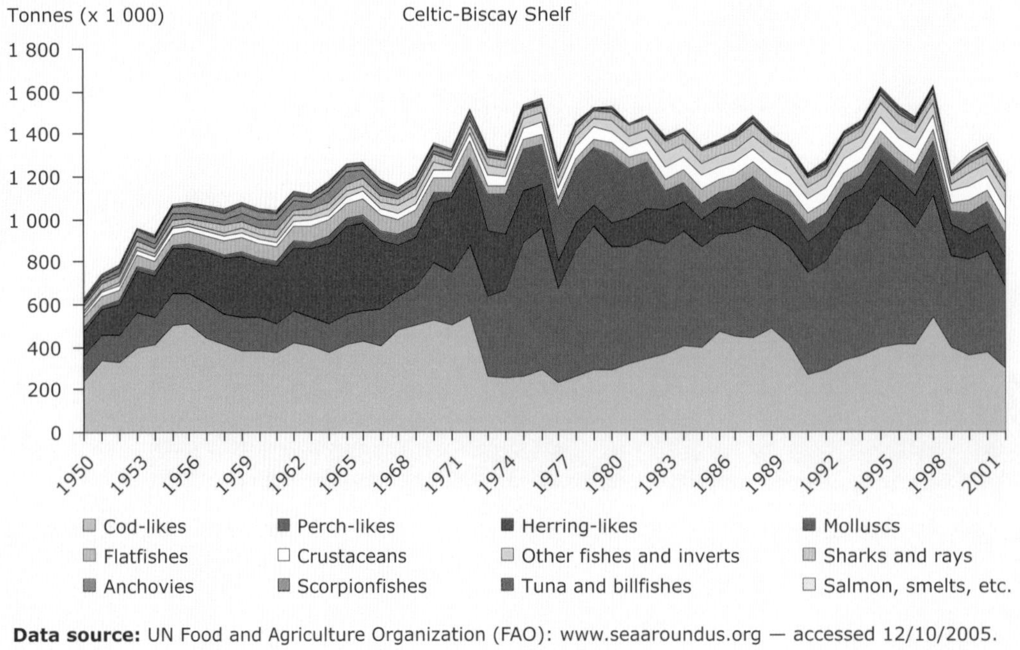

Data source: UN Food and Agriculture Organization (FAO): www.seaaroundus.org — accessed 12/10/2005.

Iberian coastal sea

The Iberian shelf region is part of the eastern Atlantic seaboard of western Europe, immediately south of the Celtic-Biscay Shelf. It stretches round the Iberian Peninsula from near the French border to Gibraltar. Much of the coastline is deeply indented with drowned river valleys. The shelf, which varies in width here from 15 to 400 kilometres, experiences intense upwelling of nutrients from the ocean depths in summer, with consequently high biological activity, a rich fishery and abundant marine mammals. This coastline was the original home of the European whaling industry in the Middle Ages.

Like the Celtic-Biscay Shelf, fierce tides and storms make it risky for shipping. The Prestige tanker disaster in 2002 happened in this region, causing massive oil pollution off the Galician coast of north-west Spain.

Commercial fish stocks are dominated by small pelagic fish such as herrings, anchovies and sardines. Landings have remained fairly constant since 1980 (Figure 6.5). Anchovies are a major catch for former whaling ports in the Basque country. The abundance of sardines and other species changes dramatically with the variable ocean conditions, mediated largely through the ocean's influence on the availability of diatoms. Thus the sardine fishery goes through natural periods of boom and bust.

Similarly, in the past, blooms of dinoflagellate algae have been caused by apparently natural variations in oceanic conditions. There is some suggestion that the recent emergence of toxic algal blooms may be a result of eutrophication and the introduction of alien species from discharged ballast water.

Figure 6.5   Landings of main commercial species in the Iberian coastal sea

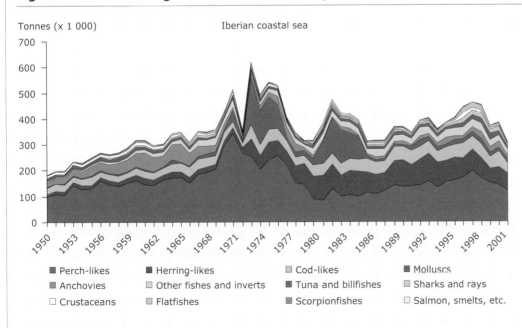

Data source: UN Food and Agriculture Organization (FAO): www.seaaroundus.org — accessed 12/10/2005.

Mediterranean Sea

The Mediterranean has been the transport hub and the fish basket of numerous civilisations; from the time of the Ancient Greeks, through the rise of Venice as the great trading port with Asia, to its modern tourist-based economy. From Spain to Greece, and from Morocco to Turkey, the Mediterranean is bordered by 20 nations. More than 130 million people live permanently along its coastline, a figure that doubles during the summer tourist season. The sea and its shores are the biggest tourist destination on Earth.

Despite covering more than 2.5 million square kilometres, and lapping at the shores of Europe, Asia and Africa, it is largely landlocked. It has a narrow upstream connection through the Bosporus to the Black Sea and almost as narrow an outflow into the Atlantic Ocean through the Straits of Gibraltar. Well-oxygenated Atlantic water flows in at the surface and flows out at depth.

Although in some respects an outsize lake, with virtually no tidal range, the Mediterranean is nonetheless a dynamic sea with wind-driven currents, big seasonal fluctuations in sea temperatures and significant local areas of upwelling that bring nutrients to the surface, especially in the Adriatic.

It also has strong sources of man-made nutrients and other pollutants delivered down rivers such as the Rhône, Po, Ebro and Nile, as well as directly from the numerous large settlements and from the fallout of air pollution over the sea. The combination of nutrient pollution from the Po and local upwelling causes serious eutrophication problems in the north Adriatic in some summers.

There are other hot spots where nutrients accumulate and cause eutrophication, mostly in estuaries and around coastal population centres. During long periods of summer calm, when the sea becomes stratified and surface water temperatures soar, toxic algal blooms can also form in these areas. In the Adriatic in particular, pollution has damaged fisheries. Toxic algal blooms and de-oxygenation cause occasional fish kills, and from time to time beaches on the Italian Adriatic coast are closed after toxic algae, such as *Ostreopsis ovata*, cause outbreaks of illness among bathers.

Elsewhere, nutrient levels, and consequently biological productivity, are low. This is especially so in the south-east Mediterranean, where natural nutrient sources in silt brought down from East Africa by the River Nile, were cut off following the damming of the river 40 years ago. This has since resulted in the collapse of fisheries in this part of the sea.

Fish catches in the Mediterranean have been fairly stable at around 1 million tonnes for a decade and more, though catches per boat have declined significantly, demonstrating that stocks are under stress (Figure 6.6). Efforts to maintain catches with high-intensity fishing equipment such as drift-nets and long-lines have caused serious problems with by-catches of marine animals such as dolphins and endangered species of turtles. Another major threat to turtles and other marine wildlife is tourism and development activities on nesting beaches.

Human activity and invasions of alien species have also damaged coastal ecosystems on which fisheries depend. A form of algae native to the Red Sea, *Caulerpa taxifolia*, has spread round the Mediterranean from the French Riviera, where it first emerged in the 1980s, obliterating sea grasses and replacing them with largely sterile algal beds.

Figure 6.6   Landings of main commercial species in the Mediterranean Sea

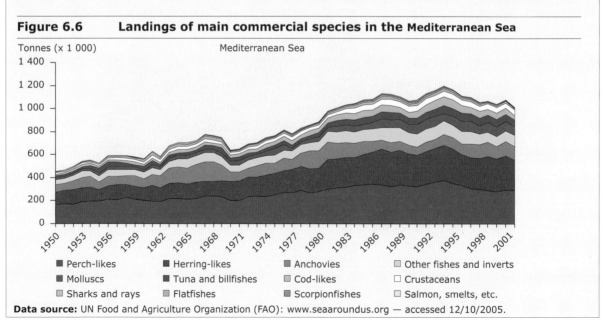

Data source: UN Food and Agriculture Organization (FAO): www.seaaroundus.org — accessed 12/10/2005.

Black Sea

The Black Sea is largely enclosed. It receives two-thirds of its water from the River Danube, and the rest from other major rivers like the Dnieper, Dniester and Don. Together these rivers drain an area of central and eastern Europe 20 times the size of the sea itself. Six countries — Bulgaria, Georgia, Romania, Russia, Turkey and Ukraine — have Black Sea shores, but a further 16 countries form part of the area that drains into the sea. The rivers bring large amounts of pollution into the ill-flushed sea, including nutrients, raw sewage, oil and heavy metals from industry. Beaches are regularly closed because they become unsafe for bathing following the formation of red tides and the build-up of sewage pathogens in coastal waters. Coastal wetlands that once filtered pollution, such as the Danube delta, have been damaged by intensive farming and the construction of navigation channels.

The sea has a low salinity, since its inflows are freshwater, and it exchanges water only slowly with the saline Mediterranean via the Bosporus. The sea is in places more than two kilometres deep, but oxygen is virtually absent below 250 metres. Below this level, comprising some 90 % of the sea's water, is the largest known volume of lifeless, anoxic water on the planet. This is essentially a natural phenomenon.

Eutrophication appears to have been a major problem only since the 1970s. Phosphates and nitrates, largely flowing into the sea from the large Danube drainage basin, have reached levels roughly double those of the also-eutrophied Baltic Sea. Eutrophication is thought to have extended the anoxic zone, which now reaches into the shallow north-west of the sea. In turn, the ever larger volume of anoxic water reduces the ability of the sea to purify itself. However, the causal link to nutrient levels may not be so simple. Evidence from a 6 000-year sedimentary record shows that the volume today is the same as then — before extensive human influence.

Combined with overfishing, eutrophication has seriously disrupted the ecosystem. It has increased the amount of plankton in the sea, boosting plankton-eating fish, while decreasing the numbers of species further up the food chain.

These changes have left the ecosystems vulnerable to invasions of alien species. In particular *Mnemiopsis*, a type of jellyfish, saw explosive growth after arriving in ships' ballast water in the late 1980s. It eventually amounted to more than 90 % of the entire biomass of the sea, and caused the collapse of anchovy and chub mackerel stocks, local oyster fisheries and even the indigenous jellyfish. Its spread was only contained with the artificial introduction of a competitor jellyfish; in the past five years there has been a modest revival of anchovy stocks, but not so far of chub mackerel (Figure 6.7).

The sea's most productive area is now the shallow Sea of Azov. However, this too has suffered from a decline in inflows of freshwater as a result of abstractions for irrigation on the River Dnieper. The crisis in fisheries in the sea has had extensive socio-economic consequences with many coastal economies undermined. Fish have also become expensive, with implications for nutrition in communities already impoverished by the collapse of the Soviet system. Meanwhile, the extensive pollution of beaches is undermining intended expansion in tourism.

Figure 6.7   Landings of main commercial species in the Black Sea

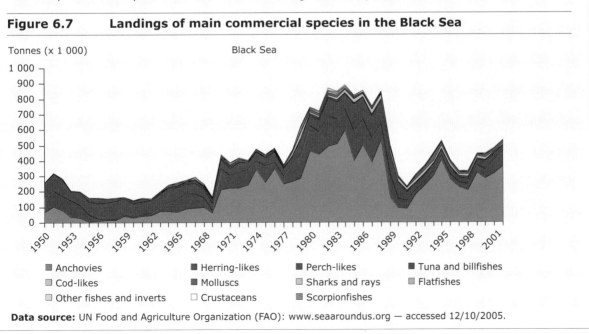

Data source: UN Food and Agriculture Organization (FAO): www.seaaroundus.org — accessed 12/10/2005.

disrupted by overfishing which has left it vulnerable to invasions, increased inflows of nutrients and pollution arising from damage to the coastal wetlands, and the extension of the anoxic zone.

6.3 The state of coastal and intertidal areas

Despite its relatively small geographic size, Europe has a very long coastline, and one that has always proven attractive for settlement. Ports have built up over the ages as centres of trade and industry, and the flat, fertile coastal plains have been the focus of agriculture and convenient land for building and transport infrastructure.

Many of Europe's capital cities are on or close to the coast, including Amsterdam, Athens, Copenhagen, Dublin, Helsinki, Lisbon, London, Oslo, Riga, Rome, Stockholm, Tallinn and Valletta. Altogether there are 280 coastal cities with populations above 50 000. In Belgium, Portugal and Spain, the population density on land within 10 kilometres of the coast is more than 50 % above that further inland. Today, about 70 million of the 455 million citizens of the enlarged EU, i.e. 16 % of the population, live in coastal municipalities, although the coastal zone represents only 11 % of the EU's total area.

In recent decades, the coasts have become a magnet for the tourist industry and for second homes based around fast-growing coastal resorts on the French and Italian Riviera, Greece, southern Spain and elsewhere. The oceans, beaches, dramatic coastlines and clean sea air have emerged as prime environmental assets. As a result, in regions such as Brittany in France, more than 90 % of the entire population lives on the coast.

Figure 6.8 Percentage of artificial coastline length by NUTS3

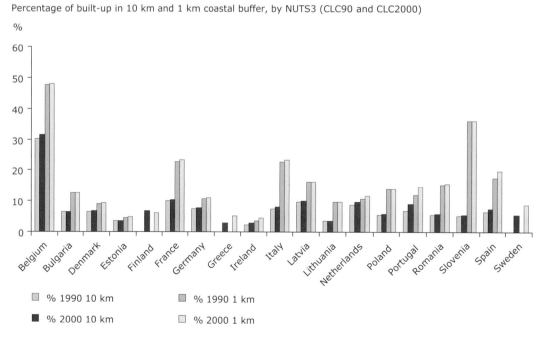

Source: Corine Land Cover 1990 and 2000; EEA, 2005.

Today, the coastal strip in many European countries is the fastest growing area, in terms of social and economic development. The Mediterranean coast of Spain, along with Ireland, has the fastest growing population in Europe, with increases of up to 50 % in the past decade. In Spain, 1.7 million houses, mostly sited along the coastal strip, are the secondary residences of Spanish city dwellers or are owned by foreigners, primarily as holiday homes. Other countries with more static populations are seeing considerable migration along the less-populated parts of the coastal zones, such as southern England, the Atlantic coast of France, as well as the coastal areas of Denmark, Sweden and Norway.

This movement of people is being accompanied by extensive development of infrastructure within the 10-kilometre coastal zones of Europe (Figure 6.8). The coastal Mediterranean region, in particular, is now one of the most densely populated regions on Earth, with more than 13 million people from the EU living near the coast. Permanent populations exceed 1 000 people per square kilometre along the French and Italian Riviera.

By one estimate, 22 000 square kilometres of the coastal zones are covered in concrete or asphalt, an increase approaching 10 % since 1990, causing habitat fragmentation and exacerbating the risks of flooding due to soil sealing.

Development is very uneven, however. Land use studies show that the greatest concentration of artificial surfaces in the coastal zone is within just 1 kilometre of the coast itself. In several parts of France, Italy and Spain, such as Andalucia, more than half of this immediate coastal strip is built upon. Two-thirds of this recent increase in artificial surfaces in the coastal zone has occurred in just four countries: France, Italy, Portugal and Spain, with most of the rest in two more — Greece and Ireland.

As a result, natural grasslands and heaths in Greece, Portugal and Spain are disappearing, and Mediterranean coastal forests are under a growing threat from fires originating on adjacent urban land. Wetlands, including marshes, coastal lagoons and estuary mudflats, have also suffered extensively from drainage to create land for development.

Traditionally, many of these intertidal and coastal areas have been regarded as having a low value — almost wasteland. Their environmental services such as nurseries for fishes, crustaceans and birds, salinas, hunting grounds, pollution filters, buffers against coastal erosion, storm surges and saltwater intrusion, absorbers of land-based nutrients and pollutants, and much else, have been ignored by developers and regulators alike. To replace these naturally fulfilled functions would impose an impossible burden on future generations of European citizens.

An estimated two-thirds of Europe's coastal wetlands have disappeared in the past century, and the loss continues. There was a net decline in European coastal wetlands of 390 square kilometres during the 1990s. Examples include peat bogs in Ireland and parts of the 200 kilometres of lagoons and salt marshes of the Languedoc-Roussillon coastline of southern France.

Another critical pressure arising from the build-up of socio-economic activities in the coastal zone is the proliferation of engineered frontage, the intensive use of natural shores for recreation and tourism, and the extraction of near-shore sand and gravel for construction purposes, leading in turn to accelerated erosion of the European coastline — one of the most visible consequences of this relentless and silent depletion of the coastal environment.

All European coastal states are to some extent affected by coastal erosion (Map 6.2, Table 6.2). About 20 000 kilometres of coast, corresponding to 20 % of the total, faced serious impacts in 2004. Most of the impacted zones, approximately 15 100 kilometres, are actively retreating, some of them despite coastal protection works along 2 900 kilometres. In addition, another 4 700 kilometres have become artificially stabilised. The area lost or seriously impacted by erosion is estimated to be 15 square kilometres per year. Within the period 1999–2002, between 250 and 300 houses had to be abandoned in Europe as a result of imminent coastal erosion risk and another 3 000 houses saw their market value decrease by at least 10 %. These losses are, however, insignificant compared with the risks of coastal flooding due to the loss of foreshore and the undermining of coastal dunes and sea defences. This threat has the potential to impact on several thousands of square kilometres and millions of people.

Beyond the tideline, Europe's sea grasses have suffered from physical destruction and pollution. Sea grass meadows are vital nurseries for fish and shellfish and provide other important ecological services such as regulating water quality and buffering the coastline from erosion. The pollution threat includes both chemical effects of eutrophication and the physical effect of reducing the penetration of light into surface waters. In addition, alien species can also impact on these habitats: one example is the arrival in the Mediterranean of the algae *Caulerpa taxifolia*, which

Map 6.2   Coastal erosion patterns in Europe

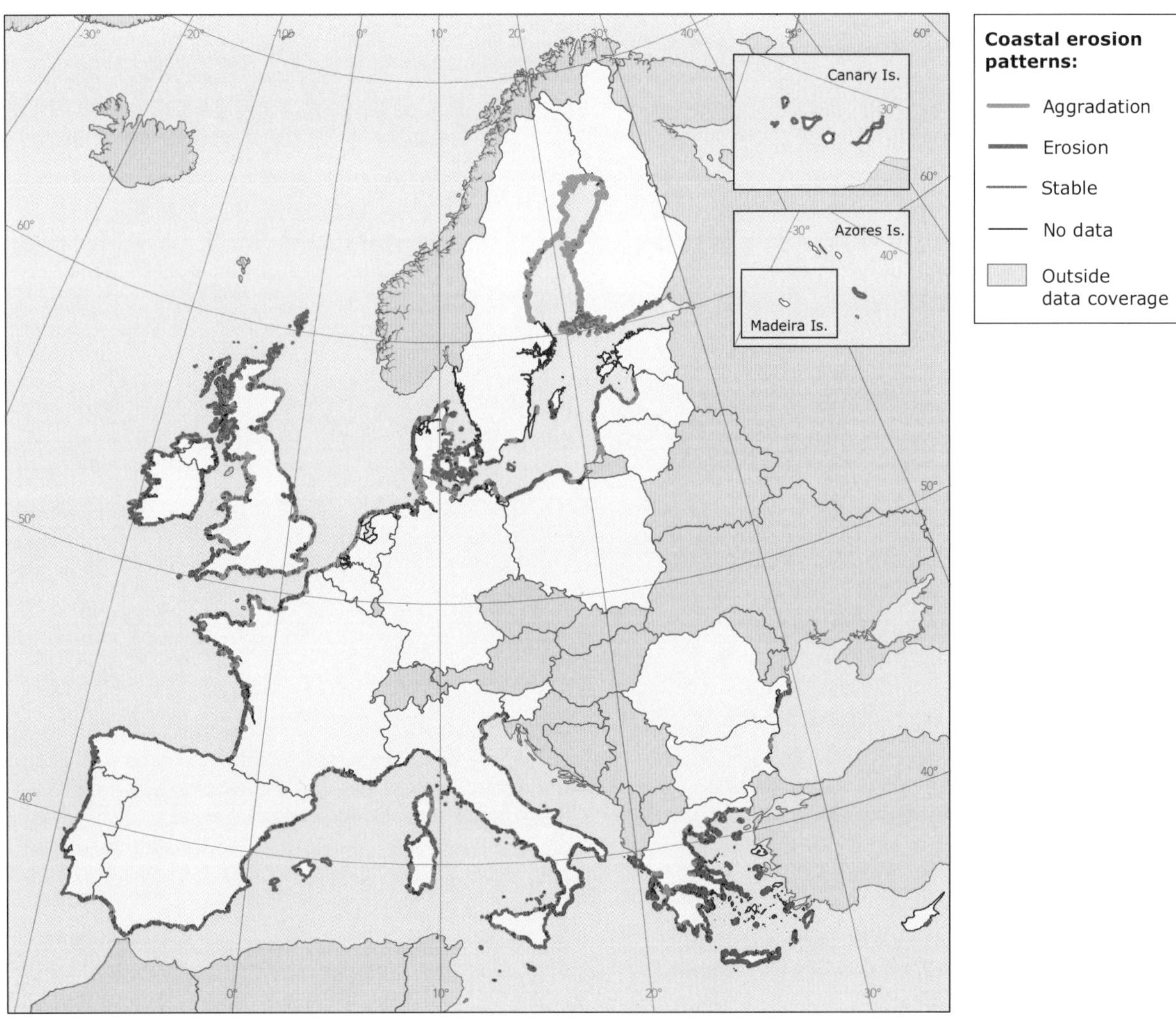

Source: Eurosion, 2004.

has spread round the shoreline, destroying sea grass meadows, since its discovery off Monaco in the 1980s.

6.4 Drivers and pressures affecting marine and coastal areas

Global drivers and pressures

The oceans surrounding Europe play a key role in controlling its climate. Through their immense heat capacity, the oceans effectively act as the 'thermostat' for the planet, moving heat between the equator and the poles; more than 80 % of the heat reaching the Earth's surface from the sun ends up in the oceans.

Chemical and biological activity in the surface waters of the oceans plays a major part in controlling the long-term composition of the Earth's atmosphere, helping to determine the Earth's response to rising levels of greenhouse gases by acting as the largest long-term sink for atmospheric carbon dioxide (CO_2).

It is estimated that gaseous exchange at the sea surface plus biological activity in shallow waters is responsible for about 85 % of the carbon removed from

Table 6.2   Extent of coastal erosion by country

	Total length of the coastline (in km)	Eroding coastline in 2001 (in km)	Artificially protected coastline in 2001 (in km)	Eroding coastline in spite of protection 2001 (in km)	Total coastline impacted by coastal erosion (in km)
Belgium	98	25	46	18	53
Cyprus	66	25	0	0	25
Denmark	4 605	607	201	92	716
Estonia	2 548	51	9	0	60
Finland	14 018	5	7	0	12
France	8 245	2 055	1 360	612	2 803
Germany	3 524	452	772	147	1 077
Greece	13 780	3 945	579	156	4 368
Ireland	4 578	912	349	273	988
Italy	7 468	1 704	1 083	438	2 349
Latvia	534	175	30	4	201
Lithuania	263	64	0	0	64
Malta	173	7	0	0	7
Netherlands	1 276	134	146	50	230
Poland	634	349	138	134	353
Portugal	1 187	338	72	61	349
Slovenia	46	14	38	14	38
Spain	6 584	757	214	147	824
Sweden	13 567	327	85	80	332
United Kingdom	17 381	3 009	2 373	677	4 705
Others (Bulgaria, Romania)	350	156	44	22	178

Source: Eurosion, 2004 (See www.eurosion.org — accessed 17/10/2005).

the atmosphere, with the remainder being taken up by terrestrial plants and soils. Ultimately, this atmospheric CO_2 is captured by the sediments of the deep ocean, but it is a slow process. Removal of the present excess levels of CO_2 in the atmosphere into seabed sediments will take more than 1 000 years.

The oceans are a litmus test of climate change and human-induced changes in the composition of the atmosphere. The effects of climate change on Europe's marine ecosystem can be seen already — in changes to the geographic distribution of species, in local and global species extinctions, through the disruption of critical planetary processes and in the degradation of important flows of goods and services from the more vulnerable ecosystems.

Already the surface waters of the oceans are 30 % more acidic than before fossil-fuel burning began, because of the increase in CO_2, and coastal waters are getting warmer and contain more freshwater as a result of inflows from melting glaciers and ice sheets and increased precipitation at high latitudes. At high latitudes, warmer air temperatures are resulting in significantly reduced sea ice cover in the Barents Sea and Arctic Ocean.

The increase in the acidity of sea water will progressively upset the oceans' chemical balance and possibly eliminate some forms of marine life. The greatest effect will be on organisms with hard shells and skeletons such as molluscs, corals and planktonic coccoliths. Even under the lowest future scenarios for carbon emissions, the cold-water corals in Europe could be virtually gone by 2050.

In European waters there is clear evidence of systematic increases in sea surface temperature, along with the periodic fluctuations associated with major natural climate cycles, such as the North Atlantic Oscillation. The net rise in surface sea water temperature will eventually reduce the oceans' ability to dissolve atmospheric CO_2, and hence the ability of the oceans to act as a sink for increased atmospheric CO_2.

With heat comes expansion, which, together with freshwater inputs from melting glaciers and ice sheets, will mean that sea levels around European coasts will rise, as will the incidence of flooding of some major capitals and cultural centres. Over the last 100 years, sea levels rose by between 0.8 millimetres a year in the western approaches of Brittany in France and Cornwall in the United Kingdom and up to 3 millimetres on the Atlantic coast of Norway. This large range is caused simply by differences in the rise and fall of the land masses.

The rising ocean temperatures also affect the composition, distribution and abundance of marine life, especially in shallow and enclosed seas such as the North Sea. There is evidence from the Sir Alistair Hardy Foundation Continuous Plankton Recorder Surveys that phytoplankton communities, the organisms most responsible for removing CO_2 and nutrients from sea water have shifted their location in relation to temperature changes. The observed changes are greatest in enclosed areas such as the North Sea, where southern species, including sub-tropical fish, have moved northwards by as much as 1 000 kilometres over the past few decades. Warm-water zooplankton such as *Calanus helgolandicus* are now twice as abundant as cold-water species such as *Calanus finmarchicus*. This warming is also believed to be inhibiting the recovery of species such as Atlantic cod, depleted through overfishing.

There is also widespread evidence of an increasing incidence of extreme concentrations of particular phytoplankton — beyond the normally occurring algal blooms— in European coastal waters. These extreme events, which can contaminate food supplies, have been observed in regions such as the Barents Sea, where they were previously unknown.

The Arctic Ocean and surrounding regions are expected to warm the most in response to increases in atmospheric greenhouse gases, with predicted warming of more than double the global mean. The extent of sea ice in the Arctic is dropping at a rate of 3 % of multi-year ice and 8 % of single-year ice per decade, suggesting that the Arctic may become ice free in summer by the end of the century.

Marine and coastal environment

The consequences of diminished Arctic sea ice cover for Europe's marine ecosystems are many and are already being observed: key amongst them are changes to the thermohaline circulation in the Arctic and Atlantic Oceans; increases in water temperatures and sunlight leading to significant alterations in primary productivity and potentially fisheries, especially in areas such as the Barents Sea; reductions in habitat for many ice-dependent species, such as polar bears, seals and some marine birds; and impacts on the distribution of marine intertidal species along the circumpolar shores.

Fisheries and aquaculture

European Commission figures show that the EU is the world's third fishing power and the first market for processed fish and aquaculture products. Catches from fisheries in the EU-25 in 2003 were 5.9 million tonnes live weight, representing about one-tenth of the world's fish catch, and from aquaculture 1.4 million tonnes. In 2004, the size of the European fleet was approximately 100 000 fishing boats, with a gross tonnage of 1.8 million.

Europe has taken steps through the common fisheries policy to help some fish stocks, notably cod, recover by reducing the overall number of vessels. However, the high level of employment in fishing — the total for just five European nations, France, Greece, Italy, Portugal and Spain is 190 000 full-time equivalent — means that there are often conflicts between the need to preserve the livelihoods of fishing communities and the recommendations of the scientific advisory bodies.

Successive efforts to rein in fleets have had only modest success in cutting catches of cod and other threatened species, and in reducing by-catches of non-target species. In 2003, the International Council for the Exploration of the Seas (ICES) reported that 61 % of Europe's demersal fish stocks were outside safe biological limits, together with 22 % of pelagic stocks, 31 % of benthic stocks and 41 % of industrial stocks. The situation today has not changed significantly. This is partly because, even though there are fewer vessels, many of them are more powerful, with more efficient fishing practices.

For many years it has been the perception that capture fisheries provide a poorer income than many other industries and occupations. One of the reasons is the peripheral geographical location of many fishing enterprises, and the fluctuations in the size of the landings. However, good returns are available in well-managed fisheries, including those where property rights have been defined in terms of a share of the catch of a species (e.g. individual transferable quotas as in Iceland and the Netherlands) or a limited area of access has been assigned.

Not all fishing enterprises are equally efficient, but the poor returns available from alternative employment in many fishery-dependent areas, and the generally low investment in peripheral local economies, have allowed a longer tail of marginal or unprofitable fishing enterprises to exist than would otherwise be the case.

The fact that the value of the whole production chain — from fishing, aquaculture, processing to marketing — is estimated to be approximately 0.28 % of the EU gross domestic product, and certainly less than 1 % in terms of contribution to the gross national product of Member States, does not reflect its highly significant role as a source of employment in areas where there are few alternatives. The number of fishermen has been declining in recent years, with the loss of 66 000 jobs in the harvesting sector, a decrease of 22 %. There has also been a 14 % decline in employment in the processing sector. In some areas these trends are threatening the viability of small coastal communities in the absence of suitable alternative employment.

The development of aquaculture in isolated coastal communities has had a positive impact on employment. On the west coast of Scotland, for example, aquaculture provides an important source of employment for local people in areas where there are very few alternatives. The EU Aqcess survey has revealed that the main reason individuals began working in fish farming was the lack of alternative employment in the local area: just under 60 % of fish farmers said there were no other job opportunities available to them. This is also the main reason why aquaculture workers stay in the industry despite relatively low pay.

Integrated assessment | Aquatic environment

Landings of fish and shellfish from European waters have declined because of the overexploited status of many stocks and tougher controls on overfished zones, especially the fishing grounds in the North Sea and the Atlantic, where stocks of cod, whiting and hake are under threat. The pressures on commercial or targeted stocks also vary considerably across the regions, largely because countries have quite different catching regimes. In Denmark, for instance, an important fraction of landings consists of 'industrial' catches of sand eels and other species for fishmeal and oil; in Spain landings are mainly for human consumption including high-value fish for sale in restaurants.

The reformulation of the common fisheries policy and the development of a European Fisheries Control Agency are intended to rebuild marine fish stocks through enhanced controls, better enforcement, local management and voluntary conservation measures.

In the meantime imbalances between domestic and external demand and domestic supply are largely being met through imports. Improved technologies for low-temperature storage and transport have created new international markets and an increased trade in fish products, with various value-added levels. These developments have also tended to suppress price responsiveness to changes in domestic supply.

The major importers by value are Norway with 21 % of the EU-15 total, Denmark with 16 %, Spain 10 % and the Netherlands and the United Kingdom each with 8 %. The figures are by value rather than quantity, given that the extent of processing varies from none from landings by foreign vessels to the sale of a final product by retailers. The major exporters by value are Spain with 16 %, France 14 %, Italy 12 %, the United Kingdom 10 % and Denmark 8 %.

One of the key drivers in fisheries is, of course, human consumption. The United Nations Food and Agriculture Organization (FAO) has estimated that consumption of fish in Europe is now about 15 % higher than in the mid-1960s. The rate of consumption per capita has remained steady for the EU-15 at 23.7 kg per year. There are wide differences in consumption per capita from country to country, reflecting European demand and widely varying culinary traditions. Overall, total consumption closely follows population size, although there are anomalies. In Turkey, which has the second largest population, consumption was only 8.0 kg per capita in 2000, while in Iceland consumption was 90 kg per capita and in Portugal 60 kg.

Shifts in consumer attitudes and preferences have been an important influence on the demand for fish. Fish is considered a 'healthy' product and has benefited from the trend towards reduced meat consumption as a requirement for a healthy lifestyle. Besides quality and price, consumers are increasingly concerned about how their food is produced. Thus, for example, farmed fish can give rise to the same concerns about the levels of antibiotics in fish products and animal welfare as any intensive livestock production system. The environmental effects of intensive fish farming may also provoke a negative consumer response when chemical additives are used for growth and disease control.

Europe's increasing consumer demand for wild fish means that imports are steadily rising. Imports to Europe increased from 6.8 million tonnes in 1990 to 9.4 million tonnes in 2003.

However, global fish catches have been stalling for some years: declining stocks are defeating the increased investment in fishing. In the longer term the prospect of compensating for the loss of European stocks with stocks from other seas is diminishing.

If European wild marine fish stocks decline, the extra demand for fish will need to be met through marine aquaculture. At present salmon is grown in the Atlantic and Baltic, turbot around Spain, sea bass and sea bream in the Mediterranean and sturgeon in the Black and Caspian Seas. Nearly 8 tonnes of fish are produced annually from aquaculture for every kilometre of coastline in the European Free Trade Agreement (EFTA) countries. Norway is the largest producer with large fish pens moored offshore, mostly holding Atlantic salmon.

Even though aquaculture can take the pressure off wild reserves of high-value fish, it also uses wild fish stocks such as capelin and sand eel to make fishmeal for the caged high-value fish. Marine aquaculture is also a significant source of additional nutrient loading in coastal waters — and disinfectants such as formalin, copper-based anti-foulants and medicines to fight sea lice infestations — and needs to be carefully managed. Average discharges of nitrogen have been calculated at 40 kilograms for every tonne of fish produced. Escapees are also a potential threat to wild populations of fish.

Tourism

The biggest driver of development in the European coastal zone in recent years has been tourism. Europe is the world's largest holiday destination, with 60 % of international tourists, and business continues to grow, by 3.8 % a year. The greatest activity is along the Mediterranean coastal zone, with France, Spain and Italy receiving respectively 75 million, 59 million and 40 million visitors a year. These represent increases of between 40 and 60 % since 1990. France and Spain are the world's top two tourist destinations.

As the big resorts of the western Mediterranean fill up, areas to the east are becoming increasingly popular, including the Greek islands, Cyprus and Malta. Malta receives more than a million tourists a year, three times its permanent population.

Tourism is the largest sector of the economy in many coastal zones, and construction of hotels, apartments and other tourist infrastructure is the dominant form of development. In French coastal regions, tourism provides an estimated 43 % of jobs, generating more revenue than fishing or shipping. This dominance of tourism is reflected in the seasonal changes in population density, with an influx of both tourists and people to work in the tourist industry each summer. Peak population densities on the Mediterranean coasts of France and Spain reach 2 300 people per square kilometre, more than double the winter populations. A further 40 % increase in peak populations is expected in the coming 20 years.

The expansion of tourism extends beyond the Mediterranean, however. The Atlantic coasts of France and Portugal, the southern Baltic coast and parts of the Black Sea coast are all seeing expansion. Other coastal areas, such as either side of the English Channel, remain popular visitor destinations and conference venues. Tourism is expected to continue to grow, though potential brakes on this could emerge from higher temperatures, fires and droughts, and a desire by tourists for emptier and less-developed resorts.

Tourism is now having a major environmental impact on many coastal areas. Besides land-grab, its demand for resources and need for waste disposal facilities cause pressure on water resources and natural coastal habitats and structures such as wetlands and sand dunes. Demand for water in Malta doubles during the tourist season; on the Greek island of Patmos, it increases sevenfold. Many regions, including Spanish resorts and Malta, are running out of water and are resorting to investment in desalination of sea water.

Tourism can, however, sometimes have a positive influence. Increasingly, tourists demand high aesthetic standards, including clean beaches, scenic beauty and amelioration of urban areas. They also provide the income for investment in clean-ups and other environmental measures.

Nature conservation

Nature conservation is an important and growing element in the coastal and marine environment. Significant areas of coastal wildlife habitat have been given protection through the EU Natura 2000 network (Figure 6.9) and there is much discussion about the efficacy of marine reserves as a tool for helping overexploited fisheries to recover.

Some countries have significantly more land designated as Natura 2000 sites in coastal zones than in the interior. They include Poland, with four times more, and Germany, Lithuania and the Netherlands, and Belgium, France and Ireland, all with at least twice as much. Habitats protected include lagoons and deltas, sandbanks and dune systems, mudflats, estuaries, reefs, sea grass meadows and small islands, as well as coastal

Integrated assessment | Aquatic environment

grasslands and forests. Countries with markedly less protected land in coastal areas than elsewhere include Greece, Italy and Spain.

As evidenced in the EU Biomare project, which documents marine sites around Europe suitable for long-term monitoring and observation, ecotourism and nature conservation are providing protection for some of Europe's more pristine areas.

Industry, energy and transport

Many industries are located on the coast, close to port facilities, with access to transport routes for supplies of raw materials, shipping of products and often large land areas. Currently, almost one in five European industrial facilities is on the coastal strip, with a third of the total clustered round the North Sea in Denmark, Germany, the Netherlands and the United Kingdom. Often these industrial complexes are built on 'reclaimed' mudflats in estuaries, replacing ecosystems of value for birdlife and other intertidal species.

Coastal zones also attract industries connected directly with marine activities, such as dredging of sand and gravel, cable laying and offshore exploration and construction. Energy facilities are also concentrated in the coastal strip. These include oil terminals, plant and pipelines connected with offshore oil installations in the North Sea, the Adriatic and elsewhere; large fossil-fuel and nuclear power plants supplied with fuel from ships or pipelines and taking advantage of sea water to provide cooling; and coastal energy plants using wave and wind power.

The conflict between visually intrusive facilities and the demand for high aesthetic standards and healthy coastal environments is growing. Some evidence of this can be seen in the demand to exploit offshore locations

Figure 6.9   Percentage of coastal surface covered by Natura 2000 designated areas

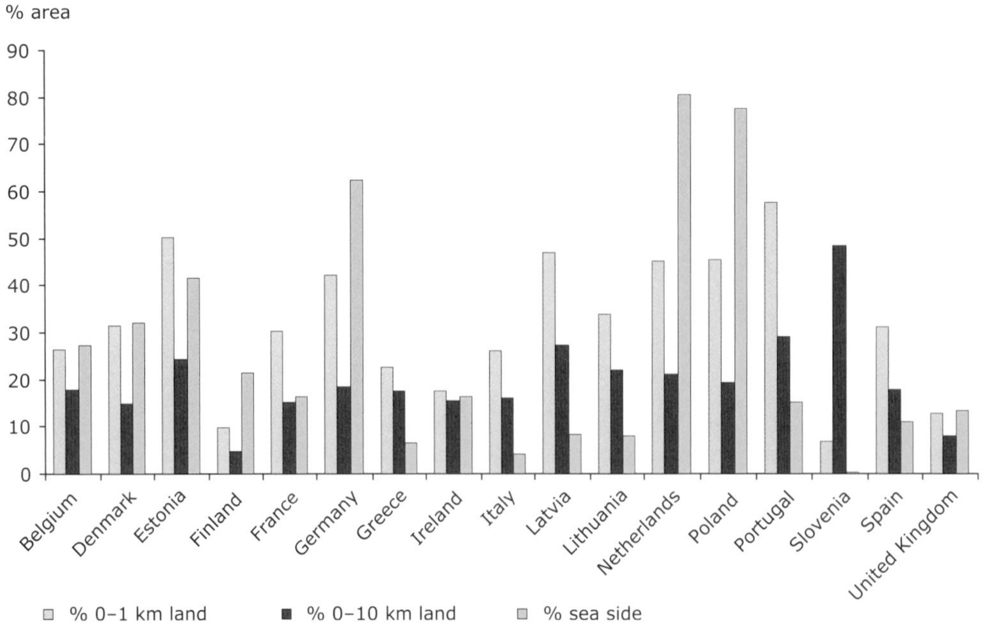

□ % 0–1 km land   ■ % 0–10 km land   □ % sea side

Note: Refers to 10 km zone for terrestrial and for marine side, respectively.
Source: EEA, 2005.

for wind farms, particularly in north-west Europe, where wind turbines can take advantage of shallow seas.

Although shipping is often ignored in national statistics and has recently been overshadowed by the growth in international air transport, during the 1990s the volume of freight moved by ship between European destinations increased by a third to about 1 270 billion tonne-kilometres, a figure similar to road freight transport. The busiest receiving ports are in Italy, the Netherlands and the United Kingdom. Passenger transport has also increased on many routes. Concerns over high-speed ferries, designed to compete with other forms of transport, are now being raised, especially in the North Sea. The European Maritime Safety Agency has been established recently to deal with this type of issue.

Despite an increase in the marine transport of oil, pollution from oil spills on a worldwide scale has been reduced by 60 % since the 1970s. The worldwide average number of accidental oil spills above 7 tonnes was estimated by the International Maritime Organization (IMO) to be 24.1 per year for the decade 1970–1979, 8.8 per year for the decade 1980–1989 and 7.3 per year for the decade 1990–1999. Nevertheless, major accidental oil tanker spills (i.e. those greater than 20 000 tonnes) still occur from time to time in European waters. In 2000 there was one spill of 250 tonnes (Germany) and in 2001 three spills totalling 2 628 tonnes, including one (Denmark) of 2 400 tonnes.

Agriculture is the sector that, whilst placing significant pressures on the coastal zone, has at the same time suffered most from coastal urbanisation and the spread of tourism. Recent EEA studies have shown that, during the 1990s, some 2 000 square kilometres of high-value farmland was lost in European coastal zones. The process has been most pronounced in Belgium, Ireland, Italy, the Netherlands and Portugal. The greatest loss has been pasture, notably in Ireland and Portugal. Nonetheless, agriculture remains a major user of (sometimes constrained) natural resources and a source of pollution in many coastal areas. In the Mediterranean coastal region, for example, where water is scarce, irrigation remains the dominant use of water, and is one of the reasons why Spain has the largest per capita use of water in Europe.

6.5 Trends in ecosystem health

One of the major difficulties in making progress in the area of coastal and marine ecosystem management and sustainable development is that indicators, targets and assessments for marine ecosystem health are currently very restricted. The European marine monitoring and assessment (EMMA) working group for the European Commission's marine strategy has recognised this. It has identified a number of issues for which a pan-European approach and set of baseline indicators and assessments need to be adopted urgently, either because of the scale of the policies involved (e.g. the common fisheries policies and the water framework directive) or because of the regional and transboundary character of the problems (e.g. invasive species and hazardous pollutants), or both. The issues include: eutrophication, hazardous substances and persistent organic pollutants, problems arising from shipping and oil discharges, overexploitation of fisheries, decline of biodiversity and habitat degradation, emergence of invasive species and threats from climate change, and extensive shoreline and coastal development.

Even without a harmonised set of baseline indicators, it is still possible to detect the early signs of trends which, by their very nature, hint at changes within the marine environment that should not be ignored.

Water quality

European efforts to clean up its surface waters have generally had a beneficial effect on coastal waters. Under the urban waste water directive, river clean-up programmes have been extended to curb discharges to estuaries. This, combined with the controls under the bathing waters directive and others to protect shellfish grounds, has reduced discharges of pathogens, organic material and nitrogen and phosphorus to coastal waters, sometimes 10-fold or more. Compliance with mandatory standards under the bathing waters directive stand at more than 95 % most years, and the more stringent guideline values have compliance greater than 85 % (Figure 6.10).

Bathing water quality is a prime example of how environmental regulation, when combined with effective monitoring and public information, has had a beneficial effect on economies. Failure to comply with the directive has demonstrably influenced tourists' choices of destination, while nominations such as the Blue Flag awards have shown clear benefits.

Concerted efforts since the 1980s have also brought about reductions in oil discharges from tankers, refineries and offshore installations. During the 1990s, Europe's refinery discharges decreased by 70 %. Nevertheless, accidents continue to happen. The break-up of the Prestige tanker off north-west Spain was a major pollution disaster that will have an impact on coastal ecosystems for years. In addition, there are indications of continuing large numbers of illegal discharges of oil from shipping in the Mediterranean and Black Seas, with consequent harm to coastal waters and shorelines.

Overall, improvements in the quality of coastal waters have been most marked in north-west Europe and least in the Mediterranean, though here the warm waters naturally destroy pathogens and hydrocarbons more quickly, and the risks of eutrophication are less compared with other badly affected areas in Europe.

Nutrient enrichment is a widespread pollution problem in coastal waters, particularly in enclosed bays and estuaries. It is predominantly the result of nitrogen pollution, and arises from a mixture of run-off of fertilisers from agricultural land, releases from coastal fish farms, air pollution fallout and sewage discharges.

Eutrophication causes changes in marine populations, with diatoms being replaced by blooms of green or blue-green algae. Intense pollution can result in the creation of 'dead zones' in which all the oxygen is removed from the water by bacteria processing the vast quantities of dead algae. Dead zones are normally seasonal, but can have major impacts on fish stocks.

There are long-standing eutrophication hot spots in the Mediterranean, for instance in the Venice area at the head of the Adriatic Sea, and the Gulf of Lion. Others occur in the Baltic Sea, Black Sea, Belt Seas, Kattegat, in the Norwegian fjords and the North Sea's Wadden Sea.

Figure 6.10   Percentage of bathing water sample points complying with Guide values (C(G)) or complying with Mandatory values (C(I)) — 2003

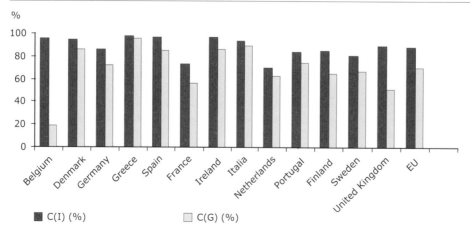

Source: European Commission-Bathing water quality database, 2005.

Marine and coastal environment   PART A

Figure 6.11   Concentrations of hazardous substances in fish from the north-east Atlantic and Baltic regions

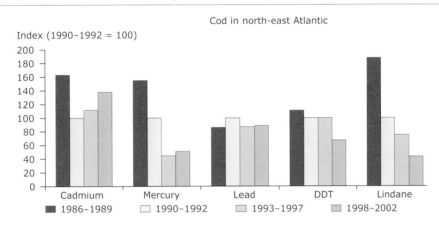

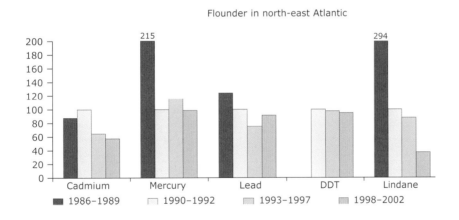

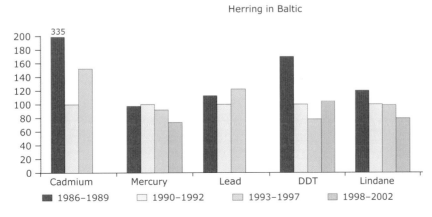

Source: EEA, 2003.

More widely, eutrophication of coastal waters also reduces the transparency of the water and causes a decline or shift in life on the seabed. Thus red algae beds disappeared from wide areas of the Black Sea, and sea grass beds from the Baltic. Eutrophication can shift the species balance, favouring shellfish that enjoy sediments rich in organic matter; filter feeders such as mussels and oysters gain over sponges and red coral that prefer clearer water.

Problems appear in most cases to be directly related to the volume of fertiliser use on land. Thus eutrophication in the Black Sea reduced during the 1990s when the economic downturn led to less fertiliser being applied. Reductions were also observed in the Baltic and North Seas following constraints on direct discharges into the Rhine.

Worsening nutrient pollution in the Mediterranean is apparently causing a deterioration in the sea grass beds that once fringed almost the entire sea, much as in the Baltic. The decline is worst around urban areas, such as Alicante, Marseilles and Venice, that discharge nutrient-rich effluents into the sea. Many fish species that use the sea grasses as nurseries are also declining. This ecological disruption has allowed an aggressive exotic weed, *Caulerpa taxifolia*, to spread, apparently after having escaped from aquaria in Monaco.

Industrial pollution
Marine transport has a direct impact on the marine environment through illegal discharges of oil and oily and other waste; the introduction of 'alien' species carried from one marine area to another in ballast water and on ship hulls; accidents resulting in oil spills or spills of dangerous chemicals; the effects of anti-fouling paints on the environment; and disturbance of sediments in coastal or shallow areas.

Environmental issues in relation to maritime transport are considered at both a global level by the International Maritime Organization and on a regional level by several of the regional marine conventions. In the Baltic there is an active programme to minimise the environmental effects of shipping, and maps of the locations of oil spillages observed by aerial surveillance are compiled annually. For the Arctic, a comprehensive assessment of Arctic marine shipping is now to be undertaken, following concerns about the opening up of the Barents Sea. Matters concerning the introduction of alien species via shipping are also reviewed annually.

Heavy metals, pesticides and hydrocarbons are entering the marine environment from air and water run-off and accumulating in marine waters and in the bodies of marine animals — especially those at the top of the food web, such as large fish, marine mammals and some bird species. Typically these substances do not kill but they have subtle effects on fertility, growth rates and health. Enclosed seas such as the Baltic and Black Seas tend to suffer most, because the pollution is not readily flushed to the open ocean. Recent studies by the EEA and the Arctic Council have shown that the problem is being amplified throughout the Arctic food chain amongst both animal and now human populations.

In most cases concentrations of these pollutants in the tissue of fish caught off Europe has fallen in the past 15 years. Cod and flounder in the north-east Atlantic, for instance, have half as much mercury, only a quarter as much lindane and marginally less cadmium than in the late 1980s (Figure 6.11). Trends for lead, the insecticide DDT (dichlorodiphenyl trichloroethane) and PCBs (polychlorinated biphenyls), however, are less clear. Some persistent organic pollutants, though often banned in Europe, continue to be widely used elsewhere and are accumulating in Arctic organisms as a result of global distillation processes.

The Helsinki Commission (Helcom) has reported that high concentrations of pollutants such as dioxins in the tissues of Baltic fish have led to intake restrictions.

Marine sediment balance
The artificial land surfaces proliferating along Europe's coastlines often extend to sea walls, harbours and other structures along the coastline itself. Around 10 % of Europe's coastline is now artificial; in Belgium, the Netherlands and Slovenia, the figure exceeds 50 %. These structures are often necessary to prevent flooding during storms, and to curb local erosion. By stopping

erosion in this way, however, the sediment balance in coastal waters is disrupted at the expense of beaches and sand spits elsewhere. Preventing coastal damage in one area can increase damage elsewhere.

Other causes of overall sediment loss in coastal waters include development of upstream dams, which trap sediments as well as water, canalisation of rivers, which reduces bank erosion, and offshore dredging of sands and gravels. For example, the Ebro delta on the Mediterranean coast of France is retreating because dams upstream on the river prevent sediment reaching the delta to maintain it against coastal erosion.

Taken together, these changes to the sediment balance have resulted in an annual loss to Europe's coastal systems of an estimated 100 million tonnes of material. In combination with rising sea levels, this has resulted in around a fifth of Europe's coastline suffering significant erosion, with coastlines retreating by an average of between 0.5 metres and 15 metres a year.

Any future rise in sea levels will dramatically increase the future risk of lost coastal land. The only solution may be to attempt to reinstate natural systems for protecting shorelines. Modern methods of 'soft' coastal engineering attempt to do this by reinforcing natural buffers against the rising tides, such as sand dunes and salt marshes, and protecting key sources of sediment and natural coastal dynamics, such as eroding cliffs, to maintain the coastal sediment balance. In some areas, for instance in parts of eastern England, coastal engineers are deliberately sacrificing land to allow 'managed' coastal retreat.

Fisheries

Overfishing in Europe's waters and the deep ocean has proved hard to tackle. Some fish stocks that have high reproductive rates, in conjunction with reduced fishing pressure, have successfully recovered from past overfishing. Most notable are the herring around Iceland and Norway and in the North Sea. Other stocks are also unlikely to recover. Sharks, skate and rays, in particular, are vulnerable because they produce few young and breed only slowly. Their recent sharp decline in the north-east Atlantic and Mediterranean is unlikely to be reversed quickly. As well as being a commercially targeted fish, these species also suffer from accidental capture, especially in drift-nets and on long-lines.

By-catches and unreported and misreported landings are all major issues that can lead to distortions in fisheries data trends. In many fisheries, between 20 and 60 % (in some even 80–90 %) of the catch is undersized or of non-targeted, non-commercial species. The average discard rate in the North Sea is 22 % of landings. Some of the highest discard rates are for crustaceans and some shrimp fisheries. Off the Portuguese coast there is a discard issue with 'verdinho' — blue whiting — that in Portugal has no commercial value; in contrast, the same fish is being landed in Spanish ports where it has a high commercial value.

Marine ecosystem structure

Fishing rarely makes species extinct, but it can easily eliminate species as significant elements in the marine ecosystem, sometimes with widespread implications for the whole structure. For example, over the past two decades the number of fish species regularly caught in nets in the Black Sea has fallen from 27 to 6.

Large fish at the top of the marine food chain are generally the most valued by consumers; and they are the first to disappear. Thus in the Black Sea the largest, top-predator species such as the swordfish, tuna and mackerel disappeared first. In the North Atlantic, the biomass of these top predators has decreased by two-thirds in 50 years.

As the big fish at the top of the food web disappear, their places in the ecosystem are taken by smaller species on which they once preyed, such as the anchovy in the Black Sea and the sprat in the Baltic. These in turn become the next target for fisheries, leading to a phenomenon known as 'fishing down the food chain'. One aspect of this is that a growing proportion of fish catches are now of plankton-eaters rather than fish eaters, a trend seen in the Atlantic, and the Mediterranean and Black Seas.

The place of fish in the food chain is measured by its 'trophic level', with the species at the top of the chain having the highest number. Research has shown a steady drop in the average trophic level of landed fish in European waters (Figure 6.12).

As fishing moves to catch second-tier species, other predatory species may emerge, such as jellyfish. These changes have knock-on effects, and can lead to entire marine systems being destabilised. Sometimes, fishing and other environmental damage provides ecological 'space' for new invasive species. The emergence of the *Mnemiopsis* jellyfish in the Black Sea is just one such case.

Other cascade effects reported in recent years by scientists include the impact of fishing pressure on sand eels in the north-east Atlantic. Sand eels are caught primarily for industrial purposes. Their disappearance deprived puffins of their main food item, causing their populations to crash in turn. In the Arctic, a decline in capelin stocks followed a revival of herring, which ate capelin larvae. The loss of capelin in turn left guillemots and several species of toothed whales hungry, causing a 50 % decline in guillemot numbers.

By-catches in fisheries are also a major threat to the survival of some endangered non-fish species round Europe's shores, including turtles and the Mediterranean monk seal. There are fewer than 500 Mediterranean monk seals left, and static fishing gear and abandoned nets are a major threat to their survival. Also in the Mediterranean, more than 50 000 turtles — including the endangered loggerhead, green and leatherback turtles — are caught in nets and by long-lines each year, with death rates in some areas as high as 50 %. Long-lines are also a major cause of seabird deaths in the Mediterranean when birds are hooked as they try to eat bait on hundreds of lines trailing from factory ships. The list of birds includes several endangered species.

Small marine mammals such as dolphins and porpoises are also caught in large numbers. Between a fifth and a half of all the cetacean strandings on the shores of England and Wales are attributed to injury during fishing. The FAO has suggested that the loss may be even greater in the Mediterranean where EU bans on drift-nets are being evaded by fishers switching to similar equipment known as anchored floating gill-nets.

By-catches of dolphins in the western Mediterranean may still exceed 3 000 animals a year, but the true extent of these by-catches and their ecological importance is often hard to establish because of a lack of data. The same applies to so-called 'ghost-fishing', in which discarded fishing gears cause fish mortality.

As catches diminish on the continental shelf round Europe, trawlers are heading into the deep waters of the Atlantic and the western Mediterranean. Here, the problems of species sustainability may be even greater. Deep-sea fish often live in fragile ecosystems where they grow and reproduce only slowly. Any recovery

Figure 6.12   Decline in mean trophic level of fisheries landings

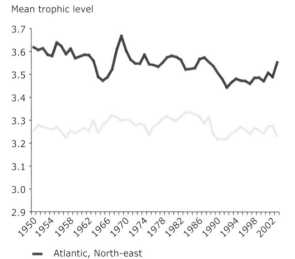

— Atlantic, North-east
— Mediterranean and Black Sea

Note: The decline in mean trophic levels results in shortened food chains, leaving ecosystems less able to cope with natural or human-induced change. The long-term sustainability of fisheries is, in turn, directly linked to human livelihoods and well-being.

Source: Adapted from Pauly *et al.*, 1998 and updated using Fishbase.

of depleted stocks will thus take much longer, often decades.

Another problem that is clearly underestimated is the by-catch of seabirds diving for food, and becoming entangled and drowning in setnets in the Baltic Sea in shallow waters of 25–30 metres. Helcom calculates this loss of seabirds to be serious, and to amount to several tens of thousands.

Biodiversity and habitats

The percentage of area protected by different conservation measures, such as marine protected areas, varies widely among the marine ecosystems of Europe. The lowest values occur in the Celtic-Biscay Shelf Sea and the Mediterranean Sea, whilst the highest are in the Baltic and the Arctic.

To establish what this means in relation to the progress that Europe is making towards meeting its 2010 target to halt biodiversity loss, an overall indicator of the trends in marine species populations has been calculated in a study for the EEA using the same approach as the WWF Living Planet Index. The indicator integrates trends in different species groups, and can be aggregated across habitats, countries and large marine ecosystems. The analysis uses more than 480 historical trends of populations of fish, marine mammals and reptiles for a total of 112 species. The results show that overall, whilst fish populations have declined, bird populations have generally improved.

Fishing technology can reduce biodiversity not only through altering the trophic dynamics, but also through damage to habitats. One major instance is trawling amongst cold-water corals in the north-east Atlantic and Arctic Oceans. Cold-water corals live around sea mounds, sometimes more than 1 000 metres down. The largest reefs, such as those in the Rockall Trough, Darwin Mounds and Porcupine Seabight areas, can cover some 100 square kilometres. They have been threatened by a move among trawling fleets since the mid-1980s to deeper waters along the edge of the continental shelf, where they catch often uncontrolled fish stocks such as orange roughy, blue ling and roundnose grenadier. Recent research has found extensive damage to cold-water coral off Ireland, Norway and Scotland. Trawling kills the coral polyps and breaks up the vital reef structures that are believed to be important fish habitat and nurseries.

The Norwegian government was the first to protect sea mounds with cold coral, and the EU introduced its own protection regime for key sites in 2003 and a Council regulation in 2004 protecting deep-water coral reefs from the effects of trawling in the area off Scotland. The Darwin Mounds are to become a special area of conservation under the habitats directive.

6.6 Future perspectives

The intense pressure on coastal ecosystems and habitats is being met with strong regulatory responses in areas such as pollution control, but far less action in others, such as curbs on inappropriate development in the coastal zone. A number of studies have shown that poor governance is often linked to a vulnerability towards ecosystem degradation and an incapacity to monitor and regulate. Good governance and harmonised, integrated policy approaches are the only solution: without these, and clear institutional arrangements and coherent management targets, the future of Europe's marine and coastal resources looks highly insecure.

Individual actions at the national level are emerging. For example, recognising that coastal development has cut citizens' access to the shoreline, the Spanish government in mid-2005 announced a plan to buy back buildings that blocked access to the coastline. However, national actions will not be sufficient to match the powerful pan-European drivers and pressures at play around Europe's coastlines and seas.

One of the major difficulties in making progress in the area of coastal and marine ecosystem management has been the general lack of coherent strategic planning at a pan-European level and no political targets, beyond those in the fisheries sector, to conserve or rebuild the health of Europe's marine ecosystems.

Integrated assessment | Aquatic environment

The significant impacts of land-based activities on the seas and coasts, and the large number of institutions and organisations involved in looking at only specific aspects of the marine system, has also meant that there has been no agreed set of baseline indicators with which to undertake an overarching integrated assessment of the health of Europe's marine environment.

Now, however, there is widespread agreement amongst all the key organisations and institutions that an ecosystem-based approach needs to be adopted to safeguard and ensure the future sustainability of the marine and coastal environments of Europe. This underpins the proposed European marine strategy, supported through the work of its working group on European marine monitoring and assessment (EMMA).

The ecosystem boundaries, indicators and future targets will be defined according to a range of criteria, including the status of the biological resources, oceanography, integrity of the adjacent catchments and land use patterns, coastal demography, goods and services, governance and political boundaries, monitoring schemes and consistency with international norms.

If approved, the marine strategy will enable Europe to develop an integrated response to the major drivers and pressures — such as coastal development, fisheries, industry, shipping, aggregates, oil and gas extraction — that act regionally and globally and are without question transboundary in nature. It will also form the natural underpinning for maritime policies, currently under preparation within the European Commission. So what are the problems that must be overcome?

Most of Europe's marine ecosystems are shared by more than one state. It is therefore essential that there are strong linkages and good governance amongst the states and amongst all the institutions, both formal and informal, which conduct or influence the management, control and regulation of the marine environment.

Over the past century, many different organisations have been established that have undertaken sectoral assessments, monitoring for the protection of the marine environment and scientific analyses of different marine resources. In many cases, the organisations have used different spatial classifications or have developed their own for data collection and assessments. For European seas alone, classifications include the economic exclusive zones (EEZs) of national territories, fisheries zones and ecological regions used by regional fisheries bodies such as the International Council for the Exploration of the Seas (ICES), the North-East Atlantic Fisheries Commission (NEAFC) and the North Atlantic Salmon Commission (NASCO), the 13 Regional Seas programmes of the United Nations Environment Programme (UNEP), and the large marine ecosystems of the Global Environment Facility, the areas applied by the Helsinki Commission (Helcom) and the Oslo and Paris Convention (OSPAR), covering other marine activities such as shipping, oil, gas and aggregate extraction and marine pollution.

Different models of assessment have also been employed, ranging from the maximum sustainable yield and spawning stock biomass models in fisheries, to indicator- and risk-based approaches for sectoral and environmental assessments.

Legally, the main treaty that deals with the management of marine resources around Europe is the UN Convention on the Law of the Sea (UNCLOS). This embodies the jurisdiction of coastal states in their EEZs and provides for broader ecosystem management in Article 92 through a general duty to protect and preserve the marine environment from pollution of all sources. UNCLOS also stipulates the duty of interested states to cooperate in the management and conservation of high seas resources.

Equally important as legally binding instruments are the UN Framework Convention on Climate Change, the Convention on Biological Diversity and the Convention on Wetlands of International Importance (Ramsar).

UNEP's Regional Seas programmes are also of significance for Europe because most have a legal framework for cooperation including conventions and appropriate protocols. So, for example, the Mediterranean Regional Seas programme adopted a

Marine and coastal environment

protocol to the Barcelona Convention on protected areas. Other regional arrangements of this nature include OSPAR and Helcom for the north-east Atlantic and Baltic respectively.

The 1995 UN Agreement on Straddling and Highly Migratory Fish Stocks explicitly calls on states to adopt measures for species belonging to the same ecosystem or associated with target stocks. The FAO Code of Conduct for Responsible Fisheries calls upon states to use responsible technologies and methods with the aim of maintaining biodiversity and conserving population structures, ecosystems and fish quality.

Over and above these, there is a wide range of ministerial, sectoral and non-governmental organisations that collate and produce information on the marine environment. Examples include the North Sea Ministerial Conference, the European Science Foundation, the Joint European Ocean Drilling Initiative, the Arctic Marine Assessment Programme and the UK Offshore Operators Association. Many of these bodies also produce periodic assessments of particular aspects of the marine environment.

It is clear from the reports of all these bodies that Europe's marine ecosystems are facing increasing pressures from an enormous range of land and marine-based activities. Yet despite the fact that at the international level there are many global and regional strategies, recommendations, binding agreements and guidelines, there is little articulation between them at the European level. In Europe there are a number of policies affecting the marine environment, such as the common fisheries policy, marine transport policy, chemicals policy, common agricultural policy, air policy and water policy, but to date none is specifically designed to protect the marine environment. There is no harmonised legislation on marine protection across the Member States. Gaps in knowledge exist because assessment and monitoring programmes are not integrated or complete, and the links between research needs and priorities remain weak.

For Europe's marine and coastal environments to continue to provide real economic benefits to its populations, remain healthy and provide food, resources and cultural support in the longer term, it is critical that a more integrated approach to management and conservation, such as the marine and maritime strategies, be adopted — one that recognises regional differences and vulnerabilities, but applies common principles and measures of progress.

6.7 Summary and conclusions

The seas and coasts around Europe are a vital resource upon which many millions of people depend, both economically and culturally. They also provide a wide range of ecosystem services that are essential to the health of Europe's environment. During the last four decades there has been a significant increase in local and regional pressures on the coastal and marine environments from urban settlement, tourism and industrial development, with the result that many of the improvements in environmental protection and clean-up are being undermined.

There are early signals that Europe's marine and coastal ecosystems are undergoing structural changes to the food chain, evidenced by the loss of key species, occurrence of large concentrations of key planktonic species in place of others and by the spread of invasive species. These are happening as a result of climate change and widespread human activities.

Different seas face both common and unique interconnected challenges, highlighting the value of integrated approaches to solutions. In the **Baltic Sea** there are continuing problems with eutrophication, overexploitation of fisheries and invasive species. In the **Barents Sea**, ecosystem-wide disturbances have been caused by overfishing and pollution from shipping, military activities and oil extraction. In the **North Sea** ecosystem damage threatens important populations of seabirds and some fish species, as a result of a wide range of pollution discharges.

In the **Celtic-Biscay Shelf Sea**, overfishing and oil drilling have damaged rich cold-water coral reefs. In the **Iberian Coast Sea** future alterations to ocean circulation due to climate change are expected to most affect future ecosystem structure. The challenges facing the **Mediterranean Sea** include coastal erosion, eutrophication, fisheries by-catches and invasive species. To the east, the structure of the **Black Sea** ecosystem has been disrupted by overfishing and by damage to coastal wetlands.

The long coastline of Europe is the site of many capital cities and internationally important ports. It is also a magnet for tourism. This has led to the coastal strip being the fastest growing area in economic and social terms. The downside is that intertidal communities of sea grass meadows and coastal wetlands, forests and heathlands have been stripped away by development and intensive foreshore construction.

On a more positive note, discharges to estuaries and coastal areas, including vital shellfish grounds, have improved with high levels of compliance under the urban waste water directive and controls under the bathing waters directive. Nevertheless, eutrophication hot spots and dead zones still remain, and worsening nutrient pollution in some areas has caused a significant deterioration in key habitats such as sea grass beds.

Looking to the future it is clear that the impacts of global warming and climate change will become widespread. They will be exacerbated by coastal development and foreshore engineering. Europe's fisheries will continue to face difficulties in balancing fishing capacity with available resources, given the modest success of common fisheries policy reforms — reducing fleet sizes, modernising vessels and deploying fishing vessels into other areas. Aquaculture, on the other hand, is having a positive impact on incomes as well as helping people to remain in rural coastal areas. Imbalances between consumer demand for fish and Europe's capacity to meet it will continue to create a 'fishy footprint' across the world as demand is met from outside the region.

The largest growing pressure on the coast and intertidal areas is coming from industrial development, tourism and coastal urbanisation. Many highly intensive industrial developments, with associated port and energy developments, are expected in the coming decades. At the same time the coasts of France, Italy and Spain receive nearly 200 million visitors per year and the number of tourists is also expected to rise. Tourism has a significant effect on foreshore development, drainage patterns and movement of sediment, with the consequence that many special conservation sites around the coast will need special attention if they are to be protected.

Often the aesthetic beauty of the sea and coast is an important aspect of tourism, so the expansion of industry along the coastal strip and into the marine area is likely to lead to many conflicts between users. The need for coherent planning is considered by many to be essential in the future development of the marine and coastal environment.

In Europe, there are a number of policies affecting the marine environment, but none is specifically designed to protect the health of its ecosystems. There is no harmonised legislation on marine protection across the Member States. Gaps in knowledge exist because assessment and monitoring programmes are not integrated or complete, and the links between research needs and priorities remain weak. The proposed ecosystem-based approach to management and sustainable development of the EU's marine strategy will enable these issues and others such as eutrophication, hazardous substances and persistent organic pollutants, discharges from shipping, the effects of fisheries, declines in biodiversity and habitat integrity, and impacts of climate change to be properly assessed.

For Europe's marine and coastal environments to continue to provide real economic benefits to its populations, remain healthy and provide food, resources and cultural support in the longer term, it is critical that a pan-European approach to management and conservation now be adopted — one that recognises regional differences and vulnerabilities, but applies common principles and measures of progress towards meeting the Lisbon agenda and other policy targets.

References and further reading

The core set of indicators found in Part B of this report that are relevant to this chapter are: CSI 21, CSI 22, CSI 23, CSI 32, CSI 33 and CSI 34.

Introduction
European Environment Agency, 2003. *Europe's environment: The third assessment*. Environmental Assessment Report No 10, Office for Official Publications of the European Communities, Luxembourg, 341 pp.

European Land Ocean Interaction Studies (ELOISE), 2004. (See www.nilu.no/projects/eloise/ — accessed 12/10/2005).

Millennium Ecosystem Assessment, 2005. *Ecosystems and human well-being: Synthesis*, Island Press, Washington, DC, 137 pp.

Sea-Search, 2004. The gateway to oceanographic and marine data and information in Europe. (See www.sea-search.net/data-access/welcome.html — accessed 12/10/2005).

Sherman, K. and Hoagland, P., 2005. *Driving forces affecting resource sustainability in large marine ecosystems*, ICES CM 2005/M:07.

Regional perspectives on the state of the marine environment
Badalamenti, F., et al., 2000. 'Cultural and socio-economic impacts of Mediterranean marine protected areas', *Environmental Conservation* 27 (2), pp. 110–125.

Black Sea Commission, 2002. *State of the environment of the Black Sea: Pressures and trends*, 1996–2000, Commission for the Protection of the Black Sea against Pollution, Istanbul, 65 pp. (See www.blacksea-commission.org/Downloads/SOE_English.pdf — accessed 12/10/2005).

Census of marine life. (See www.coml.org — accessed 12/10/2005).

European Environment Agency, 2002. *Europe's biodiversity — biogeographical regions and seas around Europe*, web report (See http://reports.eea.eu.int/report_2002_0524_154909/en — accessed 12/10/2005).

European Environment Agency, 2005. *Priority issues in the Mediterranean environment*, EEA Report No 5/2005.

Leppäkoski, E., Gollasch, S. and Olenin, S. (eds), 2002. *Aquatic invasive species of Europe — distribution, impacts and management*, Kluwer Academic Publishers, Dordrecht, Boston, London.

Meinesz, A. (translated by D. Simberloff), 1999. *Killer algae: The true tale of a biological invasion*, University of Chicago Press, Chicago, 376 pp.

Sherman, K. and Hempel, G. (eds) 2002. *Large marine ecosystems of the North Atlantic*, Elsevier, Amsterdam.

Wulff, F.V., Rahm, L.A. and Larsson, P., 2001. *A systems analysis of the Baltic Sea*, Springer-Verlag, Berlin, Heidelberg.

Zaitsev, Yu. P., 1993. 'Impacts of eutrophication on the Black Sea fauna', In: *Fisheries and environmental studies in the Black Sea system*, GFCM Studies and Reviews 64, pp. 63–85.

The state of coastal and intertidal areas
Benoit G. and Comeau A. (eds), 2005. *Sustainable future for the Mediterranean: The blue plan's environment and development outlook* (in print).

Borum, J., Duarte, C., Krause-Jensen, D. and Greve, T. (eds), 2004. *European seagrasses: An introduction to monitoring and management*, Monitoring and Managing European Seagrasses (EU project), 88 pp.

DATAR, 2004. *Construire ensemble un développement équilibré du littoral*, La Documentation Française, Paris, ISBN 2-11-005716-5, 156 pp.

European Commission, 2004. Living with the coastal erosion in Europe — Sediment and space for sustainability, Office of Official Publications of the European Communities, Luxembourg, 40 pp.

European Environment Agency, 2005. The state of the environment in Europe's coastal areas (working title), Assessment report in preparation.

JRC, 2005. Indicators on marine environment and coastal pressures: Wetland loss ME-8. (See http://esl.jrc.it/envind/meth_sht/ms_we042.htm — accessed 12/10/2005).

Drivers and pressures affecting marine and coastal areas

Aquaculture and coastal economic and social sustainability (Aqcess), 2000. EU Fifth Framework Project, Contract No. Q5RS-2000-31151. (See www.abdn.ac.uk/aqcess/. — accessed 12/10/2005).

Arctic Climate Impact Assessment (ACIA), 2004. *Impacts of a warming Arctic*, Arctic Climate Impact Assessment report, Cambridge University Press, the United Kingdom, 140 pp. (See www.amap.no — accessed 12/10/2005).

Biomare, 2003. Implementation and networking of large scale, long term marine biodiversity research in Europe, EU Contract EVR1-CT2000-20002, NIOO-CEME, Yerseke, the Netherlands, European Marine Biodiversity indicators ISBN 90-74638-14-7 and Marine Biodiversity Sites ISBN 90-74638-15-5.

Bodungen, B. von and Turner, R.K. (eds), 2001. *Science and integrated coastal zone management*, Dahlem Conference 86, Dahlem University Press.

Butler, J.R.A., 2002. 'Wild salmonids and sea louse infestations on the west coast of Scotland: Sources of infection and implications for the management of marine salmon farms', *Pest Management Science* 58, pp. 595–608.

Davies, I.M., 2000. *Waste production by farmed Atlantic salmon (Salmo salar) in Scotland*, ICES CM 2000/0.01.

Delgado, O., Ruiz, J., Perez, M. *et al.*, 1999. 'Effects of fish farming on seagrass (*Posidonia oceanica*) in a Mediterranean bay: Seagrass decline after organic loading cessation', *Oceanologica Acta* 22 (1), pp. 109–117.

DG Fisheries, 2001. European distant water fishing fleet: Some principles and some data. (See www.europa.eu.int/comm/fisheries/doc_et_publ/liste_publi/facts/peche_en.pdf — accessed 12/10/2005).

DG Fisheries, 2003. Reforming the common fisheries policy. 17 January 2003. (See www.europa.eu.int/comm/fisheries/reform/index_en.htm — accessed 12/10/2005).

DG Fisheries, 2004. Fact sheets on the common fisheries policy (Section 5.1 on structural policy and Section 5.4 on aquaculture), on the EU Online website: (See www.europa.eu.int/comm/fisheries/doc_et_publ/factsheets/facts_en.htm — accessed 12/10/2005).

Edwards, M., Licandro, P., John, A.W.G. and Johns, D.G., 2005. Ecological status report: Results from the CPR survey 2003/2004, SAHFOS Technical Report No. 2 1–6, ISSN 1744–075.

Ellett, D.J., 1993. The north-east Atlantic: a fan-assisted storage heater? *Weather* 48:118–125.

European Commission, 2000. Regional socio-economic studies on employment and the level of dependence on fishing, Lot. No 23, Coordination and Consolidation Study, Fisheries Sub Sector Strategy Paper, 53 pp.

European Commission, 2002. A strategy for the sustainable development of European aquaculture, Communication from the Commission to the Council and the European Parliament, Brussels, 19.9.2002, 24 pp., COM 2002/511 final.

European Commission, 2002. Communication from the Commission to the Council and the European Parliament on a Community action plan to reduce discards of fish, Brussels, 26.11.2002, 21 pp., COM(2002)656 final.

European Commission, 2002. Council Regulation No 2371 of 20 December 2002 on the conservation and sustainable exploitation of fisheries resources under the Common Fisheries Policy, Official Journal L358, 31/12/2002, pp. 0059–0080.

European Commission, 2002. Financial instrument for fisheries guidance — Instructions for use, ISBN 92-894-1647-5, 47 pp. (See www.europa.eu.int/comm/fisheries/doc_et_publ/liste_publi/facts/ifop_en.pdf — accessed 12/10/2005).

European Community Fisheries Register, 2003. Fishing fleet census 2003 survey.

EU fisheries policy. (See www. europa.eu.int/comm/fisheries/reform/conservation_en.htm — accessed 12/10/2005).

EU maritime transport policy. (See www.europa.eu.int/comm/transport/maritime/index_en.htm — accessed 12/10/2005).

Eurostat, 2005. (See http://epp.eurostat.cec.eu — accessed 12/10/2005).

Food and Agriculture Organization of the United Nations (FAO), 1950–. Fishstat Plus, Total production 1950–2001.

Food and Agriculture Organization of the United Nations (FAO), 2002. *The state of world fisheries and aquaculture*, SOFIA 2002, ISBN 92-5-104842-8. FAO Fisheries Department, 150 pp.

Garibaldi, L. and Limongelli, L., 2003. *Trends in oceanic captures and clustering of large marine ecosystems*, FAO Fish. Tech. Pap. 435, ISBN 92-5-104893-2, Food and Agriculture Organization of the United Nations, Rome, 71 pp.

Hansen, B., Østerhus, S., Quadfasel, D. and Turrell, W.R., 2004. Already the day after tomorrow? *Science* 305, pp. 953–954.

Intergovenmental Panel on Climate Change (IPCC), 2001. *The third assessment report of the Intergovernmental Panel on Climate Change*, Cambridge University Press, Cambridge, the United Kingdom and New York, USA.

Jurado-Molina, J. and Livingston, P., 2002. 'Climate-forcing effects on trophically linked groundfish populations: implications for fisheries management', *Canadian Journal of Fisheries and Aquatic Science* 59: 1941–1951.

Kaiser, M.J. and de Groot, S.J. (eds), 2000. *The effects of fishing on non-target species and habitats: Biological, conservation and socio-economic issues*, Blackwell Science, Oxford, the United Kingdom.

Karakassis, I., Tsapakis, M., Hatziyanni, E. et al., 2000. 'Impact of cage farming of fish on the seabed in three Mediterranean coastal areas', *ICES Journal of Marine Sciences* 57, pp. 1462–1471.

Klyashtorin, L.B., 2001. *Climate change and long-term fluctuations of commercial catches*, FAO Technical Paper 410, 86 pp.

Konsulova, T.Y., Todorova, V. and Konsulov, A., 2001. 'Investigations on the effect of ecological method for protection against illegal bottom trawling in the Black Sea. Preliminary results', *Rapp. Comm.* Int. Mer Medit. 36, p. 287.

OSPAR, 2001. Discharges, waste handling and air emissions from offshore oil and gas installations, in 2000 and 2001, ISBN 1 904426 20 4. (See www.ospar.org — accessed 12/10/2005).

OSPAR, 2002. Annual report on discharges, waste handling and air emissions from offshore oil and gas installations in 2002, ISBN 1 904426 47 6. (See www.ospar.org   accessed 12/10/2005).

OSPAR, 2003. Integrated report on the eutrophication status of the OSPAR Maritime Area based upon the first application of the comprehensive procedure, ISBN 1 904426 25 5. (See www.ospar.org — accessed 12/10/2005).

OSPAR, 2003. Liquid discharges from nuclear installations in 2003, ISBN 1 904426 62 X. (See www.ospar.org — accessed 12/10/2005).

OSPAR, 2003. Report on discharges, spills and emissions from offshore oil and gas installations in 2003, ISBN 1 904426 60 3. (See www.ospar.org — accessed 12/10/2005).

OSPAR, 2004 Environmental impact of oil and gas activities other than pollution, ISBN 1 904426 44 1. (See www.ospar.org — accessed 12/10/2005).

OSPAR, 2005. Inventory of oil and gas offshore installations in the OSPAR Maritime Area, ISBN 1 904426 66 2. (See www.ospar.org — accessed 12/10/2005).

Royal Society, 2005 Ocean acidification due to increasing atmospheric carbon dioxide. Policy document 12/05, ISBN 0 85403 6172. (See www.royalsoc.ac.uk — accessed 12/10/2005).

Seibel, B.A. and Fabry, V.J., 2003. 'Marine biotic response to elevated carbon dioxide,' *Advances in Applied Biodiversity Science* 4, pp. 59–67.

Shirayama, Y., Kurihara, H., Thornton, H. *et al.*, 2004. 'Impacts on ocean life in a high CO_2 world', SCOR-UNESCO Symposium 'The ocean in a high-CO_2 world', SCOR-UNESCO Paris.

Sir Alister Hardy Foundation for Ocean Science. www.sahfos.org.

Theodossiou, I. and Dickey, H., 2003. *Socioanalysis report, Analysis of the labour market conditions in the Aqcess study areas where fisheries and aquaculture co-exist.* Final report to the EU, DG XIV, Contract Q5RS-2000-31151.

Trends in ecosystem health

Blaber, S.J.M., Cyrus, D.P., Albaret, J.-J. *et al.*, 2000. 'Effects of fishing on the structure and functioning of estuarine and nearshore ecosystems', *ICES Journal of Marine Science* 57:590–602.

Bertrand, J.A., Gil de Sola, L., Papaconstantinou, C. *et al.*, 2002. 'The general specifications of the Medits surveys'. In: Abello, P., Bertrand, J., Gil de Sola, L. *et al.* (eds) Mediterranean marine demersal resources: The MEDITS international trawl survey (1994–1999), *Sc. Mar.* 66, pp. 9–17.

Caddy, J.F., 2000. 'Marine catchment basin effects versus impacts of fisheries on semi-enclosed seas', *ICES Journal of Marine Science* 57, pp. 628–640.

Caddy, J.F. and Garibaldi, L., 2000. 'Apparent changes in the trophic composition of the world marine harvests: The perspectives from the FAO capture database', *Ocean and Coastal Management* 43 (8–9), pp. 615–655.

Caminas, J.A. and Valeiras, J., 2001. 'Marine turtles, mammals, and sea birds captured incidentally by the Spanish surface longline fisheries in the Mediterranean Sea', *Rapp. Comm. Int. Mer. Medit.*, 36, p. 248.

Daskalov, G.M., 2002. 'Overfishing drives a trophic cascade in the Black Sea', *Marine Ecology Progress Series* 225, pp. 53–63.

De Leiva Moreno, J.I., Agostini, V.N., Caddy, J.F. and Carocci, F., 2000. 'Is the pelagic-demersal ratio from fishery landings a useful proxy for nutrient availability?' A preliminary data exploration for the semi-enclosed seas around Europe, *ICES Journal of Marine Science* 57, pp. 1090–1102.

Di Natale, A., 1995. 'Driftnet impact on protected species: Observers data from the Italian fleet and proposal for a model to assess the number of cetaceans in the by-catch', *ICCAT Collective Volume of Scientific Papers* 44, pp. 255–263.

Dolmer, P., Kristensen, P.S. and Hoffmann, E., 1999. 'Dredging of blue mussels (*Mytilus edulis L*) in a Danish sound: Stock sizes and fishery-effects on mussel population dynamics', *Fish Research* 40: 73–80.

Dosdat, A., 2001. Environmental impact of aquaculture in the Mediterranean: Nutritional and feeding aspects,

Proceedings of the seminar of the CIHEAM Network on Technology of Aquaculture in the Mediterranean, Zaragossa, 17–21 January 2000, *Cahiers Options Mediterreannes* 55, pp. 23–36.

European Environment Agency, 2004. *Arctic environment: European perspectives*. Environmental Issue Report No 38, EEA, Copenhagen.

Fiorentini, L., Caddy, J.F. and De Leiva, J.I., 1997. *Long and short term trends of Mediterranean fishery resources*, GFCM Studies & Reviews 69, Food and Agriculture Organization of the United Nations, Rome, 72 pp.

Fishbase. (See www.fishbase.org/ — accessed 12/10/2005).

Gerosa, G. and Casale, P., 1999. *Interaction of marine turtles with fisheries in the Mediterranean*, Mediterranean Action Plan-UNEP Regional Activity Centre for Specially Protected Areas.

GFCM, 2002. General Fisheries Commission for the Mediterranean, Report of the twenty-seventh session, Rome, 19–22 November 2002, Report No 27, FAO, Rome. 36 pp.

GFCM/SAC, 2002. General Fisheries Commission for the Mediterranean, Report of the fifth session of the Scientific Advisory Committee, FAO Fish. Rep. 684, 100 pp.

GFCM/ SCSA, 2002. General Fisheries Commission for the Mediterranean/Sub-Committee Meeting, Report of the fourth stock assessment, Barcelona, Spain, 6–9 May, 2002.

Gill, A.B. 2005. 'Offshore renewable energy: Ecological implications of generating electricity in the coastal zone', *Journal of Applied Ecology* 42:605–615.

Helcom *Environmental focal point information 2004 Dioxins in the Baltic Sea*, Helsinki Commission Baltic Marine Environment protection Commission, 20 pp. www.helcom.fi.

ICES, 2001. Report of the Working Group on Marine Mammal Population Dynamics and Habitats, ICES CM 2011 / ACE:01, ICES, Denmark.

ICES, 2003. Environmental status of the European seas, quality status, Federal Ministry for the Environment, Nature Conservation and Nuclear Safety, 75 pp.

ICES/ACME, 2004. Report of the ICES Advisory Committee on the Marine Environment. ICES. (See www.ices.dk/committe/acme/2004/ACME04.pdf — accessed 12/10/2005).

ICES/WGAGFM, 2003. Report of the Working Group on the Application of Genetics in Fisheries and Mariculture (See www.ices.dk/reports/MCC/2003/WGAGFM03.pdf — accessed 12/10/2005).

ICES/WGEIM, 2003. Report of the Working Group on Environmental Interactions of Mariculture, ICES. (See www.ices.dk/reports/MCC/2003/WGEIM03.pdf accessed 12/10/2005).

ICES working group reports. (See www.ices.dk/iceswork/workinggroups.asp accessed 12/10/2005).

International Maritime Organization, 2005. (See www.imo.org — accessed 12/10/2005).

Jennings, S. and Kaiser, M.J., 1998. 'The effects of fishing on marine ecosystems', Advances in Marine Biology Vol. 34, pp. 201–350.

Jennings, S., Greenstreet, S.P.R. and Reynolds, J.D., 1999. 'Structural change in an exploited fish community: A consequence of differential fishing effects on species with contrasting life histories', *Journal of Animal Ecology* 68, pp. 617–627.

Jennings, S., Kaiser, M.J. and Reynolds, J.D., 2001. Marine fisheries ecology. Blackwell Scientific Ltd, Oxford, 417 pp.

Koslow, J.A., Boehlert, G.W., Gordon, J.D.M. *et al.*, 2000. 'Continental slope and deep-sea fisheries: Implications for a fragile ecosystem', *ICES Journal of Marine Science* 57, pp. 548–557.

Laist, D.W., 1996. 'Marine debris entanglement and ghost fishing: A cryptic and significant type of bycatch?' In: Sinclair, M. and Valdimarsson, G. (eds). *Proceedings of the solving bycatch workshop: Considerations for today and tomorrow*, 25–27 September 1995, Seattle WA. Report No. 96-03, Alaska Sea Grant College Program, Fairbanks AK, pp. 33–39.

Large marine ecosystems of the world, 2003. (See www.edc.uri.edu/lme/default.htm — accessed 12/10/2005).

McGlade, J.M. and Metuzals, K.I., 2000. 'Options for the reduction of by-catches of harbour porpoises (*Phocoena phocoena*) in the North Sea', In Kaiser, M.J. and de Groot, S.J. (eds) *The effects of trawling on non-target species and habitats: Biological, conservation and socio-economic issues*, Blackwell Science, Oxford, 399 pp.

Mee, L.D., 1992. The Black Sea in crisis: A need for concerted international action, Ambi 21(4), pp. 278–286.

OECD, 2001. *Environmental outlook to 2020*, OECD.

OSPAR/QSR, 2000. *Quality status report 2000 for the north-east Atlantic*, Ospar Commission for the Protection of the Marine Environment in the North-east Atlantic. (See www.ospar.org — accessed 12/10/2005).

Pauly, D., Christensen, V., and Walters, C., 2000. 'Ecopath, ecosim, and ecospace as tools for evaluating ecosystem impact of fisheries', *ICES Journal of Marine Science* 57, pp. 697–706.

Pauly, D., Christensen, V., Dalsgaard, J. et al., 1998. 'Fishing down marine food webs', *Science* 279, pp. 860–863.

Pearson, T.H. and Rosenberg, R., 1978. 'Macrobenthic succession in relation to organic enrichment and pollution of the marine environment', *Oceanography and Marine Biology Annual Review* 16, pp. 229–311.

Pitta, P., Karakassis, I., Tsapakis, M. and Zivanovic, S., 1999. 'Natural vs. mariculture induced variability in nutrients and plankton in the Eastern Mediterranean', *Hydrobiologia* 391, pp. 181–194.

Prodanov, K., Mikhailov, K., Daskalov, G. et al., 1997. *Environmental management of fish resources in the Black Sea and their rational exploitation*, FAO Fish. Cir. 909, 225 pp.

RAC/SPA, 2003. 'Effects of fishing practices on the Mediterranean Sea: Impact on marine sensitive habitats and species, technical solution and recommendations', In Tudella S. and Sacchi, J. (eds.) *Regional activity centre for specially protected areas*, 155 pp.

Shiganova, T.A. and Bulgakova, Y.V., 2000. 'Effects of gelatinous plankton on Black Sea and Sea of Azov fish and their food resources', *ICES Journal of Marine Science* 57, pp. 641–648.

Tasker, M.L., Camphuysen, C.J., Cooper, J. et al., 2000. 'The impacts of fishing on marine birds', *ICES Journal of Marine Science* 57, pp. 531–547.

Van Dalfsen, J.A., Essink, K., Madsen, H.T. et al., 2000. Differential response of macrozoobenthos to marine sand extraction in the North Sea and western Mediterranean, *ICES Journal of Marine Science* 57, pp. 1439–1455.

Vinther, M., and Larsen, F., 2002. 'Updated estimates of harbour porpoise by-catch in the Danish bottomset gillnet fishery', Paper presented to the Scientific Committee of the International Whaling Commission, Shimonoseki, May 2002, SC/54/SM31, 10 pp.

Watling, L. and Norse, E.A., 1998. 'Disturbance of the seabed by mobile fishing gear: A comparison to forest clearcutting', *Conservation Biology* 12(6), p. 1180.

Future perspectives

Barcelona Convention. (See www.unepmap.org/ — accessed 12/10/2005).

European Commission, 2002 Communication from the Commission on the reform of the common fishery policy, 32 pp.

European Commission, 2004. *European code of sustainable and responsible fisheries practices*, Office for Official Publications of the European Communities, Luxembourg, 15 pp.

European Commission Maritime Unit. (See www.europa.eu.int/comm/fisheries/maritime/ — accessed 12/10/2005).

Froese, R., 2004. 'Keep it simple: three indicators to deal with overfishing', *Fish and Fisheries* 5: 86–91.

Gislason, H., Sinclair, M., Sainsbury, K. and O'Boyle, R., 2000. 'Symposium overview: Incorporating ecosystem objectives within fisheries management', *ICES Journal of Marine Science* 57 (3) pp. 468–475.

Grieve, C., 2001. *Reviewing the common fisheries policy: EU fisheries management for the 21st century*, Institute for European Environmental Policy (IEEP), London, ISBN 1 873906 41 2, 42 pp.

Helcom. (See www.helcom.fi — accessed 12/10/2005).

OSPAR. (See www.ospar.org/eng/html/welcome.html — accessed 12/10/2005).

McManus, E., 2005. *Biodiversity trends and threats in Europe: The marine component*, Report from Department for Environment, Food and Rural Affairs, the United Kingdom.

Pickering, H. (ed.), 2003. *The value of exclusion zones as a fisheries management tool: A strategic evaluation and the development of an analytical framework for Europe*, CEMARE Report, University of Portsmouth, the United Kingdom.

Sainsbury, K. and Sumaila, U.R., 2003. 'Incorporating ecosystem objectives into management of sustainable marine fisheries, including "Best Practice" reference points and use of marine protected areas', pp 343–362. In: Sinclair, M. and Valdimarsson, G. (eds) *Responsible fisheries in the marine ecosystem*, FAO and CABI Publishing.

Sherman, K., and Duda, A.M., 1999. 'An ecosystem approach to global assessment and management of coastal waters', *Marine Ecology Progress Series* 190, pp. 271–287.

Tasker, M.L., Camphuysen, C.J., Cooper, J. *et al.*, 2000. 'The impacts of fishing on marine birds', *ICES Journal Marine Science* 57, pp. 531–547.

United Nations Environment Programme, 2001. *Ecosystem-based management of fisheries: Opportunities and challenges for coordination between marine Regional Fishery Bodies and Regional Seas Conventions*, UNEP Regional Seas Reports and Studies No. 175, ISBN 92-807-2105-4, 52 pp.

7 Soil

7.1 Introduction

Soils are as essential to human society as air and water. They are the basis for the production of 90 % of our food, fibre and livestock food. They capture and filter rainfall, delivering it to geological formations on which millions rely for their water supplies. Properly managed soils can also absorb a significant proportion of the carbon dioxide released into the atmosphere from human activity, contributing to the moderation of climate change. A recent study, however, has suggested that rising temperatures are causing soils to unlock larger quantities of carbon dioxide than previously thought, offsetting the reductions achieved in carbon dioxide emissions from other sources.

In many parts of the continent, soils and the environmental services they provide are under threat. Human activity is triggering unsustainable levels of erosion, often combined with chemical contamination and biological degradation. Additionally, good quality agricultural soils are being sealed by the concrete and asphalt of urban and infrastructural development — indeed in some regions, such as the Mediterranean coast, soil sealing may affect large portions of the total land area.

From acid deposition to farming, from landfill seepage to mining, from highway construction to reservoir flooding, and from irrigation to overgrazing, the threats to soils are numerous. Their very resilience often means we do not perceive the damage until it is far advanced. The implications are profound for the habitability of the continent, for, while air or water pollution may disperse in a matter of days, contamination and erosion of soils can take centuries to put right.

Europe already has strategies for managing air and water quality. In line with the general recognition that soil degradation is also a serious and widespread problem, the Commission, as part of the sixth environment action programme (6EAP), established in 2002 a process towards a thematic strategy for soil protection. The soil thematic strategy (STS) identifies eight threats: contamination, erosion, decline in organic matter, compaction, salinisation, landslides, sealing, and loss of soil biodiversity. The first three are considered priorities. Five wide-ranging technical working groups were set up to examine issues of erosion, organic matter, contamination, monitoring, research and sealing, and other cross-cutting issues.

Subsidiarity and flexibility are key words for the new soil directive, which is likely to include common principles and definitions. Different 'working units' (or levels of aggregation) are proposed for different threats. For more local soil threats such as erosion, decline of organic matter, compaction and land slides, the focus of EU policy is likely to fall on the so-called 'risk-areas', to be identified by the EU Member States on the basis of common criteria. For sealing and contamination, the working unit is likely to be defined at the national and regional scale. This is because there is a need for more subsidiarity to handle these threats, due to their stronger links with the national and regional policies.

Work carried out by the technical working groups has brought into focus the paucity of available information on the geographical distribution and extent of soil-related problems, which is complicated by soils' innate heterogeneity. This chapter reflects that reality. The value of soils for sustaining many ecological functions relevant to Europe's economy, and hence competitiveness, in the face of threats such as climate change and extreme weather events is increasingly better understood. This, in turn, underlines the importance of making substantial progress on soils research, monitoring and analysis to provide a better basis for policy actions.

7.2 Erosion

Erosion of topsoil is one of the most widespread threats to the continent's soils, but there is only sparse quantitative information on actual rates and the extent of soil erosion at the European scale.

Soil erosion in Europe is primarily caused by water. It is the result of the physical impact of raindrops on exposed surfaces, combined with the ability of the subsequent run-off to dissolve nutrients and wash away soil particles. In drier areas high wind can be a threat, whipping up dust storms, particularly in finer soils.

According to a recent study called PESERA, undertaken as part of the European Commission's fifth framework programme for research, as much as a quarter of Europe's land is thought to be at some risk of erosion, with the greatest problems occurring around the Mediterranean and Black Seas, in the Balkan Peninsula and in Iceland, which has one of the highest soil erosion rates in Europe. Furthermore, the same study estimates that more than another 10 million hectares of Europe's lands are subject to high or very high risk from erosion with a further 27 million hectares under moderate risk. Countries with the largest areas at risk include Greece, Hungary, Italy, Moldova and Portugal. The PESERA results should be considered with some care. Erosion risk is overestimated in some countries (e.g. Denmark) or underestimated in some others (e.g. Spain) owing to shortcomings either for input data or modelling algorithms. Nonetheless, the results are a useful starting point and there is an opportunity to develop the methodology further so that it can provide a basis on which to build better quality results in future years.

Erosion is a natural phenomenon, of course. Indeed, it is a vital part of the functioning of the biosphere. Sediment and nutrients removed from soils by wind and rain feed life in rivers and the oceans and play an essential role in the natural carbon cycle. In the natural environment, however, those soil losses are counteracted by the formation of fresh soil as rocks beneath the soil are themselves weathered and transformed by groundwaters and the action of soil microbes. Natural factors determining the erosion potential of soil include climate, topography, vegetation and the characteristics of soil, such as how light and friable it is.

The challenge today is that human activity has dramatically accelerated the rate of soil loss. The primary causes of this escalation are the clearance of forests and dense natural vegetation, and unsustainable agriculture, including intensive arable farming and overgrazing of pastures, all of which leave soils exposed to the elements.

Erosion poses serious questions in particular about the sustainability of certain agricultural practices. Since erosion removes organic material from the soil, reducing fertility and productivity, farmers tend to apply more artificial fertilisers to maintain output. However, erosion is a process that feeds on itself as degraded soils become more vulnerable to further erosion.

Eroded soils are less efficient at filtering pollution and capturing water to replenish underground water reserves. Erosion also reduces the ability of soils to capture and store atmospheric carbon. Globally, soil loss over the centuries has reduced the amount of carbon retained by soils by about 100 billion tonnes, equivalent to some 15 years of current emissions from fossil-fuel burning.

In many areas of Europe where soils have been cultivated for long periods, organic carbon content is currently low or very low. Even modest changes in its organic carbon content can cause rapid decreases in the quality of soil structure and biodiversity. The problem is most pronounced in southern Europe, where more than 100 million hectares have an organic carbon content of less than 1 %. Across the whole of Europe nearly 230 million hectares are defined as having this low or very low content of organic carbon in the topsoil.

Soil erosion also causes impacts 'off-site'. Whilst historically deposition of eroded soil material has contributed considerably to the fertility of flood plains, in the absence of expensive dredging, it can silt up river courses and lakes, causing flooding and damaging biodiversity. When reservoirs accumulate silt, they lose water storage capacity and the potential for hydroelectricity generation. The presence of eroded soil in suspension in river systems can also significantly impact aquatic flora and fauna, with serious implications for valuable fish stocks. Erosion may also undermine man-made physical structures such as roads and bridges.

Chemically, soil erosion delivers nutrients that cause eutrophication of rivers and lakes. As improved wastewater treatment across Europe has reduced the releases of nutrients from that source, the contribution of run-off and soil erosion to eutrophication has grown. This is evident, for instance, in two UK lakes, Lough Neagh and Lough Erne, where concentrations of phosphorus have increased in spite of reduced loads from wastewater. These high concentrations were caused by a steady build-up of a surplus and continued re-application of phosphorus (arising from manures and fertilisers) in the soils in the upstream catchments.

Erosion is often seen as a process confined largely to the dry lands of southern Europe, where in extreme cases, in combination with other factors such as climate, the unsustainable use of water and a lack of vegetation, it can lead to 'desertification'. Certainly the problems are intense there. Long, dry periods leave soils vulnerable to erosion. Droughts are often broken by intense storms that can wash away large amounts of soil. Individual storms in the region have been known to remove 100 tonnes from a hectare of land, and more frequently remove 20 to 40 tonnes.

According to the desertification information system for the Mediterranean (DISMED), sensitivity to desertification is not high in Europe in comparison with its neighbouring countries. However, in the areas of the northern Mediterranean for which quantitative data are available, one-third of the territory, approximately 37 million hectares, currently shows a moderate or low sensitivity (Map 7.1). The affected areas increase to more than 70 million hectares if very low sensitivities are taken into account. Southern Portugal, southern Spain, Sicily, and parts of Greece are most seriously affected, where areas with moderate or low sensitivities range from approximately 65 % to more than 85 % of the region concerned.

In addition, the fast pace of current development in southern Europe often results in building on steep slopes that are most vulnerable to erosion when vegetation is removed. For example, this has resulted in a sharp rise in instances of landslides in Italy in the past 20 years, affecting more than 70 000 people and causing economic damage approaching EUR 11 000 million.

Map 7.1   Sensitivity to desertification in the northern Mediterranean

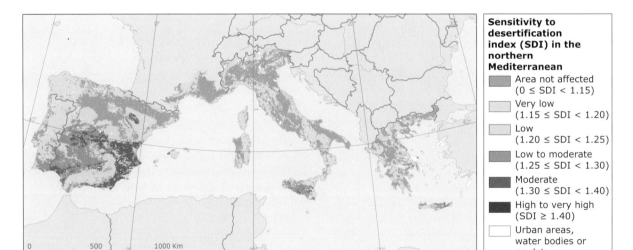

Source: DISMED project (Desertification Information System for the Mediterranean) and EEA, 2005.

Soil erosion is far from being confined to the south of the continent. In northern Europe there are wide areas of light, easily eroded soils, such as in the northern loess zone, stretching from northern France through Germany and southern Poland, and in some parts of the United Kingdom. The most obvious effects are off-site, through eutrophication and siltation of water courses.

Erosion right across the continent is expected to worsen, partly as a result of climate change, which will intensify both droughts and rainstorms. The water erosion risk is expected to increase as a result of climate change in four-fifths of Europe's agricultural areas by 2050, with the deterioration generally greatest in places that already have serious erosion problems.

All this carries major economic consequences, both on- and off-site. On-site impacts are mainly related to the loss of long-term net farm income and the cost of both restoring damage to soil structure and reversing the decline of organic matter. Off-site costs include cleaning roads and dredging eroded sediments from reservoirs used for water supply and the generation of electricity. Further costs could be incurred in restoring the aquatic environment from the effects of eutrophication and in improving the quality of water which has been damaged by eroded sediments.

The European Commission is currently preparing a quantification of the economic impacts of soil degradation. Some estimates have already been attempted which provide an indication of the magnitude of the problem across the continent. However, these do not include costs that are not related to current uses of soil or costs which may result from soil erosion but cannot be described in monetary terms, such as the loss of biodiversity or the deterioration of ecosystem health.

One such estimate suggests that the yearly economic losses in agricultural areas are in the region of EUR 53 per hectare, while the cost of off-site effects on the surrounding infrastructure, such as destruction of roads and siltation of dams, could reach EUR 32 per hectare. Data on economic losses due to soil erosion are also available for some countries and regions. In Armenia, for example, the costs of the damage from soil erosion in the past 20 years amounted to 7.5 % of national gross agricultural product.

Older, more limited studies have estimated that fertilisers needed to compensate for the loss of nutrients caused by a single wind erosion storm cost up to EUR 300 per hectare, and estimate the annual cost of short-term wind erosion damage in the Netherlands at about EUR 9 million. Other information is available on economic losses due to off-site impacts — for example, the external cost of water-induced soil erosion in Bavaria, Germany, was estimated in 1991 to be up to EUR 15 million per year.

7.3 Contamination

Soil contamination is widespread across Europe. It occurs both through localised sources of pollution, such as industrial sites, and through 'diffuse' pollution from atmospheric fallout such as acid rain, leaching of farm chemicals and even soil erosion which, as already mentioned, can liberate nutrients.

Local sources
According to the latest estimates there might be more than two million sites across Europe potentially contaminated from localised pollution sources, with an estimated 100 000 considered as needing remediation. The largest concentrations of sites are estimated to be around the old industrialised heartlands of north-west Europe, from southern United Kingdom through north-east France, Belgium and the Netherlands to the Rhine-Ruhr region of Germany. Other places with serious hot spots include the Po valley around Milan in Italy, and the old eastern European heartland of heavy industry known as the 'black triangle', which includes the Czech Republic, Slovakia, eastern Germany and parts of Poland.

Major contaminants include heavy metals, both from factory point sources, spills of mineral oil and from chlorinated hydrocarbons, and the tailings of mining and minerals processing. Cyanide leaks from metal-refining processes are a frequent problem, as are the chemical cocktails left behind by old gasworks.

The tanks of petrol filling stations are one of the most numerous and ubiquitous sources of contaminated soils. Leakage from landfills is also widespread. During the past 30 years a huge range of hazardous chemicals has been variously dumped into landfills without adequate precautions to prevent the chemicals migrating into surrounding soils, groundwaters and surface waters.

Mine drainage waters can contaminate large areas if not properly controlled. Recent examples include the Aznalcóllar mine disaster in Spain in 1998, which affected soils and the water course for 60 kilometres downstream, and the cyanide spill from a tailing treatment plant at the Baia Mare gold mine in Romania in 2000.

As former industrial land may be abandoned, problems are often hidden. The soil beneath former transport depots and railway sidings sometimes harbour a

Table 7.1   Remediation actions for soil contamination in some European countries

Country	Year	Policy or technical target
Austria	2030–2040	Essential part of the contaminated sites problem should be managed.
Belgium (Flanders)	2006	Remediation of the most urgent historical contamination. New contamination to be remediated immediately.
	2021	Remediation of urgent historical contamination.
	2036	Remediation of other historical contamination causing risk.
Bulgaria	2003–2009	Plan for implementation of Directive 1999/31/EC on landfill of waste.
Czech Republic	2010	Eliminate the majority of old ecological damage.
France	2005	Establish information system on polluted soil (BASIAS) to provide a complete scope of the sites where soil pollution could be suspected.
Hungary	2050	Handling of all sites. Government Decision No 2205/1996 (VIII.24.) adopted National Environmental Remediation Programme (OKKP).
Lithuania	2009	Waste disposal to all landfills not fulfilling special requirements should be stopped. All waste landfills not fulfilling special requirements should be closed according to approved regulations.
Malta	2004	Closure of Maghtab and il-Qortin waste disposal sites.
Netherlands	2030	All historical contaminated sites investigated and under control and remediated when necessary.
Norway	2005	Environmental problems on sites with contaminated soil, where investigation and remediation is needed, to be solved. On sites where further investigation is needed, the environmental state will be clarified.
Sweden	2020	Environmental quality objective: a non-toxic environment.
Switzerland	2025	The 'dirty' heritage of the past should be dealt with in a sustainable way within one generation.
United Kingdom (England and Wales)	2007	At a political level, the Environment Agency aims to substantially remediate and/or investigate 80 Special Sites identified under Part IIA Regime (Environmental Protection Act 1990).

Source: EEA, Eionet priority data flows, 2003.

variety of contaminants that may be hard to predict. Military installations have also often dealt with many hazardous materials, including radioactive materials, without keeping public records. The worst problems on military sites are likely to be in central and eastern Europe. In Estonia, almost 2 % of the land comprises abandoned military land operated in the past by the forces of the former Soviet Union.

In the Balkans, land has been recently contaminated by warfare, including as a result of North Atlantic Treaty Organisation (NATO) bombing during the Kosovo conflict of 1999. This left behind depleted uranium and released toxic chemicals, including mercury and dioxins, from bombed factories. However, it has often proved difficult to distinguish between pollution caused by the bombing and pre-conflict contamination. To make matters worse, large, mainly agricultural areas will remain unusable until the de-mining process has been completed.

Some national assessments have found the major sources of local soil contamination to be municipal waste landfills, industrial plants and handling losses at

Figure 7.1   Annual expenditure on contaminated site remediation by country

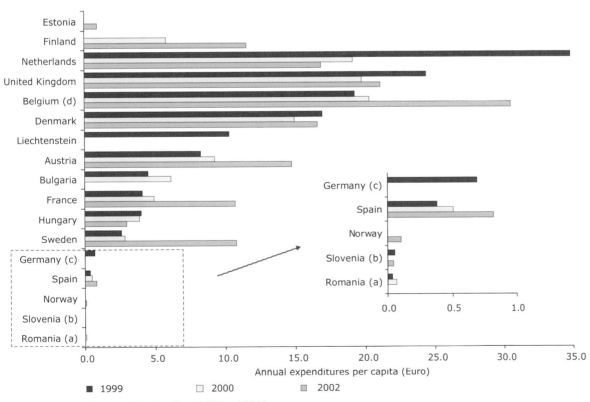

a) Romania: data from 1997 and 2000.
b) Slovenia: data from 1999 and 2001.
c) Germany: projection from estimates of expenditures from some of the 'Länder'.
d) Data for Belgium refer to Flanders.

Source: EEA, 2005.

current and former industrial facilities and distribution centres. Frequently the scale of contamination only emerges when old sites are zoned for redevelopment.

Recent EU legislation, based on the application of preventive measures, should avoid new contamination. Waste disposal is more strictly controlled, accidents and handling losses should be greatly reduced, and, in the event of mistakes, records and the chains of public accountability will be much clearer.

Nonetheless, there is a huge legacy of past contamination which is likely to extend its reach with time, as the flow of water through soils can spread the contamination both laterally beyond the site boundaries and vertically into groundwaters. Some of these contaminants are effectively permanent, though others — such as some organic pollutants and radioactive wastes — will degrade with the passage of time.

Remediation remains patchy and European targets are yet to be developed, although the majority of European countries have established national actions to deal with the problems (Table 7.1). Some have taken a proactive stance, mapping former industrial and waste sites and investing large amounts in cleaning up or containing leakage — often linked to policies of redeveloping 'brownfield' former industrial sites in preference to annexing farmland. National annual expenditure on remediation also varies, with national expenditures ranging from a low of EUR 2 to a high of EUR 35 per capita (Figure 7.1).

Most countries have now also put in place legislative instruments which apply the 'polluter pays' principle to cleaning up contamination. However, in many cases the polluters have long since disappeared, so in practice a considerable share of remediation is paid for with public money, on average about 25 % of total costs. Still, the amount of money spent on remediation is relatively small (8 %) compared with the estimated overall costs. New techniques for remediation, such as 'bioremediation' — in which micro-organisms are used to biodegrade organic compounds or hyper-accumulating plants are used to reduce the heavy metal content of soils — hold out the prospect of reducing costs. Nevertheless, the applicability of these techniques is expected to be limited and so the legacy of contaminated sites will remain dauntingly large for some time to come.

Diffuse sources

Diffuse pollution of soils, although probably not as critically widespread as local contamination, presents an even larger problem of accountability and clean-up. Nonetheless, few densely populated areas are without their contamination hot spots. In Lithuania, a country of 6.5 million hectares, almost half the land is contaminated by heavy metals.

Acidification

The most widespread form of diffuse contamination in Europe arises from acid deposition, especially in northern and central Europe (see Chapter 4). Some soils can neutralise the acidity, but many, particularly the thinner and naturally acid soils of northern Europe, cannot. Acid rain leaches vital soil ingredients such as calcium and magnesium from soils and can liberate toxic metals such as aluminium which can then build up elsewhere to toxic levels.

Overall, acid deposition has been reduced in recent years by more than 50 % across Europe. Although sulphur emissions may be much reduced, those of nitrogen remain high, not only increasing acidification in places but also adding to the ecological damage from 'over-fertilisation' of soils, often resulting in the eutrophication of water courses. Soil erosion and run-off from fertilisers often intensify this effect.

Critical loads for acidification and eutrophication are exceeded across the Benelux countries, the Czech Republic, Germany, Hungary, Poland and Slovakia, as well as in northern France, southern Scandinavia and parts of the United Kingdom. Acidified soils are often all but impossible to rehabilitate. Application of lime will reduce acidity, but the wider geochemical damage will remain. Natural recovery can take hundreds or even thousands of years. Hence, the decline of acid deposition will have only a limited impact in those areas already heavily affected.

Farmland

In parts of Europe such as Belgium, Denmark, the Netherlands and northern France, contamination from aerial spraying of farm chemicals such as pesticides is also a problem, particularly if they percolate through soils into groundwaters.

A study produced for the European Commission as part of the process to develop a thematic strategy on the sustainable use of pesticides identifies the current legal situation on aerial spraying in Europe as very heterogeneous, ranging from a total ban in some countries (e.g. Slovenia and Estonia) and a ban with few exceptions (e.g. Italy) to comparatively weak restrictions (e.g. Spain) and no regulation (e.g. Malta). The study proposes strict minimum requirements for the application of certain pesticides to reduce drift problems, which can affect the health of operators and

Map 7.2   Soil contamination by heavy metals

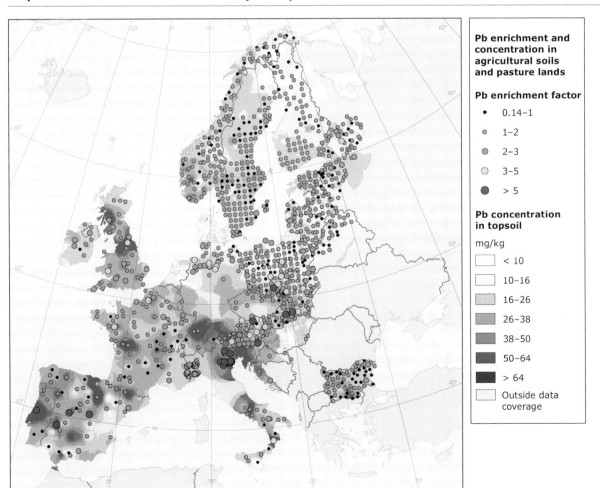

Note: Only randomly selected enrichment value dots shown for Austria, Bulgaria and Slovakia.
Source: Baltic Soil Survey (BSS), the Foregs Geochemical Baseline Mapping Programme and Eionet, 2003.

bystanders, and reduce water contamination, with no significant socio-economic impacts expected.

This proposed measure together with others such as mandatory checking of spraying equipment, integrated pest management and pesticide-free (or pesticide-reduced) zones on land such as Natura 2000 sites, could deliver a reduction of up to 16 % in pesticides use in the mid to long term with consequent reductions in the risks of environment and human health impacts. Farmers would also be expected to benefit economically from these measures with savings on pesticide use more than offsetting the additional costs of maintaining spraying equipment.

Heavy metals from industrial plants are sometimes applied to soils in sewage sludge taken from wastewater treatment works handling effluent from factories. The nutrients in this sludge can improve soil fertility in the short term where nutrients are in short supply, but the heavy metals may accumulate, potentially damaging long-term fertility (Map 7.2). The precise impacts will generally depend on the extent of heavy metal contamination of the sludge. This is limited by the EU sewage sludge directive which prohibits the use of untreated sludge on agricultural land. The directive also limits the rates and longevity of application of treated sludge in areas where fruit and vegetable crops are grown and where animals are grazing.

Less than 5 % of EU farmland is currently treated with sewage sludge, and most sludge contains only tiny amounts of heavy metals. However, the requirements of EU legislation such as the urban waste water treatment directive and the landfill directive, which limit other disposal options for sewage sludge, may tend to increase their application to land. Currently the heavy metal content of sewage sludge tends to be higher in southern Europe.

Other threats
In parts of the Balkans a new form of land contamination has emerged in recent years: landmines. By one estimate, a quarter of Bosnia's ploughed land is mined as a result of the recent conflict. Meanwhile, nuclear power stations, research facilities and weapons manufacturing plants have caused some contamination of European soils with radionuclides.

Most cases are highly localised and a result of spills. The major exception is the fallout from the Chernobyl disaster in 1986, which rained large quantities of radioactive isotopes on to parts of Belarus and the Ukraine. As a result, human habitation is still banned from a 30-kilometre-radius exclusion zone around the accident site because of extensive contamination of soils and ecosystems. It will be many decades before people can return.

Smaller amounts of fallout also fell in rain on Poland, north-east Scandinavia and the United Kingdom, where — 20 years later and more than 2 000 kilometres from the scene of the accident — livestock reared on some hillsides are still checked before sale for radioactivity picked up by the animals from grass grown on soils that remain contaminated.

7.4 Sealing

As soils are sealed off, compacted and deprived of air and water, so most biological activity ceases. There are no precise figures but, across the EU-15, as much as a fifth of the land is used for settlements, industry and infrastructure. In the Ruhr region of Germany that proportion rises to 80 %. Often it is the continent's best soils that are sealed: most population centres and infrastructure in Europe are built on fertile valley soils and around estuaries, typically taking the soils most productive for either agriculture or natural vegetation. Yet soil sealing by infrastructures and urban development is increasing at a faster rate than population, mostly at the expense of arable land and permanent crops, a clear indication of development which is not sustainable.

Between 1990 and 2000 about 50 000 hectares a year were used for housing services and recreation. Overall this represents around half the land area sealed across Europe. This rate of uptake of land for residential purposes varies from more than 70 % in Ireland and

Luxembourg to 16 % in Greece and 22 % in Poland, where urban development is mainly driven by the expansion of economic activities.

Sealing soils increases run-off by eliminating percolation of rainwater underground. It thus contributes to the widely recognised problem of increased storm run-off and flood risks, including mudflows and landslides. It also reduces recharge rates for groundwaters. Furthermore, by reducing the amount of time that moisture spends at the surface before being diverted into drains, soil sealing can also reduce evaporation, thereby influencing local climates.

Some countries have sought to limit the rate of sealing of soils by policies to redevelop existing abandoned sites such as old factories, so-called 'brownfield' development. This can, however, lead to increased localised problems within urban areas as the new developments often result in greater areas of sealed soil than the facilities or derelict land they replace.

Despite such initiatives, soil sealing continues. Typically this is due to changes in human lifestyles, such as suburbanisation and the development of tourist activity, rather than growing populations. Between 1990 and 2000, the built-up area of Europe expanded by about 12 %, while population increased by just 2 % (Figure 7.2). While not all urban land is sealed, it seems likely that more land is sealed for each European inhabitant than ever before. Delving deeper, we can see that the vast majority of this land uptake that results in soil sealing is for housing and recreation, with transport networks also making a contribution.

In Germany, for instance, an average of around an extra 100 hectares of land is converted to settlements and infrastructure each day. Settlements make up 80 % of this, and roads and other transport infrastructure most of the rest. While some of the land remains open — being converted from fields into suburban gardens or roadside verges — around half is permanently sealed. The German Government, mindful of this loss, has set a target to reduce the land lost to settlements and infrastructure to 30 hectares a day by 2020.

Rates of urbanisation have been greatest in recent times around the Mediterranean coast, including France, Italy, Spain and the islands, and on the French Atlantic coast. Often this is linked to expansion of tourism. High rates of future urbanisation are also expected in Finland, Ireland and Portugal.

Urbanisation and transport infrastructure are not the only causes of the sealing of soils. Others include reservoirs, which flood land, and even mechanised agriculture, which can so compact the soil surface that it becomes impermeable, effectively sealing off what lies beneath.

Recent research from Slovakia has highlighted compaction as the most widespread source of physical soil degradation in central and eastern Europe, affecting more than 60 million hectares. Most prevalent in areas where heavy machinery is used in agriculture

Figure 7.2   Built-up land and population trends

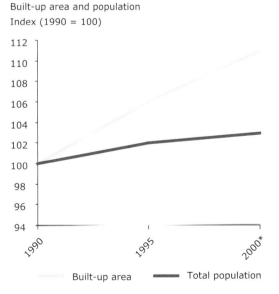

* Data for 2000 or latest available year

Source: EEA, 2004.

and forestry, compaction reduces the air-filled porosity and permeability of soils, increases its strength and partly destroys soil structure. The area affected by compaction is growing as wheel loads in agriculture continue to increase.

7.5 Salinisation

Salinisation of soils is another common diffuse contamination problem. It is caused by the accumulation of salts on or near the surface of the soil and may result in completely unproductive soils.

Evaporation of saline groundwater, groundwater extraction itself and industrial activities can contribute, but salinisation is most frequently the result of poor irrigation practice. Poor drainage and evaporation concentrate salts on irrigated land — even good quality irrigation water contains some dissolved salt and can leave behind tonnes of salt per hectare per year. Additionally, irrigation can raise groundwater to within a metre of the surface, bringing up more dissolved salts from the aquifer, subsoil and the root zone. Unless the salts are washed away below root level, soil salinity stunts growth and eventually kills off all but the most resistant plants. Salinisation has a strong impact on a range of physico-chemical properties of soil and, above certain thresholds, restoration is very expensive if not impossible to carry out. In extreme cases, salinisation becomes a perverse form of desertification brought about through the application of water.

Estimates of the extent and severity of salinisation are not easy to make due to the progressive nature of the process and the difficulty of detecting it in its early stages. However, as much as 16 million hectares or 25 % of irrigated cropland in the Mediterranean may be affected.

7.6 Summary and conclusions

Europe's soil is uniquely varied — more than 300 major soil types have been identified across the continent. Lost soil may eventually be replaced through natural processes of weathering rocks, a process that can take as little as 50 years to produce a few centimetres of new soil in areas with ample rain and organic inputs but thousands of years in mountainous areas such as the Alps. Soil is thus, on the timescales of normal environmental interest, a non-renewable resource.

There are many threats to soil — erosion, sealing, contamination, salinisation. These have proven difficult to tackle up to now and are expected to continue to be a challenge in line with expected future developments in Europe on urbanisation, intensive agriculture and industrialisation/deindustrialisation.

Countries have been taking more and more action, especially on the issue of contaminated sites. Many of the threats to soil, however, are interlinked through the main socio-economic developments (e.g. erosion, compaction, diffuse contamination and salinisation all result from agriculture), and so more integrated and coordinated actions in the future would deliver many positive effects, in a cost-effective way.

There are no overall estimates of the cost of soil erosion, contamination and sealing in Europe. One estimate puts the annual loss to farmers alone at EUR 53 per hectare from erosion, with another EUR 32 per hectare from off-site impacts of erosion, such as damage to infrastructure and siltation of reservoirs. That suggests a cost to non-Russian Europe of around EUR 15 billion a year.

These cost estimates are not insubstantial. In addition, the ecological services provided by soil are under further threat from climate change — desertification, extreme weather events — so the costs can be expected to get higher in the future. This could, in time, have implications for Europe's food security as recognised through the Global Monitoring of Environment and Security initiative established by the European Commission and Member States in 2003.

What is being done? Directives on nitrates, sewage sludge and others will help, as will recent reforms to the common agricultural policy that remove most subsidies from production and move them to other services, including protection of biodiversity and soils. Further, it is expected that the thematic strategy for soil protection and the soil framework directive will facilitate the coordination and implementation of existing but different policies related to soil.

Many data on soil have already been collected by the wide range of organisations which support the numerous different 'users' of soil. Nonetheless, there remain important gaps in the data, and access to them is difficult — few can be directly used for policy purposes and most of them cover small geographical areas.

Progress is being made towards closing these gaps and producing better information to support policy-making, for example by the collaboration on the development of the European data centre, led by the Joint Research Centre in collaboration with the EEA and its Eionet partners, and with the support of other parts of the European Commission. Recognition of the importance of a coherent framework for the monitoring and assessment of Europe's soils, and the streamlining of existing activities, is a significant step towards the success of the thematic strategy and framework directive.

References and further reading

The core set of indicators found in Part B of this report that are relevant to this chapter are: CSI 14, CSI 15, CSI 25 and CSI 26.

Introduction
Bellamy, P.H. et al., 2005. *Nature*, Volume 437, pp. 245–248.

EEA-UNEP, 2000. *Down to earth: Soil degradation and sustainable development in Europe. A challenge for the 21st century*. Environmental Issues Series No 6, EEA/United Nations Environment Programme, Luxembourg.

European Commission, 2001. *The sixth environment action programme*, COM(2001) 31 final, 2001/0029 (COD), Brussels.

European Commission, 2002. *Towards a strategy for soil protection*, COM(2002) 179 final. (See www.europa.eu.int/comm/environment/soil/index.htm — accessed 14/10/2005).

European Commission, 2004. *Final reports of the thematic working groups*. (See http://forum.europa.eu.int/Public/irc/env/soil/library — accessed 14/10/2005).

European Environment Agency, 1999. *Environment in the European Union at the turn of the century*, Environmental Assessment Report No 2, Office for Official Publications of the European Communities, Luxembourg.

Erosion
Doleschel, P. and Heissenhuber, A., 1991. *Externe Kosten der Bodenerosion*. Landw. Jahrbuch 68 Jahrg. — H 2/91.

European Commission, 2002. *Soil erosion risk in Europe*, European Commission Joint Research Centre, Brussels.

European Environment Agency, 2000. Final report on Task 6 of the Technical Annex for the 1999 subvention to the European Topic Centre on Soil (working document prepared by BGR), EEA, Copenhagen.

European Environment Agency, 2002. *Assessment and reporting on soil erosion*, Background and workshop report, Technical Report No 94, EEA, Copenhagen.

European Environment Agency, 2003. *Europe's environment: the third assessment*, Environmental Assessment Report No 10, EEA, Copenhagen.

European Environment Agency, 2003. *Europe's water: An indicator-based assessment*, Topic Report No 1/2003, EEA, Copenhagen.

García-Torres, L. et al., 2001. 'Conservation agriculture in Europe: Current status and perspectives'. In: *Conservation agriculture, a worldwide challenge*, I World Congress on Conservation Agriculture, Madrid, 1–5 October 2001, ECAF, FAO, Córdoba, Spain.

Gross, J., 2002. 'Wind erosion in Europe: Where and when?' In Warren, A. (ed.) *Wind erosion on agricultural land in Europe*, EUR 20370 EN, 13-28, Office for the Official Publications of the European Communities, Luxembourg.

Intergovernmental Panel on Climate Change, 2001. *Climate change 2001: impacts, adaptation, and vulnerability*, Summary for policymakers, A Report of Working Group II of the IPCC.

Neemann, W., Schäfer, W. and Kuntze, H., 1991. 'Bodenverluste durch winderosion in Norddeutschland — erste quantifizierungen' (Soil losses by wind erosion in north Germany — first quantifications), *Z.f. Kulturtechnik und Landentwicklung* 32, pp. 180–190.

Oldeman, L.R. *et al.*, 1991. GLASOD world map of the status of human-induced soil degradation, ISRIC, Wageningen and UNEP, Nairobi.

Van Lynden, G.W.J., 2000. *Soil degradation in central and eastern Europe: The assessment of the status of human-induced degradation*, FAO Report 2000/05, FAO and ISRIC.

Zdruli, P., Jones, R. and Montanarella, L., 2000. *Organic matter in the soils of southern Europe*, Expert Report prepared for DG ENV/E3 Brussels, mentioned in EEA-UNEP, European Commission Joint Research Centre, European Soil Bureau.

Contamination

European Commission, 2004. *Final reports of the thematic working groups*. (See http://forum.europa.eu.int/Public/irc/env/soil/library — accessed 14/10/2005).

European Commission, 2004. Assessing economic impacts of the specific measures to be part of the Thematic Strategy on the Sustainable Use of Pesticides. Executive Summary of the Final Report.

European Environment Agency, 2003. *Europe's environment: the third assessment*, Environmental Assessment Report No 10, EEA, Copenhagen.

European Environment Agency, 2005. No14 *Core set of indicators guide*, Technical Report 1/2005, EEA, Copenhagen.

Sol, V.M. *et al.*, 1999. *Toxic waste storage sites in EU countries*, A preliminary risk inventory R-99/04, WWF, Institute for Environmental Studies of the Vrije University, Amsterdam.

Van Lynden, G.W.J., 2000. *Soil degradation in central and eastern Europe: The assessment of the status of human-induced degradation*, FAO Report 2000/05, FAO and ISRIC.

Sealing

EEA-UNEP, 2000. *Down to earth: Soil degradation and sustainable development in Europe. A challenge for the 21st century*, Environmental Issues Series No 6, EEA, United Nation Environment Programme, Luxembourg.

European Environment Agency, 2004. EEA signals 2004, EEA, Copenhagen.

European Environment Agency, 2005. No 14 *Core set of indicators guide*, Technical Report 1/2005, EEA, Copenhagen.

European Environment Agency, 2005.: *Sustainable use and management of natural resources*, EEA, Copenhagen (in print).

Salinisation

EEA-UNEP, 2000. *Down to earth: Soil degradation and sustainable development in Europe. A challenge for the 21st century*, Environmental Issues Series No 6, EEA, United Nation Environment Programme, Luxembourg.

European Environment Agency, 2003. *Europe's environment: the third assessment*, Environmental Assessment Report No 10, EEA, Copenhagen.

FAO ,2000. *Global network on integrated soil management for sustainable use of salt-affected soils*. (See http://fao.org/ag/AGL/agll/spush — accessed 14/10/2005).

Soil   PART A

8 Biodiversity

8.1 Europe's biodiversity: the background

'Biological diversity' is defined by the United Nations Convention on Biological Diversity as the variability among living organisms from all sources including, among others, terrestrial, marine and other aquatic ecosystems and the ecological complexes of which they are part; this includes diversity within species, between species and of ecosystems (Article 2 of the United Nations Convention on Biological Diversity, 1992).

The countries of the European Union are home to a wide range of biomes (the basis for ecosystem services) that host around 1 000 species of vertebrate animals, some 10 000 plant species and maybe 100 000 different invertebrates, not including marine species. These are significant levels of species diversity, and yet, in comparison to many other parts of the world, the numbers are relatively small.

This is mostly a reflection of the geological history of Europe. Repeatedly over the past 2 million years, great ice sheets have spread across northern and central Europe, removing soil and vegetation and sanitising the land. Every time, life has had to start again, colonised from warmer areas to the south. The last of these glaciations only ended around 10 000 years ago.

While the glaciations stripped Europe of many of its species, the continent has nonetheless developed a variety of ecosystems. Extending from the Arctic Circle to the Mediterranean and from the Caucasus to the Canary Islands, it is home to permafrost and deserts, dry forests and alpine mountains, semi-tropical lagoons and Arctic fjords, steppe and peat bog. This variety in itself is an important resource and a buffer against climate change, geological disturbances and human disruption of the landscape.

There is a substantial variety of wildlife habitats in Europe. Some habitats harbour endemic species, that is, species that can be found nowhere else on Earth. Some mountain regions of southern Europe, in particular, as well as islands under the macaronesian bio-geographic region (Azores, Madeira and Canary Islands), are rich in endemic plants. Amid the natural conifer forests of the Baetic and sub-Baetic Mountains in southern Spain, for instance, there are more than 3 000 plant species — one of the richest troves in Europe. In parts of the mountains, 80 % of the plants are unique to the area. Almost as rich are the Gudar and Javalambre mountains near Valencia.

Other biodiversity rich spots with more than 1 000 plant species, many of them endemic, include the Pyrenees and the Alps. The highest number of plant and animal species in Europe is hosted in the Mediterranean basin, which has been identified by Conservation International as one of the world's 34 biodiversity hot spots. Particularly rich are the mountains of the Balkans and southern Greece, as well as 5 000 or so Mediterranean islands. These last include the Greek island of Crete, and Cyprus where the Troodos Mountains are particularly rich, with 62 unique species of plants. At a smaller scale, a large number of areas have been identified in Europe as of special importance for particular groups of species such as birds, butterflies and plants.

Most of Europe's land surface has been used for centuries to produce food and timber or provide space for living. Less than a fifth can be regarded as not directly managed at present. Much of that is under pressure.

The most habitat-relevant changes in land across the continent during the 1990s were the increases in artificial habitats (5 %) and in inland surface water (some 2.5 %), due to the creation of dams. Losses were identified for heath, scrub and tundra (some 2 %), and wetland mires, bogs and fens, which diminished by 3.5 %. Many of these wetlands have been lost to coastal development, mountain reservoirs and river engineering works. These changes have in some cases caused dramatic changes in landscape character and biodiversity richness.

Biodiversity

8.2 The changing countryside: intensive farmland and urban expansion

Europe is unique in global terms because the diversity of its species is to a large extent dependent upon landscapes created by human influence. More than on any other continent, Europe's biodiversity has been shaped by agriculture since the last glaciations. Remarkably few areas of even the highest conservation value are truly natural. A continuation of traditional methods of land management is essential to species survival in these areas.

Europe has some of the oldest and most enduring agricultural landscapes, from the woodlands and olive groves of the south to the reindeer pastures of Scandinavia. Areas defined by ecologists as 'semi-natural' farmland, forest and grassland habitats are home to many of Europe's most valued species.

Map 8.1 Share of targeted agricultural habitat types (dependent on extensive farming practices) within Natura 2000 sites

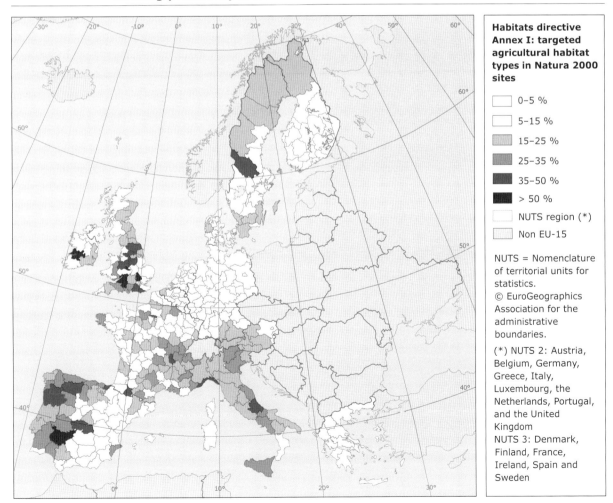

Source: EEA, 2004.

The largest semi-natural areas are in eastern and southern Europe. They include semi-natural pastures, steppe and dehesas (grasslands with scattered oak trees, typical in parts of Iberia), and mountain pastures. Many of these are under threat and have been given protected status. Central to this protection is the network of Natura 2000 areas designated for conservation under the EU birds and habitats directives. The network is designed to ensure the long-term conservation of the most typical and most threatened species and habitat types and now covers almost 18 % of the EU-15 and is being extended to the new Member States. According to current estimates, 17 % of the sites on the adopted lists are 'agri-ecological' landscapes that depend on the continuation of existing, usually extensive, farming practices to survive (Map 8.1).

The structure and functions of the European countryside are in many places also under threat from a range of developments. The urbanisation of Europe and the intensification of its agriculture, as well as forest management over the past half-century, have caused profound changes to the traditional agri-ecological landscapes and the species that live in them. Emerging new threats include the spread of transport networks and tourism infrastructure, the abandonment of agriculture and climate change.

Urbanisation remains a major threat to habitats across Europe. Suburban sprawl, highways, minerals extraction and industry are spreading across former rural areas. Some 800 000 hectares of Europe's land, an area more than three times the size of Luxembourg, was covered by concrete and asphalt during the 1990s, an increase in built-up area of 5 %.

One feature of this trend is that the traditional sharp divide between urban and rural areas is progressively breaking down. Urban areas are becoming less densely populated as people prefer living in semi-rural and suburban areas — an aspiration more easy to fulfil when households have one, two or even more private cars. The spread of transport infrastructure, as well as directly taking land, has also fragmented the natural and semi-natural areas through which it passes, disrupting migration routes and spreading air pollution and noise.

As suburban zones expand, they become greener, with parks and gardens and golf courses. Equally, in many rural areas, farming ceases to become the dominant economic activity as tourist accommodation, horse stabling, market gardening, theme parks and other land-hungry activities move in. Many farm labourers' houses are taken over by city dwellers as second homes. Even farming areas look very different, with large tracts of soil under glass and plastic.

Coastal areas are being subject to especially intense development, partly as a result of mass tourism. The coastal zones and islands of the Mediterranean, which are especially rich in species diversity, are under particular pressure. Urban sprawl is growing in all countries but most intensively in the Benelux countries, northern Italy, much of Germany, Portugal and Ireland, and around Paris and Madrid. In some cases this process has been stimulated by EU regional development policies.

This process is likely to continue as prosperity grows. The more prosperous EU countries have a larger built-up area per inhabitant than poorer countries. Demographic and social changes are generating a decrease in average household size. Unless development policies change, new EU Member States, which generally have less suburban sprawl, can be expected to develop similarly, consuming large areas of natural and agro-ecological landscapes as they do.

Meanwhile, planned extensions of the motorway network, especially in new Member States, will see more than 12 000 kilometres of new motorways built within the next decade.

In some European countries where intensification of farming is most pronounced ecologists are attaching an increased importance to conserving wildlife in urban areas. Even mammals such as foxes invade urban areas to take advantage of the abundant food available, much of it food thrown away by humans. Cities, especially those with old industrial areas, often provide a variety of unique wildlife habitats — some polluted and some simply abandoned — where unusual species of plants and insects congregate. Many such urban 'brownfield' sites contain more species than intensively managed farms in the countryside nearby.

Clearly the requirements of conservation are changing, and the maintenance of Europe's biodiversity will depend on action on a wide range of policy areas from agriculture and forestry through regional development, tourism and energy to land use and transportation.

The development of policies to ensure the preservation of ecosystems and habitats in Europe requires different approaches than in other parts of the world where nature is more pristine. In Europe, classical conservation methods, such as the creation of national parks, can protect only a fraction of the continent's biodiversity. Protecting Europe's species, habitats and ecosystems therefore requires broader support for the social and economic systems that developed and sustain them.

8.3 Major ecosystems across Europe

This section addresses major terrestrial and freshwater ecosystems; marine ecosystems are addressed in Chapter 6 and landscapes are assessed more comprehensively in Chapter 2.

European landscapes can be described in terms of the species and habitat types present in them. Their richness is essential when considering present and future ecosystem services, in particular in relation to potential adaptations to climate change. Maintaining the variety inherent in landscapes in terms of their health and connectivity is no longer a stand-alone target of nature conservation but a main challenge for society. Across Europe, landscapes differ, but most are under pressure and experiencing rapid changes which give cause for concern.

Farmland

Farmland, including arable land and permanent grassland, is one of the dominant land uses in Europe, covering more than 45 % (180 million hectares) of the EU-25. It has been estimated that 50 % of all species in Europe depend on agricultural habitats. Consequently, some of the most critical conservation issues today relate to changes from traditional to modern farming practices on habitats such as hay meadows, lowland wet grasslands, heathlands, chalk and dry grasslands, blanket bogs, moorlands and arable land.

The most significant pressures currently affecting farmland biodiversity are the loss and fragmentation of semi-natural habitats, the introduction of invasive species, the direct effects of pesticide or mechanical treatments and water consumption for irrigation, as well as the loss of crop varieties and livestock breeds.

There are two key trends leading to the loss and fragmentation of semi-natural habitats in agriculture in Europe today. One is the intensification of agriculture. The other is the abandonment of farmland. The latter happens when intensification is not possible or is uneconomic and when farmers and their families move out of farming. Both changes often cause a decline in biodiversity.

The intensification and mechanisation of agriculture is the most obvious threat. It results in numerous physical, chemical and biological changes to the landscape. Stone and earth terraces on steep hillsides are abandoned; hedgerows are degraded; small irregular fields with different crops are converted into large monoculture fields; pastures, ponds and other wet areas are drained; rivers are canalised and numerous small streams disappear; cattle are kept indoors while their pastures are turned over to growing fodder; crop rotations are lost; pastures are converted to arable land; farm woodlands, including coppiced and pollarded trees, are converted to agriculture.

At the same time, more intensive use of fertilisers, pesticides and water, coupled with the use of modern machinery, are all changing landscapes by reducing plant diversity and sometimes poisoning wildlife. Pesticides reduce the abundance of many insects and invertebrates and can poison the birds and mammals that feed on them. Nitrate fertilisers impact widely on soils and aquatic ecosystems. An example is provided within the experimental project Biodepth which covers a variety of grasslands across Europe: it shows that crop productivity, as expressed by hay yield, declines in line with reductions in plant diversity.

There are still areas of Europe, however, where soil and climatic constraints have meant that it has not been possible to intensify farming practices to the same extent as elsewhere. Such areas not only generally contain more of a patchwork of semi-natural and natural habitats but also the farmland is more varied and subject to a greater range of management intensities.

Although such high nature value (HNV) farmland occurs in association with traditional cropping systems in southern Europe, the majority of remaining HNV farmland is now largely associated with livestock grazing systems on semi-natural habitats in the mountains and other remote areas of this and other parts of Europe. These areas host habitats of relatively high biodiversity value (Figure 8.1). Approximately 15–25 % of the European countryside can be categorised as HNV farmland.

Due to the relatively small remaining area of undisturbed natural habitats, the so-called 'semi-natural farmland habitats' and in particular semi-natural grasslands, have become relatively more important for European biodiversity. Depending on biogeographic contexts or local situations, these habitat types often have higher levels of biodiversity than undisturbed areas, as is the case for vascular plants in semi-natural grasslands in Sweden.

Figure 8.1   General relationship between agricultural intensity and biodiversity

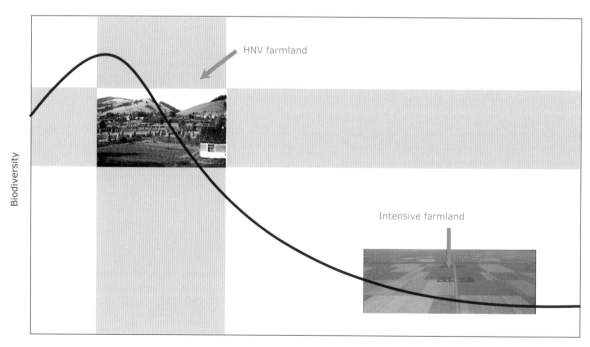

Source: After Hoogeveen *et al.*, 2001.
Photos: Peter Veen (left); Vincent Wigbels (right).

Often the cessation of farming is scarcely better for biodiversity than intensification. Farmers give up their land because the soils are poor, because it is too remote from markets or labour, or simply because it is surplus to requirements. Mountain regions have suffered especially from abandonment. Traditional transhumance pastoral systems are almost gone in many areas. Mediterranean areas at risk of drought and forest fires are also being abandoned on a large scale, as are parts of eastern and central Europe, where economic conditions no longer make farming viable. About 30 % of Estonia's farmland is currently out of production.

In places, other economic activities take over. Alpine shepherds and their flocks give way to skiers and hikers, for instance. Tourist resorts take over around the Mediterranean coast and islands. Often, however, the land is simply abandoned.

On the face of it, abandoning farmland to nature sounds good for biodiversity. But in practice this tends not to be true, or is double edged. In Latvia, where large farms growing cereals and sown grasslands were abandoned in the 1990s, bird species such as white storks and corncrakes increased their numbers on abandoned land, but grassland plants such as the marsh gentian and marsh dandelion that depend on grazing to create their ideal habitat have declined.

Abandonment often leaves behind a simplified and transient ecosystem, populated by fast-growing, opportunistic and invasive species. This results from the loss of land management practices that have boosted biodiversity, often for many hundreds of years. These practices include the mowing of meadows and the grazing of chalk grasslands, and the maintenance of micro-features such as walls, hedges and ponds.

So abandonment generally reduces the diverse patchwork of extensive agricultural habitats. Many species of plants and animals disappear. In Estonia it is the most biologically valuable farmland that is being lost. More than 50 % of permanent grasslands that are rich in plant species and need mowing or grazing to survive have been abandoned.

Intensification and abandonment can take place in the same region. Where abandonment dominates, the result can be a cycle of depopulation and further land abandonment, as young people leave in search of work. The situation is particularly worrying in central and eastern Europe, where economic changes in the past 15 years have already impoverished rural areas, and where privatisation of collective farms has reduced job opportunities.

The problem is likely to become even more severe in the coming years among the new EU nations, which currently have the largest share of extensively used farming areas. In future, economic restructuring may increase the magnet of urban areas as centres of economic regeneration. Economic pressures

---

Biodiversity and biotechnology

Developments in technology pose opportunities as well as challenges for biodiversity policy and the chances of achieving the 2010 targets. New biotechnology techniques have the potential to deliver improved food quality and environmental benefits through agronomically enhanced crops, leading to more sustainable agricultural practices in both the developed and developing worlds.

However, the development of biotechnology, and genetically modified organisms (GMOs) in particular, has also raised concerns about the possible impacts on human health and the environment, including biodiversity. The European Community is a signatory party to the Cartagena Protocol on Biosafety, which seeks to protect biological diversity from the potential risks posed by living modified organisms resulting from modern biotechnology.

The EU has been legislating on GMOs since the 1990s and has the toughest adoption procedures in the world. Only GMOs that have been positively assessed through strict authorisation procedures can be placed on the market in the European Union. Directive 2001/18/EC is concerned with the experimental release of GMOs into the environment, for example in connection with field tests, and the cultivation, import and transformation of GMOs in industrial products.

on the farming sector for either intensification or abandonment will probably become more intense.

The mid-term review of the EU common agricultural policy in 2003 placed environmental concerns at the heart of the debate. Consequently, from 2005 farmers are receiving a single farm payment based on their historic level of support, provided they undertake to comply with a suite of EU directives (including the birds and habitats directives) and keep their land in good agricultural and environmental condition.

Although a wide array of measures can be funded under the rural development heading, it is anticipated that this modification of the policy will release funds to encourage more farmers to join agri-environment schemes, thereby helping to strengthen the preservation of ecologically valuable agricultural land. However, much will depend on the total budget available for rural development, and on the manner in which Member States apply agri-environment and other instruments under the CAP.

Biological diversity is fundamental to agriculture and food production. A rich variety of cultivated plants and domesticated animals serves as the foundation for agricultural biodiversity. Yet people depend on a mere 14 mammal and bird species for 90 % of their food supply from animals. Just four species — wheat, maize, rice and potato — provide half of our energy from plants. When food producers focus on this limited range, however, less-commercial species, varieties and breeds may die out, along with their specialised traits.

A wide range of species dependent upon farmland habitats has been affected by the increasing intensification of farming practices, thus becoming threatened. For example, more than 400 species of vascular plants in Germany have declined because of habitat loss or fragmentation due to agricultural intensification, while in the United Kingdom there has been a greater decrease in recent decades in plant diversity in arable habitats than in any other habitat. Farmland invertebrates have also suffered, with total insect abundance, including moths, butterflies, sawflies,

Figure 8.2 Trends in EU farmland bird populations in some EU countries between 1980 and 2003 based on 24 characteristic bird species

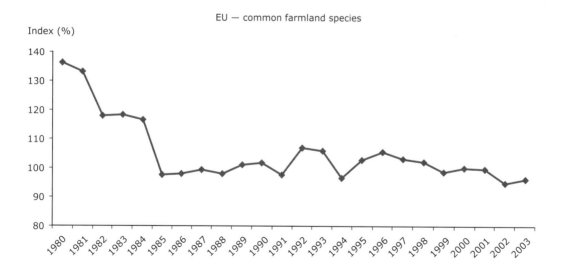

Source: EEA, 2005 based on data from BirdLife International.

parasitoid wasps and aphids, decreasing in both numbers and range.

Changes in the populations of individual farmland bird species have been particularly well documented (Figure 8.2). For example, the red-backed shrike (*Lanius collurio*) has shown a widespread decline in Europe. It is thought that the application of inorganic nitrogen fertiliser and the use of insecticides has reduced the abundance of food for this species.

The marsh fritillary butterfly (*Euphydryas aurinia*) is declining in almost every European country. The United Kingdom (and Ireland) are believed to be the major remaining strongholds for the species, but even here it has declined substantially over the last 150 years. The main factors contributing to the decline are agricultural improvement of marshy and chalk/limestone grasslands, afforestation and changes in livestock grazing practices.

Europe is home to a large proportion of the world's domestic livestock diversity, with more than 2 500 breeds registered in the United Nations Food and Agriculture Organization (FAO) breeds database. A large number of European breeds are threatened because of their perceived lack of economic competitiveness. In nearly all EU-15 countries about 50 % of all livestock breeds have been categorised as extinct, or of endangered or critical status (Figure 8.3).

Europe's HNV pastoral grazing systems depend on hardy old livestock breeds adapted to natural conditions and to practices such as transhumance. For

Figure 8.3 Distribution of the endangered risk status of national main livestock breeds (cattle, pig, sheep, goat and poultry) in EU-15

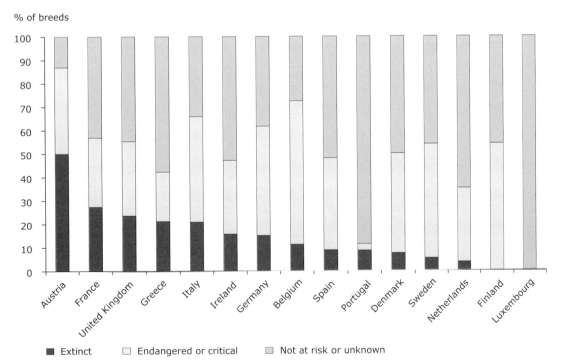

Source: EEA, 2005. Prepared by IRENA from data within FAO's Domestic Animal Diversity Information System.

example, Avileña negra cattle in central Spain can walk 20–40 kilometres a day on the journey to their summer mountain pastures. Modern breeds — which can produce a lot of milk and meat — need large quantities of rich grass and supplementary feeds, and cannot cope with the same conditions. This switch of breeds has therefore led to the abandonment of remote pastures in many areas, and the loss of biodiversity that depends on grazing impacts.

Forests

Despite Europe's high population density, roughly 30 % of the continent's land area is still covered by forest, which remains a key ecosystem for biodiversity. Most of these forests are semi-natural. During the 20th century concern at the sustainable supply of timber and pulp encouraged most governments to pass laws on enhancing the productive function of the forests.

Recent estimates show a slight overall increase in the extent of European forests, by about 0.5 % per year. Most of this has taken place on abandoned farmland, in equal measure through spontaneous regrowth and deliberate plantation — the latter often with funding support from the European Union. Afforestation has been highest in Ireland, Iceland and the Mediterranean countries, in particular Spain, France, Portugal, Turkey, Greece and Italy.

Most forests in Europe are, to some extent, economically productive and about 25 % of the forest area is subject to more or less extensive protection. These forests cover some 37 million hectares and are designated for the protection of biodiversity, soil or water supply. In the Natura 2000 network, forests currently cover almost half of the total number of designated areas.

The bulk of Europe's surviving 'natural' forests, unaffected by humans, are concentrated in a few, mainly northern, boreal regions. Scattered relicts of undisturbed forests also occur in the mountainous areas of the Balkan, Alpine and Carpathian regions. Natural forests often contain a diverse range of tree species, usually accompanied by a wide range of non-tree species. However, all forests, even monoculture plantations, are reservoirs of biodiversity.

Tree species composition is a key factor to consider when assessing the development of biodiversity conditions in forests. Unfortunately it is not possible to present European-level data on the long-term development of the overall tree species composition in the main types of European forest. Data reported by countries on forest-related vascular plants (including the trees) provide an insight into the situation of threatened species of this group in European countries (Map 8.2).

Agriculture and biodiversity management issues

The main policy instruments for site protection at EU level are the birds and habitats directives (79/409/EEC, 92/43/EEC). Annex I of the habitats directive lists 198 natural and semi-natural habitat types that must be maintained in a favourable conservation status. Of these, 65 have been shown to be threatened by the intensification of agriculture practices, whilst 26 grazed pasture habitats and 6 mown grassland habitats are threatened by the abandonment of pastoral management practices. The Natura 2000 network is building on special protection areas (SPAs) and proposed sites of Community interest (pSCIs) that will safeguard these habitats. Despite the importance of farmland across Europe for biodiversity, agricultural habitats only form about 35 % of the total area listed as pSCIs in the EU-15. Only Greece, Portugal and Spain have a higher proportion of such habitats within the pSCIs they have listed.

The current reform of the EU agricultural policy represents a radical change in the system of farm support provided within the EU, by decoupling support payments from production. The subsequent effects on farming practices and land use patterns are largely unknown. The likely impacts on farmland biodiversity are also currently unclear.

The increased use of agri-environment schemes in rural development measures is good in principle. However, the reforms to date have done little to address the question as to whether or not the programmes themselves have been effective in achieving their biodiversity objectives to protect biological features that have evolved as integral functional components of farming systems.

Biodiversity

In contrast to much of the rest of the world, forestry in Europe today extracts timber at a rate slower than or equal to the rate of regrowth. For the EEA member countries as a whole, average rates of felling are only two-thirds of the amount of regrowth. Afforestation can be achieved either naturally, by seeds from remaining or neighbouring trees, or by planting. Natural regeneration conserves the genetic diversity and, if the original stand is suitable, maintains the natural species composition of the forest. In practice, however, planting is often preferred because it creates homogeneous stands and can be tailored to need, often with the use of 'improved' genetic material.

In other respects forestry practice in Europe is developing in a way that can be considered good for biodiversity. For instance, with felling rates lower than growth rates, Europe's forests of all sorts are growing older. Bigger, older trees are typically of greater value for moss and other plants that grow on the trees themselves, and they may contain dead and hollow parts that are important for a number of plants, fungi, animals and insects. Nowadays, in many European countries, forestry practices aim at increasing the amount of dead wood in the forests.

Map 8.2   Total number of endangered vascular plant species and the share of endangered tree species and other endangered vascular plant species in forests

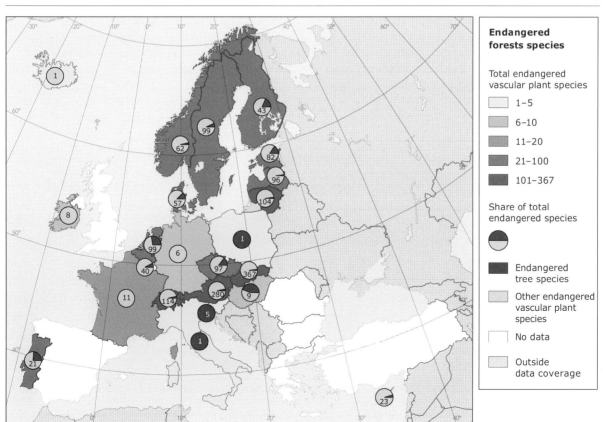

Source: UN-ECE/FAO, 2000 and updates.

Forest fires, especially in the Mediterranean region, pose a threat to the productive potential of forests and to surrounding land. At the same time, they are also a natural feature of most forests and a vital part of their dynamics, creating clearings and new habitats. From a biodiversity perspective, therefore, fire suppression may threaten species dependent on habitats formed by fire, especially in boreal forest. Moreover, fire suppression runs the risk of increasing the standing stock of timber ready to be burned in future, thus 'priming' the forest for a future, larger conflagration.

On the other hand, many fires are far from natural, since they are caused by people. They also cause significant economic, social and ecological losses. Thus forest fire management needs an integrated approach, taking account of ecosystem needs and long-term fire suppression strategies rather than simply operating short-term fire prevention regimes.

Freshwater ecosystems

Few of Europe's large freshwater systems are close to what can be considered their natural ecological state. Many have lost numerous species because of pollution and alterations to natural flow and flood regimes. Nevertheless, a marked improvement in the water quality of many rivers and lakes in recent decades has made water suitable again for the return of some lost species.

Pollution clean-up has contributed to this improved prospect and is discussed in Chapter 5. Better management practices, such as the construction of ponds and the provision of fish ladders through dams and weirs, have also contributed to this improvement. Still, much remains to be done to restore the quality of water, riverine habitats and biological communities in many areas. In addition, new threats are emerging. Climate change will change water temperatures, quantity and flow characteristics; while invasive non-native species represent a growing threat to freshwater biodiversity.

Europe has approximately 1.2 million kilometres of rivers. Most are, by global standards, small. Only about 70 of Europe's rivers have a catchment area exceeding 10 000 square kilometres. Along these rivers are around 600 000 lakes larger than 0.01 square kilometres, mostly in Finland and Sweden. As with rivers, there are many more small lakes than larger ones. Size matters: small lake and river water bodies are rich in biodiversity but often extremely sensitive to anthropogenic pressures, such as agricultural activities.

The EU water framework directive (WFD) is now the prime legislative instrument for the protection of the water environment of Europe. It covers all surface and groundwater bodies. One of its principal objectives is to achieve good chemical and biological water status

Regulation of the Danube — Europe's largest river

There have been major modifications to the Danube's flow since the 19th century, as communities along the river sought to control floods and improve navigation. This involved the construction of dykes along the river that reduced inundation of the floodplains. In Hungary in the Middle Danube, for instance, the area of floodplain that is seasonally inundated has fallen by 93 % from 22 000 square kilometres to 1 800 square kilometres.

Other changes have reduced river length, which has accelerated the passage of flood peaks. As a result, river flows have became more extreme, with higher floods and worse fluvial droughts. The straightening and dredging of the river bed has also increased channel erosion, deepening river beds, lowering water levels and breaking the river's contact with its backwaters. This in turn has led to falling water tables in surrounding aquifers and extensive siltation of surviving water bodies on the floodplain.

The annual inundation of the Danube floodplains has historically been a crucial event for maintaining the reproduction and productivity of fish populations, especially in the middle reaches. Dykes along the River Tisza, one of the largest tributaries of the Middle Danube, have caused an enormous loss of fish-spawning habitat, and a 99 % reduction in fish catches.

by 2015. The only exception should apply to water bodies designated by their governments as 'heavily modified', and where over-riding socio-economic reasons prevent the necessary improvements. The WFD is directly relevant to the management of Natura 2000 sites, for the conservation of those habitats and species dependent on water.

Most rivers in Europe have been subject to extensive damming for hydroelectric power, channelisation to facilitate transport and drainage of riparian habitats to provide agricultural land. Such modifications have led to widespread losses of aquatic habitats and biodiversity, with thousands of small lakes, ponds and streams lost entirely to drainage for agricultural land. Today, very few unregulated waters remain.

There is an increased awareness of the conservation importance of riverine and wetland habitats, and their role in buffering dry land against floods. In traditional farming systems, riverine and lake-shore habitats were often grazed or mowed, but allowed to flood. These areas offered valuable habitats for many rare species. Recreating and restoring these habitats is one of the greatest challenges for current and future actions on nature conservation.

The common otter, *Lutra lutra*, is found in European rivers, lakes and marshes, as well as in coastal waters. The species was once widespread, but the inland water populations in particular decreased dramatically during last century in countries such as France, although the otter still thrives in Ireland (Map 8.3). Destruction of habitat, pollution of watercourses and trapping have all contributed to its decline. There are now signs of recovery in, for instance, Denmark and the United Kingdom. Nevertheless, otters are still absent or sparse in many other countries, for example France.

Salmon, *Salmo salar*, is widely seen as an indicator of the health of rivers. Once widespread in northern and middle Europe, salmon requires good water quality and natural riffles and other features to support breeding and maintain stocks. Furthermore, the fish must be able to swim from the sea to upstream river spawning areas. Since the 1970s there has been a general decline in Europe's salmon.

There have been similar declines in other fish stock, such as eels and sturgeon, in many European rivers in response to dams, other river modifications and pollution. Many European countries have also seen declines in a wide range of species of freshwater plants,

Map 8.3 Otter populations in France in 1930 and 1990

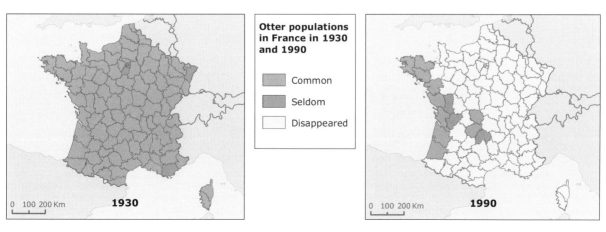

Source: www.cigogne-loutre.com/html/dispaloutre.html — accessed 13/10/2005.

animals and invertebrates such as mayflies, dragonflies, stoneflies and caddis flies, with hardy generalists and some new invasive species surviving, while local specialist species disappear.

Wetlands

Freshwater ecosystems are more than just rivers and lakes. Among the most biologically productive freshwater areas are wetlands, including lagoons, estuaries, riparian forests, grazed wet meadows and farm ponds. Although varying in size, often only seasonally wet and rarely focused on, wetlands are vital for a wide range of biodiversity.

Modifications to rivers, combined with intensified agriculture, urban development and changes to agricultural drainage and run-off and water abstraction have caused a massive decline in these ecosystems. In north and western Europe, for instance, 60 % of wetlands disappeared during the 20th century, and that decline continues. The EEA countries have seen a 3.5 % loss of large wetland areas since 1993; estimates of loss would rise to 10 % if changes in small wetlands were included. Traditional uses of wetlands are being abandoned throughout Europe.

This continued decline has happened at the same time as a growing concern to protect surviving wetlands, and serious efforts at conservation management. All EEA countries are parties to the Ramsar Convention on Wetlands, and have designated about 19 % of their total wetland area. According to national reports, there has been an overall negative change in ecological status of Ramsar sites (Figure 8.4).

Additionally, in the EU Member States, important wetlands are given strong protection through the birds and habitats directives. Other positive features include

Figure 8.4 **Ecological change status of Ramsar sites within the EEA member countries according to national reports to the Ramsar Convention**

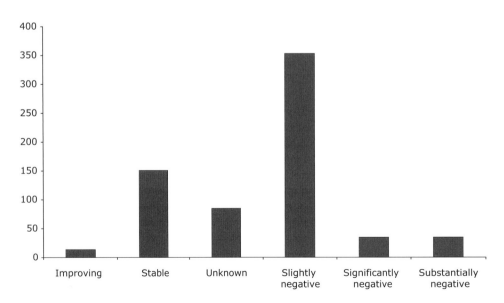

Note: There is no objective measure in place for countries to report changes in actual wetland area or ecological status. The data behind this figure are uncertain, for example Nivet and Frazier (2002, 2004) concluded that only 16 countries have adequate national wetland inventory information available.

Source: Ramsar Sites Database, 2004.

reformed agricultural policy, which now aims to avoid adverse effects on wetlands. There is also increased public and local community awareness of the value of wetlands, including their value for local traditions and culture. Wetland ecotourism activities are of growing significance.

In an independent report by the World Bank with WWF, Ramsar site designation was considered to be a significant factor in increasing conservation success. While conservation success was considered to be relatively high in the Ramsar Europe region generally over a period from 1993–1995 to 1999, eastern Europe reported a slight downturn.

The outlook for wetlands of international importance, such as those included in the Natura 2000 network, those with Ramsar designation and those with potential for ecotourism, appears to be quite good, at least in the medium term, thus contributing to the 2010 target of halting the loss of biodiversity in Europe. Nevertheless, for the majority of wetlands without protection or recognition, the outlook remains at best mixed.

Mountainous areas

Mountain environments in Europe are among the continent's most valuable natural areas, rich in biodiversity. They are also among the most vulnerable. European mountains host many endemic species, attracted by their isolation and special climate conditions. For example, more than 2 500 out of Europe's 11 500 vascular plant species are found mainly above the tree line.

Though often apparently stable, mountain regions are experiencing unprecedented change. Large-scale industrial projects, such as damming for hydroelectric power, mining and development of transport infrastructure, invade the mountains, often with drastic consequences for nature and biodiversity. Many European mountain areas are also important tourist areas, with an increased pressure especially from ski resorts. Meanwhile, the abandonment of farming and livestock grazing is influencing mountain vegetation as well as species diversity.

Despite the increasing pressure, some successful measures have been taken to bolster biodiversity in Europe's mountains. There has been widespread designation of mountain areas for Natura 2000 protection. A number of other EU programmes and directives recognise mountain areas in need of special attention, for example the common agricultural policy, the European Regional Development Fund, the directive on less favoured areas and the water framework directive.

The populations of several large herbivores have increased in the Alps in recent years, partly as a result of direct human actions such as reintroductions. The southern chamois, *Rupicapra pyrenaica*, nearly became extinct because of intensive hunting and poaching. In the past 40 years, regulation of hunting has led to an increase in numbers from a few thousand to 50 000 individuals in the Pyrenees, the Cantabrian Mountains and the Apennines.

Other large mammals are experiencing difficulties or are in decline. The wolverine, *Gulo gulo*, is the only large mammal predator in Europe that is naturally confined to mountains, where it lives on semi-domesticated reindeer. Long-term hunting and persecution has reduced wolverine populations and the total population in northern Europe is now less than 1 000 individuals, but apparently steady.

The brown bear, *Ursus arctos*, originally a widespread species in Europe, is today largely confined to mountains, and is now among the rarest large mammals in Europe. Like the wolf, it is seldom appreciated by local people, because it induces fear and attacks farm animals. The western European populations in the Pyrenees, the Cantabrian Mountains, the Trentino Alps and the Apennines are very small and fragmented. Nonetheless, the bear lives on in Finland and Sweden, where around 2 000 survive, in the Carpathian Mountains of Romania and Slovakia, and in the mountain ranges of the Balkan Peninsula, where a substantial number of bears can still be found.

The Pyrenean ibex, *Capra pyrenaica pyrenaica*, has for centuries been in decline due to hunting. The small

Spanish residual population recently faced new threats from habitat destruction, human disturbance, poaching and its own faltering genetic diversity. These led to a serious decline in numbers, and eventually to extinction when the last individual succumbed to a falling tree in 2000.

8.4 Invasive alien species

Invasive alien species are species introduced outside their natural habitats where they have the ability to out-compete native species. They are widespread in the world and found in all types of ecosystems: plants, insects and other animals comprise the most common types in terrestrial environments. Their threat to biodiversity is considered second only to habitat loss. Invasions are expected to increase because of the growing globalisation of trade, tourism and business travel.

Alien species also threaten our economic and societal well-being. Weeds reduce crop yields, increase control costs and decrease water supply, thereby degrading freshwater ecosystems. Pests destroy plants and increase control costs; and dangerous bugs continue to kill or disable millions of people every year.

Considerable uncertainties surround the economic costs of invasive species, but estimates of the impact of particular species on different sectors indicate the magnitude of the problem. The international trade in birds, for instance, in which the EU is an important actor, exposes populations to infectious diseases such as Asian bird influenza. The recent avian flu outbreaks in Belgium and the Netherlands resulted in 30 million poultry being killed, and cost industry and taxpayers hundreds of millions of euro.

The majority of non-native species in inland waters have been introduced accidentally, for aquaculture or for angling purposes. For many species the ecological effects are unknown but where the impact is known, the effects on the ecosystem have mainly been adverse, i.e. the species are invasive.

Despite decades of research, knowledge of the ecological and human dimensions of invasive species remains incomplete. Only some 20 % of the world's species have been scientifically described, so we are unable to predict either which species are likely to become invasive or the economic and social impacts they may have. This would suggest taking a precautionary approach to mitigate the occurrence of invasions through increasing globalisation of markets.

8.5 Climate change and biodiversity

Large uncertainties remain about the capacity of ecosystems to resist, accommodate or even sometimes benefit from climate change. Nevertheless, there is a strong probability that climate change will become the dominant force in changes to the continent's biodiversity, overwhelming the forces of habitat destruction, pollution and overharvesting, whether for good or ill.

Climate change will impact almost every aspect of Europe's biological life. Growing seasons and flowering times will alter; so will migration times and destinations. Species unable to move will decline or die out; others will take advantage of climatic space that opens up. Pests will change their domains. Carbon dioxide in the atmosphere will fertilise some plants, while drought will undermine others.

Often ecosystems are shaped less by average conditions and more by large natural disturbances such as fires, floods, high winds and droughts. Climatologists suggest that the probability and intensity of such extreme events may change even more than average conditions.

The one certainty is that a changing climate will put pressure on many species and ecosystems. It is thus of paramount importance to protect as much as possible of the natural landscape to improve the chances of a smooth transition to new climatic conditions. As climate zones shift, species will need to move. For some, this may be easy enough but for others it could be very hard. Species need habitats in which to live, and if the habitat as a whole cannot move, then the migrant may be left homeless.

Some regions of Europe have been identified as probably more vulnerable to climate change. In the Arctic, higher temperatures have already brought a greater variety of plants to Arctic lakes, and new niches may open up as permafrost thaws, glaciers retreat and temperatures warm. However, some endemic Arctic plants will be lost. Moreover, as sea ice conditions change, there will be threats to marine mammals, particularly polar bears which need sea ice from which to hunt in the cold Arctic waters.

Mountain species are able to cope with extreme conditions and may handle moderate warming quite well. Migrating up hillsides to keep pace with moving climate zones will involve much smaller distances than migrating on flatter lands. On the other hand, many plants in mountain areas occupy small niches with very localised climatic conditions; if these conditions change there may be nowhere suitable for them to grow.

The most extreme case will be near mountain summits. As temperature zones move up mountainsides, cold-loving species, having retreated to even higher altitudes, may find nowhere left to go. Plants, insects and mammals alike could be stranded. At the same time, other species, including trees, will be migrating from lower slopes, creating a botanical gridlock in which the delicate specialist endemic species will be most at risk. So, for instance, around the summits in the Alps, there could be a profusion of species but also the significant disappearance of local endemics.

One study suggests that a 1 °C warming in the Alps will result in the loss of 40 % of local endemic plants, while a 5 °C warming would produce a 97 % loss. Another study confirms the trend, suggesting a 90 % loss with a warming of 3 °C. Specific mountain plant species under threat include the mountain bladder fern (*Crysopteris montana*).

Coastal zones will suffer complex changes as rising sea waters invade freshwater ecosystems, storms become more intense, water quality changes in the warm

Projected impacts of climate change on European flora

Following earlier Euromove surveys, a study by the advanced terrestrial ecosystem analysis and modelling (Ateam) project of projected changes in the late 21st century distribution of 1 350 European plant species under seven climate change scenarios came up with the following conclusions:
- Even under the least severe scenario (mean European temperature increase of 2.7 °C), the risks to biodiversity appear to be considerable.
- More than half the species studied could be vulnerable or threatened by 2080.
- Different regions are expected to respond differently to climate change, with the greatest vulnerability in mountain regions (approximately 60 % species loss, including many endemic species) and the least in the southern Mediterranean and Pannonian regions.
- The boreal region is projected to lose few species, although gaining many others from immigration.
- The greatest changes, with both loss of species and a large turnover of species, are expected in the transition between the Mediterranean and Euro-Siberian regions.

The results of the study cannot be taken as precise forecasts given the uncertainties in climate change scenarios, the coarse spatial resolution of the analysis and uncertainties in the modelling techniques used. In particular the relatively crude grid scale of the study may hide potential refuges for species and environmental heterogeneity that could enhance species survival, especially in mountain areas where the risk of extinctions could be overestimated. On the other hand, impacts of land use change, which were not taken into consideration, could increase the vulnerability of these refuges to fire or other disturbances, which, in combination with the lack of propagule flow, could compromise the survival of remnant populations.

temperatures, and flows of sediments and freshwater down rivers change. Wetlands, already under grave threat from development, will suffer further damage from climate change.

Some Atlantic coastal wetlands may cope well with sea inundations because they are adapted to a wide tidal range. They have evolved protective features such as sand spits. Both the Mediterranean and Baltic Seas, however, are virtually tideless and thus have no coping strategies for inundation. Several predictions put the likely loss of coastal wetland habitat in these two seas, under a 2–3 °C warming, at greater than 50 %. Several large river deltas in the Mediterranean, such as those of the Ebro and Po rivers and the lagoons within them, are thought to be particularly at risk.

The Mediterranean region as a whole, while prone to coastal changes, is also likely to face more droughts and fires, land degradation due to desertification and spreading salinity in newly irrigated areas, and loss of wetlands.

Several studies have concluded that the Mediterranean is probably the part of Europe most vulnerable to climate change. Much of the region's biodiversity is already close to its climatic limit, and particularly vulnerable to the droughts that climate models suggest will become ever more frequent. Even small changes in temperatures and rainfall could have severe consequences for some tree species most typical of the Mediterranean landscape. In practice, increased fire risk may become the most serious threat. Fire is already

Figure 8.5   Special protection areas (SPAs) established under the EU birds directive (EU-25)

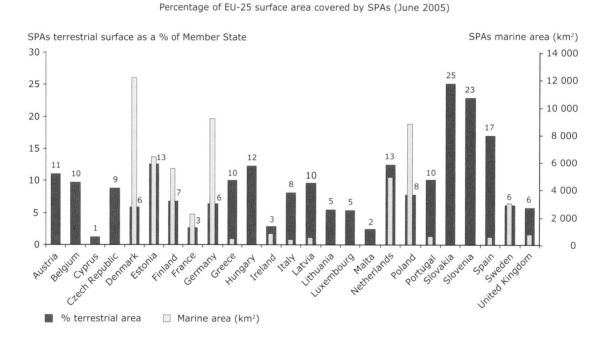

Note: Although there is no agreed percentage of terrestrial or marine areas that require SPA designation by individual Member States, it is clear that some countries need to conserve larger areas if the intended network is to be realised.

Source: EEA, 2005.

the crucial survival determinant for a number of tree and shrub species in the region as, each year, an area the size of Corsica is scorched.

8.6 Main biodiversity policy responses

European countries have long made commitments to protect their nature through joining international conventions, including the Ramsar Convention on the Conservation of Wetlands of International Importance (1971); the Helsinki Convention on the Baltic Sea (1974); the Barcelona Convention on the Mediterranean (1976); the Bonn Convention on Migratory Species (1979); the Bern Convention on European Wildlife and Natural Habitats (1979); and the Convention on the Protection of the Alps (1991). At the same time, the EU has been developing its own strategy for protecting its critical wildlife habitats, the wider landscape and the biosphere.

EU action began with protected area programmes under the 1979 birds directive and the 1992 habitats directive. In 1998, the Community adopted a biodiversity strategy, which was developed in accordance with the United Nations Convention on Biodiversity (CBD) signed at the Earth Summit in 1992. Under the strategy, a series of biodiversity action plans on natural resources, agriculture, fisheries, and development and economic cooperation followed in 2001. Additionally, commitments under the CBD have been carried forward into the EU sixth environment action programme and its thematic strategies, which

Figure 8.6 EU habitats directive: sufficiency of Member State proposals for designated sites (EU-15, September 2004)

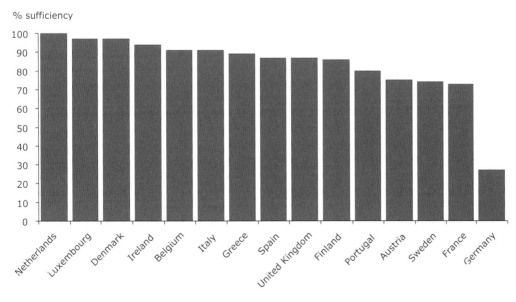

Note: As shown in the 'sufficiency' indicator, some countries need to strengthen their contribution to Natura 2000 under the habitats directive. Bars show the degree to which Member States have proposed sites that are considered sufficient to protect the habitats and species mentioned on the habitats directive Annex I and II (situation of September 2004). Marine species and habitats are not considered.

Source: Natura 2000 database.

cover such issues as the marine environment, soil protection, air pollution, the sustainable use of pesticides, and the urban environment, all of which touch upon biodiversity concerns.

At the centre of the EU biodiversity strategy is the creation of a coherent ecological network of protected areas, Natura 2000, made up of special protection areas (SPAs) to conserve 194 bird species and sub-species, as well as migratory birds, and special areas of conservation (SACs) to conserve the 273 habitat types, 200 animal species and 724 plant species listed under the habitats directive.

By February 2005, 4 169 SPAs, covering nearly 382 000 square kilometres, had been classified across the EU-25, of which 325 000 square kilometres are terrestrial (approximately 8 % of the Community's land area) and 56 000 square kilometres are marine (Figure 8.5).

The establishment of a list of sites of Community interest (SCIs), as a prelude to selecting special areas of conservation, has not been as rapid as initially hoped. Nonetheless, 19 516 sites covering nearly 523 000 square kilometres have been proposed as SCIs across the whole of the EU-25, covering almost 14 % of its land area as well as 65 000 square kilometres of marine area. These sites cover four of the six biogeographic regions identified by the habitats directive — Alpine, Atlantic, Continental and Macaronesian. The sufficiency of EU-15 Member State proposals for designated sites under the EU habitats directive is quite high with the exception of Germany (Figure 8.6).

Member States have six years following the adoption of the lists of SCIs to establish the measures necessary to protect and manage designated sites, and in doing so designate them as special areas of conservation.

The Natura 2000 network must be ecologically coherent both within individual Member States and between Member States and neighbouring countries in order to provide species and habitats the best possible chances of survival in the face of climate change.

The habitats directive also recognises the need to tackle conservation of species and habitats within and beyond designated protected areas, and to integrate management plans into broader landscapes and seascapes, contributing to the practical implementation of the 'ecosystem approach' promoted by the CBD.

Progress is being made towards the implementation of the Natura 2000 network. Almost 18 % of the EU's land area is protected, and a significant part is a net addition to the total area of nationally designated sites in Europe. Because many SPAs and SCIs overlap, the total protected area is less than the summed area of SPAs and SCIs.

Some findings of the 2003–2004 EU biodiversity policy review

At the World Summit on Sustainable Development in Johannesburg, South Africa, in 2002, nations agreed to significantly reduce the rate of loss of biodiversity in the EU by 2010. The EU had already gone further, committing itself to halting biodiversity decline by 2010. To plan its approach to meeting these ambitious targets, the EU began a review of its biodiversity strategy in 2003. Some of the findings are presented here.

Many species remain threatened in Europe: 43 % of European avifauna has an unfavourable conservation status; 12 % of the 576 butterfly species are very rare or declining seriously on the continent; up to 600 European plant species are considered extinct in the wild or critically rare; 45 % of reptiles and 52 % of freshwater fish are threatened. Some species such as the Iberian lynx, the slender-billed curlew and the Mediterranean monk seal are on the verge of extinction in the wild. Even once-common species such as skylarks have seen their populations drop dramatically in recent years.

These trends are not surprising given the generally low rate of implementation of both the strategy and the action plans in Member States, and the extent of natural habitat loss outside protected areas. The strategy itself underlined, however, that much of Europe's wildlife is to be found outside protected areas. Wider efforts are therefore needed to protect landscapes, especially traditional extensive farming systems, as suitable for wildlife.

More recently, and in response to the development of the strategic plan of the CBD, the countries of the EU endorsed the 'Malahide Message' in 2004. The message contains 18 concrete targets on how to work towards reaching the EU goal to stop the loss of biodiversity by 2010.

Meanwhile market forces are encouraging farmers to produce more organic crops. Although organic production may not necessarily lead to a reduction in intensity, it will mean fewer inputs, including no artificial pesticides and fertilisers. The reliance on animal manure and crop rotation to maintain soil fertility and combat pests and diseases reduces the risk of eutrophication of freshwaters and, by removing direct toxins, generally promotes more wildlife. In 2003 organic farming represented 4 % of the total farmed area in the EU-15, a doubling in just five years. In the 10 new Member States, where consumer demand and state help for organic farming are both lower, the proportion remains below 1 %.

Besides the organic movement, certification, also often market-led, is helping to promote both quality products and awareness of biodiversity issues. Two EU regulations associated with the origin and processing of agricultural and food products have played a role in this development.

Nonetheless it is recognised that additional efforts are needed, particularly to conserve HNV farmland and improve the biodiversity value of intensively managed farmland.

The EU forestry strategy, adopted in 1998, considers biodiversity as an element of sustainable forest management. Most European countries have made significant efforts to reduce the threats to, and enhance forest biological diversity within protected forest areas and through more environmentally sustainable and close-to-nature management practices in the countryside. This includes the increasing reintroduction in the last 10 years of native tree species in forest areas, the diversity of which had been affected by monospecific plantations of exotic species.

The development of certification initiatives, such as that of the Forest Stewardship Council, which define and encourage sustainable forestry regimes, is expected to have a positive effect. So, too, is the emergence of consumer-led demand through buyers' groups within the retail industry for sustainably produced wood and wood products, even though this is not directly targeted at the preservation of biodiversity.

However, action is still needed to mitigate threats to the forest ecosystems by long-range pollution and alien invasive species, to ensure the long-term survival of threatened species, and to establish an ecologically adapted fire regime. Additionally, consideration should be given to how forest management for carbon dioxide sequestration may affect biodiversity.

A range of general issues is in need of further consideration to help steer future action:

- the long-range damage to biodiversity from transboundary pollution such as acid rain and climate change;

- the failure to break the common perception that conservation and economic development are incompatible;

- the continued abandonment of traditional wildlife-friendly extensive farming methods; and

- gaps between theory and practice in Europe's management of forests and fisheries.

The broad objectives, set at the Community level, to protect nature and manage natural resources according to principles of sustainability could benefit from getting closer to local practice. In part this points to opportunities to improve the coherence of governance between different levels of administration in countries and at the EU level. The implementation of policies, strategies and directives has been relatively slow, with the Natura 2000 process already 15 years in development. Subsidies remain that encourage landowners to undermine ecological goods and services, though recent reforms to the common agricultural policy point the way forward. Nevertheless, the external costs to biodiversity have not yet been fully internalised in the sectors that have most impact.

The EU's biodiversity policy review culminated in a conference on 'Biodiversity and the EU' held under the Irish Presidency at Malahide in May 2004. The resulting 'Message from Malahide' achieved a broad degree of consensus on priorities towards meeting the 2010 targets. The Message contains 18 objectives with a set of targets relating to each. The Commission is now developing a new communication on biodiversity which will provide its response to Malahide. It is expected to provide a road-map of priority measures for the EU to 2010.

8.7 The global picture: how biodiversity underpins society

Healthy ecosystems deliver an abundance of life-sustaining services, often at no cost (Figure 8.7). Some we instantly recognise for their economic value. Ecosystems provide wild crops such as timber, fruit, nuts and medicinal herbs. In more heavily managed landscapes, soils and the microbial populations within them maintain a life-support system for arable crops, grazing animals and managed forests, from which modern societies gain most of their food, fibre and timber.

Other ecological services of biodiversity are more indirect and often less well recognised. Natural vegetation maintains insects that pollinate crops and control pests. Soils and vegetation store and filter water, watering crops, filling underground water reserves and protecting against floods. Evapotranspiration from vegetation and soils creates rain and cools the land, while gas exchanges between the atmosphere and vegetation maintain atmospheric chemistry. Among the services so provided is the moderation of climate change by trapping carbon dioxide that would otherwise remain in the atmosphere. Ecosystems also act as sinks for waste products, absorbing and oxidising them. They also contribute to landscapes that are valued for tourism and their cultural and psychological value.

Nature still provides direct genetic resources. A quarter of all modern medicines, though mostly synthesised, have their origins in traditional plant remedies. Drug companies are among the most assiduous corporate 'bio-prospectors' in rainforests and elsewhere, seeking out the active ingredients developed by nature and often already discovered and used by local communities.

Every lost forest is a risk to such enterprises. In 1987, a crucial chemical for fighting HIV was discovered in leaves and twigs sampled from a tree called *Calophyllum langierum*. Unfortunately, when scientists returned to find more material, they found that the original tree was gone and no more could be found. A similar gene has since been identified in a related tree, but it is not as active as the original. Meanwhile, the genetic variety present in wild precursors of major food crops remains a valuable resource for plant breeding to fight pests and increase yields. Most of these services are simply impossible for humanity to replicate. Therefore future well-being is dependent on the maintenance of the planet's ecological services, through protecting its biodiversity.

Biological and ecological systems are in a constant state of natural flux, so conservation need not be about preserving every habitat intact or keeping every endangered species untouched. Species are constantly becoming extinct — probably about one in every million each year.

However, conservation works best when it is about preserving those basic life-support systems on which we depend. What is worrying about the current situation is the scale of change triggered by human activity — a scale that undermines the ecosystems and the services they provide. Whether market instruments can be used to protect biodiversity and the ecosystem services that it underpins remains an open question.

Biodiversity

Figure 8.7   Ecosystem services and their links to human well-being

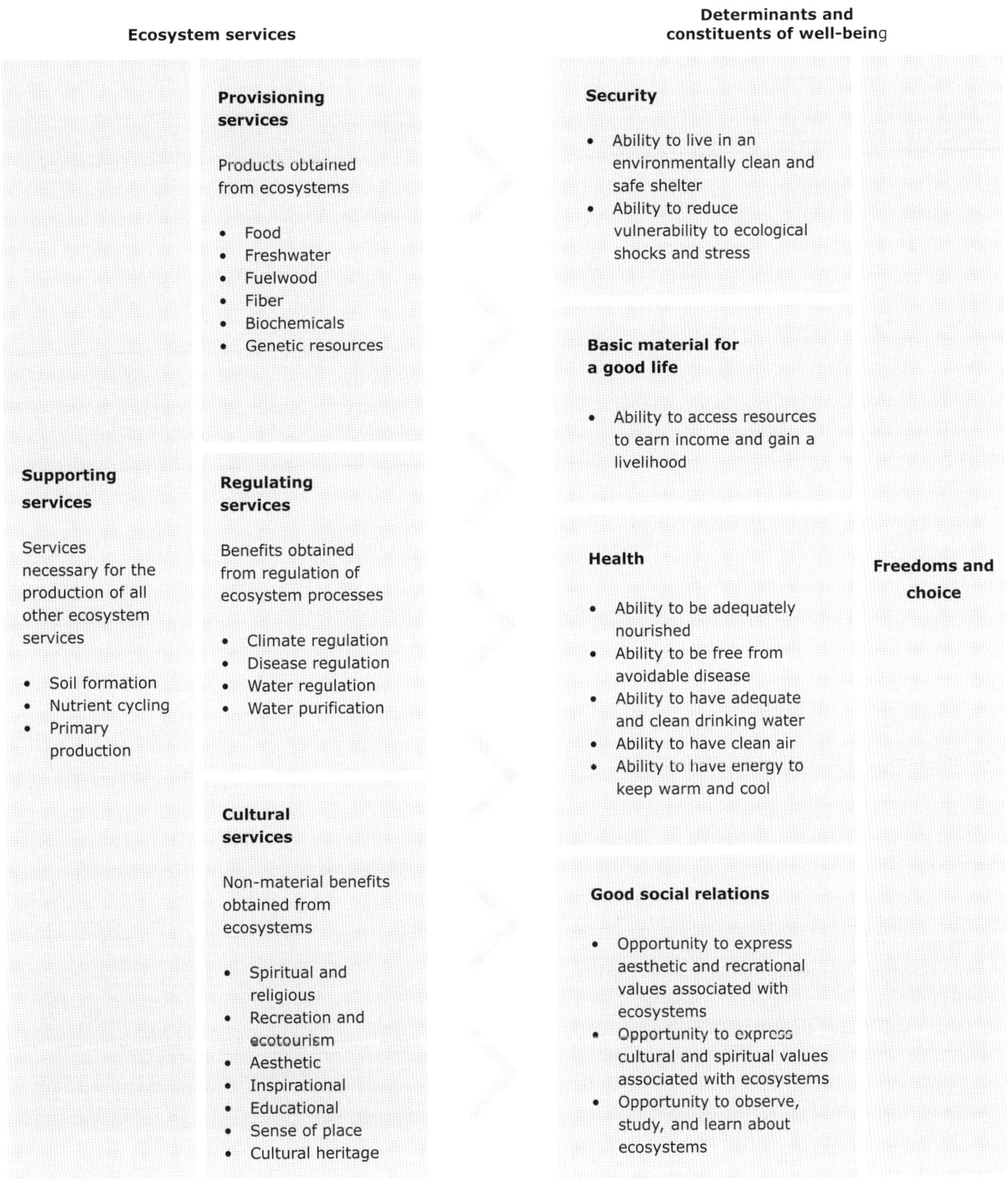

Ecosystem services

Determinants and constituents of well-being

Provisioning services

Products obtained from ecosystems

- Food
- Freshwater
- Fuelwood
- Fiber
- Biochemicals
- Genetic resources

Supporting services

Services necessary for the production of all other ecosystem services

- Soil formation
- Nutrient cycling
- Primary production

Regulating services

Benefits obtained from regulation of ecosystem processes

- Climate regulation
- Disease regulation
- Water regulation
- Water purification

Cultural services

Non-material benefits obtained from ecosystems

- Spiritual and religious
- Recreation and ecotourism
- Aesthetic
- Inspirational
- Educational
- Sense of place
- Cultural heritage

Security

- Ability to live in an environmentally clean and safe shelter
- Ability to reduce vulnerability to ecological shocks and stress

Basic material for a good life

- Ability to access resources to earn income and gain a livelihood

Health

- Ability to be adequately nourished
- Ability to be free from avoidable disease
- Ability to have adequate and clean drinking water
- Ability to have clean air
- Ability to have energy to keep warm and cool

Good social relations

- Opportunity to express aesthetic and recreational values associated with ecosystems
- Opportunity to express cultural and spiritual values associated with ecosystems
- Opportunity to observe, study, and learn about ecosystems

Freedoms and choice

Source: Millennium Ecosystem Assessment, 2005.

It may be that legal instruments will, as now, remain the main method of protection. What is clear is that many new instruments of all kinds are likely to be needed if the huge task of maintaining ecosystems and biodiversity is to be achieved.

The current rate of species extinctions is around a thousand times higher than the natural rate. Between 10 and 30 % of all mammal and bird species are currently threatened with extinction, and the geographical extent of human transformation of the planet's landscape is unprecedented. A study by the Wilderness Conservation Society defined areas of the Earth's land surface as being influenced by humans if:

- human population density was above 1 person per square kilometre;

- there was a road or major river within 15 kilometres;

- the land was used for agriculture or was within two kilometres of a settlement or railway; and

- it produced enough light to be visible to a space satellite at night.

By that measure, 83 % of the Earth's land surface is under human influence. The Millennium Ecosystem Assessment (MEA) have attempted to capture the extent to which we have degraded natural ecosystems and the price we are paying for it. It found that more land has been converted to agricultural use in the last 50 years than in the 18th and 19th centuries combined. More than half of all the synthetic nitrogen fertilisers ever used on the planet have been applied since 1985.

Overall the MEA concluded that 60 % of ecosystem services that support life on Earth — the services that purify and regulate water, deliver fisheries, regulate air quality, climate and pests — are being degraded or used unsustainably. As most of that damage has been done in the last 50 years, it may be too soon to be sure of the lasting impacts of our abuse.

It is far from clear that natural systems can cope with this without widespread collapse of those ecological services. Many of the systems and services are in evident decline — including ocean fisheries and freshwater supply, regulation of air quality and climate, protection against soil erosion and timber production. Meanwhile, ecosystem loss such as deforestation is causing epidemics of diseases such as malaria, a disease that came close to elimination 35 years ago but now kills three million people a year, mostly children. It may also be related to the spread from the natural world to humans of viruses such as Ebola and HIV.

Ecosystem damage is increasing human vulnerability to a range of natural disasters. Storms, tsunamis and high tides rip through coastal communities because mangroves and coral reefs have been destroyed. Floods engulf communities inland because deforestation has destabilised soils and reduced their ability to absorb heavy rains. Elsewhere the loss of forests allows wildfires to spread across the landscape.

Figure 8.8   Ecological overshoot 1961–2002

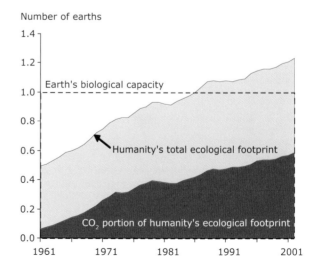

Source: Global Footprint Network, 2004.

Human influence does not necessarily lead to degradation. People can thrive in a landscape while maintaining its rich biodiversity. Nature can cope with a certain level of human pressure. Surviving agri-ecological landscapes, even in densely populated Europe, illustrate this.

Nevertheless, it is clear that the world is too populous for us to return to a relationship with nature based on hunter-gatherer economies, or even on traditional agricultural economies. However, developing technologies for living in very large numbers at a high standard does not mean we can forgo the natural resources on which all our wealth and health depends. We need to conserve and nurture the planet's ecosystems to ensure our own survival.

8.8 Tracking Europe's ecological footprint

Europe's impact on biodiversity extends far beyond its own shores. We use materials from across the globe to feed, clothe, house and transport ourselves. And our waste is spread around the world — on the winds and via ocean currents. Europe's high per capita consumption and waste production means that its impact on ecosystems is felt well beyond its own borders.

One attempt to capture that is the 'ecological footprint' — a measure of how much of the ecological capacity of the Earth we use up to grow our food and fibre, dispose of our waste, create room for our cities and infrastructure, and provide other ecological services such as sequestration of our carbon dioxide pollution. It has been developed by WWF, the global conservation organisation, and the Global Footprint Network, among others.

Figure 8.9   EU-25 and Switzerland — footprint versus population

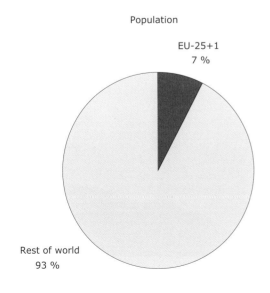
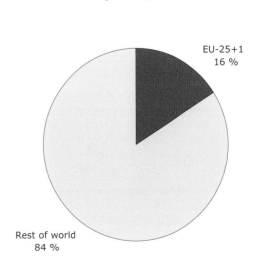

Source: Global Footprint Network, 2004.

Integrated assessment | Terrestrial environment

By these measures the global footprint of humanity was 2.5 times greater in 2002 than it was in 1961. We are now overusing the planet's resources by about 20 % (Figure 8.8).

The ecological footprint is normally measured in terms of hectares of land and productive ocean take-up to provide the goods and ecological services of a country's citizens. This can then be compared with the actual area available, the planet's biocapacity. According to these calculations, the planet's available biocapacity is between 1.5 and 2 hectares per person, though less than half the world lives at this level. North Americans require around 9 hectares to maintain their lifestyles, western Europeans 5 hectares, central and eastern Europeans 3.5 hectares and Latin Americans 3 hectares. The EU share of the world's footprint is more than twice its share of the global population (Figure 8.9).

Such calculations are inevitably crude, and not without controversy. Nonetheless, they can act as a warning about how we manage and share the planetary resources and ecological services on which we all depend.

Some countries, because they have low population densities, can reasonably claim that, while they consume more than their share of the planet's resources, they also contribute more. Not so for Europe, however. The continent is running up a large ecological deficit with the rest of the world. The difference between its footprint and its domestic biologically productive capacity is large and growing.

In 1961, the EU-25's global footprint was around 3 hectares per person, which was virtually the same as the continent's biocapacity. By 2001, Europe's global ecological footprint had risen to more than twice its internal biocapacity. Effectively it requires two continents with the size and fecundity of modern-day Europe to maintain the continent in the style to which it has become accustomed.

Europe achieves that by using its wealth to import the biocapacity of others. In effect, Europe exports many of its environmental problems, by buying products that are created through the depletion of natural capital elsewhere, including the poor, developing world.

Europe's footprint across the globe

So how has Europe's footprint grown, and what impact does it have on the rest of the planet? Europe's demand for fish is a potent case. Fish is the last wild source of animal protein available to Europe in and around its territory. Demand is increasing, while most of the fisheries of Europe are seriously overexploited. Despite growing production of fish from aquaculture, Europe has increasingly turned to foreign waters to maintain supplies. In 1990, the EU-15 imported some 6.8 million tonnes of fish products; by 2003, that had increased by almost 40 % to 9.4 million tonnes.

Analysing Europe's footprint

The list of the global 'top 20' countries with the biggest per capita ecological footprints is headed by the United Arab Emirates, USA, Kuwait and Australia. But European countries feature strongly. The European countries with the highest ecological footprint, as calculated by WWF, are Sweden and Finland, at around 7 hectares per person each. They occupy fifth and sixth places. Overall, European countries occupy more than half of the top 20 places.

Europe's footprint in other countries is created in part by its imports of a range of crops such as coffee, tea, bananas and other fruit, soy and palm oil, wood and fish. However, carbon dioxide emissions from burning fossil fuels are alone responsible for half Europe's total footprint.

Some countries have begun to decouple economic growth from their ecological footprint. One is Germany, which has not increased its footprint since about 1980 — even though it remains more than twice the country's biocapacity. Much of this has been achieved by reducing coal-burning, and reducing its footprint from both acid rain and carbon dioxide emissions. Poland's footprint fell dramatically after the collapse of the former Soviet Union, but it has not grown as its economy has recovered, probably as a result of the closure of much heavy industry. By contrast, the ecological footprints of France and Greece have continued to grow.

EU fleets work in the territorial waters of 26 foreign countries where the EU has negotiated access. Half of these are in Africa. While the deals are open and legal and contain clauses on sustainable harvesting, there are criticisms that, particularly in Africa, some EU fleets are depleting fish stocks and depriving local artisan fishers of their traditional catches.

Europe also imports large quantities of shrimp. Most shrimps in international trade are the products of aquaculture, so there is little direct loss to wild shrimp populations. However, particularly in Asia, shrimp farmers create their ponds by clearing coastal mangrove forests. The increase in shrimp farming over the past two decades has been a major cause of the destruction of around a quarter of the world surviving mangroves.

Mangroves are one of the most biodiverse tropical forest ecosystems. They provide other ecological services, too. The tsunami in Asia in 2004 showed how they protect against storms and tidal waves. Areas of India and neighbouring countries that had cleared their mangroves for shrimp farms generally suffered more from the tsunami than those that still had their mangroves, because the mangroves provided a buffer against the lethal tidal wave.

Timber is another critical natural resource widely exported to Europe, often from poor developing countries where the sustainability of the trade has been widely questioned.

While European countries produce enough timber to supply much of our needs for wood, paper and board, a large part of the remainder comes from tropical countries where illegal logging is often rampant, and ecologists warn of the ecological and social effects of deforestation. Half of Belgium's plywood imports come from the tropics, along with 30 % of French log imports, 50 % of Portuguese sawn wood imports and 30 % of UK veneer imports.

Forest resources are critical in most developing countries, both for national economies and for the subsistence lifestyles of inhabitants of the forests themselves. The World Bank estimates that more than a billion of the world's poorest inhabitants are in some measure dependent on forest resources for their livelihoods. Sustainably managed and harvested, the forests should benefit the people.

The volume of timber imported by the EU is less than by some other continents. Europe is responsible for about 4 % of world trade in timber, but the trade is concentrated in certain areas. European companies dominate the trade in timber from the countries of Central Africa, for instance, taking 64 % of timber exports from the region. Timber makes up a fifth of the EU's total trade with Central Africa. Within the EU, France is the largest importer, followed by Spain, Italy and Portugal.

It is often not easy to establish if imported timber comes from legal or illegal sources, especially when the supply chains are complex and the imported products have been processed along the way. In Asia, there are strong indications that large volumes of wood are harvested illegally in countries such as Cambodia, Indonesia and Myanmar, with some of this undoubtedly reaching Europe.

The World Bank estimates that around a half of all logging in Indonesia may be illegal. This means loggers are removing timber from someone else's land — often that of native forest inhabitants — or at an ecological or social cost unacceptable to the government. Among species threatened by this destruction are the last orang-utans of Borneo and Sumatra. Besides the environmental destruction and the loss of livelihoods for forest dwellers, the Bank calculates that the illegal trade results in a loss of revenue to the government of more than EUR 500 million a year.

Europe is also a major importer of vegetable oil products, especially soybean oil and meal and palm oil that are produced in the tropics on forest land cleared for the purpose. Soybean products come primarily from South America, and palm oil from South-East Asia.

Globally, the EU is the second biggest importer of soy products and, after efforts were stepped up to eliminate animal protein in animal feed, it has become the world's largest importer of soybean meal.

Europe's biggest source of soybean products is Brazil; in 2004 Europe imported almost half of Brazil's 19 million tonnes of exported soy products. This comes at a major ecological cost. Soybean is now probably the largest cause of the destruction of natural habitats in Brazil. Besides rainforests, large areas of dry savannah, known in Brazil as *cerrado*, are being cleared for soybean plantations. The *cerrado*, mostly in the Mato Grosso region of the country, receives much less protection than rainforests, but it is home to more than 4 000 endemic plant species as well as endangered animals such as the giant armadillo and the giant anteater. Seeing Brazil's success in selling to Europe, both Argentina and Paraguay have ambitious plans to expand soybean production in their own Chaco and Atlantic forests.

Palm oil exports to Europe come primarily from South-East Asia. Palm oil finds its way into a huge number of food products, from margarine and cooking oils to confectionery, ice cream, noodles and bakery products. The EU is one of the world's top importers, with 17 % of world trade. The two largest producers are Malaysia and Indonesia: combined, they have 85 % of global production. Expanded production, much of it to meet growing markets in Europe, is a main driver for forest clearance in both countries, as well as exacerbating social conflicts over the ownership of forest resources.

Europe's global ecological footprint also extends to water. While Europe does not directly import water, it does import large volumes of crops that have been grown using scarce irrigation water in other lands. Economists have characterised this as 'virtual water'. Three commodities — wheat, rice and soybean products — make up almost two-thirds of the world trade in virtual water.

The volumes of water involved are huge. It takes between 2 000 and 5 000 litres of water to grow 1 kilogram of rice and 7 500 litres to grow the 250 grams of cotton needed to make a single t-shirt. More and more countries come under water stress, and as the cost of providing water for irrigation grows, there is increasing discussion of how sustainable such trade in virtual water is.

European countries are among the world's largest importers of virtual water, with annual imports estimated at around 400 billion cubic metres. Typical imports of virtual water come in the form of tomatoes and oranges from Israel, cotton from Egypt and Australia and rice from South-East Asia. The Netherlands alone imports some 150 billion cubic metres of virtual water. Germany, Italy and Spain are also in the world top ten importers, with more than 60 billion cubic metres each.

The EU also has a large footprint in the live animal trade. The EU imports 92 % of all internationally traded wild birds, for instance. The leading importers are Italy, the Netherlands and Spain. Many of the birds are listed as endangered by the Convention on International Trade in Endangered Species of Wild Flora and Fauna (CITES). A study by non-governmental organisations found that over the past four years the EU imported three million birds listed under CITES. The trade could have been the route of introduction of Asian bird influenza to Europe in 2003.

8.9 Giving biodiversity a cash value

We live in a world where value is usually measured in cash terms. The problem for protecting biodiversity is that, however much we sense its worth or comprehend its importance in maintaining ecological services, that worth is hard to put a price on. Often, economic enterprises do not pay the cost of the damage they do to ecosystems. Equally, there is often no benefit or incentive for those who take the trouble to preserve those assets. The world economic system has yet to find a satisfactory way to internalise these losses of natural capital on which the system itself ultimately depends.

A new generation of economists is attempting to put a price on biodiversity and to evaluate the benefits of the services that ecosystems provide. They believe that the process of evaluation will help policy-makers appreciate the value of natural assets. It will allow society to assess better who wins and who loses when natural forests are logged out, wetlands drained and coral reefs wrecked, and to consider alternative economic strategies and whether they provide a better return in protecting ecosystem services. Ultimately the new economists hope that the value of ecosystem services can be routinely incorporated into mainstream market mechanisms.

Biodiversity can appear to many a rather abstract concept. So what specifically are the economists trying to value? There are four categories:

- Direct use values. These include things we harvest, such as timber, food and plant medicines, together with features of nature that we use without consuming, such as landscapes we visit.

- Indirect values. These are the ecological services provided by nature. Wetlands, for instance, purify water; forests maintain wildlife and capture and store carbon, so moderating climate change; mangroves protect coastlines from storms and tsunamis.

- Option values. These are both direct and indirect values that are not used now, but might be in future. Thus mangroves might be worth protecting because they will in future provide a barrier against a rise in sea levels. A forest may be kept because one day it might yield a cure for a disease.

- Existence values. These are largely cultural or spiritual. Europeans might see a value in a rainforest even if we never expect to use or visit it, or gain any services from it. We just like to know that it is there.

The first two of these values can, in theory at least, be measured. Directly used resources have a cash value in the market place. We can measure, for instance, the value of a harvest that would be lost if a rainforest were clear-felled. Indirect values can be indirectly measured, too, by assessing the cost of a replacement for the ecological service, whether purifying water, cooling the air or preventing floods.

Option and existence values may be no less important to society, but they are harder to assess. Conventionally, economists would 'discount' future value, thus giving little credence to option value, but is that acceptable when, through the United Nations, governments have agreed to the proposition that we should maintain the planet's ecosystems in a state fit for the use of future generations?

The trouble is that the cash value of a rainforest may be most readily realised by harvesting things of direct value with little regard to things of indirect value or to option or existence values, by clear-felling the forest for timber, for instance. If, however, these other values are included, it would be more economic to harvest the forest it a way that allows it to regenerate, and maintain the value of its asset for other uses. Similarly coral reefs might best be protected from destructive fishing, and mangroves from conversion to shrimp farms.

That is the theory; turning it into practice is harder. A private owner of land will generally only be able to 'harvest' the direct value of the resource. The indirect value has a wider constituency of beneficiaries who do not in legal terms have any ownership or control over the resource. Governments may have to intervene, either to establish economic instruments to allow the owner to benefit from the indirect value of the resource, or to enact laws on behalf of the wider community to prevent loss of those indirect values.

How market instruments can be used to protect biodiversity and the ecosystem services that it underpins remains an open question. It may be that legal instruments will, as now, remain the main method of protection. What is clear is that many new instruments of all kinds are likely to be needed if the huge task of maintaining ecosystems and biodiversity is to be achieved.

8.10 Summary and conclusions

Europe is home to around 1 000 species of animals, birds and fish, some 10 000 plant species and maybe 100 000 different invertebrates. This richness of European biodiversity and ecosystems is essential when considering present and future ecosystem services, in particular in relation to potential adaptations to climate change. Maintaining the variety of ecosystems in terms of their abundance, health and connectivity is not a stand-alone target of nature conservation but a main challenge for society. Across Europe, most large ecosystems exhibit worrying signs of rapid changes.

Most of Europe's land surface is in productive use — less than a fifth can be regarded as unproductive, and most of that is just formerly productive land that has, possibly temporarily, been abandoned. The largest losses of habitats and ecosystems for biodiversity across the continent during the 1990s were in heath, scrub and tundra, and wetland mires, bogs and fens. Many wetlands have been lost to coastal development, mountain reservoirs and river engineering works. Similarly, although more of Europe is covered by trees today than in the recent past, many forests are harvested more intensively than before.

These losses are impacting individual species. Although almost 18 % of the Community's land area is protected as part of the European strategy to conserve its critical wildlife habitats, many species remain threatened, including 42 % of native mammals, 15 % of birds, 45 % of butterflies, 30 % of amphibians, 45 % of reptiles and 52 % of freshwater fish.

Europe's high rates of consumption and waste production impact biodiversity far beyond its own borders and shores. We use materials from across the global to feed, clothe, house and transport ourselves. Our waste, too, is spread around the world — on the winds and via ocean currents. In 1961, the EU-25's global footprint was around three hectares per person, which was virtually the same as the continent's biocapacity. By 2001, Europe's footprint had risen to more than twice its internal biocapacity.

While uncertainties remain about the capacity of ecosystems to resist, accommodate or possibly even benefit from it, climate change will affect almost every aspect of Europe's biological life. Growing seasons and flowering times will alter; so will migration times and destinations. Species unable to move will decline or die out; others will take advantage of the climatic space that opens up. Pests will change their domains. Carbon dioxide in the atmosphere will fertilise some plants, while drought or floods will undermine others.

The European Union and its Member States have agreed an ambitious target to halt biodiversity loss by 2010, recognising the seriousness of the threat to the planet's ecological resources and our well-being. Progress, albeit slow, is being made on several fronts, and awareness is being raised among key stakeholders. This is despite the complexities surrounding biodiversity and our limited understanding of the interplay between genes, species, habitats, ecosystems, biomes and landscapes.

Conservation is not just about preserving special habitats and threatened species. It is about preserving those basic life-support systems on which life on Earth depends. Whether market instruments can be used to protect biodiversity and the ecosystem services or whether legal instruments will, as now, remain the main method of protection is an open question. What is clear is that much more effort is needed to implement to best effect the policy instruments already available for the benefit of biodiversity, and that new instruments of various kinds are likely to be needed if the huge task of maintaining ecosystems and biodiversity, on which our standards of living depend, is to be achieved.

References and further reading

The core set of indicators found in Part B of this report that are relevant to this chapter are: CSI 07, CSI 08, CSI 09, CSI 14, CSI 26 and CSI 34.

Europe's biodiversity: the background

American Museum of Natural History, 2005. The current mass extinction. (See www.well.com/user/davidu/extinction.html — accessed 13/10/2005).

Blondel, J., 2005. 'La biodiversité sur la flèche du temps', Presentation made at the first international conference on 'Biodiversity, science and governance', held in Paris on 24–28 January 2005. (See www.recherche.gouv.fr/biodiv2005paris/. — accessed 13/10/2005).

Mittermeier, R. et al., 2005. *Hot spots revisited: Earth's biologically richest and most endangered terrestrial ecoregions*, Conservation International, Washington.

Thomas, J.A., Telfer, M.G., Roy, D.B. et al., 2004. 'Comparative losses of British butterflies, birds, and plants and the global extinction crisis', *Science* 303, pp. 1879–1881.

The changing countryside: intensive farmland and urban expansion

European Environment Agency, 2005, CLC database (See http://dataservice.eea.eu.int/dataservice — accessed 13/10/2005).

European Environment Agency, 2004. *High nature value farmland-characteristics, trends and policy challenges*, EEA Report 1/2004, Copenhagen.

European Environment Agency, 2002. *Towards an Urban Atlas: assessment of spatial data on 25 European cities and urban areas*, EEA Issue Report 30, Copenhagen.

EuroGeoSurveys, 2004, European Landscapes for Living (See: www.gsi.ie — accessed 13/10/2005).

Major ecosystems across Europe

Andres, C. and Ojeda, F., 2002. 'Effects of afforestation with pines on woody plant diversity of Mediterranean heathlands in southern Spain', *Biodiversity and Conservation*, Vol. 11, No 9, September 2002, pp. 1511–1520, Springer Science+Business Media B.V., formerly Kluwer Academic Publishers B.V.

Birdlife, 2004. *Birds in Europe: Population estimates, trends and conservation status*, Birds Conservation Series No 12, Birdlife International. (See www.birdlife.org/action/science/indicators/pdfs/2005_pecbm_indicator_update.pdf — accessed 13/10/2005).

Bradshaw, R. and Emanuelsson, U., 2004. 'History of Europe's biodiversity', Background note in support of a report on 'Halting biodiversity loss', EEA, Copenhagen (unpublished).

Bruszik, A. and Moen, J., 2004. 'Mountain biodiversity', Background note in support of a report on 'Halting Biodiversity Loss', EEA, Copenhagen (unpublished).

Council of Europe, 2001. European rural heritage. *Naturopa*, Issue No 95, Strasbourg.

Council of Europe, 2002. Heritage and sustainable development. *Naturopa*, Issue No 97, Strasbourg.

Delanoe, O., de Montmollin, B. and Olivier L. (eds), 1996. *Conservation of Mediterranean island plants: Strategy for action*, 106 pp., IUCN Publications, Cambridge, the United Kingdom and Covelo CA, USA.

Diaci, J. (ed.), 1999. *Virgin forests and forest reserves in central and eastern European countries*, Proceedings of the invited lecturers' reports presented at the COST E4 Management Committee and Working Group meeting in Ljubljana, Slovenia 25–28 April 1998, University of Ljubljana. 171 pp. (includes country reports on Bosnia and Herzegovina, Croatia, Czech Republic, Poland, Romania, Slovenia and Switzerland).

Diaci, J. and Frank, G., 2001. 'Urwälder in den Alpen: Schützen und Beobachten, Lernen und Nachahmen', In: Internationale Alpenschutzkommission (ed.), *Alpenreport*, Vol. 2, Verlag Paul Haupt, Stuttgart, pp. 253–256.

Dufresne, M. et al., in print. *Vieux arbres et bois mort: des composantes essentielles de la biodiversité forestière*, Proceedings of the workshop on 'Gestion forestière et biodiversité' held in Gembloux (BE) on 23 March 2005, Faculté des sciences agronomiques de Gembloux, Plateforme biodiversité.

Edwards, M. et al., 2003. Fact sheet on phytoplankton, submitted to ETC/Air and Climate Change, EEA, Copenhagen.

European Bird Census Council, Royal Society for the Protection of Birds, BirdLife and Statistics Netherlands, 2005. *A biodiversity indicator for Europe: Wild bird indicator update 2005*.

European Environment Agency, 1998. *Europe's environment: The second assessment*, EEA, Copenhagen.

European Environment Agency, 1999. *Environment in the European Union at the turn of the century*, EEA, Copenhagen.

European Environment Agency, 2004. *Agriculture and the environment in the EU accession countries — Implications of applying the EU common agricultural policy*, Environmental Issue Report No 37, EEA, Copenhagen.

European Environment Agency, 2004. *High nature value farmland: Characteristics, trends and policy challenges*, EEA Report No 1/2004, Luxembourg, Office for Official Publications of the European Communities.

European Environment Agency, 2004. *Impacts of Europe's changing climate: An indicator-based assessment*, EEA Report No 2/2004, Luxembourg, Office for Official Publications of the European Communities

European Environment Agency, 2004. IRENA indicator fact sheet, IRENA 15: Intensification/extensification (See http://themes.eea.eu.int/IMS_IRENA/Topics/IRENA/indicators/IRENA15 %2C2004/index_html — accessed 13/10/2005).

European Environment Agency, 2004. *The state of biological diversity in the European Union*, Report prepared by the European Environment Agency for the Stakeholders' Conference 'Biodiversity and the EU — Sustaining life, sustaining livelihoods', held on 25–27 May 2004 in Malahide, Ireland.

European Topic Centre on Nature Protection and Biodiversity (ETC/NPB), 2002. *Identification of introduced freshwater fish established in Europe and assessment of their geographical origin, current distribution, motivation for their introduction and type of impacts produced*.

Eurostat, 2005. Fishery statistics (1990–2003). (See http://epp.eurostat.cec.eu.int/cache/ITY_OFFPUB/KS-DW-04-001/EN/KS-DW-04-001-EN.PDF — accessed 13/10/2005).

Food and Agriculture Organization of the United Nations, 2000. *World watch list for domestic animal diversity* (3rd edition), FAO, Rome.

Food and Agriculture Organization of the United Nations, 2001. Global forest resources assessment 2000 — Main report, FAO Forestry Paper No 140, FAO, Rome. (See www.fao.org/forestry/site/fra2000report/en — accessed 13/10/2005).

Food and Agriculture Organization of the United Nations, 2005. *The state of the world's forests 2005*.

Hallanaro, E.-L. and Pylvänäinen, M., 2002. *Nature in northern Europe — Biodiversity in a changing environment*, Nord 2001:13, Nordic Council of Ministers, Copenhagen.

Hoogeveen, Y.R., Petersen, J.E., Gabrielsen, P., 2001, *Agriculture and biodiversity in Europe*. Background report to the High-Level European Conference on Agriculture and Biodiversity, 5–7 June, Paris. STRA-CO/AGRI (2001) 17. Council of Europe/UNEP

IUFRO, INRA, 2005. Proceedings of the conference on 'Biodiversity and conservation biology in plantation forests', held in Bordeaux, France (in print).

Lazdinis, M. et al., 2005. 'Afforestation planning and biodiversity conservation: Predicting effects on habitat functionality in Lithuania', *Journal of Environmental Planning and Management*, Volume 48, Number 3/May 2005, pp. 331–348, Routledge, part of the Taylor & Francis Group.

Loreau, M., 2000. 'Loss of biodiversity decreases biomass production in European grasslands', *GCTE News*, 15, 3–4.

Ministerial Conference for Protection of Forests in Europe, 2003. MCPFE work programme, Pan-European follow-up of the Fourth Ministerial Conference on 'The protection of forests in Europe' 28–30 April 2003, Vienna, Austria, adopted at the MCPFE Expert Level Meeting 16–17 October 2003, Vienna, Austria.

Nivet, C. and Frazier, S., 2002. *A review of European wetland inventory information*, Wetlands International.

Nixon, S., Tren, Z., Marcuello, C. *et al.*, 2003. Topic Report 1/2003, EEA, Copenhagen.

RIVM, 2004. Environmental data compendium. (See www.rivm.nl/milieuennatuurcompendium/en/index.html — accessed 13/10/2005).

UNECE/FAO, 2000. *Forest resources of Europe, CIS, North America, Australia, Japan and New Zealand* (TBFRA 2000), Main report, UNECE/FAO contribution to the Global Forest Resources Assessment 2000, United Nations, New York and Geneva.

United Nations Economic Commission for Europe, 2003. *The condition of forests in Europe*, Executive Report 2003, Federal Research Centre for Forestry and Forest Products (BFH), UNECE, Hamburg.

United Nations Economic Commission for Europe, 2004. *The condition of forests in Europe*, Executive Report 2004, Federal Research Centre for Forestry and Forest Products (BFH). UNECE, Hamburg.

Van Swaay, C.A.M., 2004. *Analysis of trends in European butterflies*, Report VS2004.041, De Vlinderstichting, Wageningen.

Van Swaay, C.A.M and Warren, M.S., 1999. *Red Data Book of European butterflies (Rhopalocera)*, Nature and Environment, No 99, Council of Europe Publishing.

Invasive alien species

Nixon S., Kristensen P., Fribourg-Blanc, B. *et al.*, 2004. Pressures on freshwater biodiversity, Background note in support of a report on 'Halting biodiversity loss', EEA, Copenhagen (unpublished).

Zenetos, A., Todorova, V. and Alexandrov B., 2002. *Marine biodiversity changes in the Mediterranean and Black Sea regions*, Report to the European Environment Agency. (See www.iasonnet.gr/abstracts/zenetos.html — accessed 13/10/2005).

Climate change and biodiversity

Grabherr, G., 2003. 'Overview: Alpine vegetation dynamics and climate change — a synthesis of long term studies and observations', In: Nagy, L., Grabherr, G., Körner, C. and Thompson, D.B.A. (eds), Alpine biodiversity in Europe, *Ecological Studies* 167, pp. 399–409.

Lehner, B., Henrichs, T., Döll, P. and Alcamo, J., 2001. *EuroWasser: Modelbased assessment of European water resources and hydrology in the face of global change*, Kassel World Water Series No 5, Centre for Environmental Systems Research, University of Kassel.

Theurillat, J.P. and Guisan, A., 2001. Potential impact of climate change on vegetation in the European Alps: A review. *Climatic Change* 50, pp. 77–109.

Thomas, C.D., Cameron, A., Green, R.E. *et al.*, 2004. Extinction risk from climate change, *Nature* 427, pp. 145–148.

Thuiller, W., Lavorel, S., Araújo, M.B. *et al.*, 2005. *Climate change threats to plant diversity in Europe*, Proceedings of the National Academy of Sciences of the United States of America, June 7, 2005 , Vol. 102, No 23, pp. 8245–8250.

Main biodiversity policy responses

Bennett, H., 2005. *Cross-compliance in the CAP: Conclusions of a Pan-European project 2002–2005*, IEEP, London.

Buord, S., Lesouef, J.-Y. and Richard, D., in print. 'Consolidating knowledge on plant species in need of urgent attention at European level', In: *Proceedings of the 4th Planta Europa Conference held in Valencia, Spain , 17–20 September 2004*.

Davis, S., Heywood, V.H. and Hamilton, A.C. (eds), 1994–1997. *Centres of plant diversity* (three vols), World Wide Fund for Nature and International Union for Conservation of Nature and Natural Resources, Gland, Switzerland.

De Heer, M., Kapos, V., Ten Brink, B.J.E., 2005. Biodiversity trends in Europe: Development and testing of a species trend indicator for evaluating progress towards the 2010 target, *Phil. Trans. R. Soc. Lond. B.* (in print).

European Commission, 2001. *Environment 2010: Our future, our choice* — Sixth Environment Action Programme, 2001, COM(2001)31; OJ L242.

European Commission, 2005. Communication from the Commission to the Council and the European Parliament — reporting on the implementation of the EU forestry strategy, COM(2005) 84 final. (See www.europa.eu.int/comm/agriculture/publi/reports/forestry/com84_en.pdf — accessed 13/10/2005).

European Platform for Biodiversity Research Strategy, 1999–2005. (See www.epbrs.org/epbrs_library.html — accessed 13/10/2005).

European Topic Centre on Biological Diversity (ETC/BD), 2005. EUNIS database on species. (See http://eunis.eea.eu.int/ — accessed 13/10/2005).

IUCN, 2004. Resolutions made at the Third World Conservation Congress. (See www.iucn.org/congress/members/submitted_motions.htm — accessed 3/2005).

IUCN, 2004. *The 2004 IUCN Red List of threatened species*. (See www.redlist.org — accessed 13/10/2005).

The global picture: how biodiversity underpins society

Brashares, J., Arcese, P., Sam, M. *et al.*, 2004. 'Bushmeat hunting, wildlife declines, and fish supply in West Africa', *Science* 306, p. 1180.

Chivian, E. (ed.), 2002. *Biodiversity: Its importance to human health*, Interim Executive Summary, Center for Health and the Global Environment, Harvard Medical School. (See www.med.harvard.edu/chge/Biodiversity.screen.pdf — accessed 13/10/2005).

Pisupati, B. and Warner, E., 2003. *Biodiversity and the Millennium Development Goals*, IUCN, Regional Biodiversity Programme Asia, Sri Lanka.

Reid, W. *et al.*, 2005. Millennium Ecosystem Assessment synthesis report, pre-publication final draft approved by MA Board on March 23, 2005.

Starke, L. (ed.), 2004. *The state of the world 2004*, Special focus: The consumer society, Worldwatch Institute. (See www.worldwatch.org — accessed 13/10/2005).

Ten Brink, P., Monkhouse, C. and Richartz, S., 2002. Promoting the socio-economic benefits of Natura 2000, Background report for European Conference on 'Promoting the socio-economic benefits of Natura 2000', Brussels 28–29 November 2002, IEEP. (See www.ieep.org.uk — accessed 13/10/2005).

Tilman, D., 2005. 'Biodiversity and ecosystem services: Does biodiversity loss matter?' Presentation made at the first international conference on 'Biodiversity, science and governance', held in Paris on 24–28 January 2005. (See www.recherche.gouv.fr/biodiv2005paris/ — accessed 13/10/2005).

UNECE/FAO, 2000. *Forest resources of Europe, CIS, North America, Australia, Japan and New Zealand* (TBFRA 2000), Main report, UNECE/FAO contribution to the Global Forest Resources Assessment 2000, United Nations, New York and Geneva.

UN/World Bank, 2005. *Millennium Ecosystem Assessment*.

World Bank, 2004. Sustaining forests — a development strategy. (See http://lnweb18.worldbank.org/ESSD/ardext.nsf/14ByDocName/ForestsStrategyandOperationalPolicyForestsStrategy — accessed 13/10/2005).

World Health Organization, 2003. Fact Sheet No 134: Traditional medicine. (See www.who.int/mediacentre/factsheets/fs134/en/ — accessed 13/10/2005).

WWF India, 2004. Tsunami's aftermath: On Asia's coasts, progress destroys natural defences. (See http://wwfindia.org/tsunami1.php — accessed 13/10/2005).

Tracking Europe's ecological footprint

Brown, J. and Ahmed, 2004. *Sustainable EU fisheries — facing the environmental challenges, Consumption and trade of fish*. IEEP, London.

FAO, 2005. *The state of world fisheries and aquaculture*, FAO, Rome.

Halwell, B., 2002. Home grown: The case for local food in a global market, *Worldwatch Paper* 163.

Hoekstra, A.Y., Hung, P.Q., 2004. *Virtual water trade — A quantification of virtual water flows between nations in relation to international crop trade*. IEEP, London.

IIED, 2002. Drawers of water II. (See www.iied.org/sarl/dow/pdf/uganda.pdf — accessed 13/10/2005).

ITTO, 2003. *Annual review and assessment of the world timber situation 2003*, International Tropical Timber Organization.

Pauly, D., Christensen, V., Dalsgaard, J. *et al.*, 1998. Fishing down marine food webs, *Science* 279, pp. 860–863.

Picard, O. *et al.*, 2001. *Evaluation of the Community aid scheme for forestry measures in agriculture of Regulation No 2080/92*, Final Report, Institut pour le Développement Forestier, Auzeville, France.

UNEP/Grid Arendal, 2004. Poverty-biodiversity mapping applications, Discussion paper prepared for the IUCN World Congress, November 2004. (See www.povertymap.net/publications/doc/iucn_2004/stunting.cfm — accessed 13/10/2005).

USDA, 2005. *Brazil oilseeds and products soybean update 2005*, GAIN Report Number BR5604. (See www.fas.usda.gov/gainfiles/200502/146118775.pdf — accessed 13/10/2005).

USDA, 2005. Oilseeds: World markets and trade. (See www.fas.usda.gov/oilseeds/circular/2005/05-03/toc.htm — accessed 13/10/2005).

WWF, 2004. *Living planet report 2004*. (See www.panda.org/downloads/general/lpr2004.pdf — accessed 13/10/2005).

Giving biodiversity a cash value

Scottish Parliament, 2002. SPICe Briefing: Rural tourism, 21 August 2002. (See www.scottish.parliament.uk/whats_happening/research/pdf_res_brief/sb02-92.pdf — accessed 13/10/2005).

Seafood choices alliance. (See www.seafoodchoices.org/ — accessed 13/10/2005).

World Bank, IUCN and The Nature Conservancy, 2004. *How much is an ecosystem worth? Assessing the economic value of conservation*, International Bank for Reconstruction and Development/World Bank, Washington.

9 Environment and economic sectors

9.1 Introduction

The economy depends on the environment. The natural environment provides invaluable ecological services, including forests that moderate local climate, wetlands that absorb floods, and soils that purify water and buffer pollution. It also provides sources of materials, water, medicines and energy, as well as sinks for our wastes and pollution, recycling toxic materials into benign and sometimes useful forms. Finally, it offers space for people's homes and their leisure pursuits, and room for other species. Particularly in the developed world, economic prosperity is needed to deliver effective environmental management.

The assessment of realistic values for ecological services — values that reflect their true place within modern economies — is still in its infancy. This is, perhaps, one reason why we are still eroding the planet's natural resources faster than may be viable. As the World Business Council for Sustainable Development said in the Millennium Ecosystem Assessment: 'Business cannot function if ecosystems and the services they deliver — like water, biodiversity, fibre, food and climate — are degraded or out of balance…'.

It took the whole of human history to the year 1900 for the world economy to grow to a gross domestic product (GDP) of EUR 1.7 trillion (USD 2 trillion) at 1990 prices. Fifty years later this had grown to EUR 4.1 trillion (USD 5 trillion) and by 2001 to EUR 31 trillion (USD 37 trillion), more than seven times the 1950 amount. It is the speed and scale of this economic development that threatens the integrity of the ecological services which underpin economic activity. It is now generally accepted that there are physical limits to continuing economic growth based on resource use (Figure 9.1).

The current rates of change in economic growth and population make it more difficult than before for ecosystems and their associated services to adapt.

Figure 9.1 World economic growth 1900–2001 and links to the use of environmental services

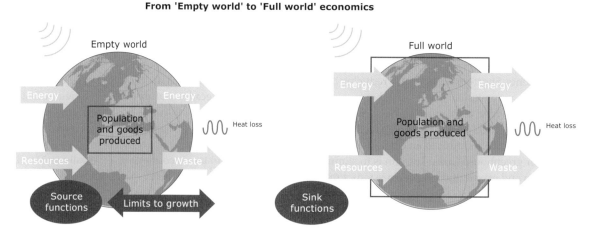

1900: USD 2 trillion by EUR 1.7 trillion.   2001: USD 37 trillion by EUR 31 trillion.

Source: EEA based on data from OECD.

Together with rapidly rising personal consumption patterns, demographic changes and economic transformation, the increasing economic use of environmental assets allows relatively little time for ecological adaptation. Worryingly, analyses of trends suggest that we should expect the intensity of use of ecological services to increase in the future.

9.2 The changing state of Europe's environment

The overall picture of the state of Europe's environment remains complex. On the plus side, there have been substantial reductions in emissions of substances that deplete the ozone layer, reductions in air emissions that cause acidification and air pollution, and cleaner water as a result of reductions in point source pollution. Protection of biodiversity, through the designation and protection of habitats, has provided some improvements in maintaining ecosystem productivity and landscape amenity. Such progress, overall, has been brought about mainly through 'traditional' measures such as regulating products and production processes, and protecting important nature sites. These policy areas are supported through well-established European Union legislation and in many cases are also directly or indirectly framed within international conventions.

Trends in other environmental pressures such as greenhouse gases and waste generation have been upward, in line with broader socio-economic developments. Short-term targets for reducing greenhouse gas emissions are expected to be met by 2008–2012 provided all planned polices and measures are implemented. As part of its effort to reach its target, the EU has in 2005 introduced an emissions trading system for greenhouse gases. The aim is to stimulate innovation and give reductions in emissions a market value. Long-term targets for emissions reductions, though, established to prevent harmful climate change, are not expected to be reached without substantial changes to the energy mix. Many countries are already developing adaptation strategies in recognition of the need to act on expected future long-term impacts.

Climate change is already visible. Increasing temperatures across Europe, changing precipitation patterns in different regions, melting glaciers and ice sheets, increased frequency of extreme weather events, rising sea levels and increasing stress on terrestrial and marine ecosystems are among the most visible impacts on the environment.

The EU has made substantial progress in reducing the environmental impacts of waste disposal and will make further progress as recently adopted legislation on landfill and incineration is implemented. Nevertheless, the volume of most waste streams continues to rise in step with growth in GDP — by 2020 we can expect to be producing nearly twice as much waste as today if current trends continue.

At the same time, air quality in urban areas continues to have adverse effects on people's health, and in rural areas on ecosystems. The impacts in rural areas are expected to decline substantially on the basis of existing policies and measures; however, negative impacts are expected to remain significant in highly populated areas up to 2020.

Much remains to be done on point source emissions to water especially in the EU-10, whilst there has been relatively little progress in reducing nitrates in water across the EU-25. Implementation of the urban waste water treatment directive should reduce point source emissions substantially in the EU-10, but nutrient discharges from rural populations and agriculture are expected to remain a major water pollution problem in coming decades. Future prospects suggest that eutrophication of Europe's freshwater and marine waters will remain a challenge.

Biodiversity loss is continuing, especially on farmland. Gains and losses of plant species are expected in the future in some countries as a consequence of climate change. Soil remains a resource under pressure with sealing and contamination of soils in and around urban areas being of particular concern. Exceedances of critical loads for soils from nitrogen deposition are expected to decline for most areas of Europe in the coming decades.

Integrated assessment | Integration

Table 9.1 The sixth environment action programme (6EAP) — Are we on track?

Action on tackling climate change

Target	Outlook	Region
Kyoto Protocol commitment of an 8 % reduction in GHG emissions in the EU as a whole by 2008–2012 compared with 1990 levels (Art. 5.1)	-> With existing domestic policies and measures alone (as of mid-2004), a reduction in emissions of less than 3 % is expected in the EU	EU-25
	-> However, taking into account the latest policy developments and all the additional policies, measures and third-country projects planned so far, the EU-15 is likely to meet its target	EU-15
Long-term objective of a maximum global temperature increase of 2 °C over pre-industrial levels (Art. 2)	-> Global temperature to increase by more than 3 °C by 2100	EU-25
	-> Potential to reach target by a long-term deep reductions of global and EU GHG emissions	EU-25
Use of renewable energy sources [...] meeting the indicative target of 12 % of total energy use by 2010 (Art. 5.2 (ii (c)))	-> Renewable energy sources in total energy use expected to be about 7.5 % by 2010	EU-25
Doubling the overall share of combined heat and power to 18 % of total gross electricity production (Art. 5.2 (ii (d)))	-> Combined Heat and Power in total gross electricity production is expected to be about 16 % by 2030	EU-25
Promote the development and use of alternative fuels in the transport sector (Art. 5.2 (iii (f)))	-> Biofuels in transport final energy demand are expected to be 1 %, 2 % and 4.5 % by 2005, 2010 and 2030	EU-25
Decoupling economic growth and the demand for transport (Art. 5.2 (iii (h)))	-> Relative decoupling from GDP is expected over the next 30 years for both passenger and freight transport demand	EU-25

Action on nature and biodiversity

Target	Outlook	Region
Halting biodiversity decline with the aim of reaching this objective by 2010 (Art. 6.1)	-> Losses in the number of plant species are expected as a consequence of climate change in some European countries	EU-25
Protection and appropriate restoration of nature and biodiversity from damaging pollution (Art. 6.1)	-> On the basis of existing policies and measures, air pollution and its impacts on health and ecosystems are expected to decline significantly up to 2030	EU-25
Encouraging more environment-responsible farming, such as extensive, integrated, and organic farming (Art. 6.2 (f))	-> Moderate expansion of good farming practices expected	EU-25

Environment and economic sectors

PART A

Action on environment and health and quality of life

Target	Outlook	Region
Ensure that the rates of extraction from water resources are sustainable over the long term (Art. 7.1)	-> Total water withdrawals are expected to decrease by 2030, but water stress may remain in southern Europe	EU-25
Achieve levels of air quality that do not give rise to significant negative impacts on and risks to human health and the environment (Art. 7.1)	> On the basis of existing policies and measures, all emissions of land-based air pollutants (except ammonia) are expected to decline significantly up to 2030	EU-25
	-> The EU as a whole is expected to comply with the 2010 targets of the NEC directive	EU-25
	-> Impacts on human health and ecosystems are expected to diminish substantially, although large differences across Europe persist	EU-25
Sustainable use and high quality of water, ensuring a high level of protection of surface and groundwater, preventing pollution (Art. 7.2 (e))	-> The urban waste water directive is expected to significantly reduce the overall discharge of nutrients	EU-25
	-> Agricultural nutrient surpluses are expected to be moderately reduced in 2020	EU-15
	-> Pressures are expected to increase significantly in the New-10, due to mineral fertiliser use	New-10

Action on the sustainable use and management of natural resources and waste

Target	Outlook	Region
Indicative target to achieve 22 % of electricity production from renewable energies by 2010 (Art. 8.1)	-> Electricity production from renewable energy expected to be about 15 % in 2010	EU-25
Significant overall reduction in the volumes of waste generated (Art. 8.1)	-> Waste generation continues to grow across Europe. In the New-10 relative decoupling from GDP growth is expected (but not in the EU-15)	EU-25
Establishment of goals and targets for resource efficiency and the diminished use of resources (Art. 8.2 (i (c)))	-> Resource productivity in the New-10 is expected to remain about 4 times lower than in the EU-15	EU-25

Although anti-pollution measures taken over the past half-century have dramatically reduced the presence of many known toxins, the number of toxic substances in consumer products, pharmaceuticals and the wider environment has grown. Individual chemicals, such as endocrine disrupters, are likely to be detrimental to human health and reproduction, while there is growing scientific concern about the effects of the cocktail of chemicals each of us is exposed to every day.

Many of Europe's commercial fish stocks are overharvested and some are in danger of collapse. As a result, an increasing proportion of fish for European consumption is caught outside European waters, by either foreign or licensed European vessels. Europe's environmental footprint on the world's fisheries is unsustainably large and, besides questions of equity, forms part of a threat to the survival of the resource itself.

There has been a decline in the health of Europe's forests, attributable at different times to air pollution and drought, with a quarter of the continent's trees currently rated as damaged. This damage has particularly serious implications in Europe's remaining old-growth forests, where biodiversity is at its richest.

As indicated in Chapter 1, the EU's sixth environment action programme provides the main framework for action to 2012. The programme identifies key environmental problems within which are embedded various objectives and targets relevant to these priorities and the economic sectors that have most impacts. Outlooks for the future suggest that full implementation of existing environmental policies would deliver significant improvements in the coming years in several fields and help the EU meet its targets in a number of areas. Nevertheless, progress is expected to be limited towards targets for greenhouse gases, renewable energies and transport (Table 9.1).

New and more integrated actions are therefore needed that reflect the strong relationship between environmental problems and socio-economic developments, over space and over time. What Europe is now facing is a series of mainly diffuse source issues that require both actions across a number of well-established sectors, whether agriculture, transport, manufacturing or energy production, and actions which engage with social factors such as urbanisation, personal consumption and waste production.

A look at recent developments and prospects in four of the main sectors — transport, agriculture, energy and households — and their impacts on the environment helps to provide some clues as to where the focus for such future integrated actions could lie. A fifth sector, industry, which has major environmental impacts, directly influences trends in the other four: for example, the metals and materials industries in transport, the chemicals industry in agriculture, the minerals industry in the energy supply sector and the construction industry in households. This sector, especially the manufacturing part, is addressed further in the section on eco-innovation in the next chapter.

9.3 Developments in four socio-economic sectors

Transport

An efficient and flexible transport system is essential for our economy and our quality of life. The current European system poses significant and growing threats to the environment, to human health, and to the economy, through, for example, increasing congestion. Passenger and freight transport, by road, air and sea, are either growing at the same rate as or faster than the economy overall, implying that the eco-efficiency of transport in the EU economy, and the decoupling of growth in transported passenger or tonne from growth in GDP, are not improving. Trends to 2020 suggest that decoupling will continue to be a challenge overall (Figure 9.2).

Transport volumes in the EU-25 have increased steadily over the past decade: about 30 % for freight transport and almost 20 % for passenger transport. This growth is strongly linked to infrastructural development that, in turn, contributes to air pollution, the sealing of soil and the fragmentation of habitats across many parts of Europe, as well as exposing a significant proportion of the population to high noise levels. Freight transport has increased as a result of changed procurement and distribution strategies of companies (outsourcing, just-in-time delivery) and the development of the internal market as companies exploit the competitive advantages of different European regions.

Causes of the growth in passenger transport include an increase in the number of households and in the number of cars per household, as well as a lengthening of the average journey. This last trend is influenced by such factors as urban sprawl, together with the location of services including schools, shops and medical facilities; the availability and pricing of public transport; and changes in lifestyle fuelled by two incomes per household and a wider choice of leisure activities.

Unsurprisingly, transport is the fastest growing consumer of energy, currently accounting for 31 % of Europe's final energy consumption. Greenhouse gas emissions are also growing rapidly — by more than

20 % between 1990 and 2003 — and they are expected to be 50 % higher by 2030 than they were in 1990. Aviation, as the fastest growing mode of transport, and marine transport account for an increasing share of these emissions while remaining outside the coverage of environmental policies such as the Kyoto Protocol and fuel taxation. On the road, increasing traffic volumes and a rising number of larger, heavier and more powerful vehicles travelling ever further has more than offset progress in improving energy efficiency, stimulated by the industry's voluntary commitments to reduce average CO_2 emissions from new passenger cars to 140 grams/kilometre by 2008/2009.

The rapid increase in passenger and freight demand projected over the next 30 years, together with the difficulties in replacing oil as the fuel on which the sector depends, suggests that transport will be one of the most difficult sectors in which to reduce carbon dioxide (CO_2) emissions. Even increases in fuel prices, possibly through such measures as the introduction of carbon permits, seem unlikely to substantially alter this picture, unless appropriate policies for new fuels are developed alongside such measures.

Technological developments, including catalytic converters and other technical abatement measures on road vehicles, have resulted in marked decreases of some other pollutants such as ozone precursors and acidifying substances. Emissions of these regulated pollutants fell by about a third between 1990 and 2002 across EEA countries, with further improvements expected as stricter limits come into force and the vehicle fleet is renewed.

Developments in vehicle technology go hand in hand with improved fuel quality standards. Lead has been banned in the EU-25 and new standards for sulphur content have been set at 50 parts per million (ppm) by 2005, falling to 10 ppm by 2009. There is, however, increasing evidence that standardised test cycles used for the type approval of road vehicles do not necessarily represent 'real world' driving conditions. The issue of 'chip-tuning' of diesel vehicles to boost power at the expense of fuel efficiency and low emissions is a further cause for concern.

Figure 9.2   Transport — decoupling outlooks to 2020 for key environmental resources and pressures

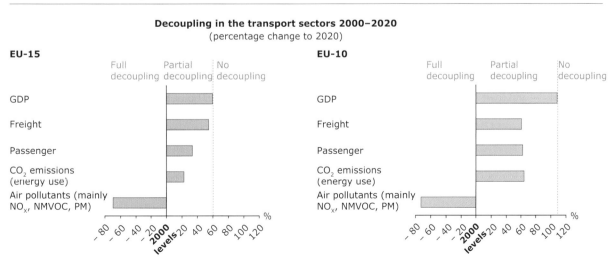

Source: EEA, 2005.

Technical improvements in vehicles and fuels can be supported by economic incentives such as taxation tied to CO_2 performance, road pricing policies or environmental zoning. The introduction of mandatory CO_2 emission limits might be considered, too. There is also a need to raise public awareness about how much car parameters such as size, weight and engine power, and energy-consuming equipment such as air conditioning, influence CO_2 emissions.

Any emission control policy needs to be complemented by other measures aimed at controlling road transport volumes. If the forecast growth in road transport is not to undermine current and expected achievements, focus needs to be put on user behaviour. Options include improving spatial planning to reduce distances to and between key services and providing settlements with improved access to better public transport. Given the slow rate of change in housing and infrastructure stock, and the fact that decisions are seldom based on considerations of what is best for the environment, these measures would inevitably take some time to produce benefits. Investment in public transport and pricing mechanisms could, however, also strengthen a shift to more environmentally sound transport and improve incentives to higher load factors.

Thus, a sustainable road transport policy that guarantees social inclusion and economic development with a high level of environmental quality and safety has to combine a number of different approaches, instruments and strategies that aim to:

- improve efficiency by reducing the number and average distance of journeys;

- shift transport to more environmentally benign modes;

- use existing vehicle capacity and infrastructure more effectively; and

- improve the environmental performance of vehicles.

Some instruments, such as road user charges or fuel taxes, can contribute to several or all strategies simultaneously, while others — for example setting emission standards for vehicles or the provision of public transport — basically influence one or two approaches.

Emissions of air pollutants from aviation and marine transport, which are not subject to international regulation, and from rail and inland shipping have not reduced substantially. In the case of aviation and marine transport, they have grown significantly due to increasing volumes allied to a lack of strict and mandatory standards. It is expected that emissions of sulphur dioxide and nitrogen oxides from maritime activities will surpass land-based emissions within the next 20 to 30 years.

Agriculture

Across Europe, highly developed agricultural land patterns and their functions have evolved over centuries to ensure that the population is fed and rural landscapes are maintained. Current agricultural activity has substantial environmental impacts in terms of greenhouse gas and air pollutant emissions, contributing to climate change and acidification; pollution of water by nitrates, phosphorus, pesticides and pathogens; habitat degradation and species loss; and the over-abstraction of water for irrigation. Looking ahead to 2020, in the EU-15, a partial decoupling of water and mineral fertilizers uses is expected, and full decoupling for nutrients surpluses and greenhouse gas emissions. Partial and full decoupling are also expected in the EU-10 for water use and greenhouse gas emissions, but no decoupling might characterise the development of mineral fertilizers use and nutrients surpluses (Figure 9.3).

Farmland boasts a wide range of habitats and species that depend to a large extent on continued (extensive) agricultural use. However, depopulation is occurring in many rural areas, profoundly affecting the countryside and the environment. Low and variable incomes, hard working conditions and a lack of social services and leisure activities in many areas make traditional farming a less attractive option for young people living in a predominantly urban Europe — the proportion of the elderly is already very high amongst Europe's farmers. Depopulation is a phenomenon all over Europe, whether from hill farms in the Alps or traditional small farms from Poland to Portugal. The trend is particularly worrying in central and eastern Europe, where recent political and economic changes during the 1990s negatively affected the conditions for farming. As a result, further land abandonment is to be expected.

Agriculture's share of the total national land area ranges from 30–60 % in the new Member States. Here many private farmers, with limited formal agricultural training, rely on relatively outdated machinery and buildings. Economic restructuring and lack of capital caused a sudden drop in agricultural investment in the 1990s. This resulted in lower pesticide and fertiliser inputs, with a consequent reduction of pollution, and, in most EU-10 countries, the abandonment of biodiversity-rich grassland systems.

The reduced investment in erosion mitigation and in manure storage facilities poses significant environmental risks if, as expected, agriculture in these countries intensifies in the future. Indeed, fertiliser use in the new Member States is expected to increase by up to 50 % by 2020, while in the EU-15 fertiliser use is expected to remain stable. The increased use of inputs will be a key factor behind the expected increased yields and agricultural production in the EU-10, and brings with it risks of environmental pollution requiring careful management.

Over many decades, in response to greater demand driven by improved standards of living, population growth and urbanisation, large-scale rationalisation and industrialisation of agricultural production has taken place. This has led, amongst many outcomes, to pastures and semi-natural grasslands being converted to intensive farmland, with the consequent destruction of habitats such as hedgerows and ponds that have supplied, over at least the past 250 years, niches for a wide range of species. Moreover, conversion of marginal land to agriculture has taken place in parts of Portugal and Spain and to a smaller extent in the south-west of France. Withdrawal of farming has occurred in some mountain areas in southern Europe, as well as in many new Member States.

Figure 9.3 Agriculture — decoupling outlooks to 2020 for key environmental resources and pressures

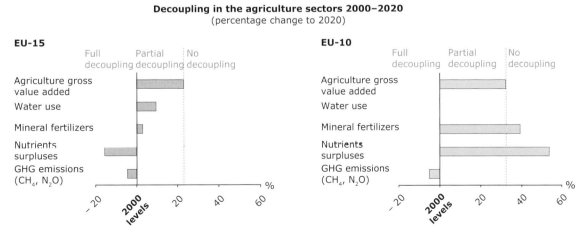

Source: EEA, 2005.

Agricultural intensification has brought about a rapid decline in semi-natural vegetation such as hedgerows and field borders. Wild-living species of both fauna and flora rely for their survival on habitats and the corridors that connect them — for example, roughly two-thirds of the currently endangered bird species depend on agricultural habitats. These have become increasingly fragmented, making the maintenance of viable species populations more difficult. As a result, over the last few decades, biodiversity on farmland has declined. Farmland species of particular conservation concern occur throughout Europe, but many of them are associated with high nature value (HNV) farmland, particularly in southern Europe.

A realisation that the regional identity of European landscapes — testimony of the continent's combined natural and cultural heritage — is at risk has placed the conservation of biodiversity on agricultural land high on the political agenda. Of the many relevant conservation efforts at European level, the most important are the habitats and birds directives and the biodiversity action plan for agriculture. In the sixth environment action programme, the EU has committed itself to halting the decline of biodiversity by 2010.

Conserving HNV farmland is essential to achieving this target. Under the EU common agricultural policy (CAP), agri-environment schemes are being used as a tool to give farmers compensation for taking specific environmental measures that can support HNV areas. However, the rate of uptake varies greatly: it is particularly low in southern European countries, including Portugal and Spain, where the share of HNV farmland is relatively high. Thus, the challenge for agri-environment schemes is specifically to target those areas that could benefit most from conservation.

Nitrates from agriculture continue to damage the environment, contributing to eutrophication of coastal and marine waters and pollution of drinking water, especially where groundwaters have become contaminated. Problematically, substantial time lags can occur before changes in agricultural practices are reflected in groundwater quality. The length of these lags, which may be measured in decades, varies according to soil type and the specific hydro-geological conditions of the groundwater body and overlying substrate.

It is generally cheaper to prevent nitrates reaching the water in the first place. A review of the possible costs to farmers comes to an initial estimate of EUR 50–150 per hectare per year to alter farming methods to comply with standards set by the EU nitrates directive. This is considerably cheaper than the estimated costs of removing nitrates from polluted waters. Moreover, changing farming practices puts the responsibility on the farmers who have caused the pollution, rather than on the consumer.

The nitrogen (N) surplus in the farmland soils of the EU-15 declined from 65 to 55 kilograms N/hectare between 1990 and 2000. In some European hot spots, surpluses as high as 200 kilograms/hectare can be found. These surpluses are the overriding contributor to continuing high nitrate levels in Europe's rivers. Looking forward, the good news is that such surpluses are expected to decouple completely from agricultural production growth in the EU-15 and partially decouple in the EU-10. Nonetheless, forecasts suggest they will continue to increase in absolute terms.

Currently nitrate levels in surface and groundwaters are lower in the EU-10 than in the EU-15. However, if agriculture intensifies in the EU-10, as expected, good implementation of the EU nitrates directive, supported by the CAP cross-compliance rules that tie funding to legislation and other measures, will be essential to avoid creating extensive, expensive and long-lived water pollution problems in the coming years.

Withdrawal for agricultural irrigation is the largest source of water abstraction in southern Europe and will continue to be so in the future. Technological developments have led to some improvements in efficiency — and there is scope for much greater uptake of these new technologies — but these have been more than offset by increases in the area of irrigated land. The hotter drier summers predicted as a result of future climate change will further increase pressures on water use in the next 20–30 years. In northern Europe, abstractions for irrigation are relatively small, and may decrease further in future as a result both of improved technologies and of the expected wetter conditions. For the EU-10, as well as for southern Europe, savings in the future from more efficient irrigation systems are likely to be cancelled out by increases in the need to irrigate as a result of anticipated climate change.

Changing climatic conditions will, in all likelihood, have a range of favourable and unfavourable impacts on agriculture. For example, the annual growing season for plants, including agricultural crops, lengthened by an average of 10 days between 1962 and 1995 and is projected to continue getting longer. In most parts of Europe, particularly the middle and northern Europe, agriculture could also potentially benefit from a limited temperature rise. However, while Europe's cultivated area could expand northwards, agricultural productivity in some parts of southern Europe may be threatened by water shortages. More frequent extreme weather, especially heatwaves, could bring more bad harvests. Agriculture's capacity to adapt will be a key factor in response to expected climate change in Europe.

Energy

Energy services provide all of us with comfort and mobility, and underpin economic competitiveness and security. Despite reductions of some air emissions, the energy supply sector (including electricity and heat production, refineries, etc.) is a prime contributor to environmental concerns such as climate change, air pollution and water stress. In particular, it continues to be the major source of greenhouse gas emissions (around one-third of total emissions) and emissions of acidifying substances such as sulphur dioxide and nitrogen oxides (about 30 % of total emissions). Future developments thus depend to a large extent on progress in decoupling environmental pressures from production and consumption.

Energy consumption is expected to continue increasing over the coming decades but to partially decouple from GDP, consolidating past reductions in energy intensity (Figure 9.4). At the same time, the policy targets for increasing sources of renewable energy are not expected to be met across the EU-25 without additional policies and measures. As a consequence, the energy sector is expected to contribute to increasing greenhouse gases and climate change in coming decades, while reductions in emissions of acidifying substances are expected to continue.

Past measures to reduce air emissions from power stations have been hugely successful. In the EU-15, between 1990 and 2002, emissions of sulphur dioxide and nitrogen oxide from public electricity and heat production fell by 64 % and 37 % respectively, despite a 28 % increase in the amount of electricity and heat produced. This success has been attained through strict regulations setting clear emission standards based on available technological abatement measures.

The introduction of flue gas desulphurisation and the use of lower sulphur coal and oil contributed some two-thirds of the sulphur dioxide reductions; another major contributor was the switch in the fuel mix away from coal and oil towards lower sulphur fuels such as natural gas, prompted by the liberalisation of energy markets and, to a lesser extent, improvements in the efficiency of the conversion process. Some of these developments produced one-off benefits, however, and will not contribute to any further decoupling of environmental pressures from production and consumption.

The development of the electricity sector over the 1990s demonstrates that new technologies can be introduced. Electricity produced from gas doubled in both the EU-15 and the new Member States between 1995 and 2002 as competition favoured gas use due to the high efficiencies and low capital costs associated with some gas-based technologies, in particular combined cycle gas turbines (CCGT).

Overall, the CO_2 emissions intensity of power production fell by about a fourth between 1990 and 2002 in the EU-25, but increases in demand meant that CO_2 emissions from power production declined only slightly, by around 5 %. For CO_2 emissions, end-of-pipe abatement technologies are not yet available. This may change in the future with the planned use of CO_2 capture and storage. This technology separates CO_2

from the flue gas or a process gas before burning. It can substantially reduce CO_2 emissions from the burning of fossil fuels. However, the process is expensive and requires a substantial extra amount of energy; the long-term safe storage potential, and even feasibility, is not yet fully known.

With CO_2 capture and storage not yet being commercially available, reducing CO_2 emissions requires a lower consumption of fossil fuels (coal, oil, gas). Since the majority of electricity — and over three-quarters of total energy consumption — is produced from fossil fuels, this requires deeper changes in electricity generation. Technologies are available to reduce the CO_2 emissions of electricity. These include the increased use of non-fossil fuels such as renewable and nuclear energy, improving the efficiency of the conversion process, or using less carbon-intensive fossil fuels such as natural gas. The use of combined heat and power plants, which produce not only electricity but also make use of the heat that would otherwise be lost, can also contribute to substantial CO_2 emissions reductions.

Many of these measures imply investing in new plants and infrastructure as opposed to applying abatement technologies in existing plants. Combined heat and power plants need a heat distribution infrastructure to the end-user, while some renewable technologies — such as wind energy — face the problem of fluctuating electricity production. Nevertheless, the difficulties in achieving such structural changes are primarily due to socio-economic barriers, not the lack of technical solutions. If long-term targets and appropriate incentives are set, such changes can be realised within the ongoing renewal of the European power system.

The importance of further penetration of new, less carbon-intensive technologies and fuels in electricity generation can be demonstrated by the results of scenarios which have been developed for the EEA. If no additional policies and measures were implemented to mitigate expected climate change, the share of coal in electricity production would decrease in the short term but increase after 2015, before returning to its current level in 2030. Despite a further penetration of gas-fuelled technologies in the short term, their rate

Figure 9.4 Energy — decoupling outlooks to 2020 for key environmental resources and pressures

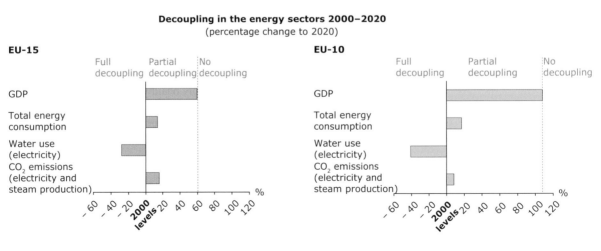

Source: EEA, 2005.

of growth is expected to decline as a result of higher natural gas import prices, enhanced by concerns about security of supply. The share of electricity from technologies such as renewable energy sources and combined heat and power plants would increase by only a few percentage points until 2030. This would lead to CO_2 emissions from electricity and steam production being some 15 % above 1990 levels in 2030.

The scenarios also highlight the important emissions reduction potential of low-carbon technologies that already exist but have yet to be fully mobilised. The scenarios suggest that the introduction of a carbon price alone would not be sufficient to reach high shares of renewable energies but would have to be complemented by specific policies and measures. These include direct price support, subsidies and loans or market-based mechanisms — for example, calls for tender for electricity from renewable sources, trading of 'green certificates' or voluntary payments of premium rates for renewable electricity by consumers.

Large reductions in water withdrawal for electricity production are expected in coming decades as newer power plants operating with tower cooling systems replace older plants using once-through systems (Figure 9.4). Tower cooling systems commonly require only a twentieth of the water per MWh for cooling purposes. These reductions can be achieved despite an expected near doubling of electricity production in Europe by 2030.

The future of nuclear power remains unclear across the Community apart from in, for example, Finland and France. Some believe that, as the current generation of nuclear power stations reaches the end of its useful life, so the share of electricity generated in this way will dwindle. Others suggest that to mitigate the effects of climate change and avoid possible future shortages or massive price increases, nuclear power must remain a significant option. The debate is likely to continue.

Households and demography

Important drivers of Europe's changing environmental pressures are demographics and increasingly affluent lifestyles. The environmental pressures of personal consumption are generally lower than those of the production they drive, but are expected, as in the recent past, to grow substantially faster than overall GDP and in line with increased house building, transport use and tourism.

Europe's population has stabilised for now. Over the next 30 years, the overall population of the EU-25 is expected to remain broadly at around 455 million. Current projections suggest there will be 7 % fewer people in the EU-10 by 2030, with declines particularly concentrated in rural areas. Furthermore, in line with trends throughout the developed world, the Europe of 2030 is likely to have a substantially higher proportion of older people.

Assuming a continuance of current working lives and retirement patterns, which is far from certain, this ageing of the population means that the proportion of Europe's population that is economically active is set to drop substantially, placing a greater importance on each of those working to generate more wealth. Leaving aside issues of immigration policy which are beyond the scope of this report, this calls for innovative thinking about the structure of taxation and benefits, including the possibility of moving some of the tax burden away from labour and on to resource use and pollution.

An older Europe may also bring about changes in consumption patterns. More old people will mean an increasing proportion of the national income being spent on health. It is also conceivable that, as the number of old people unable or unwilling to drive increases, demand for public transport will also increase. Additionally it has been suggested that, as the number of reasonably healthy and relatively wealthy older people grows, so too will the demand for tourism and second homes. However, with the exception of the increasing demand for health services, these are as yet uncharted territories.

Europe, again in concert with much of the developed world, is also experiencing a reduction in the size of the average household. By 2030 this will have fallen from more than 3 in 1990, through the current figure of around 2.75, to around 2.4. Driven by a variety of factors, led by increasing personal wealth but including the ageing population, high divorce rates, and the increasing number of adults who choose either to live alone or not to marry, the number of households in Europe is expected to rise by approaching one-fifth. In general, more households result in net increases in demand for energy and water and generate greater volumes of waste.

More goods, including computers, stereo systems, mobile phones, household appliances and air-conditioning systems, are being bought. Although new equipment is sometimes less wasteful of resources, this is not always the case. For example, many electronic goods run on stand-by mode when not in use, and so use substantially more electricity than their predecessors. The recent Green Paper on Energy Efficiency states that according to available studies as much as 20 % of energy savings could be realised in a cost-effective way by 2020. Demand-side improvements in efficiency will probably be more dependent on awareness-raising among end-use consumers and on providing incentives to change behaviour as well as on regulations that foster higher technical standards.

Within the EU-25, water withdrawals for household consumption are expected to increase at a rate less than expected household expenditure growth up to 2020 (Figure 9.5). Demand-side measures such as more efficient homes and appliances linked to taxes and charges explain this trend. Nevertheless, household water withdrawals are expected to increase substantially in the EU-10 as these countries approach average consumption levels in the EU-15 in the coming decades.

In the 1990s the EU set a target of reducing the municipal waste stream to below 300 kilograms per person per year by 2000. Unfortunately this has not been achieved and waste production continues to rise. Landfill remains the most common route for its disposal but the implementation of the EU landfill directive is cutting the use of this method for biodegradable municipal waste. The directive was intended to reduce the generation of carbon dioxide, methane and nitrous oxide, all of which are greenhouse gases controlled under the Kyoto Protocol, and it is adding pressure on manufacturers, retailers and local authorities to find new and innovative ways of reducing the waste stream — for example, by using biodegradable waste for all sorts of energy production.

Figure 9.5   Households — decoupling outlooks to 2020 for key environmental resources and pressures

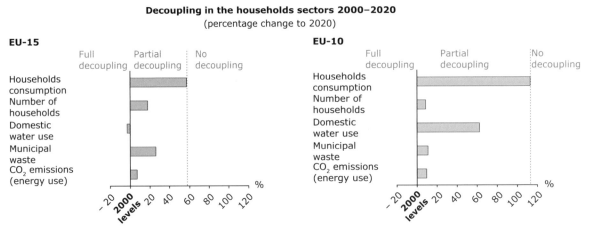

Source: EEA, 2005.

Experiences for packaging waste show both the extent to which Europe has tackled the problem — and the distance it has to go. Consumers and industry seem happy to recycle their waste packaging, but extremely reluctant to take steps to avoid producing it in the first place. Most packaging waste policies relate to recycling and recovery, rather than reduction.

In most EU countries packaging waste production still increases in line with GDP. Absolute rates range from 217 kilograms per person per year in Ireland to 87 kilograms in Finland, but the trend everywhere is continuing upwards. Analysts expect paper and cardboard waste generation within the EU-15 to grow by more than 60 % between 2000 and 2020, roughly in line with growth in GDP — dreams of a paperless office induced by changes in information technology have proved false.

In contrast, most countries have easily exceeded their targets for recycling packaging waste. Although the EU target was to recycle 25 % by 2001, the overall recycling rate of packaging in the EU-15 is now above 50 %. This reflects the relative ease of adopting 'end-of-pipe' solutions rather than effecting structural changes that reduce materials or energy flows. It is also an example of the management dictum that what gets measured gets done. In this case, specific targets concern recovery and recycling, while the real challenge — that of waste reduction — remains an aspirational objective.

Looking ahead, municipal waste volumes are expected to decouple partially from GDP growth up to 2020, with most progress expected in the EU-10 where economic recovery is expected to provide opportunities for adopting better, more up-to-date technologies (Figure 9.5).

Most Europeans live in urban areas, which are generally connected to sewerage systems. In northern Europe the majority of housing is connected to the most efficient treatment facilities for removing pollutants from wastewater, while in western European countries only around half of the wastewater is treated in this way. In southern European countries and the EU-10, just 50–60 % of the population is connected to wastewater treatment plants of any kind. There remains, still, substantial scope for wider application of tertiary treatment facilities in many parts of Europe. There also remains substantial scope for combining investments in treatment plants with charging levies to reduce pollution at source and hence treatment costs; currently countries focus primarily on investment in treatment plants.

At the same time, increasing wealth enables many Europeans to invest their savings in second homes. These will often increase development pressures in environmentally vulnerable areas already subject to tourism pressures, such as the Mediterranean coastal areas. The arrival of second-home owners, including significant numbers of retirees, from northern Europe is already the largest cause of construction in parts of Spain. Such investors may, however, help rural economies, particularly in more remote, marginal or mountain areas. They may also allow the continuation of low-intensity agro-ecosystems as part-time activities.

Personal car travel has risen by more than 3 % a year for the past three decades. During 2001, the average European travelled 14 000 kilometres across all transport modes. On current trends, each of us is expected to travel 7 000 kilometres further in 2030. This puts pressure on the land and inevitably has a deleterious effect on urban air quality. During the 1990s, although Europe's motorway network grew by a quarter, extra traffic filled the new roads as fast as governments could build them. This 'generated traffic' generally fills 50 % to 90 % of available road capacity within a year or so. This is, in part, consumer choice, but studies now indicate that the development of out-of-town shopping centres and the spatial pattern of medical and educational facilities also play significant roles.

Air travel's share of the total distance travelled is expected to double to more than 10 % by 2030. Recent changes such as cheap flights and on-line booking are making it more attractive to travel by plane across Europe, rather than by car or train. This considerable growth in air travel is driven by rising demand both from Europeans and from foreign travellers wishing to come to Europe. The travel and tourism industry is now a major economic force, producing 11 % of the continent's GDP and accounting for 12 % of its employment, a major consumer of goods, water and land, and significant producer of waste and emissions of greenhouse gases.

Europe's growing demand for housing, food, consumer goods, transport, tourism and waste disposal are putting pressure on its land, water and air quality, as well as causing the loss and fragmentation of its wildlife habitats. In the coming years, these pressures are expected to be particularly strong along the Mediterranean and Atlantic coasts of southern Europe, and may be widely felt across rural Europe, as more people travel beyond their urban existence into the countryside to improve their quality of life and relax.

9.4 Summary and conclusions

The regulatory regime developed in Europe over the past 30 years has delivered an impressive list of achievements. It has provided the stable basis for the development of technologies that have decoupled some environmental pressures from economic growth, especially pressures from point sources. Nevertheless, it is recognised that environmental regulation of this kind can only achieve so much. The sectoral activities that lie behind many of today's ongoing environmental problems have multiple sources which often require behavioural change and therefore cannot be addressed through command and control regulations alone. Rather, a combination of regulatory standards, technological changes, financial measures, economic instruments, voluntary agreements and information provision provide a more effective mix of actions. Different combinations are appropriate for different problems and sectors.

For the transport sector, regulation and voluntary agreements have provided the stable basis for the auto industry to innovate, and economic instruments, especially taxes and charges, have contributed to making explicit the hidden costs of pollution and changing consumer behaviour to some degree.

For the energy supply sector, regulation has also provided a stable basis for innovation. In the area of renewables, recent policies have provided the basis for increases in venture capital to fund company start-ups. Economic instruments and financial measures have been dominated by subsidies for fossil fuels. More recently, tradeable permits have been used to encourage cost-efficient reductions in emissions of air pollutants.

The agriculture sector is shaped by financial measures taken under the CAP. There have been substantial reforms in recent years shifting from environmentally damaging subsidies for production to incentives that protect the environment and promote economic growth and social cohesion. The cross-compliance measures linking CAP payments to action by farmers on reducing nitrates is an innovative example of integrated action that could be applied more widely, for example the use of Cohesion Funds and recycled charges for building wastewater treatment plants and reducing pollution at source. Innovation has been dominated by production rather than eco-efficiency objectives so there remains substantial scope to increase the use, for example, of more efficient irrigation technologies.

The household sector is different; it is not as homogeneous as the other sectors, nor is it backed by well-defined policy objectives and measures. Changing public behaviour is difficult and often more politically sensitive. Economic instruments, especially taxes and charges, are used extensively in countries to internalise the costs of environmental services such as water provision, sewerage treatment and waste collection. There is huge scope to increase the use of already developed eco-efficient technologies, but financial incentives and awareness-raising activities are relatively absent.

As some major environmental problems are interlinked, and with many sectoral activities contributing to the same environmental problems, there is substantial potential through more integrated approaches to deliver benefits beyond those which could be achieved through unilateral approaches. Examples include reductions in emissions of sulphur dioxide to deal with acidification that at the same time deliver secondary benefits for climate change; the switch from subsidies in agriculture, transport and energy that contribute to environmental degradation towards incentives that change behaviour; and investments in new technologies that reduce diffuse environmental pressures such as hydrogen and carbon sequestration, and at the same time create jobs and contribute to improving Europe's overall competitiveness. The concluding chapter looks ahead by assessing three interlinked approaches that could form the basis for making progress on integration in the future.

References and further reading

All of the core set of indicators found in Part B of this report are relevant to this chapter. The most relevant are: CSI 11, CSI 14, CSI 16, CSI 17, CSI 18, CSI 20, CSI 24, CSI 27, CSI 28, CSI 29, CSI 30, CSI 31, CSI 32, CSI 35 and CSI 36.

Introduction
European Environment Agency, 1999. *Environment in the European Union at the turn of the century*, Environmental Assessment Report No 2, EEA, Copenhagen.

European Environment Agency, 2005. *European environment outlook*, EEA Report No 4/2005, Copenhagen.

Maddison, A., 2004. *The world economy: historical statistics*, Organisation for Economic Co-operation and Development, Paris.

Millennium Ecosystem Assessment, 2005. *Ecosystems and human well-being. Opportunities and challenges for business and industry*.

The changing state of Europe's environment
European Environment Agency, 2005. *European environment outlook*, EEA Report No 4/2005, Copenhagen.

European Environment Agency, 2005. *Environment and health*. EEA Report, Copenhagen (in print).

Food and Agriculture Organization of the United Nations, 2005. *State of the world's fisheries 2004*, FAO, Rome.

Developments in four socio-economic sectors
European Commission, 2001. *The sixth environment action programme*, COM(2001) 31 final, 2001/0029 (COD).

European Commission, 2004. EU common agricultural policy explained. www.europa.eu.int/comm/agriculture/publi/capexplained/cap_en.pdf.

European Council, 1999. Directive 1999/31/EC of 26 April 1999 on the landfill of waste, Official Journal L182, 16/07/1999.

European Environment Agency, 2002. *Corine land cover update 2000: Technical guidelines*, Technical Report No 89, EEA, Copenhagen.

European Environment Agency, 2004. *EEA signals 2004*, Copenhagen.

European Environment Agency, 2004. *Ten key transport and environment issues for policy-makers*, EEA Report No 3/2004, Copenhagen.

10 Looking ahead

10.1 Introduction

Europe faces several interconnected challenges in coming decades. These include greater global competition for natural resources and markets; pressures on social and territorial cohesion from an ageing population and from a decreasing size of families; and environmental problems from climate change, biodiversity loss, use of land and water resources, over-fishing and marine ecosystem impacts, soil loss, and air pollution and health impacts from day-to-day living, and the widespread use and generation of chemical substances.

Europe is well placed to meet these challenges. It has some of the world's most competitive companies, a quality of life that is one of the best in the world, a long history of industrial and institutional innovation, and a wide range of people and cultures that can stimulate a diversity of economic and social activities. It also has a rich and varied environment that, if nurtured, can maintain and sustain a high quality of life in the face of rapid change.

Both the challenges that Europe faces and its capacities to manage them are interlinked by webs of ecological, economic and social networks. Cost effective measures need to be similarly interlinked via more coherent and integrated responses.

The integration of environmental policies into economic activities is one key response. In addition, environmental measures should be designed to achieve high environmental standards whilst contributing to or at least not inhibiting, innovation, social integration and the reform of markets and governance. Recent debates on environmental policies have shown that unless such policies are seen to contribute to these wider issues they can easily be relegated to 'luxury' items that must await some future prosperity.

Three main interlinked approaches could help Europe to make further environmental and economic progress. Firstly, stronger and more coherent environmental policy *integration* to ensure that environmental issues are fully reflected in all policy-making. This is particularly needed in the economic sectors that contribute most to environmental problems i.e. transport, agriculture and energy. Secondly, the *internalisation* of the environmental costs of energy and resource use into more realistic market prices via environment taxes, charges, tradeable permits and tax and subsidy reform. And thirdly, more efficient use of renewable and non-renewable resources, via measures that stimulate **eco-innovation**.

10.2 Integration

Institutional and financial integration

Article 6 of the EU Treaty states that 'environmental protection requirements must be integrated into the definition and implementation of the Community policies and activities, … in particular with a view to promoting sustainable development'.

Two types of institutional integration are necessary: the horizontal level, that makes links across government between ministries, and Parliamentary committees, at Member State and EU level; and the vertical integration between regional, national, city and local governments.

Environmental policy integration is a feature of the EC Treaty, the sixth environment action programme, the Cardiff integration process and the EU sustainable development strategy. It is promoted, indirectly, in the White Paper on European governance. Environmental objectives are, in principle, also embedded in the Lisbon process, which is the ten-year strategy to make the EU the world's most dynamic and competitive economy.

The role of governments in establishing goals, regulatory frameworks, incentives and information flows whilst encouraging more environmentally responsible activities by corporations, investors, consumers and citizens is an overall feature of these various initiatives.

Progress in sectoral integration has been slow over the last five years due in part to the failure to adequately address institutional integration. However, a closer look reveals some signs of positive change. The Cardiff process, initiated in 1998 to stimulate sectoral integration at the EU level, has encouraged a gradual breakdown of some administrative walls between sector and environment departments; the establishment of environmental units in sector directorates-general of the Commission; and a reorientation of some departments to address more integrated issues, e.g. rural development.

The development of the thematic strategies under the sixth environment action programme further supports new cross-departmental and multi-stakeholder engagement. Increasing the institutional capacity to support environmental policy integration in terms of human and financial resources could offer additional rewards.

Meanwhile, there has been a quiet revolution in the strategic management and coordination of the European Commission and Council activities. The potential of the EU's move towards multiannual and annual planning offers the potential for putting environmental integration into practice. This is also applicable to budgetary planning cycles and auditing, both of which could be used to promote environmental integration.

The European Parliament has used its budgetary role to advance the integration of environment into other policy areas, such as the Structural and Cohesion Funds. This process of greening the EU's budget could be further encouraged by regular and comprehensive reporting on the environmental impacts of EU spending programmes and on progress with environmental policy integration.

New forms of governance are also emerging, such as the 'open method of coordination', aiming at better linking countries and stakeholders in policy processes. In the new EU Member States, environment ministries have used the high priority that the EU has given to environment protection as a means to raise their profile in government. In some of the old Member States, the shift of environmental responsibilities to other ministries has increased opportunities for better policy integration.

National governments have made good progress in terms of developing and agreeing high-level political commitments to environmental policy integration and sustainable development. Most of the 25 EU Member States (EU-25) have established national sustainable development strategies. There is, however, little evidence so far of these strategies being implemented, and considerable opportunities exist for greater cross-country learning.

Since the early 1990s, many countries have developed committees to address environmental integration. Germany's committee of state secretaries for sustainable development is one such example. Other countries, such as Austria and Belgium, have established inter-ministerial commissions to support the implementation of sustainable development commitments. A large number of countries now have environment or sustainable development advisory councils, with councils in Finland, Latvia and Lithuania also serving inter-ministerial coordination functions.

Few countries have exploited opportunities to link their regular strategic planning, budgeting and auditing with the delivery of overarching environmental or sustainable development commitments, although some useful examples are emerging in the Netherlands, Sweden and the United Kingdom. Few countries have explicitly allocated responsibilities for environmental policy integration throughout all key departments, although some countries have established environmental units in some sectoral ministries.

In the new Member States, the transposition and implementation of EU legislation is improving the quality of the environment and decreasing transboundary pollution. There is an opportunity to reorganise governance structures in many countries to bring together policy decision processes (e.g. under IPPC directive) and strengthen cooperation in international networks (e.g. IMPEL).

However, the priority given to economic development has been seen to put at risk the implementation of necessary environmental protection measures. There is therefore a need to ensure sufficient financial resources for implementation of EU legislation. There is also a unique opportunity to decouple environmental from economic pressures especially for the energy, transport and industry sectors. Funds from the EU could be better targeted to local, more sustainable solutions in this respect. The new Member States' extensive spatial planning experience could also be utilised to strengthen further transboundary and cooperative planning initiatives, e.g. for new roads that have already demonstrated the ability to deliver improved environmental outcomes.

The importance of vertical integration is illustrated by the EEA studies into the effectiveness of urban wastewater treatment systems and packaging waste systems in selected EU countries. Packaging waste management is complex, including industry, retailers, consumers, and local and national governments. Institutional arrangements, incentives and governance become every bit as important as the policy itself. Pre-existing institutional arrangements can make effective implementations easier — or more difficult — to achieve.

For urban wastewater, clear lines of responsibility and financing were important in Denmark and the Netherlands for achieving full or near-full compliance through implementation of the urban wastewater treatment directive. In contrast, the overlap of responsibilities in France and Spain between authorities at the national, regional and local levels, together with large investment needs and bottlenecks in financing, appear to have been important reasons for greater implementation difficulties.

The corporate social responsibility movement is bringing additional pressure to bear on the environmental performance of companies, particularly when their performance can be monitored by common approaches to indicators, as with the global reporting initiative. At sector level, corporate initiatives within the chemicals, food, fisheries and forests industries are encouraging more responsible environmental activity, including certification schemes that encourage informed consumer choice.

Investors are increasingly looking towards the environmental performance of their funds and of the corporations within them. Initiatives such as the Green Fund system in the Netherlands, which includes tax incentives for green investments and a partnership with the financial sector, illustrates the potential of such market-based instruments to influence flows of capital towards more sustainable activities, and in doing so help promote over the long term the internalisation of environmental costs into the prices of goods and services.

Evaluating progress

Building on previous work by the Organisation for Economic Co-operation and Development (OECD) and others, and reflecting the national and EU practice summarised here, a possible framework for evaluating progress with environmental policy integration has been developed by EEA (Figure 10.1).

The framework focuses on the following six main areas: political commitment, vision and leadership; administrative culture and practices; assessments and information for decision-making; policy instruments such as market-based instruments that promote internalisation; monitoring progress towards objectives and targets; and eco-efficiency. The evaluation of progress in these six areas is supported by a checklist of relevant criteria.

The framework serves two purposes: firstly, it helps to show how integration can be promoted; and secondly, it provides a single framework for evaluating progress towards environmental policy integration in a consistent manner and across very different economic sectors. It can also be used at all levels of governance, from EU institutions to national, regional and local governments, and even within large companies.

10.3 Internalisation using market-based instruments

Purpose and progress

Market-based instruments can help to realise environmental and economic policy objectives simultaneously in a cost-effective way by taking account of the hidden costs of production and consumption to our health and the environment.

Currently, the prices of goods and services do not fully reflect the environmental costs of their provision, use and disposal — so-called 'environmental externalities'. The greater reflection of the costs of environmental replacement, recovery and reparation in market prices is becoming increasingly urgent.

For example, the price of coal, oil and natural gas does not fully include the costs that will be incurred by climate change and other environmental degradation created by burning them; the price of a hardwood table does not fully include the cost of the lost biodiversity in the forests from which its timber was taken or the increased risk of flooding from logged land; water bills do not always include a tariff for depleted and polluted aquifers; the price of food in supermarkets does not fully include the environmental impacts of the

Figure 10.1   Framework for evaluating integration of environment into sector policies

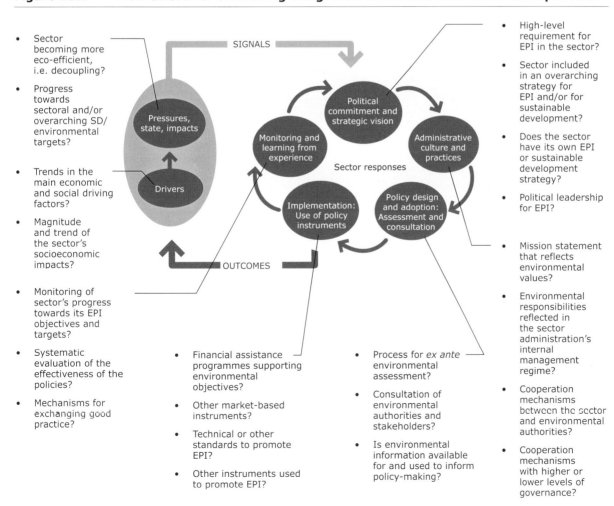

Figure 10.2 Development of environmental tax bases in EU-15, Iceland and Norway since 1996

Source: EEA, 2005.

agricultural systems that produced it, nor the health effects and noise of traffic fumes from the trucks that brought the food to the store.

All environmental policy tools can help to internalise environmental costs by encouraging companies and consumers to pay for their pollution by meeting environmental standards. However, once environmental targets are reached under regulation, for example, there is often no continuing incentive to go further.

Figure 10.3 Overview of environmental tax bases in EU-10 and other countries, 2004

	Cyprus	Czech Republic	Estonia	Hungary	Latvia	Lithuania	Malta	Poland	Slovenia	Slovakia	Bulgaria	Croatia	Romania	Turkey
Air/energy														
CO_2*			X						X					
SO_2		X	X	X	X	X		X	X	X		X		
NO_x		X	X	X	X	X		X	X	X		X		
Other air pollutants		X	X	X	X	X		X	X	X		X		
Fuels	X	X	X	X	X	X	X	X	X	X	X	X	X	X
S in fuels													X	
Transport														
Car sales				X	X	X	X	X		X	X		X	X
Annual circuation tax	X	X	X	X	X	X	X	X	X	X	X	X	X	X
Water														
Water effluents		X	X	X	X	X		X	X	X	X		X	
Waste														
Waste taxes		X	X		X			X		X		X	X	
Noise														
Aviation noise		X												
Products														
Tyres				X	X	X				X	X			
Beverage cont.		X		X										
Packaging			X	X	X	X		X		X				
Bags														
Pesticides														
CFCs		X		X	X					X			X	
Batteries				X	X					X	X			
Light bulbs					X	X				X				
PVC/phtalates										X				
Lubrication oil		X		X						X				
Fertilisers														
Paper, board					X	X				X				
Solvents														
Resources														
Raw materials			X	X	X	X					X	X		

Source: EEA, 2005.

Market-based instruments, on the other hand, use the more realistic pricing of goods and services to provide continuous incentives for European producers and consumers to reduce tax by producing and using more eco-efficient innovations. In addition, such instruments also provide more flexibility for companies with different technologies and cost structures to respond to the need for environmental improvements. However, the net impacts of such instruments are not as predictable as direct regulations, and a mix of policy tools may be required for the sake of environmental effectiveness and equity.

There are several forms of market-based instruments. These include taxes and charges on products and processes regarded as environmentally damaging; deposit schemes that provide a refund when the product or packaging is returned for recycling; and tradeable permits for pollution or some other activity such as fishing that needs to be limited. Permits are growing in popularity because they combine flexibility of response with the reasonable certainty that targets will be met.

A number of more recent EU environmental legislation includes specific provisions to allow governments to use these instruments to meet the targets, including the water framework and packaging waste directives. The 2005 EU emissions trading scheme for greenhouse gases aims at helping the EU to meet its shared commitment to the Kyoto Protocol targets and is the first major market-based instrument to be implemented at the EU level.

Market-based instruments in the Member States have mostly taken the form of environmental taxes and charges — for instance as a differential on fuel tax to encourage certain fuels such as low-sulphur diesel or lead-free petrol, or alternative fuels such as ethanol. Market-based instruments have also been widely used by the new EU Member States, particularly to mitigate air pollution. A number of European countries have also introduced taxes on non-renewable raw materials such as sand, gravel and limestone, or on products such as plastic bags. Many are aimed at encouraging the recycling of materials.

In the EU-15, the development of environmental taxes between 1996 and 2004 shows that progress has been made in the application of taxes across a wide range of areas (Figure 10.2). Interestingly, the new EU-10 Member States have made substantial progress on putting in place environmental taxes over a relatively short period of time, in particular, with regard to air pollutants, products and raw materials (Figure 10.3).

The effectiveness of market-based instruments

Evidence suggests that instruments work better if they are well-designed in themselves for the long term and as part of a wider package of instruments, if the reasons for having them and the way revenues will be used are clearly communicated to the public and if the levels at which 'prices' are set reflect both an incentive to producers and consumers to change behaviour and a realistic analysis of affordability.

Europe has a long tradition of imposing high taxes on motor fuels. Taxes, excluding VAT, make up more than half of the pump price of petrol in almost all of the EU-15. Partly as a result of the tax, the European car fleet is much more fuel efficient than that of USA, and has much lower emissions of carbon dioxide for every kilometre driven. New passenger cars in the EU need on average 6–7 litres per 100 kilometres; this figure is 10–11 litres per 100 kilometres for USA.

Several countries have introduced CO_2 taxes as an additional tool to achieve the objectives of climate change policy. In Denmark, industry reduced its CO_2 intensity by 25 % in seven years from 1993–2000; analysis has shown that at least 10 percentage points resulted from the CO_2 tax. The impact came about through both fuel switches and energy efficiency, each accounting for about half the CO_2 reduction.

A variety of other pay-as-you-go charging systems for road transport have been variously introduced in Europe. Traditionally powered cars have to pay a congestion charge for driving in the central London area. It imposes a flat rate for travelling in the city during daytime and has cut traffic volume by 15 %; it speeds traffic flow and generates revenue to improve the city's public transport system. Switzerland has had an environmental standard-dependent, kilometre-based charge for heavy vehicles since 2001. Austria and Germany have introduced similar charges for the use of road infrastructure, but these do not internalise environmental costs.

Per kilometre road pricing is likely to emerge more widely now that efficient satellite-based and computerised systems for monitoring and charging vehicles are becoming available. Increasingly, economy-damaging congestion is a main driver, but the environment will benefit too. Such systems would, advocates argue, be transparent, equitable, economically efficient and environmentally effective. Similar thinking is happening in other fields. Many European countries are switching from conventional fixed rates for water supply, or those based on the value of the property, to water metering. The evidence is that metering cuts general water use, typically by around 10 %.

The pragmatic mixing of market-based instruments with other measures is well illustrated by the water sector. It has rarely proved possible to introduce full market pricing for water where it might have had most effect, e.g. for agricultural irrigation. For example, during the drought in the summer of 2005 in southern Europe, where irrigation is generally the largest user of water, its use was primarily controlled through bans rather than charges. Paying more for a previously 'free' or cheap product or service was deemed to be too unpopular to be practicable. However, realistic waste water charges in the Netherlands, and their use to help companies reduce their polluted wastewater, was more cost-effective in complying with the urban waste water treatment directive than in countries that only built wastewater treatment plants.

Some market-based instruments raise revenues. The revenues from environmental taxes commonly go into the public coffers and can be used to offset other taxes, or to help finance government programmes and other actions that are beneficial for the environment. The revenues from environmental charges are usually meant to finance collective services from which the charge-payer benefits. Emissions trading systems raise revenue if the credits are auctioned, although giving them away without cost is the favoured option in practice. Finally, the reform of harmful subsidies may yield savings in the government budget, or provide revenues to finance incentives that can support more environmentally friendly technologies such as organic farming or renewable energy.

European and national subsidies to agriculture, fisheries, transport and energy production do not efficiently balance economic needs with long-term environmental integrity. Local subsidies can also encourage less environmentally friendly options. For example, when German cities such as Bremen, Dresden and Stuttgart investigated the size of the subsidies associated with the free use of their infrastructure given to car transport, they found that this averaged EUR 128 per citizen, much higher than the municipal subsidies for more environmentally sustainable public transport.

Anomalies in the tax system can accentuate environmental damage. For instance, aviation fuel and fuel for shipping are free from the heavy taxes that make up most of the cost of fuelling European road transport and, in some cases, trains. This international subsidy has, among other things, helped stimulate the boom in aviation. If taxes on all fuels were introduced, their impacts on the environment would be more transparent and, in time, reduced.

In some countries, the revenue from environmental taxes is used to lower other taxes, primarily those on labour. The Swedish programme over the period 2001–2010 will switch EUR 3.3 billion from labour taxes to environmental taxes. This environmental tax reform focuses on shifting the tax burden from the welfare-negative taxes on labour, capital and consumption to welfare-positive taxes on environmental externalities.

At the EU-15 level, energy tax revenues that make up almost 80 % of all environmental tax revenues have risen, and the average effective tax rate on labour (measured by implicit tax rate (ITR) which equals social security contributions of employers and wage earners plus other non-wage personal taxes on wages and salaries, divided by total pre-tax labour income) has dropped, indicating a small shift of the tax burden from labour to energy. Moreover, the overall energy efficiency in the EU has improved in parallel with increased energy taxation.

Equity, competitiveness and innovation concerns

Nevertheless, the energy-tax burden is unevenly spread over target groups, with the bulk resting on consumers. In the Nordic countries, for example, households consume about 20 % of all energy, but pay about 60 % of all energy taxes. By far the biggest contribution comes from taxes on motor fuels (petrol and diesel). Energy carriers such as coal, and heavy and light oil, typically used in manufacturing, are taxed at a much lower level.

This potential tax shift would not be large in the short term as long as energy taxes, which tend to impact the consumer more highly than other sectors, were not raised considerably. Options that might be more equitable include transport taxes that have a share of slightly more than 1 % in total tax revenues, and pollution and resource taxes which make up only 0.2 % of the total tax take in the EU-15. In considering these options, however, it should be remembered that revenue-raising, market-based instruments are primarily a tool of environmental policy; other tools are available to implement labour market policy.

Some environmental taxes can be socially inequitable, since the poorer members of society generally spend a greater proportion of their income on basic needs such as food, water and energy. Denmark, which has the largest range of environmental taxes in Europe, raising 10 % of government revenues this way, has found that energy taxes, and particularly the tax on electricity, impact the poor harder, although less than existing taxes on alcohol and tobacco, and VAT. On the other hand, transport taxes are relatively benign for the poor, and pollution taxes are neutral in their distributional impact.

The latest and most innovative form of market-based incentive is the tradeable 'pollution permit' aimed at limiting resource use and emissions. The EU greenhouse gas emissions trading scheme allocates permits for major companies in certain sectors to emit greenhouse gases. By limiting the allocation of permits to less than projected emissions, it creates a market in the permits. Companies that fall short of permits needed to cover their emissions can either reduce their emissions themselves, or, if that would be cheaper, buy from those who have permits to spare, perhaps because of investment in clean technologies. The scheme provides the auction option to Member States to a limited extent, but, currently, very little use is made of it.

The initial allocation of permits, for the period 2005–2007 and covering carbon dioxide only, is seen as a dress rehearsal for the next five years when Europe must meet its legally binding emissions targets under the Kyoto Protocol. It was introduced relatively seamlessly — in marked contrast to past efforts to introduce an EU-wide carbon dioxide and energy tax, which were abandoned after concerted opposition from several quarters.

There is no evidence that market-based instruments damage the competitiveness of the economy or of specific sectors. This is due to the design of the instruments; to exemption possibilities that avoid unacceptable cost impacts; and to measures that compensate those affected by recycling revenues. Such instruments can maintain or improve competitiveness by encouraging cost-effective and innovative responses to environmental demands.

There is evidence that the venture capital needed to connect technology development with market penetration is lacking for environmental innovations. Environmental technology is seen as more risky, and less of a niche market than biotechnology, computer software and telecommunications. Incentives may therefore be needed to stimulate the design and marketing of innovative and more eco-efficient technologies.

Most of the barriers to implementation of market-based instruments can be overcome by: progressive removal of subsides and regulations that contribute to environmental damage; recycling of saved revenues to provide incentives for eco-innovation; better design of instruments and mitigation measures to deal with inequities; progressive implementation to build up trust and confidence in the measures over time; and integration of market-based instruments for environmental policy into those for economic and social policy so that revenues can be used to support broader tax reforms.

10.4 Resource productivity and eco-innovation

Different resources require different approaches

Some 75–90 % of resources currently used are non-renewable, at least within timescales that are relevant to humans and many ecosystems. This compares with 50 % at the beginning of the last century. A better overall balance between using the stocks of non–renewables and the flows of renewable resources — mainly from bio-based and recycled sources — is essential for maintaining ecosystem services and can provide a strong incentive for eco-innovations.

There are several reasons to focus on improving non-renewable resource productivity over the coming decades. Some of the main ones are the changing nature of environmental pressures; a growing disparity in global non-renewable resource use; rising prices and competition for raw materials; increasing international security risks; and the need to boost EU competitiveness.

One example of a better balance towards renewable resources is the increased use of biomass to produce electricity, heating and transport fuels. This can provide both environmental benefits and an alternative source of income for those living in rural areas. However, biomass production might create additional pressures on biodiversity, soil and water resources, as well as taking land that could be used for food or other production. Bio-energy crops therefore need to be developed that can reduce soil erosion and compaction, minimise nutrient inputs into ground and surface waters, and use fewer pesticides and less water.

If these crops are then to be converted into biofuels for transport, new conversion technologies will have to be used, such as the biomass-to-liquids technology. The increased use of biomass and other renewable energies can also contribute to the reduction of Europe's dependence on energy imports, which otherwise is forecast to grow from 50 % in 2005 to 70 % in 2030.

Improving non-renewable and renewable resource productivity can help strengthen the synergies between environmental protection and growth. The 'clean, clever and competitive' initiative of the Dutch government in 2004 identified many ways in which European companies could achieve significant increases in resource productivity and at the same time reduce environmental pressures. Other studies in many Member States and at EU level have demonstrated that large potential economic and environmental gains can be made at the sector, company and household levels by reducing resource use.

Nevertheless, too much focus on reducing the overall use of resources can hide 'hot' flows of particularly damaging materials which require different approaches than for other materials. For example, the extraction of some metals or the handling of hazardous substances requires specific regulatory attention, although the complexities of estimating and regulating the environmental impacts of a single substance at different stages of its life-cycle are formidable. More research of the life cycle of such small resource volumes that have high environmental impacts would help improve understanding of how innovation can help mitigate these impacts.

Non-renewable resource productivity gains — a mixed picture

Trends in global non-renewable resource use suggest that the current European economic model cannot be followed by emerging economies since this would increase global consumption by between two and five times. Reports such as the Millennium Ecosystem Assessment suggest this would be simply unsustainable given the Earth's finite ecological capacity.

Effectively, amongst other measures, Europe, along with other parts of the developed world, needs to reduce its overall resource consumption by increasing its resource productivity, if it is to be better placed to adapt to future changes.

The average material productivity — raw materials consumed per unit of gross domestic product (GDP) — in the EU-25 is 1 kg/EUR, slightly less than in USA but twice as high as in Japan. The picture is similar for energy productivity, where the difference in the efficiency of the Japanese economy is even more pronounced, suggesting there is room for learning from this country's experiences in particular and others in general.

Over the past decades in Europe, there has been less focus on materials and energy productivity than on labour productivity. For example, between 1960 and 2002 labour productivity in Europe rose by 270 %, compared with 100 % for materials and just 20 % for energy. These trends are largely the result of a shift towards automated production (leading to more energy use and so offsetting energy productivity gains) and to structural changes in the economy. Earlier and full internalisation of environmental costs could have helped to further improve energy and resource productivities.

The cost structure of manufacturing in Germany, and probably throughout the larger economies of the EU, shows that materials and energy costs more than double labour costs. In this respect, the European economy could also be seen as over-consuming natural resources and under-consuming labour. Adjusting this imbalance could also reduce the degradation of the global environment, whilst contributing to Europe's long-term competitiveness and employment.

During the past decade Europe has achieved a relative decoupling of materials and energy use from GDP, but absolute resource use has remained steady. There are large differences between EU countries — in part depending on the modernity, type and level of the predominant industries — with material intensity varying from 11.1 kg/EUR of GDP in Estonia to 0.7 kg/EUR in France. Nonetheless, resource and energy productivity in western Europe is, on average, four times higher than in the EU Member States of central and eastern Europe. This indicates substantial opportunities to achieve greater parity in resource use between the EU-15 and EU-10 through technology transfer and other measures.

Outlooks to 2020 show a partial decoupling of water use, material flows, and waste from economic growth in the industry sector (Figure 10.4). This is expected to be achieved in part by continuing structural changes in the European economy away from resource intensive industries and towards the services sector. These structural changes will, however, allow Europe to continue exporting its environmental pressures by shifting the production of the goods we consume to developing countries.

These countries will also suffer from being the source of increasing greenhouse gas emissions resulting from the transport of goods back to Europe for our consumption. This makes it more difficult for developing countries to meet their emissions reductions targets while allowing Europe to meet some of its targets without having to significantly change its current consumption and productions patterns.

Developments in European manufacturing industry demonstrate the potential for reducing energy use at the same time as increasing economic output. Between 1990 and 2003, the sector's final energy consumption fell by almost 8 % while its added value rose by 17 %. Projections suggest that substantial improvements in industrial energy intensity could continue both under baseline assumptions and in a climate-mitigation scenario. Under these, the energy needed to produce one unit of economic added value in 2030 would be almost half 1990 levels.

The decrease in energy intensity can be explained partly by structural changes to the economy. But it is also the result of improvements in energy efficiency, influenced by technological innovation. Looking ahead, a recent Commission green paper outlines how energy efficiency measures could, by 2020, improve the energy consumption in EU-25 by more than 20 %, saving EUR 60 billion and creating, directly or indirectly, a million new jobs. This would translate into citizens each saving EUR 200–1 000 per year.

Some of the savings would be achieved if the EU directive on energy performance of buildings (2002/91/EC) were fully implemented. If the energy certification provisions of the directive were improved and extended to the renovation of older buildings, the savings could be almost doubled and some 250 000 skilled jobs created. In turn, this could stimulate innovation in the development of new and sustainably produced products and materials.

Recent studies also indicate that the adoption of energy efficiency policies have accelerated gains in efficiency, by fostering new, energy-efficient technologies. For example, refrigerators significantly improved after the introduction of energy efficiency labels and standards. Additionally, it has been observed that countries which introduce strict regulatory standards bring new technologies to world markets faster than their competitors.

The data are less robust for gains in materials efficiency because of the relative lack of interest in resource productivity compared with labour and energy productivity. However, a recent German study shows the potential for savings of EUR 5–10 billion on material input costs in small- and medium-sized companies across just four sectors — metals manufacturing, construction, electricity generation and distribution, chemicals and synthetic products.

Furthermore, a UK study shows that waste minimisation in manufacturing produced savings in annual operating costs of EUR 3–5 billion. Other studies indicate that once the process of identifying materials and energy savings begins, other, often large, eco-efficiency gains are also identified and implemented, providing a stream of unintended secondary benefits that are rarely captured by initial estimates of cost savings.

In general, a lack of awareness of the real costs of obtaining, using and disposing of materials and energy is a significant barrier to the wider implementation of many eco-innovations. Both the internal costs to companies and the external costs to societies of energy and materials consumption are generally hidden from the minds of decision-makers. For example, at a company level, savings as a result of waste minimisation are usually identified as reductions in waste disposal costs. However, the total savings available include the reduced purchasing and processing costs of not handling 'unnecessary'

Figure 10.4 Industry — decoupling outlooks to 2020 for key environmental resources and pressures

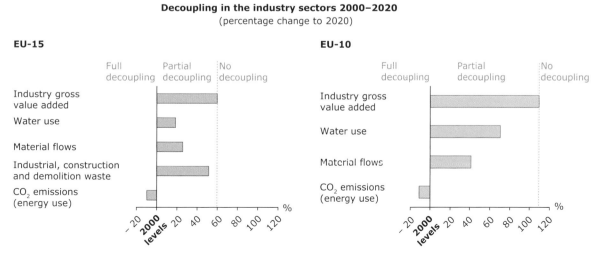

Source: EEA, 2005.

materials, which can be more than double the waste disposal costs.

Creating the conditions for future eco-innovations

The EU can move towards a more balanced economic development that is underpinned by eco-innovation and the recognition of the economy's dependence on the environment. Under the relaunched Lisbon agenda, the contribution that eco-innovation can make to economic growth and employment is fully recognised. Already the EU's eco-industries, which employ more than two million people and are growing by around 5 % a year, account for about one-third of the global market. Exports grew by around 8 % in 2004, producing a trade surplus for the EU of some EUR 600 million.

Equally important to the encouragement of eco-innovation is the promotion of a culture favourable to research and development. There is a proportionally lower number of annual patent applications in the EU. In 2002, research and development expenditure as a proportion of GDP in the EU-25 (at 1.93 %) lagged behind both Japan (3.12 %) and USA (2.76 %). Within the EU-25, research and development expenditure in the EU-15 outstrips that in the EU-10 (Figure 10.5).

Figure 10.5 Public and private sector expenditure on research and development in EU-25

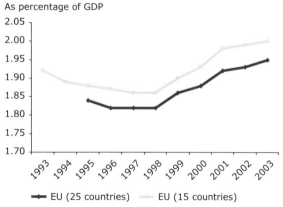

Source: Eurostat, 2005.

The strategic importance of investment in research and development under both the Lisbon and the Sustainable Development strategies was recognised at the 2002 Barcelona European Council, where it was agreed that overall EU spending on research and development should gradually increase and reach 3 % of GDP by 2010.

At the same time, the Commission has proposed the development of an action plan to tackle obstacles to the development, take-up and use of environmental technologies, and the Parliament agreed to this proposal. This has resulted in the EU environmental technologies action plan (ETAP) which also provides a framework for action by Member States. The EU's new research framework programme (FP7) for the period 2007–2013 includes some EUR 2.5 billion for the environment, an increase of nearly 60 % over FP6. In addition, the European Commission has proposed a competitiveness and innovation framework programme for 2007–2013 totaling EUR 4.2 billion, of which some EUR 500 million will be dedicated to supporting eco-innovation initiatives.

There are also benefits to be realised by recycling savings realised from resource productivity improvements towards investment in innovation. A recent German study that modelled the effects of dematerialisation on economic growth and the state budget concluded that if materials and energy savings were reinvested in research and development and engineering strategies, it would lead to 2.3 % GDP growth, the creation of an additional 750 000 jobs, and decreased public spending on social welfare.

Public authorities can also encourage more eco-efficient procurement policies. The World Summit on Sustainable Development in Johannesburg in 2002 called on 'authorities at all levels to promote public procurement policies that encourage [the] development and diffusion of environmentally sound goods and services'.

Public authorities in the EU spend an estimated EUR 2 trillion on goods, works and services every year providing significant opportunities for encouraging a large and stable market for eco-innovations. For example, public administrations could contribute to 18 % of Europe's Kyoto obligations by switching to renewable energy sources. A survey carried out in the EU in 2003 found that almost a fifth of public authorities claimed to have adopted environmental procurement policies in one or more areas, whether buying organic produce for their canteens or using environmentally certified timber for construction. Many said they would do more if they were provided with better advice on best practice.

Many municipal authorities have adopted policies for renewing their vehicle fleets with cars that use low-emissions fuels, and for reducing their own greenhouse gas emissions by investing in renewable energy generation or combined heat- and powerplants for their buildings, including authority-owned housing stock. Some have joined global networks of cities, such as ICLEI (International Council for Local Environment Initiatives), formed under Article 28 of Agenda 21 agreed at the Earth Summit in Rio de Janeiro, Brazil, in 1992. European cities also joined hundreds of others in San Francisco in USA on World Environment Day 2005 to sign 'urban environmental accords' covering energy, waste production, urban design and other goals.

Awareness and information campaigns specifically targeted at individual sectors can help overcome the lack of information on the real costs of both wastes and pollution, and on how to reduce them at source. For example, Europe's chemical industries, particularly those concerned with the production and use of fertilisers and pesticides, are working with farmers to achieve more eco-efficient use of their products. At the same time they are developing such innovations as biocleaning plants, which can replace septic tanks, and the use of living cells from moulds, yeasts and bacteria, that can be used as 'cell factories' to produce enzymes for industry, as well as to make antibiotics, vitamins, vaccines and proteins for medical use. In this they have sometimes been encouraged by regulations and taxes on the overuse of their products as well as by incentives to develop less environmentally damaging chemicals.

Environmental labelling and other consumer initiatives are part of the policy focus on information to the public.

Labelling schemes related to energy efficiency have been particularly successful. Given the choice, consumers will often buy energy- and water-efficient white goods, perceiving a benefit both in their fuel bills and to the environment. Initiatives, such as the forest and marine stewardship councils that guide consumers on sustainable products, also help in this respect.

Policies to enhance eco-innovation could also usefully address the financial, institutional and behavioural factors that 'lock in' current patterns of consumption and production. Innovation studies show that a stable policy framework, guided by long-term overall targets and stimulated by flexible policy packages which address interrelated economic realities, are needed to interact with the dynamics of many actors and stakeholders. The Dutch transition approach illustrates one way of tackling this.

Making progress in eco-innovation will be a complex process. However, it could be greatly helped by an increasing involvement of the public in establishing the acceptable risks of innovation balanced against the dangers of inertia in the face of climate change and other environmental threats. Eurobarometer confirms that citizens are concerned about the environment, and understand that environmental protection is often an incentive to innovation rather than an obstacle to economic performance. This provides support for increased, transparent, innovation and contributes more and better jobs to a sustainable future.

10.5 Summary and conclusions

The United Nations Millennium Ecosystem Assessment defines the natural environment in terms of the services, sources, sinks and spaces it provides. There is a growing body of influential opinion, expressed by business leaders, scientists and opinion-formers that environmental concerns and economic growth are far from mutually exclusive; rather they are intrinsically linked. However, the recognition of the real economic value of the natural world, and of our dependence on it for our continued prosperity, is still poorly understood, largely because the connections are relatively invisible.

Environmental policy measures have served the European society, its economy and its environment well over the last 30 years. Much has been achieved across Europe in the past decades to improve the quality of the air we breathe and the water we drink, and to dispose of many of the wastes that we generate. Policies have so far mainly addressed environmental deterioration from easily visible point sources of pollution. In doing so, Europe has encouraged technological advances and developed internationally recognised expertise in several eco-technologies and in environmental policy-making.

Current environmental challenges are more complex, diffuse and less visible than in the past, and increasingly robust science has demonstrated that environmental deterioration is continuing. Our consumption patterns are driving a rapidly rising use of natural resources in Europe and globally. As a result, our health continues to be damaged, the pollution of our waters continues, our biodiversity is still in decline, and our emissions of greenhouse gases have not fallen sufficiently to avoid climate change.

Our analyses of these issues suggest that we now have to consider tackling diffuse sources of pollution, whether, for example, from the cars we drive or from the way farmers respond to the increasing demand for cheap and plentiful foods. Taking action to address these diffuse sources will require integrated measures across economic sectors — agriculture, transport, manufacturing and energy production — and actions that engage with socio-economic factors such as household size, urbanisation, personal consumption and waste production.

Three interlinked approaches could help to realise the benefits available to Europe from tackling these recent or emerging realities: stronger and more coherent environmental policy integration, particularly through institutional and financial reform; the internalisation of the real costs of our use of the natural world into market prices which will contribute to the more efficient use of renewable resources, energy and materials; and the more efficient use of renewable and non-renewable resources via measures that stimulate eco-innovation.

There are sometimes trade-offs between economic and environmental priorities but they can be exaggerated. Many costs are short term (from two to five years), and can be eliminated through dynamic efficiency gains from innovation. EU citizens and businesses recognise that well designed environmental regulations can encourage innovation, particularly when they are predictably phased in over the longer term. Recent, more integrated policy approaches, especially using market-based instruments supplemented with regulations and information campaigns, are more cost-effective and can stimulate innovation better than most policy measures of the 1970s and 1980s.

There are great opportunities in Europe to make better use of the latest technologies in energy, transport and materials use. These can help achieve the eco-efficiency gains needed to avoid breaching environmental thresholds and to provide emerging economies with the ecological space to expand. However, substantial barriers to the exploitation of these opportunities remain, especially environmentally damaging subsidies and the absence of financial incentives to eco-innovate.

Ecological tax reform, alongside a shift to eco-friendly incentives, can help protect the environment, boost innovation and employment, and help deal with problems posed by an ageing population. Such reforms could include a gradual shift over 20–30 years of much of the tax-base away from income (which is at risk due to a dwindling workforce) and from capital (which often discourages investment and innovation), towards taxes on consumption, pollution and inefficient use of energy and materials — thus providing a wider tax-base from an ageing population and lifetime consumption.

The timescale to put in place effective policy measures can be 5–20 years, whilst the harmful impacts, and the time to reverse them, can take up to 100 years or more. Policy actions taken now to prevent the costly consequences of inaction later. Past examples show that inaction can be both very costly and long-term as the histories of asbestos, acid rain, the ozone hole, polychlorinated biphenyls (PCBs), and diminished fish stocks illustrate. In contrast, where action has been taken, all evidence indicates that the costs are usually overestimated while the benefits are underestimated.

The EU has already taken action towards greater coherence and integration of environmental and economic considerations. For instance, the development of the thematic strategies under the sixth environment action programme supports new cross-departmental and multi-stakeholder engagement. Meanwhile there has been a quiet revolution in the strategic management and coordination of the European Commission and Council activities. The potential of the EU's move towards multiannual and annual planning offers the potential for putting environmental integration into practice. Furthermore, Member States are individually taking their own action in support of such integration and internalisation.

The concepts of institutional and financial reforms are, of themselves, significant drivers of innovation. Some of the resulting changes may be painful, such as the reform of possibly outdated and environmentally harmful systems of subsidies. Nonetheless, there are proven cases and studies to indicate that environmental care and management creates economic opportunities and jobs — confirming that by being clever and clean, Europe can also be competitive, as eco-efficient innovations contribute to the broader social and economic goals of the Lisbon agenda.

This assessment of the state of the European environment demonstrates that the current and future challenges to our environment and the services it provides are long-term and strongly linked. They can best be managed by using similarly interlinked policy measures. These solutions often require behavioural changes from social and economic agents that are encouraged and facilitated by government actions. Progress will often be gradual and over several decades. This timeframe can provide space for policy learning and for attracting the broad support of both economic activities and citizens.

Eurobarometer polls show that citizens understand the importance of the environment for Europe's future welfare and are willing to take action, but only if others do. This provides an opportunity to engage with the public on how to tackle together the long-term environmental challenges we face. Their support is essential for the success of more integrated and innovative policy measures. Actions are needed now. The health, social and economic costs of inaction can be very large as experiences illustrate. And Europe is well placed to lead the way by creating smarter, cleaner, more competitive and more secure European societies.

References and further reading

Introduction
European Environment Agency, 2005. *European environment outlook*. EEA Report No 4/2005.

United Nations/World Bank, 2005. *Millennium Ecosystem Assessment*.

VROM, 2004. *Clean, clever and competitive*, Knowledge document for Dutch informal environmental council.

Integration
European Environment Agency, 2005. *Environmental policy integration in Europe — Administrative culture and practices*. EEA Technical report No 5/2005.

European Environment Agency, 2005. *Environmental policy integration in Europe — State of play and an evaluation framework*. EEA Technical report No 2/2005.

Green Funds. (See www.sustainablebusiness.com — accessed 24/10/2005).

Internalisation using market-based instruments
European Environment Agency, 2005. *Market-based instruments for environmental policy in Europe*. EEA report, Copenhagen (in print).

European Environment Agency, 2005. *Climate change and a European low-carbon energy system*. EEA Report No 1/2005, Copenhagen.

European Environment Agency, 2005. *Effectiveness of packaging waste management systems in selected countries: an EEA pilot study.* EEA Report 3/2005, Copenhagen.

European Environment Agency, 2005. *Effectiveness of urban wastewater treatment policies in selected countries: an EEA pilot study.* EEA Report 2/2005, Copenhagen.

European Environment Agency, 2005. *Environmental policy integration in Europe — State of play and an evaluation framework.* Technical report No 2/2005, Copenhagen.

European Environment Agency, 2005. *Household consumption and the environment.* EEA report, Copenhagen (in print).

European Environment Agency, 2004. *Impacts of Europe's changing climate.* EEA Report No 2/2004, Copenhagen.

European Environment Agency, 2005. *European environment outlook.* EEA Report No 4/2005, Copenhagen.

European Environment Agency, 2004. *Ten key transport and environment issues for policy-makers, TERM 2004 — Indicators tracking transport and environment integration in the EU,* Copenhagen.

European Environment Agency, 2004. *Agriculture and the environment in the EU accession countries — Implications of applying the EU common agricultural policy.* Environmental issue report No 37, Copenhagen.

European Commission, 1998. Towards Sustainability — *The fifth environment action programme (1992–2000).* Decision 2179/98. 10.10.1998 OJ L275/1.

UNDP, 2004. *Human Development Report 2004 —* Indicator 12 Technology: Diffusion and creation. http://hdr.undp.org/statistics/data/pdf/hdr04_table_12.pdf.

Resource productivity and eco-innovation
Arthur D. Little, FHI ISI, Wuppertal Institute, 2005. *Studie zur Konzeption eines Programms für die Steigerung der Materialeffizienz in mittelständichen Unternehmen, Abschlussbericht für das BMWA.*

Cambridge Econometrics and AEA Technology, 2003. *The benefits of greener business — the cost of unproductive use of resources.* Unpublished. A report submitted to the European Environment Agency.

Enerdata, ISI-FhG, ADEME, 2001. *Energy efficiency in the European Union 1990–2000,* SAVE-ODYSSEE project on energy efficiency indicators.

Environmental Technologies Action Plan, 2005. Conclusions of the ETAP working conference 'Financial instruments for sustainable innovations', 21–22 October 2004, Amsterdam.

European Commission, 2001. *A sustainable Europe for a better world: A European Union strategy for sustainable development* (Commission's proposal to the Gothenburg European Council), COM(2001)264 final.

European Commission, 2002. *Towards a European strategy for the security of energy supply,* Green Paper COM (2002)769 final.

European Commission, 2005. *Doing more with less,* Green paper on energy efficiency.

European Commission, 2005. *Integrated guidelines for growth and jobs (2005–2008),* Communication from the President, in agreement with Vice-President Verheugen and Commissioners Almunia and Spindla, COM(2005)141 final, 2005/0057 (CNS).

European Council, 1991. Directive 91/271/EEC on urban waste water treatment.

European Environment Agency, 2005. *Environmental policy integration in Europe — Administrative culture and practices.* EEA Technical report No 5/2005, Copenhagen.

European Environment Agency, 2005. *Environmental policy integration in Europe — State of play and an evaluation framework,* EEA Technical report No 2/2005, Copenhagen.

European Environment Agency, 2005. *Sustainable use and management of resources*, EEA Report, Copenhagen (in print).

European Environment Agency, 2005. Briefing: *How much biomass can Europe use without harming the environment?* EEA Briefing series, Copenhagen.

European Commission, 2001. *European governance — a White Paper* COM(2001) 428 final 25.07.2001.

European Parliament and Council, 1994. Directive 94/62/EC of 20 December 1994 on packaging and packaging waste.

European Parliament and Council, 2002. Directive 2002/91/EC of 16 December 2002 on the energy performance of buildings.

European Parliament and Council, 2000. Directive 2000/60/EC establishing a framework for Community action in the field of water policy also known as the water framework directive (WFD).

Fischer, H. *et al.*, 2004. Wachstums- und Beschäftigungsimpulse rentabler Materialeinsparungen. In: *Hamburgisches Welt-Wirtschafts-Archiv*. 84. Jahrang, Heft 4.

International Energy Agency, 2004. *Oil crises and climate challenges: 30 years of energy use in IEA countries.*

International Energy Agency, 2005. *The experience with energy efficiency policies and programmes in IEA countries, Learning from the critics*, IEA Information Paper.

Joest, F., 2001. 'An evolutionary perspective on structural change and the role of technology', In Binder, M., Jaenicke, M., Petschow, U. *Green industrial restructuring: International case studies and theoretical interpretations*, Springer.

Lapillonne, B. and Eichammer, W., 2004. *Energy efficiency trends in industry in the EU-15*, Assessment based on Odyssee-Indicators.

United Nations, 1992. *Agenda 21* — report of the Earth Summit in Rio de Janeiro, New York.

United Nations, 2002. *Report of the World Summit on Sustainable Development in Johannesburg*, New York. www.johannesburgsummit.org/.

Van der Voet, *et al.*, 2004. *Policy Review on Decoupling: Development of indicators to assess decoupling of economic development and environmental pressure in the EU-25 and AC-3 countries*. CML report 166, Leiden: Institute of Environmental Sciences (CML), Leiden University — Department Industrial Ecology.

VROM, 2004. *Clean, clever and competitive*, Knowledge document for Dutch informal environmental council.

Summary and conclusions

Europen Environment Agency, 2001. *Late lessons from early warnings: the precautionary principle 1896–2000*. Environmental issue report No 22.

European Commission, 2005. Communication from the Commission to the Council and the European Parliament on Thematic Strategy on air pollution. COM (2005) 446 final.

Forest Stewardship Council. (See www.fscus.org/ — accessed 19/10/2005).

Marine Stewardship Council. (See www.msc.org/ — accessed 19/10/2005).

B Core set of indicators

B

Core set of indicators

Setting the scene	255
Air pollution and ozone depletion	
01 Emissions of acidifying substances	256
02 Emissions of ozone precursors	260
03 Emissions of primary particles and secondary particulate precursors	264
04 Exceedance of air quality limit values in urban areas	268
05 Exposure of ecosystems to acidification, eutrophication and ozone	272
06 Production and consumption of ozone depleting substances	276
Biodiversity	
07 Threatened and protected species	280
08 Designated areas	284
09 Species diversity	288
Climate change	
10 Greenhouse gas emissions and removals	292
11 Projections of greenhouse gas emissions and removals	296
12 Global and European temperature	300
13 Atmospheric greenhouse gas concentrations	304
Terrestrial	
14 Land take	308
15 Progress in management of contaminated sites	312
Waste	
16 Municipal waste generation	316
17 Generation and recycling of packaging waste	320
Water	
18 Use of freshwater resources	324
19 Oxygen consuming substances in rivers	328
20 Nutrients in freshwater	332
21 Nutrients in transitional, coastal and marine waters	336
22 Bathing water quality	340
23 Chlorophyll in transitional, coastal and marine waters	344
24 Urban wastewater treatment	348
Agriculture	
25 Gross nutrient balance	352
26 Area under organic farming	356
Energy	
27 Final energy consumption by sector	360
28 Total energy intensity	364
29 Total energy consumption by fuel	368
30 Renewable energy consumption	372
31 Renewable electricity	376
Fisheries	
32 Status of marine fish stocks	380
33 Aquaculture production	384
34 Fishing fleet capacity	388
Transport	
35 Passenger transport demand	392
36 Freight transport demand	396
37 Use of cleaner and alternative fuels	400

PART B Core set of indicators

Setting the scene

Part B of the report presents a four-page summary of each of the 37 indicators in the EEA core set based on data available mid-2005. For each indicator we give the key policy question, the key message and an assessment. This is followed by information on the indicator definition, the rationale behind the indicator, the policy context and a section on uncertainty.

On top of being an important source of information in its own right, the core set underpins the integrated assessment in Part A and also the country analysis in Part C. References to the indicators and how they have been used can be found in those parts.

The complete indicator specifications, technical explanations, caveats and assessments are available at EEA's website (currently at www.eea.eu.int/coreset). The assessments will be updated on a regular basis as new data becomes available.

The EEA identified a core set of indicators in order to:

- provide a manageable and stable basis for indicator-based assessments of progress against environmental policy priorities;

- prioritise improvements in the quality and coverage of data flows, which will enhance comparability and certainty of information and assessments;

- streamline contributions to other indicator initiatives in Europe and beyond.

The establishment and development of the EEA core set of indicators was guided by the need to identify a small number of policy-relevant indicators that are stable, but not static, and that give answers to selected priority policy questions. They should, however, be considered alongside other information if they are to be fully effective in environmental reporting.

The core set covers six environmental themes (air pollution and ozone depletion, climate change, waste, water, biodiversity and terrestrial environment) and four sectors (agriculture, energy, transport and fisheries).

The indicators in the core set were selected from a much larger set on the basis of criteria widely used elsewhere in Europe and by the OECD. Particular attention was given to relevance to policy priorities, objectives and targets; the availability of high-quality data over both time and space, and the application of well-founded methods for indicator calculation.

The core set, and particularly its assessments and key messages, is targeted mainly at policy makers at the EU and national level who can use the outcomes to inform progress with their policies. EU and national institutions can also use the core set to support streamlining of data flows at the EU level.

Environmental experts can use it as a tool for their own work by using the underlying data and methodologies to do their own analysis. They can also look at the set critically, give feedback and so contribute to future EEA core set developments.

General users will be able to access the core set on the web in an easily understandable way, and use available tools and data to do their own analyses and presentations.

Core set of indicators | Air pollution and ozone depletion

01 Emissions of acidifying substances

Key policy question

What progress is being made in reducing emissions of acidifying pollutants across Europe?

Key message

Emissions of acidifying gases have decreased significantly in most EEA member countries. Between 1990 and 2002, emissions decreased by 43 % in the EU-15 and by 58 % in the EU-10, despite increased economic activity (GDP). For all EEA member countries, excluding Malta, emissions decreased by 44 %.

Indicator assessment

Emissions of acidifying gases have decreased significantly in most EEA member countries. In the EU-15, emissions decreased by 43 % between 1990 and 2002, mainly as a result of reductions in sulphur dioxide emissions, which contributed 77 % of the total reduction. Emissions from the energy, industry and transport sectors have all been significantly reduced, and contributed 52 %, 16 % and 13 % respectively of the total reduction in weighted acidifying gas emissions. This reduction is due mainly to fuel switches to natural gas, economic restructuring of the new Länder in Germany and the introduction of flue gas desulphurisation in some power plants. So far, the reductions have resulted in the EU-15 being on track to reaching the overall target for reducing acidifying emissions in 2010.

Emissions of acidifying gases have also decreased significantly in the EU-10 and candidate countries (CC-4). Emissions in the EU-10 Member States decreased by 58 % between 1990 and 2002, also mainly as a result of the large reduction in sulphur dioxide emissions, as in the EU-15 countries.

The reduction in emissions of nitrogen oxides is due to abatement measures in road transport and large combustion plants.

Indicator definition

The indicator tracks trends since 1990 in anthropogenic emissions of acidifying substances: nitrogen oxides, ammonia, and sulphur dioxide, each weighted by their acidifying potential. The indicator also provides information on changes in emissions by the main source sectors.

Indicator rationale

Emissions of acidifying substances cause damage to human health, ecosystems, buildings and materials (corrosion). The effects associated with each pollutant depend on its potential to acidify and the properties of the ecosystems and materials. The deposition of acidifying substances still often exceeds the critical loads of ecosystems across Europe.

The indicator supports assessment of progress towards implementation of the Gothenburg Protocol under the 1979 Convention on Long-range Transboundary Air Pollution (CLRTAP) and the EU Directive on National Emissions Ceilings (NECD) (2001/81/EC).

Policy context

Emission ceiling targets for NO_x, SO_2 and NH_3 are specified in both the EU National Emission Ceilings Directive (NECD) and the Gothenburg Protocol under the United Nations Convention on Long-range Transboundary Air Pollution (CLRTAP). Emission reduction targets under NECD for the EU-10 have been specified in the treaty of accession to the European Union 2003.

The NECD generally involves slightly stricter emission reduction targets for 2010 than the Gothenburg Protocol for the EU-15 countries.

Emissions of acidifying substances   PART B

Figure 1	Emission trends of acidifying pollutants (EEA member countries), 1990–2002

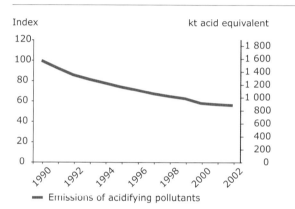

— Emissions of acidifying pollutants

Note: Data not available for Malta.

Data source: Data from 2004 officially reported national total and sectoral emissions to UNECE/EMEP Convention on Long-range Transboundary Air Pollution.

Figure 2	Emission trends of acidifying pollutants (EU-15), 1990–2002

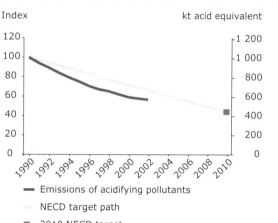

— Emissions of acidifying pollutants
··· NECD target path
■ 2010 NECD target

Note: Data source: Data from 2004 officially reported national total and sectoral emissions to UNECE/EMEP Convention on Long-range Transboundary Air Pollution.

Figure 3	Change in emission of acidifying substances (EFTA-3 and EU-15) compared with 2010 NECD targets (EU-15 only), 1990–2002

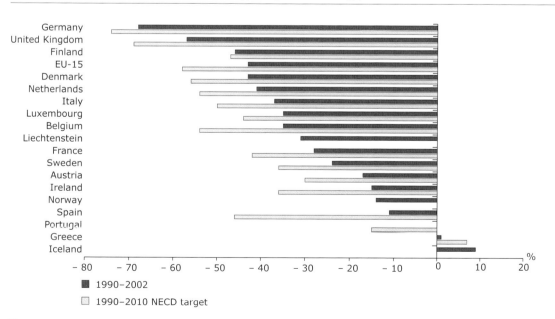

■ 1990–2002
□ 1990–2010 NECD target

Note: Data source: Data from 2004 officially reported national total and sectoral emissions to UNECE/EMEP Convention on Long-range Transboundary Air Pollution (Ref: www.eea.eu.int/coreset).

Core set of indicators | Air pollution and ozone depletion

Figure 4 Change in emission of acidifying substances (CC-4 and EU-10) compared with 2010 NECD targets (EU-10 only), 1990–2002

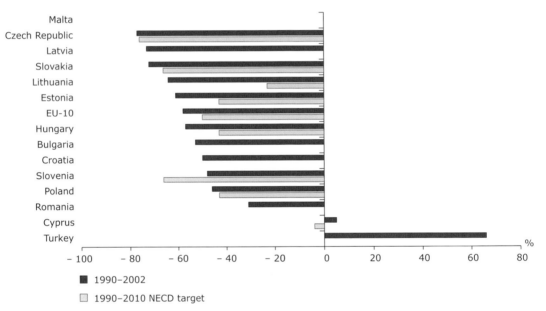

Note: Data not available for Malta.

Data source: Data from 2004 officially reported national total and sectoral emissions to UNECE/EMEP Convention on Long-range Transboundary Air Pollution (Ref: www.eea.eu.int/coreset).

Indicator uncertainty

The use of acidifying potential factors leads to some uncertainty. The factors are assumed to be representative of Europe as a whole; different factors might be estimated on the local scale.

The EEA uses data officially submitted by EU Member States and other EEA member countries which follow common guidelines on the calculation and reporting of emissions for air pollutants.

NO_x, SO_2 and NH_3 estimates in Europe are thought to have an uncertainty of about ± 30 %, 10 % and 50 % respectively.

Emissions of acidifying substances   PART B

Figure 5 Contribution to total change in acidifying pollutant emissions for each sector and pollutant (EU-15), 2002

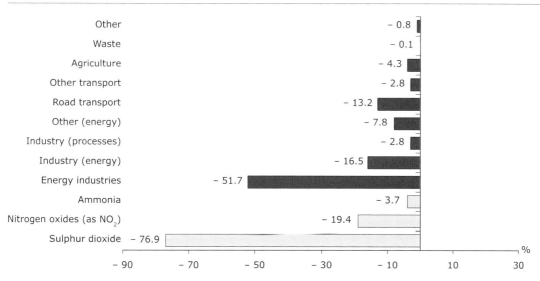

Note: 'Contribution to change' plots show the contribution to the total emission change between 1990–2002 made by a specified sector/pollutant.

Data source: Data from 2004 officially reported national total and sectoral emissions to UNECE/EMEP Convention on Long-range Transboundary Air Pollution (Ref: www.eea.eu.int/coreset).

02 Emissions of ozone precursors

Key policy question

What progress is being made in reducing emissions of ozone precursors across Europe?

Key message

Emissions of ozone-forming gases (ground-level ozone precursors) were reduced by 33 % across the EEA member countries between 1990 and 2002, mainly as a result of the introduction of catalysts in new cars.

Indicator assessment

Total emissions of ozone precursors were reduced by 33 % across the EEA member countries between 1990 and 2002. For the EU-15 countries, emissions were reduced by 35 %.

Emission reductions in the EU-15 since 1990 are due mainly to the further introduction of catalytic converters for cars and increased penetration of diesel, but also as a result of the implementation of the solvents directive in industrial processes. Emissions from the energy and transport sectors have both been significantly reduced, and contributed 10 % and 65 % respectively of the total reduction in weighted ozone precursor emissions. Emission reductions of the ozone precursors covered by the national emission ceilings directive (non-methane volatile organic compounds, NMVOCs, and nitrogen oxides, NO_x) have resulted in the EU-15 being on track towards reaching the overall target for reducing these emissions in 2010.

Emissions of non-methane volatile organic compounds (38 % of total weighted emissions) and nitrogen oxides (48 % of total weighted emissions) contributed the most to the formation of tropospheric ozone in 2002. Carbon monoxide and methane contributed 13 % and 1 % respectively. The emissions of NO_x and NMVOC were reduced significantly between 1990 and 2002, contributing 37 % and 44 % respectively of the total reduction in precursor emissions.

In the EU-10 ([1]), total ozone precursor emissions were reduced by 42 % between 1990 and 2002. Emissions of non-methane volatile organic compounds (32 % of the total) and nitrogen oxides (51 % of the total) were the most significant pollutants contributing to the formation of tropospheric ozone in EU-10 countries in 2002.

Indicator definition

This indicator tracks trends since 1990 in anthropogenic emissions of ozone precursors: nitrogen oxides, carbon monoxide, methane and non-methane volatile organic compounds, each weighted by their tropospheric ozone-forming potential. The indicator also provides information on changes in emissions by the main source sectors.

Indicator rationale

Ozone is a powerful oxidant and tropospheric ozone can have adverse effects on human health and ecosystems. The relative contributions of ozone precursors can be assessed on the basis of their tropospheric ozone-forming potential (TOFP).

Policy context

Emission ceiling targets for NO_x and NMVOCs are specified in both the EU National Emission Ceilings Directive (NECD) and the Gothenburg Protocol under the United Nations Convention on Long-range Transboundary Air Pollution (CLRTAP). Emission reduction targets for the EU-10 under NECD have been specified in the treaty of accession to the European

---

([1]) Data from Malta not available.

Emissions of ozone precursors

Figure 1 Emission trends of ozone precursors (ktonnes NMVOC-equivalent) for EEA member countries, 1990–2002

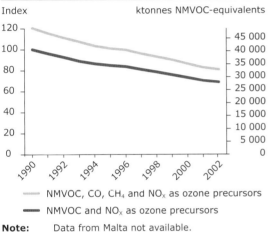

- NMVOC, CO, CH₄ and NOₓ as ozone precursors
- NMVOC and NOₓ as ozone precursors

Note: Data from Malta not available.
Data source: Data from 2004 officially reported national total and sectoral emissions to UNECE/EMEP Convention on Long-range Transboundary Air Pollution and the UNFCCC.

Figure 2 Emission trends of ozone precursors (ktonnes NMVOC-equivalent) for EU-15, 1990–2002

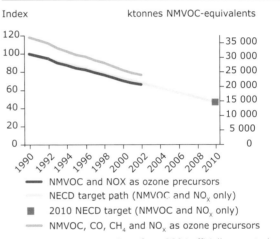

- NMVOC and NOX as ozone precursors
- NECD target path (NMVOC and NOₓ only)
- 2010 NECD target (NMVOC and NOₓ only)
- NMVOC, CO, CH₄ and NOₓ as ozone precursors

Note: Data source: Data from 2004 officially reported national total and sectoral emissions to UNECE/EMEP Convention on Long-range Transboundary Air Pollution and the UNFCCC.

Figure 3 Change in emission of ozone precursors (EFTA-3 and EU-15) compared with 2010 NECD targets (EU-15 only), 1990–2002

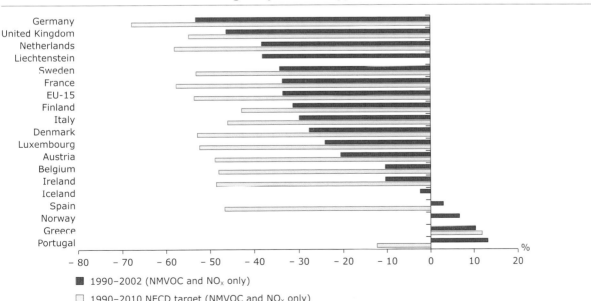

■ 1990–2002 (NMVOC and NOₓ only)
□ 1990–2010 NECD target (NMVOC and NOₓ only)

Note: Data source: Data from 2004 officially reported national total and sectoral emissions to UNECE/EMEP Convention on Long-range Transboundary Air Pollution and the UNFCCC (Ref: www.eea.eu.int/coreset).

Core set of indicators | Air pollution and ozone depletion

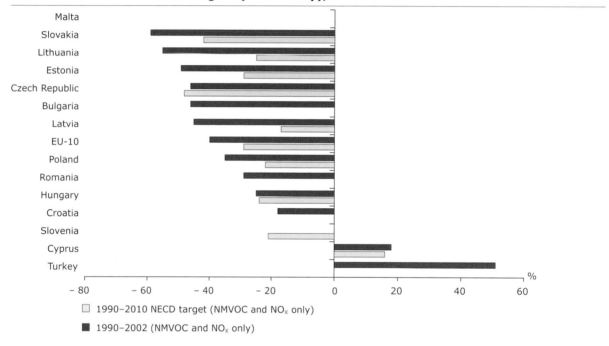

Figure 4 Change in emission of ozone precursors (CC-4 and EU-10) compared with 2010 NECD targets (EU-10 only), 1990–2002

☐ 1990–2010 NECD target (NMVOC and NO_x only)
■ 1990–2002 (NMVOC and NO_x only)

Note: Data from Malta not available.

Data source: Data from 2004 officially reported national total and sectoral emissions to UNECE/EMEP Convention on Long-range Transboundary Air Pollution and the UNFCCC (Ref: www.eea.eu.int/coreset).

Union 2003. There are no specific EU emission targets set for carbon monoxide (CO) or methane (CH_4).

The NECD generally involves slightly stricter emission reduction targets than the Gothenburg Protocol.

Indicator uncertainty

The EEA uses data officially submitted by EU Member States and other EEA member countries which follow common guidelines on the calculation and reporting of emissions for the air pollutants NO_x, NMVOC and CO, and IPCC for the greenhouse gas CH_4.

NO_x, NMVOC, CO and CH_4 emission estimates in Europe are thought to have an uncertainty of about ± 30 %, 50 %, 30 % and 20 % respectively. The use of ozone formation potential factors leads to some uncertainty. The factors are assumed to be representative for Europe as a whole; uncertainties are larger and other factors are more relevant on the local scale. Incomplete reporting and resulting intrapolation and extrapolation may obscure some trends.

Emissions of ozone precursors | PART B

Figure 5   **Contribution to change in ozone precursors emissions for each sector and pollutant (EU-15), 1990–2002**

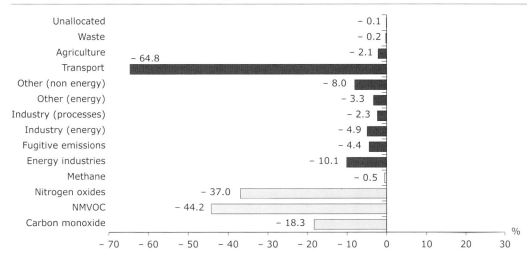

Note: Data not available for Malta.

Data source: Data from 2004 officially reported national total and sectoral emissions to UNECE/EMEP Convention on Long-range Transboundary Air Pollution and the UNFCCC (Ref: www.eea.eu.int/coreset).

03 Emissions of primary particles and secondary particulate precursors

Key policy question

What progress is being made in reducing emissions of fine particles (PM_{10}) and their precursors across the EU-15?

Key message

Total EU-15 emissions of fine particles were reduced by 39 % between 1990 and 2002. This was due mainly to reductions in emissions of the secondary particulate precursors, but also to reductions in primary PM_{10} emissions from energy industries.

Indicator assessment

EU emissions of fine particles were reduced by 39 % between 1990 and 2002. Emissions of NO_X (55 %) and SO_2 (20 %) were the most important contributing pollutants to particulate formation in the EU-15 in 2002. The reductions in total emissions between 1990 to 2002 were due mainly to the introduction or improvements of abatement measures in the energy, road transport, and industry sectors. These three sectors contributed 46 %, 22 % and 16 % respectively to the total reduction.

Indicator definition

This indicator tracks trends in emissions of primary particulate matter less than 10 µm (PM_{10}) and secondary precursors, aggregated according to the particulate formation potential of each precursor considered.

The indicator also provides information on changes in emissions from the main source sectors.

Indicator rationale

In recent years scientific evidence has been strengthened by many epidemiological studies that

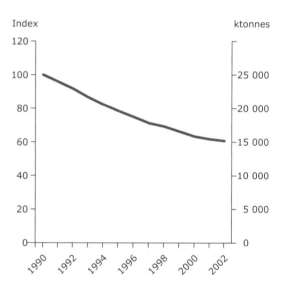

Figure 1 Emissions of primary and secondary fine particles (EU-15), 1990–2002

— Particulate emissions (primary and secondary)

Note: Data source: Data from 2004 officially-reported national total and sectoral emissions to UNECE/EMEP Convention on Long-range Transboundary Air Pollution. Where emissions of primary PM_{10} were not reported by countries, estimates have been obtained from the RAINS model (IIASA) (Ref: www.eea.eu.int/coreset).

indicate an association between long- and short-term exposure to fine particulate matter and various serious health impacts. Fine particles have adverse effects on human health and can be responsible for and/or contribute to a number of respiratory problems. Fine particles in this context refer to the sum of primary PM_{10} emissions and the weighted emissions of secondary PM_{10} precursors. Primary PM_{10} refers to fine particles (defined as having an aerodynamic diameter of 10 µm or less) emitted directly to the atmosphere. Secondary PM_{10} precursors are pollutants that are partly transformed into particles by photo-chemical reactions in the atmosphere. A large fraction of the urban population is exposed to levels of fine particulate

Figure 2 Changes in emissions of primary and secondary fine particles (EFTA-3 and EU-15), 1990–2002

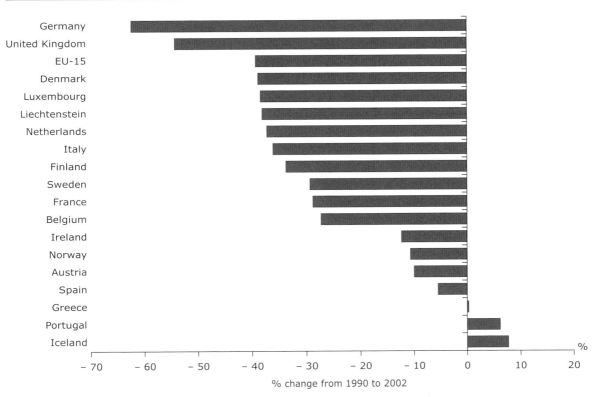

Note: Data source: Data from 2004 officially-reported national total and sectoral emissions to UNECE/EMEP Convention on Long-range Transboundary Air Pollution. Where emissions of primary PM_{10} were not reported by countries, estimates have been obtained from the RAINS model (IIASA) (Ref: www.eea.eu.int/coreset).

matter in excess of limit values set for the protection of human health. There have been a number of recent policy initiatives that aim to control particulate concentrations and thus protect human health.

Policy context

There are no specific EU emission targets set for primary PM_{10}. Measures are currently focused on controlling emissions of the secondary PM_{10} precursors. However, there are several directives and protocols that affect emissions of primary PM_{10}, including air quality standards for PM_{10}, in the first daughter directive to the framework directive on ambient air quality and emission standards for specific mobile and stationary sources for primary PM_{10} and secondary PM_{10} precursors.

For the particulate precursors, emission ceiling targets for NO_x, SO_2 and NH_3 are specified in both the EU National Emission Ceilings Directive (NECD) and the Gothenburg Protocol under the United Nations Convention on Long-range Transboundary Air Pollution (CLRTAP). Emission reduction targets for the EU-10 have been specified in the Treaty of Accession to

Core set of indicators | Air pollution and ozone depletion

Figure 3   Contributions to the changes in emission of primary and secondary fine particles (PM_{10}), per sector and per pollutant (EU-15), 2002

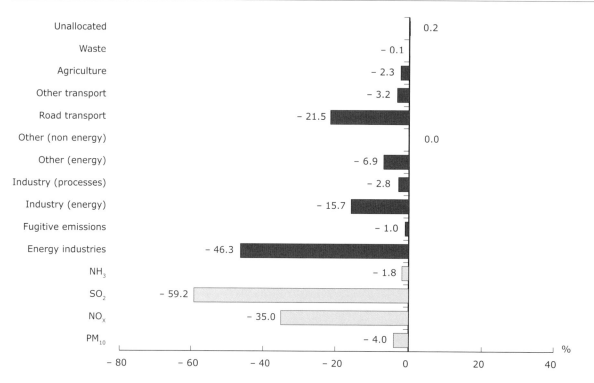

Note: 'Contribution to change' plots show the contribution to the total emission change between 1990–2002 made by a specified sector/pollutant.

Data source: Data from 2004 officially-reported national total and sectoral emissions to UNECE/EMEP Convention on Long-range Transboundary Air Pollution. Where emissions of primary PM_{10} were not reported by countries, estimates have been obtained from the RAINS model (IIASA) (Ref: www.eea.eu.int/coreset).

the European Union 2003 in order that they can comply with the NECD. In addition, the treaty of accession also includes emission targets for the EU-25 region as a whole.

Indicator uncertainty

The EEA uses data officially submitted by EU Member States and other EEA countries which follow common guidelines on the calculation and reporting of emissions for air pollutants.

NO_x, SO_2 and NH_3 estimates in Europe are thought to have an uncertainty of about 30 %, 10 % and 50 % respectively.

The primary PM_{10} emission data are generally more uncertain than the emissions on secondary PM_{10} precursors.

The use of generic particulate formation factors leads to some uncertainty. The factors are assumed to be representative of Europe as a whole; different factors may be estimated at the local scale.

Emissions of primary particles and secondary particulate precursors

04 Exceedance of air quality limit values in urban areas

Key policy question

What progress is being made in reducing concentrations of air pollutants in urban areas to below the limit values (for SO_2, NO_2 and PM_{10}) or the target values (for ozone) defined in the air quality framework directive and its daughter directives?

Key message

Large fractions of the urban population are exposed to concentrations of air pollutants in excess of the health-related limit or target values defined in the air quality directives. Exposure to SO_2 shows a strong downward trend but no clear downward trend is observed for the other pollutants.

PM_{10} is a pan-European air quality issue. The limit values are exceeded at urban measuring stations for background concentrations in nearly all countries.

Ozone is also a widespread problem, although the health-related target values are less frequently exceeded in north-western than in southern, central and eastern Europe.

NO_2 limit values are exceeded in the densely populated areas of north-western Europe and in large agglomerations in southern, central and eastern Europe.

Exceedances of SO_2 limit values are observed only in a few eastern European countries.

Indicator assessment

PM_{10} particles in the atmosphere result from direct emissions (primary PM_{10}) or emissions of particulate precursors (nitrogen oxides, sulphur dioxide, ammonia and organic compounds) which are partly transformed into particles (secondary PM) by chemical reactions in the atmosphere.

Although monitoring of PM_{10} is limited, it is clear that a significant proportion of the urban population (25–55 %) is exposed to concentrations of particulate matter in excess of the EU limit values set for the protection of human health (Figure 1).

Figure 2 shows a downward trend in the highest daily mean PM_{10} values until 2001.

Although reductions in emissions of ozone precursors appear to have led to lower peak concentrations of **ozone** in the troposphere, the health-related target value for ozone is exceeded over a wide area and by a large margin. About 30 % of the urban population was exposed to concentrations above the 120 µg O_3/m³ level during more than 25 days in 2002 (Figure 3).

Data from a consistent set of stations over the period 1996–2002 show hardly any significant variation for the 26th highest maximum daily 8-hour mean (Figure 4).

Figure 1   Exceedance of air quality limit value of PM_{10} in urban areas (EEA member countries), 1996–2002

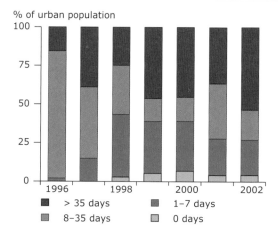

Note: Representative monitoring data were not available before 1997. Over the period 1997–2002 the total population for which exposure estimates are made increased from 34 to 106 million as a result of an increasing number of monitoring station reporting air quality data. Year-to-year variations in exposure classes might be caused partly by meteorological variability and partly by the changes in spatial coverage.

Data source: Airbase (Ref: www.eea.eu.int/coreset).

Figure 2   Highest daily PM_{10} concentration (36th highest daily 24h-mean) observed at urban stations (EEA member countries), 1997–2002

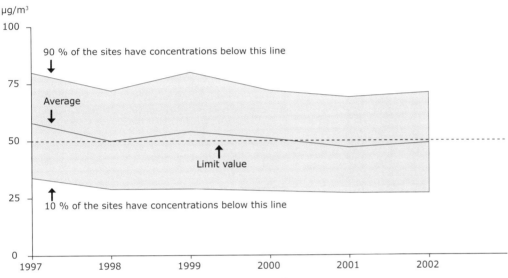

Note: Data source: Airbase (Ref: www.eea.eu.int/coreset).

About 30 % of the urban population live in cities with urban background concentrations in excess of the annual limit value of 40 µg/m³ of nitrogen dioxide. However, limit values are probably also exceeded in cities where the urban background concentration is below the limit value, in particular at hot spots in locations with a high traffic density.

The main source of emissions of nitrogen oxides (NO_x) to the air is the use of fuels: road transport, power plants and industrial boilers account for more than 95 % of European emissions. Enforcement of current EU legislation (large combustion plants and IPPC directive, auto-oil programme, NEC directive) and CLRTAP protocols have resulted in a reduction in emissions. This reduction is not yet reflected in the annual average concentrations observed at the urban monitoring stations measuring background concentrations.

Sulphur in coal, oil and mineral ores is the main source of emissions of sulphur dioxide to the atmosphere. Since the 1960s, the combustion of sulphur-containing fuels has largely been removed from urban and other populated areas, first in western Europe and now also increasingly in most central and eastern European countries. Large point sources (power plants and industries) remain the predominant source of sulphur dioxide emissions. As a result of the significant reductions in emissions achieved in the past decade, the percentage of the urban population exposed to concentrations above the EU limit value has been reduced to less than 1 %.

Indicator definition

The indicator presents the percentage of the urban population in Europe potentially exposed to ambient air concentrations (in µg/m³) of sulphur dioxide, PM_{10}, nitrogen dioxide and ozone in excess of the EU limit or target value set for the protection of human health. Where there are multiple limit values (see section on policy context) the indicator presents the most stringent case.

The urban population considered is the total number of people living in cities with at least one monitoring station.

Figure 3   Exceedance of air quality target values for ozone in urban areas (EEA member countries), 1996–2002

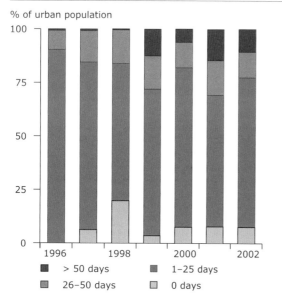

% of urban population

- ■ > 50 days
- ■ 26–50 days
- ■ 1–25 days
- □ 0 days

Note: Over the period 1996–2002 the total population for which exposure estimates are made increased from 50 to 110 million as a result of an increasing number of monitoring station reporting under the EoI Decision. Data prior to 1996 with coverage of less than 50 million people are not representative of the European situation. Year-to-year variations in exposure classes might be caused partly by meteorological variability and partly by the changes in spatial coverage.

Data source: Airbase (Ref: www.eea.eu.int/coreset).

Indicator rationale

Epidemiological studies have reported statistically significant associations between short-term, and especially long-term, exposure to increased ambient PM concentrations and increased morbidity and (premature) mortality. PM levels that may be relevant to human health are commonly expressed in terms of the mass concentration of inhalable particles with an equivalent aerodynamic diameter equal to or less than 10 μm (PM_{10}). Health effect associations for the fine fraction ($PM_{2.5}$) are even more clearly evident. Although the body of evidence concerning the health effects of PM is increasing rapidly, it is not possible to identify a concentration threshold below which health effects are not detectable. There is therefore no recommended WHO air quality guideline for PM, but the EU has set a limit value.

Exposure to high ozone concentrations for periods of a few days can have adverse health effects, in particular inflammatory responses and reduction in lung function. Exposure to moderate ozone concentrations for longer periods may lead to a reduction in lung function in young children.

Short-term exposure to nitrogen dioxide may result in airway and lung damage, decline in lung function, and increased responsiveness to allergens following acute exposure. Toxicology studies show that long-term exposure to nitrogen dioxide can induce irreversible changes in lung structure and function.

Sulphur dioxide is directly toxic to humans, its main action being on the respiratory functions. Indirectly, it can affect human health as it is converted to sulphuric acid and sulphate in the form of fine particulate matter.

Policy context

This indicator is relevant information for the Clean Air for Europe (CAFE) programme. The Air Quality Framework Directive (96/62/EC) defines basic criteria and strategies for air quality management and assessment for a set of health-relevant pollutants. In four 'daughter' directives, it establishes the framework under which the EU has set limit values for SO_2, NO_2, PM_{10}, lead, CO and benzene and target levels for ozone, heavy metals and polyaromatic hydrocarbons to protect human health.

Emission reduction targets for national emissions have been set in the Gothenburg Protocol by the CLRTAP, and by the EU National Emission Ceiling Directive (NECD; 2001/81/EC). This is intended to address, simultaneously, pollutant-specific ambient air quality problems affecting human health, as well as ground

Figure 4 Peak ozone concentration (26th highest maximum daily 8h-mean) observed at urban background stations (EEA member countries), 1996–2002

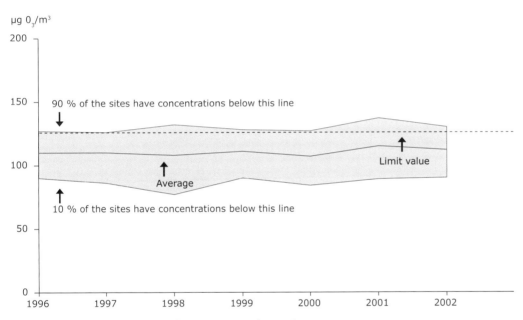

Note: Data source: Airbase (Ref: www.eea.eu.int/coreset).

level ozone, acidification and eutrophication affecting ecosystems.

The targets used for these indicators are the limit values set by Council Directive 1999/30/EC for sulphur dioxide, nitrogen dioxide, particulate matter and lead in ambient air and the target value and long term objective for ozone for the protection of human health set by Council Directive 2002/3/EC.

Indicator uncertainty

It is assumed that the air quality data submitted officially to the European Commission under the exchange-of-information decision have been validated by the national data supplier. Station characteristics and representativeness is often insufficiently documented. The data are generally not representative of the total urban population in a country. In a sensitivity analysis, the indicator has been based on the most exposed station in a city. In this worst-case calculation, the highest number of exceedance days observed at any of the operational stations (classified as urban, street, other or not defined) is assumed to be representative of the whole city. Locally, the indicator is subject to year-to-year variations due to meteorological variability.

PM_{10} data have been considered from monitoring stations using the reference method (gravimetry) and other methods. Documentation is incomplete whether countries have applied correction factors for non-reference methods, and if they have, which ones. Uncertainties associated with this lack of knowledge may result in a systematic error of up to 30 %. The number of data series available varies considerably from year to year and is insufficient for the period before 1997.

05 Exposure of ecosystems to acidification, eutrophication and ozone

Key policy question

What progress is being made towards the targets for reducing the exposure of ecosystems to acidification, eutrophication and ozone?

Key message

There have been clear reductions in the acidification of Europe's environment since 1980, but with some tailing off in improvement since 2000. Continued attention and further action is needed to ensure that the targets set for 2010 are achieved.

Eutrophication has declined slightly since 1980. However, only limited further improvement is expected by 2010 with current plans.

Most agricultural crops are exposed to ozone levels that exceed the EU long-term objective set for their protection, and a significant fraction are exposed to levels above the target value to be attained by 2010.

Indicator assessment

There have been substantial reductions in the area subjected to **deposition of excess acidity** since 1980 (see Figure 1) ([1]).

Data on a country basis indicate that already by 2000 all but six countries had less than 50 % of their ecosystem areas in exceedance of critical loads of acidity. Further substantial progress is anticipated for virtually all countries in the period 2000–2010.

Eutrophication of ecosystems shows less progress (Figure 1). There have been limited improvements at the European level since 1980, and very little further improvement is expected in individual countries between 2000 and 2010. The broader European continent continues to have a lesser problem than the countries of the EU-25.

The target value for **ozone** is exceeded in a substantial fraction of the arable area of the EEA-31: in 2002, about 38 % of the total area of 133 million ha (Figure 2 and Map 1). The long-term objective is met in less than 9 % of the total arable area, mainly in the United Kingdom, Ireland and the northern part of Scandinavia.

Indicator definition

The indicator (Figures 1 and 2) shows the ecosystem or crop areas which are subject to deposition or ambient concentrations of air pollutants in excess of the so-called 'critical load' or level for the particular ecosystem or crop.

'The critical load or level is defined as the estimated quantity of pollutant deposited or ambient concentration below which exposures to the pollutant are such that significant harmful effects do not occur according to present knowledge.'

Thus the critical load is an indication on how large a burden an ecosystem, or crop, can withstand in the long term without suffering from harmful effects.

The percentage of the area of ecosystem, or crop, exceeded indicates the extent of possible significant harmful effects in the long term. The magnitude of exceedance is thus an indication of the significance of future harmful effects.

The critical load of acidity is expressed in acidifying equivalents (H^+) per hectare per year (eq $H^+.ha^{-1}.a^{-1}$).

Ozone exposure, critical level, EU target value and the long-term objective are expressed as accumulated exposure to concentrations of over 40 ppb (about 80 µg/m^3) of ozone (AOT40) in the following unit: (mg/m^3)h.

([1]) It is difficult to assess the quantitative improvements since 1990 as acidification status in this base year (1990) remains to be reassessed using the latest critical loads and deposition calculation methodology.

Exposure of ecosystems to acidification, eutrophication and ozone

Figure 1   EU-25 and Europe-wide ecosystem damage area (average accumulated exceedance of critical loads), 1980–2020

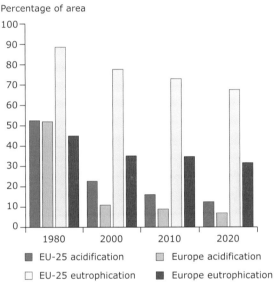

Figure 2   Exposure of crops to ozone (exposure expressed as AOT40 in (mg/m³)h in EEA member countries, 1996–2002 (²)

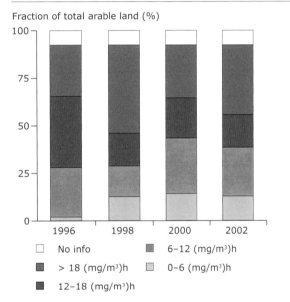

Note: Data source for deposition data used to calculate exceedances: EMEP/MSC-W.

Data source: UNECE — Coordination Center for Effects (Ref: www.eea.eu.int/coreset).

Indicator rationale

Deposition of sulphur and nitrogen compounds contributes to acidification of soils and surface waters, leaching of plant nutrients and damage to flora and fauna. Deposition of nitrogen compounds can lead to eutrophication, disturbance of natural ecosystems, excessive algal blooms in coastal waters and increased concentrations of nitrate in groundwater.

The estimated capacity of a location to receive depositions of acidifying or eutrophying pollutants without harm ('critical load') may be thought of as the threshold total quantity of air polluting compounds deposited, which should not be exceeded if ecosystems

Note: The target value for the protection of vegetation is 18 (mg/m³)h while the long-term objective is set at 6 (mg/m³)h.

The fraction labelled 'no information' refers to areas in Greece, Iceland, Norway, Sweden, Estonia, Lithuania, Latvia, Malta, Romania, and Slovenia for which either no ozone data from rural background stations or no detailed land cover data are available. Bulgaria, Cyprus, and Turkey are not included.

Data source: Airbase (Ref: www.eea.eu.int/coreset).

are to be protected from risk of damage, according to present knowledge.

Ground-level ozone is seen as one of the most prominent air pollution problems in Europe, mainly because of its effects on human health, natural ecosystems and crops. Threshold levels set by the EU for the protection of human health and vegetation, and critical levels agreed under the LRTAP Convention

---

(²) The sum of the differences between hourly ozone concentration and 40 ppb for each hour when the concentration exceeds 40 ppb during a relevant growing season, e.g. for forest and crops.

Core set of indicators | Air pollution and ozone depletion

Map 1   **Exposure above AOT40 target values for vegetation around rural ozone stations (EEA member countries), 2002**

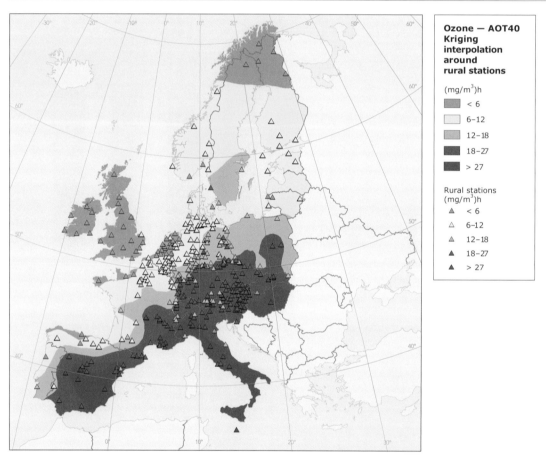

Note: Reference period: May–July 2002 (Kriging interpolation around rural stations).

Data source: Airbase (Ref: www.eea.eu.int/coreset).

for the same purpose, are exceeded widely and by substantial amounts.

Policy context

This indicator is relevant information for the Clean Air for Europe (CAFE) programme. A combined ozone and acidification abatement strategy has been developed by the Commission, resulting in an Ozone Daughter Directive (2002/3/EC) and a National Emission Ceiling Directive (2001/81/EC). In this legislation, target values have been set for ozone levels and precursor emissions for 2010. The long-term objectives of the EU are largely consistent with the long-term objectives of no exceedances of critical loads and levels as defined in the UN-ECE CLRTAP protocols to abate acidification, eutrophication and ground-level ozone.

Negotiation of emission reduction agreements has been based on model calculations, and the reporting of emission reductions in accordance with these agreements would indicate the improvement in environmental quality required by the policy objectives:

National Emission Ceilings Directive 2001/81/EC, Article 5

Acidification: Reduction in areas exceeding critical loads for acidification by 50 % (in each 150 km resolution grid square) between 1990 and 2010.

Vegetation-related ground-level ozone exposure: By 2010 the ground-level ozone load above the critical level for crops and semi-natural vegetation (AOT40 = 3 ppm.h) shall be reduced by one-third in all grid cells compared with the 1990 situation. In addition, the ground-level ozone concentrations shall not exceed an absolute limit of 10 ppm.h expressed as an exceedance of the critical level in any grid cell.

UNECE CLRTAP Gothenburg Protocol (1999)

The protocol sets emission limits with target dates to abate acidification, eutrophication and ground-level ozone. While environmental quality objectives are not specified, full attainment of emission targets is intended to result in an improvement in the state of the environment.

EU Ozone Daughter Directive (2002/3/EC)

The ozone directive defines the target value for the protection of vegetation as an AOT40-value (calculated from hourly values from May to July) of 18 (mg/m^3)h, averaged over five years. This target value should be attained in 2010 (Article 2, indent 9). It also defines a long-term objective of 6 (mg/m^3)h as AOT40.

Indicator uncertainty

The exceedance of deposition of critical loads for acidification and eutrophication presented in this indicator is itself a calculation derived from reported air emissions. Model estimates of pollutant depositions are used rather than observed depositions on account of their larger spatial coverage. Computer modelling uses officially reported national pollutant emission totals and their geographical distributions using documented procedures. Temporal and spatial coverage is imperfect, however, as a number of annual national totals and geographical distributions are not reported according to time schedules. The resolution of the computer estimates has improved recently to 50 km grid averages. Local pollutant sources or geographical features below this scale will not be well resolved. The meteorological parameters used for modelling pollutant supply are mainly computations, with some adjustment towards observed conditions.

The critical load estimates are reported by official national sources, but face difficulties of geographical coverage and comparability. The latest reporting round in 2004 supplied estimates for 16 of the 38 EEA participating countries. For a further nine countries, earlier submissions were reported as being still valid. Those reporting did so for a variety of ecosystem classes, although reported ecosystems typically covered less than 50 % of their total country area. For other countries the most recently submitted critical loads data are used.

Methodology uncertainty in the indicator for **ozone** is due to uncertainty in mapping AOT40 based on interpolation of point measurements at background stations. The different definitions of AOT40 values (accumulation during 8.00 to 20.00 CET following the Ozone Directive or accumulation during daylight hours following the definition in NECD) is expected to introduce minor inconsistencies in the data set.

At the data level it is assumed that the air quality data officially submitted to the Commission under the exchange of information decision and to EMEP under the UNECE CLRTAP have been validated by the national data supplier. Station characteristics and representativeness are often not well documented and coverage of territory and i time is incomplete. Yearly changes in monitoring density will influence the total monitored area. The indicator is subject to year-to-year fluctuations as it is sensitive mainly to episodic conditions, and these depend on particular meteorological situations, the occurrence of which varies from year to year.

06 Production and consumption of ozone depleting substances

Key policy question

Are ozone depleting substances being phased out according to the agreed schedule?

Key message

The total production and consumption of ozone depleting substances in the EEA-31 decreased significantly until 1996 and has since stabilised.

Indicator assessment

The production and consumption of ozone depleting substances (ODS) has decreased significantly since the 1980s (Figures 1 and 2). This is a direct result of international policies (the Montreal Protocol and its amendments and adjustments) to phaseout the production and consumption of these substances. Production and consumption in the EEA-31 is dominated by the EU-15 countries, which account for 80–100 % of total ODS production and consumption. The overall decline is in accordance with the international regulations and the agreed schedule.

Indicator definition

This indicator tracks the annual production and consumption of ozone depleting substances (ODS) in Europe. ODS are long-lived chemicals that contain chlorine or/and bromine and that destroy the stratospheric ozone layer.

Figure 1 Production of ozone depleting substances (EEA-31), 1989–2000

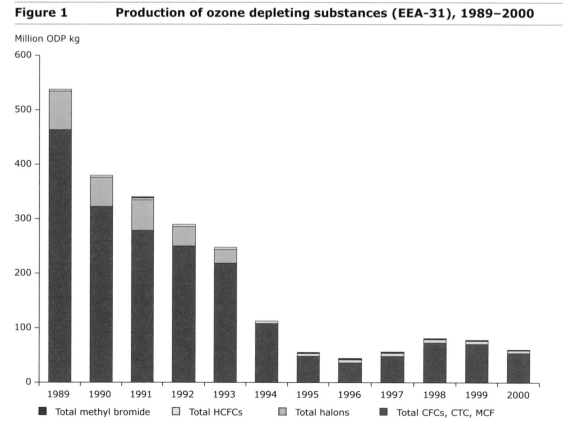

Note: Data source: UNEP (Ref: www.eea.eu.int/coreset).

Production and consumption of ozone depleting substances PART B

Figure 2   **Consumption of ozone depleting substances (EEA-31), 1989–2000**

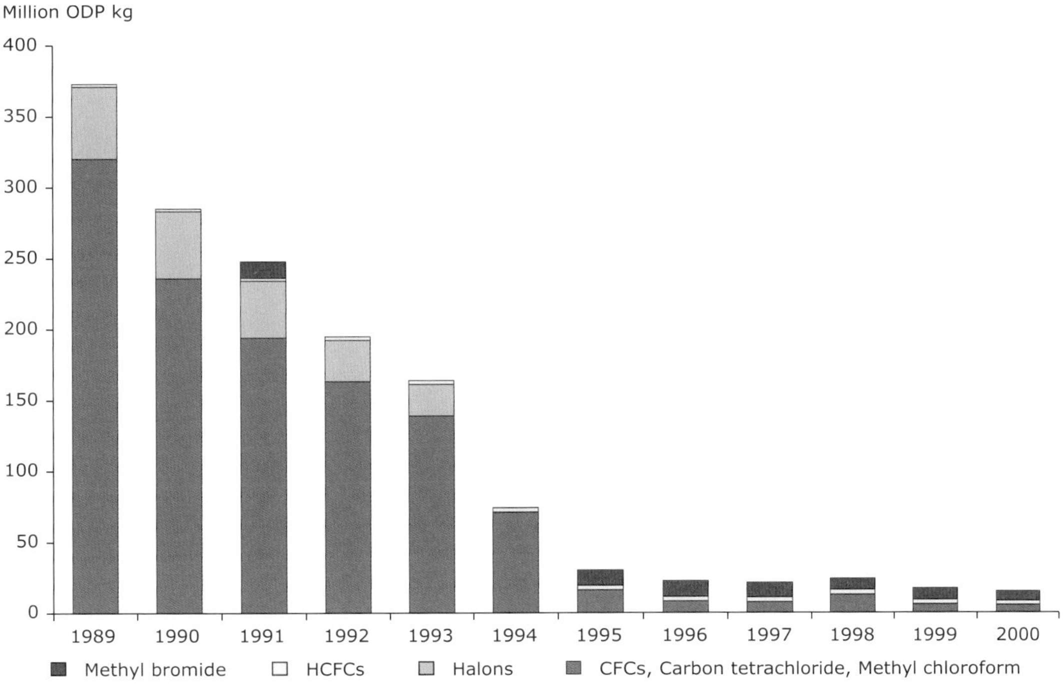

Million ODP kg

Legend: Methyl bromide · HCFCs · Halons · CFCs, Carbon tetrachloride, Methyl chloroform

Note: Data source: UNEP (Ref: www.eea.eu.int/coreset).

Developed countries have not been allowed to produce or consume halons since 1994, and CFCs, carbon tetrachloride and methyl chloroform since 1995. A limited production of ODS is still allowed for designated essential uses (e.g. metered dose inhalers) and for developing countries to meet their basic domestic needs.

The indicator is presented as million kg of ODS weighted by their ozone depletion potential (ODP).

Indicator rationale

Policy measures to limit or phaseout the production and consumption of ozone depleting substances (ODS) have been taken since the mid 1980s in order to protect the stratospheric ozone layer from depletion. This indicator tracks progress towards this limiting or phasing-out of production and consumption.

Policies focus on the production and consumption rather than emissions of ODS. This is because emissions

Table 1   **Article 5(1) and Non-article 5(1) countries of the Montreal Protocol**

Montreal Protocol	EEA member countries
Article 5(1)	Cyprus, Malta, Romania and Turkey
Non-article 5(1)	All other EEA member countries

The European environment | State and outlook 2005   277

Table 2   **Summary of phaseout schedule for Non-article 5(1) countries, including Beijing adjustments**

Group	Phaseout schedule for Non-article 5(1) countries	Remark
Annex-A, group 1: CFCs (CFC-11, CFC-12, CFC-113, CFC-114, CFC-115)	Base level: 1986 100 % reduction by 01.01.1996 (with possible essential use exemptions)	Applicable to production and consumption
Annex A, group 2: Halons (halon 1211, halon 1301, halon 2402)	Base level: 1986 100 % reduction by 01.01.1994 (with possible essential use exemptions)	Applicable to production and consumption
Annex B, group 1: Other fully halogenated CFCs (CFC-13, CFC-111, CFC-112, CFC-211, CFC-212, CFC-213, CFC-214, CFC-215, CFC-216, CFC-217)	Base level: 1989 100 % reduction by 01.01.1996 (with possible essential use exemptions)	Applicable to production and consumption
Annex B, group 2: Carbon tetrachloride (CCl_4)	Base level: 1989 100 % reduction by 01.01.1996 (with possible essential use exemptions)	Applicable to production and consumption
Annex B, group 3: 1,1,1-trichloroethane (CH_3CCl_3) (= methyl chloroform)	Base level: 1989 100 % reduction by 01.01.1996 (with possible essential use exemptions)	Applicable to production and consumption
Annex C, group 1: HCFCs (HydroChloroFluoroCarbons)	Base level: 1989 HCFC consumption + 2.8 % of 1989 CFC consumption Freeze: 1996 35 % reduction by 01.01.2004 65 % reduction by 01.01.2010 90 % reduction by 01.01.2015 99.5 % reduction by 01.01.2020, and thereafter consumption restricted to the servicing of refrigeration and air-conditioning equipment existing at that date. 100 % reduction by 01.01.2030	Applicable to consumption
	Base level: Average of 1989 HCFC production + 2.8 % of 1989 CFC production and 1989 HCFC consumption + 2.8 % of 1989 CFC consumption Freeze: 01.01.2004, at the base level for production	Applicable to production
Annex C, group 2: HBFCs (HydroBromoFluoroCarbons)	Base level: year not specified. 100 % reduction by 01.01.1996 (with possible essential use exemptions)	Applicable to production and consumption
Annex C, group 3: Bromochloromethane (CH_2BrCl)	Base level: year not specified. 100 % reduction by 01.01.2002 (with possible essential use exemptions)	Applicable to production and consumption
Annex E, group 1: Methyl bromide (CH_3Br)	Base level: 1991 Freeze: 01.01.1995 25 % reduction by 01.01.1999 50 % reduction by 01.01.2001 75 % reduction by 01.01.2003 100 % reduction by 01.01.2005 (with possible essential use exemptions)	Applicable to production and consumption

Production and consumption of ozone depleting substances

from multiple small sources are much more difficult to monitor accurately than from industrial production and consumption. Consumption is the driver of industrial production. Production and consumption can precede emissions by many years, as emissions generally occur after the disposal of products in which ODS are used (fire-extinguishers, refrigerators, etc.).

Release of ODS to the atmosphere leads to depletion of the stratospheric ozone layer which protects humans and the environment from harmful ultra-violet (UV) radiation emitted by the sun. Ozone is destroyed by chlorine and bromine atoms which are released in the stratosphere from man-made chemicals — CFCs, halons, methyl chloroform, carbon tetrachloride, HCFCs (all completely anthropogenic) and methyl chloride and methyl bromide. Depletion of stratospheric ozone leads to increases in ambient ultra-violet radiation at the surface, which has a wide variety of adverse effects on human health, aquatic and terrestrial ecosystems, and food chains.

Policy context

Following the Vienna Convention (1985) and the Montreal Protocol (1987) and their amendments and adjustments, policy measures have been taken to limit or phaseout the production and consumption of ozone depleting substances

The international target under the Ozone Convention and Protocols is the complete phase-out of ODS, according to the schedule below.

Countries falling under Article 5, paragraph 1 of the Montreal Protocol are considered as developing countries under the protocol. Phaseout schedules for Article 5(1) countries are delayed by 10–20 years compared with Non-article 5(1) countries (Table 1).

Indicator uncertainty

Two data sets are used in the fact-sheet: (1) UNEP data, as reported by the countries to the UNEP Ozone Secretariat (data provided for production and consumption), and (2) DG Environment data, as reported by companies to DG Environment (data provided for production, consumption, import and export). Generally, production data are reported only when individual company performance cannot be recognised in the statistics. So if one or two companies within a country or group of countries only produce a substance, data may be missing because of privacy protection of companies.

The uncertainty in the statistics is unknown, as an uncertainty estimate is not reported by the companies. Production figures are generally better known than consumption, because production occurs only in a few factories, while use of ODS (consumption) takes place in many factories.

Emissions are more uncertain than consumption figures, because emissions take place when products in which ODS are used (e.g. fire-extinguishers, refrigerators) are discarded. The time when these products are discarded is unknown, and hence when the corresponding emissions will occur.

The definition of production in the DG Environment and the UNEP data is different. In the DG Environment data, production is real production without subtracting ODS recovered and destroyed or used as feedstock (intermediate products that are used to produce other ODS).

An estimate of the uncertainty for the EU-15 can be obtained by comparing the DG Environment data with the UNEP data.

07 Threatened and protected species

Key policy question

What measures are being taken to conserve or restore biodiversity?

Key message

Identifying and establishing lists of protected species at national and international levels are important first steps in conserving species diversity. European countries have agreed to join efforts to conserve threatened species by listing them for protection in EU directives and/or the Bern Convention. Some, but not all, of the globally endangered species of wild fauna occurring in Europe in 2004 are currently under European protection status. The responsibility of the EU towards the global community for the conservation of these species is high.

Indicator assessment

According to IUCN (2004), 147 vertebrate (mammals, birds, reptiles, amphibians and fish) and 310 invertebrate species (crustaceans, insects and molluscs) that occur in the EU-25 are considered to be globally threatened, since they have been categorised as critically endangered, endangered, and vulnerable.

The overall assessment shows that specific protection status under EU legislation and the Bern Convention exists for all globally threatened bird species, and for a fair percentage of the reptiles and mammals. However, most of the globally threatened amphibians and fish, as well as invertebrate species occurring in the EU-25 are not protected at the European level. Information on whether these receive protection at the national level, where they occur, is not readily available.

All 20 globally threatened bird species occurring in EU-25 are protected either under the EU birds directive (which, while protecting all bird species, lists a number of species in its Annex I for which strict habitat management is needed) or the Bern Convention (Annex II).

Up to 86 % of reptile and mammal species have been protected at the European level so far: 12 out of 14 globally threatened reptile species and 28 out of 35 mammal species have been included in the EU habitats directive (Annexes II and IV), or the Bern Convention (Annex II).

Less than half of the amphibian and fish species have been protected under European legislation so far; 7 out of 15 amphibian species and 24 out of 63 fish species have been included in the legislative lists.

The gap for invertebrate species is vast. Only 43 out of 310 species have been included in the lists.

The indicator in its present form cannot directly assess the effectiveness of EU biodiversity policies. It can only confirm the extent of European responsibility to the global community and show the extent to which global responsibilities are covered by European legislation.

Indicator definition

This indicator depicts the number and percentage of the globally endangered species of wild fauna occurring in the EU-25 in 2004 that are granted European protection status through the EU birds and habitats directives or the Bern Convention. The indicator takes into account modifications to the respective legislative lists of species resulting from EU enlargement.

Threatened and protected species PART B

Figure 1 **Percentage of inclusion of globally threatened species occurring in EU-25 in protected species lists of EU directives and the Bern Convention**

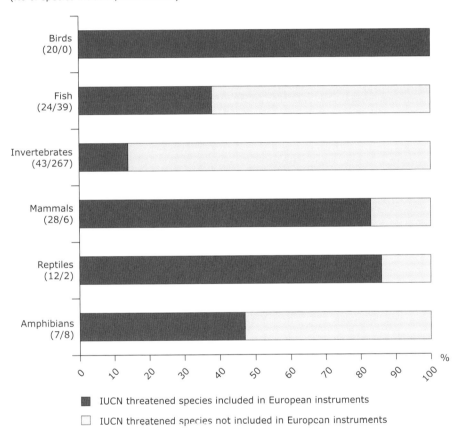

Note: Data source: 2004 IUCN list, Annexes of EU birds and habitats directives and Bern Convention (Ref: www.eea.eu.int/coreset).

Indicator rationale

There are a number of ways of assessing progress towards the target of halting the loss of biodiversity in Europe by 2010.

The International Union for Conservation of Nature (IUCN) has been monitoring the extent and rate of biodiversity degradation for several decades by assigning species to red-list categories through detailed assessment of information against a set of objective, standard, quantitative criteria. This assessment is made at the global level and the most recent one was published in 2004.

Globally threatened species are present in and also outside Europe, and some of them may not be classified as threatened at the regional or national levels within the EU. How far European legislation, which is further linked to European policies on nature and biodiversity, takes the EU responsibility to the global community into consideration is shown by the information that the indicator provides on the number of globally threatened species that are protected at the European level.

Indicator uncertainty

The indicator does not currently identify how many species of wild fauna listed as globally endangered are found only in Europe. It also does not consider the protection of species that do not occur in the global red lists but which are endangered in Europe. Finally it does not include data on plants.

Policy context

Halting the loss of biodiversity by 2010 is a target expressed by the 6EAP and the European Council at Gothenburg, and re-enforced by the Environment Council in Brussels in June 2004.

The Council also emphasises 'the importance of monitoring, evaluating and reporting on progress towards the 2010 targets, and that it is absolutely vital to communicate biodiversity issues effectively to the general public and to decision-makers in order to provoke appropriate policy responses'.

Targets

There are no specific quantitative targets for this indicator.

The target to 'halt the loss of biodiversity by 2010' implies not only that species extinction must be stopped but that threatened species must be shifted to a better status.

Threatened and protected species   PART B

08 Designated areas

Key policy question

What measures are being taken to ensure the *in situ* conservation of biodiversity components?

Key message

In situ conservation of species, habitats and ecosystems entails the establishment of protected areas. The increase in the cumulative area of sites within the European Ecological Natura 2000 network during the past ten years is a good sign of commitment to the conservation of biodiversity. Some of the Natura 2000 sites include areas that have not already been designated under national laws, thus contributing to a direct increase in the total area designated for *in situ* conservation of biodiversity components in Europe.

Indicator assessment

Worldwide, countries use the designation of protected areas as a means of conserving biodiversity components (genes, species, habitats, ecosystems), each country applying its own selection criteria and objectives. A common EU perspective was defined by the birds and habitats directives. On the basis of these, EU Member States have classified and/or proposed sites for establishing the European Natura 2000 network.

The indicator shows that there has been a steady increase in the cumulative area of sites designated to the Natura 2000 network over the past ten years, from approximately 8 to 29 million ha under the birds directive (as special protection areas) and from 0 to approximately 45 million ha under the habitats directive (as sites of Community importance). Some countries have greater representation of species and habitats listed in the two directives than others. Therefore these countries have designated larger parts of their territory, as is the case with countries of southern Europe as well as the large countries of the north. Spain leads by contributing more than 10 million ha, followed by Sweden with about 5 million ha.

The second part of the indicator demonstrates the extent to which nationally designated sites that already exist are fulfilling the criteria of the European directives. It also provides a snapshot of the significance of the contribution of European legislation to *in situ* conservation in Europe.

Indicator definition

The indicator comprises two parts:

- the cumulative surface area of sites designated over time under the birds and habitats directives by each EU-15 Member State;

- the proportion of the area coverage of the sites designated by a country only under the EC birds and habitats directives, protected only by national instruments, and covered by both.

Indicator rationale

There are a number of ways of assessing progress towards the target of halting the loss of biodiversity in Europe by 2010.

The indicator aims to assess progress of *in situ* conservation of biodiversity components, which entails the establishment of protected areas. Progress is shown at the EU level, namely with the establishment of the Natura 2000 network. Quantitative information on the cumulative area comprising the Natura 2000 network over time in the EU-15 is broken down by country in the first part.

The second part of the indicator assesses whether the establishment of the Natura 2000 network is likely to increase the overall surface of protected areas in

Designated areas

Figure 1 Cumulative surface area of sites designated for the habitats directive over time (sites of Community importance — SCIs)

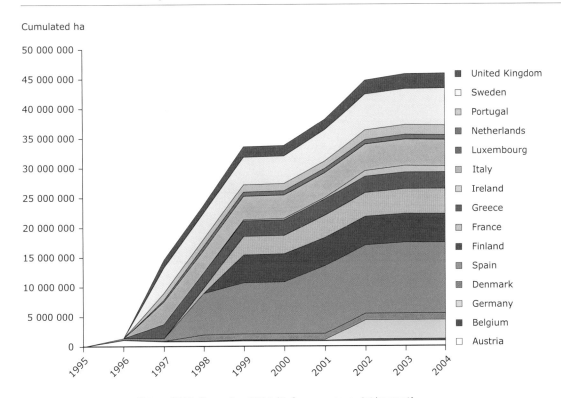

Note: Data source: Natura 2000, December 2004 (Ref: www.eea.eu.int/coreset).

Europe, by examining the proportion of the nationally designated areas included in the Natura 2000 network by each Member State, at a given point in time.

Policy context

Halting the loss of biodiversity by 2010 is one target expressed by the EU 6th environment action plan and the European Council at Gothenburg (2001). This target was fully endorsed at the Pan-European level in 2003. The European Council has also urged the Commission and Member States to implement the new programme of work on protected areas, adopted in the context of the Convention of Biological Diversity in 2004. This programme includes the need to update information on the status, trends and threats to protected areas.

At the EU level, policy on nature conservation is essentially made up of two pieces of legislation: the birds directive and the habitats directive. Together, they establish a legislative framework for protecting and conserving the EU's wildlife and habitats.

Targets

At the global level, the Convention on Biological Diversity (CBD) has set relevant targets to be achieved by 2010: Target 1.1 is the effective conservation of at least 10 % of each of the world's ecological regions and Target 1.2 is the protection of areas of particular importance to biodiversity.

Figure 2 Cumulative surface area of sites designated for the birds directive over time (special protection areas — SPAs)

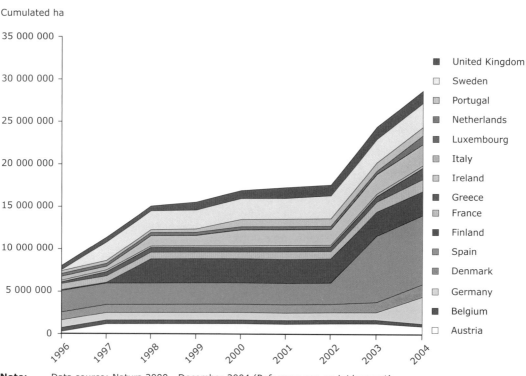

Note: Data source: Natura 2000, December 2004 (Ref: www.eea.eu.int/coreset).

At the Pan-European level, the target is full establishment of the Pan-European Ecological Network, of which Natura 2000 is a part, by 2008.

At the EU level, Member States should contribute to the establishment of Natura 2000 in proportion to the representation within their territories of the natural habitat types and the species mentioned in the directives.

With regard to time, the Natura 2000 network should be completed on land by 2005, implemented for marine sites by 2008, and management objectives for all sites should be agreed and instigated by 2010.

Indicator uncertainty

The indicator does not currently address all the targets set, especially sufficiency and evaluation of the management of sites. The EU-10 have not been assessed.

Designated areas   PART B

Figure 3 Proportion of total surface area designated only for the habitats directive, protected only by national instruments, and covered by both (sites of Community importance — SCIs)

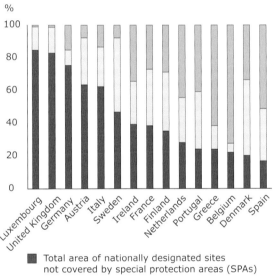

■ Total area of nationally designated sites not covered by special protection areas (SPAs)
□ Total area of designated special protection areas (SPAs) covered by national designations
▣ Total area of special protection areas (SPAs) not covered by national designations

Note: Data source: CDDA, October 2004; Proposed sites of Community importance database, December 2004 (Ref: www.eea.eu.int/coreset).

Figure 4 Proportion of total surface area designated only under the birds directive, protected only by national instruments, and covered by both (special protection areas — SPAs)

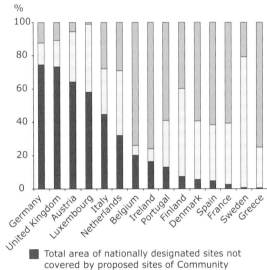

■ Total area of nationally designated sites not covered by proposed sites of Community importance (pSCIs)
□ Total area of proposed sites of Community importance (pSCIs) covered by national designations
▣ Total area of proposed sites of Community importance (pSCIs) not covered by national designations

Note: Data source: CDDA, October 2004; Special protection areas database, December 2004 (Ref: www.eea.eu.int/coreset).

The European environment | State and outlook 2005   287

09 Species diversity

Key policy question

What is the state and trend of biodiversity in Europe?

Key message

The populations of selected species in Europe are falling. Since the early 1970s, butterfly and bird species linked to different habitat types across Europe show population declines of between 2 % and 37 %. The declines may be linked to similar trends in the land cover of specific habitats between 1990 and 2000, especially certain wetland types as well as heaths and scrubs.

Indicator assessment

The indicator links population trends of species belonging to two groups (birds and butterflies) to the trends in the extent of different habitat types deriving from land cover change analysis for 1990–2000.

The assessment is based on 295 butterfly species and 47 bird species linked to 5 different habitat types across several European countries. Results vary among species/habitats groups, but it is striking that both birds and butterflies, linked to different habitat types, show a decline in all the habitats examined.

The declines in the populations of wetland bird and butterfly species can be explained by direct habitat loss as well as habitat degradation through fragmentation and isolation. Mires, bogs and fens, which are specific wetland habitats, declined most in area (by 3.4 %) across the EU-25 between 1990–2000, a result based on detecting changes bigger than 25 hectares.

Heaths and scrubs have a particularly high diversity of butterfly species, up to at least 92 species in the habitats surveyed. Direct habitat loss (by 1.6 %) as well as habitat degradation through fragmentation and isolation also play a role in the very substantial decline (28 %) observed among butterfly species.

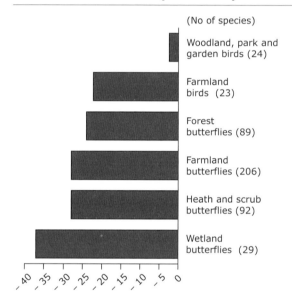

Figure 1 Trends in birds and butterfly populations in the EU-25 (% decline)

■ Trend from 1980 to 2002 (%)

Note: The numbers in brackets show the number of species taken into account for each habitat type. The bird trends reflect the period 1980–2002. The butterfly trends reflect the period 1972/73–1997/98.

Data source: Pan-European Common Bird Monitoring project (EBCC, BirdLife Int, RSPB), Dutch Butterfly Conservation (Ref: www.eea.eu.int/coreset).

The highest number of species assessed, namely 206 butterfly species and 23 bird species, occur in the farmland habitat. These species are typical of open grassy areas such as extensively farmed areas, grasslands, meadows and pastures. The two species groups show very similar trends of decline: 28 % and 22 % respectively. The main pressures related to this decline are loss of extensive farmland with a low or no input of nutrients, herbicides and pesticides, and an increase in agricultural intensification, which leads, among other factors, to loss of marginal habitats and hedgerows and a higher input of fertilisers, herbicides and insecticides.

Species diversity

Figure 2 Land cover change from 1990 to 2000 expressed as % of the 1990 level, aggregated into EUNIS habitat level 1 categories

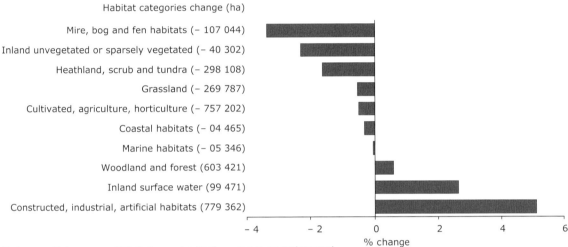

Note: Data source: EEA Data service (Ref: www.eea.eu.int/coreset).

The area of woodland and forest habitats has increased by 0.6 % since 1990, which in absolute terms is about 600 000 hectares. However, the species linked to the woodland and forest habitats have declined. The 89 butterfly species occurring in this habitat show a decline of 24 % and birds occurring in woodland, parks and gardens show a 2 % decline. Nearly all forests in Europe are managed to some extent and the various management schemes surely have an impact on species diversity. For example, the presence of dead wood and old-growth trees are of importance to birds for nesting and feeding, and clearing of forests is an important factor for forest butterflies.

Indicator definition

This indicator comprises two parts:

- Population trends of species and species groups. Currently the species groups considered are: birds, namely those species occurring in farmland, woodland, parks and gardens, and invertebrates, namely butterflies. The time reference of the species data sets used is also given.

- Change in area of the 10 main EUNIS habitat types, calculated on the changes in land cover between two points in time.

Indicator rationale

The indicator presents information on the state and trends of biodiversity in Europe, addressing species and their habitats in an interlinked way. In order to approach the issue, the trends of widely distributed taxonomic groups may be assessed through a range of habitats over the whole of Europe. Given the data availability on a European level, birds and butterflies were selected as a proxy for species and habitat biodiversity in general. Species from both groups can be linked to a range of different habitats, and their trends may also be considered as representative of the quality of a habitat with regard to other species.

In the case of birds, the species assessed are all common (numerous and widespread) breeding birds, with large distribution areas over Europe, linked to farmland, woodland, park and garden habitats.

Figure 3 Temporal coverage for the three data sets

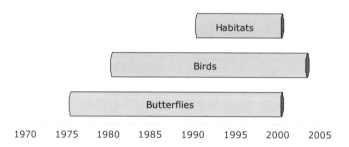

In the case of butterflies, the species assessed are not necessarily present in all countries, nevertheless each can be related to one of four major EUNIS habitat types, namely farmland, forest, heath, and scrub and wetlands.

An interpretation of the resulting species population trends per habitat type requires the assessment of trends in habitat area. For this indicator, the approach taken is to analyse land cover changes of the different habitat types between 1990 and 2000.

Future development of the indicator will clearly involve extending the concept to other species and species groups, while also defining common criteria for inclusion or deletion of species and by improving the selection of species in relation to habitats.

Policy context

'Halting the loss of biodiversity by 2010' is an objective of the European strategy for sustainable development, adopted in 2001 and further endorsed at the Pan-European level in 2003 by the Kiev resolution on biodiversity. Other relevant European Community policies include the 6th environment action programme and the European Community biodiversity strategy and action plans.

At the global level, the Convention on Biological Diversity (CBD) in 2002 committed the parties to achieving a significant reduction in the current rate of biodiversity loss at the global, regional and national level by 2010.

Targets

The overall target is to halt the loss of biodiversity by 2010.

No specific quantitative target is identified.

Indicator uncertainty

At present the indicator is prone to uncertainty on various levels. The major uncertainty is in the general lack of data from other species groups, and the incomplete geographical and temporal coverage of the data. In addition, the data are based on voluntary work by NGOs, which are dependent on continued funding and resources.

Farmland, woodland, park, and garden birds: since the species selection has been based on expert judgment and not on statistical evidence of the occurrence of each species, it is anticipated that links to habitats may be not as strong. The same list of bird species was used for all countries.

Butterflies: only very few countries have butterfly monitoring (the United Kingdom, the Netherlands, and

Belgium) but the network is building up. The butterfly trends used for this assessment are therefore based on trends in distribution as a proxy for population trends.

Data sets — geographical and time coverage at the EU level
Specifically on farmland, woodland, park, and garden birds: Data are available for 16 of the EU-25 Member States for 1980–2002 (unavailable for Cyprus, Finland, Greece, Lithuania, Luxembourg, Malta, Portugal, Slovenia and Slovakia). Data reflect different monitoring periods among countries.

Specifically on butterflies: monitoring data are not available for all species; distribution data are used.

Data sets — representativeness of data at the national level
Farmland, woodland, park, and garden birds: the representativeness of the data at the EU level is high because the selected species are widely distributed in Europe. At the national level, however, some of the selected species may be less representative, and other species not selected for this indicator may be more representative of the farmland or forest ecosystems of a country.

Butterflies: good representativeness since the data derive from questionnaires filled out by national experts.

Data sets — comparability
Farmland, woodland, park, and garden birds: overall comparability for the EU-25 is good. Data collection is based on a Pan-European monitoring scheme using a standardised methodology across countries.

Butterflies: comparability is good.

Core set of indicators | Climate change

10 Greenhouse gas emissions and removals

Key policy question

What progress is being made in reducing greenhouse gas (GHG) emissions in Europe towards the Kyoto Protocol targets?

Key message

Total EU-15 GHG emissions in 2003 were 1.7 % below base-year levels. Increases in carbon dioxide emissions were offset by reductions in nitrous oxide, methane and fluorinated gas emissions. Carbon dioxide emissions from road transport increased whereas emissions from manufacturing industry decreased.

Total EU-15 GHG emissions (including Kyoto Protocol flexible mechanisms) in 2003 were 1.9 index points above the hypothetical linear EU target path. Many EU-15 Member States were not on track to meet their burden-sharing targets. Total GHG emissions in the EU-10 decreased considerably (by 32.2 %) between the aggregate base year and 2003, due mainly to the economic restructuring transition process towards market economies. Most EU-10 Member States are on track to meet their Kyoto targets.

Indicator assessment

Total EU-15 GHG emissions in 2003 were 1.7 % below base-year levels. Four EU-15 Member States (France, Germany, Sweden and the United Kingdom) were below their burden-sharing target paths excluding Kyoto mechanisms. Luxembourg and the Netherlands were below their burden-sharing target paths including Kyoto mechanisms. Nine Member States were above their burden-sharing target paths: Greece and Portugal (excluding Kyoto mechanisms), Austria, Belgium, Denmark, Finland, Ireland, Italy, the Netherlands and Spain (including Kyoto mechanisms). Considerable emissions cuts have occurred cin Germany and the United Kingdom, the EU's two biggest emitters, which together account for about 40 % of total EU-15 GHG emissions; the 1990 to 2003 reductions were 18.5 % in Germany and 13.3 % in the United Kingdom. Compared with 2002, EU-15 emissions in 2003 increased by 1.3 %, due mainly to increases from energy industries (by 2.1 %), because of growing thermal power production and a 5 % increase in coal consumption in thermal power stations. From 1990 to 2003, EU-15 transport CO_2 emissions (20 % of total EU-15 GHG emissions) increased by 23 % due to road transport growth in almost all the Member States. CO_2 emissions from energy industries increased by 3.3 % due to increasing fossil fuel consumption in public electricity and heat plants, but Germany and the United Kingdom reduced their emissions by 12 % and 10 %, respectively. In Germany this was due to efficiency improvements in coal-fired power plants and in the United Kingdom to the fuel switch from coal to gas in power production. Reductions were achieved in EU-15 CO_2 emissions from manufacturing industries and construction (by 11 %), due mainly to efficiency improvements and structural change in Germany after reunification. CH_4 emissions from fugitive emissions decreased the most (by 52 %), due mainly to the decline of coal mining, followed by the waste sector (by 34 %), due mainly to reducing the amount of biodegradable waste in landfills and installing landfill gas recovery. Industrial N_2O emissions decreased by 56 %, due mainly to specific measures at adipic acid production plants. N_2O emissions from agricultural soils reduced by 11 %, due to a decline in fertiliser and manure use. HFC, PFC and SF_6 emissions from industrial processes, which account for 1.6 % of GHG emissions, decreased by 4 %. All EU-10 Member States that joined the EU in 2004 have to reach their Kyoto targets individually (Cyprus and Malta have no Kyoto target). Total emissions have declined substantially since 1990 in almost all EU-10, due mainly to the introduction of market economies and the consequent restructuring or closure of heavily polluting and energy-intensive industries. Emissions from transport started to increase in the second half of the 1990s. However, emissions in almost all EU-10 were well below their linear target paths — thus they were on track to meet their Kyoto targets.

Based on their emission trends until 2003, the EU accession countries Romania and Bulgaria, as well as the EEA member country Iceland, were on track to meet their Kyoto targets. Based on their emission trends up

Greenhouse gas emissions and removals

Figure 1 **Development of EU-15 greenhouse gas emissions from base year to 2003 and distance to the (hypothetical) linear EU Kyoto target path (excluding flexible mechanisms)**

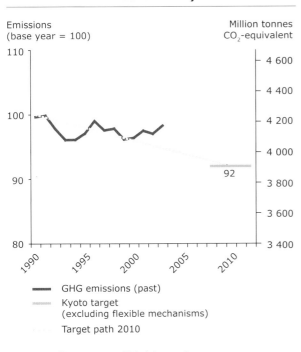

GHG emissions (past)
Kyoto target (excluding flexible mechanisms)
Target path 2010

Note: Data source: EEA data service (Ref: www.eea.eu.int/coreset).

to 2003, the EEA member countries Liechtenstein and Norway are not on track to achieve their Kyoto targets.

Indicator definition

This indicator illustrates current trends in anthropogenic GHG emissions in relation to the EU and Member State targets. Emissions are presented by type of gas and weighted by their global warming potentials. The indicator also provides information on emissions from sectors: energy industries; road and other transport; industry (processes and energy); other (energy); fugitive emissions; waste; agriculture and other (non-energy). All data are in million tonnes CO_2-equivalent.

Indicator rationale

There is growing evidence that emissions of greenhouse gases are causing global and European surface air temperatures to increase, resulting in climate change. The potential consequences at the global level include rising sea levels, increased frequency and intensity of floods and droughts, changes in biota and food productivity and increases in diseases. Efforts to reduce or limit the effects of climate change are focused on limiting the emissions of all greenhouse gases covered by the Kyoto Protocol. This indicator supports the Commission's annual assessment of progress in reducing emissions in the EU and the individual Member States to achieve the Kyoto Protocol targets under the EU Greenhouse Gas Monitoring Mechanism (Council Decision 280/2004/EC concerning a mechanism for monitoring Community GHG emissions and for implementing the Kyoto Protocol).

Policy context

The indicator analyses the trend in total EU GHG emissions from 1990 onwards in relation to the EU and Member State targets. For the EU-15 Member States, the targets are those set out in Council Decision 2002/358EC in which Member States agreed that some countries be allowed to increase their emissions, within limits, provided these are offset by reductions in others. The EU-15 Kyoto Protocol target for 2008–2012 is a reduction of 8 % from 1990 levels for the basket of six greenhouse gases. For the EU-10, accession countries and other EEA member countries, the targets are included in the Kyoto Protocol. For an overview of the national Kyoto targets see the IMS website.

Core set of indicators | Climate change

Figure 2   **Distance to target for the EU-15 in 2003 (EU Kyoto Protocol and EU Member State burden-sharing targets)**

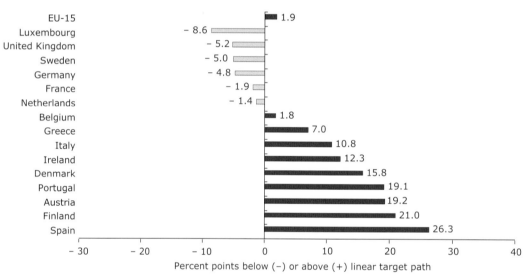

Note: Data source: EEA Data service (Ref: www.eea.eu.int/coreset).

Figure 3   **Development of EU-10 greenhouse gas emissions from base year to 2003**

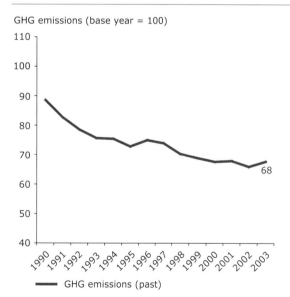

Note: Excluding Malta and Cyprus, which do not have Kyoto Protocol targets.

Indicator uncertainty

The EEA uses data officially submitted by EU Member States and other EEA countries which perform their own assessments of the uncertainty of reported data (good practice guidance and uncertainty management in national GHG inventories: Intergovernmental Panel on Climate Change (IPCC)). The IPCC suggests that the uncertainty in the total GWP-weighted emission estimates, for most European countries, is likely to be less than +/– 20 %. Total GHG emission trends are likely to be more accurate than the absolute emission estimates for individual years. The IPCC suggests that the uncertainty in total GHG emission trends is +/– 4 % to 5 %. This year for the first time uncertainty estimates were calculated for the EU-15. The results suggest that uncertainties at EU-15 level are between +/– 4 % and 8 % for total EU-15 GHG emissions.

For the EU-10 and EU candidate countries, uncertainties are assumed to be higher than for the EU-15 because of data gaps. The GHG emission indicator is an established indicator and is used

Figure 4 Change in EU-15 emissions of greenhouse gases by sector and gas 1990–2003

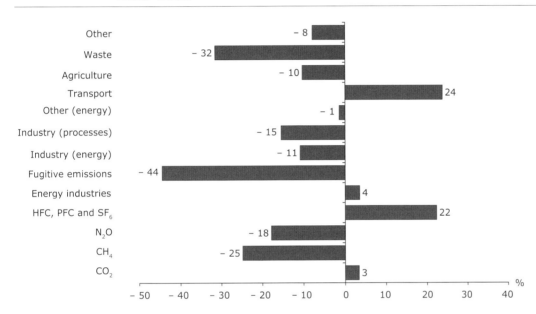

Note: Data source: EEA Data service (Ref: www.eea.eu.int/coreset).

regularly by international organisations and at the national level. Any uncertainties involved in the calculation and the data sets need to be accurately communicated in the assessment, to prevent erroneous messages influencing the political process.

11 Projections of greenhouse gas emissions and removals

Key policy question

What progress is projected towards meeting the Kyoto Protocol targets for Europe for reducing greenhouse gas (GHG) emissions to 2010: with current domestic policies and measures, with additional domestic policies and measures, and with additional use of the Kyoto mechanisms?

Key message

The aggregate projections for the EU-15 for 2010, based on existing domestic policies and measures, show emissions falling to 1.6 % below base-year levels. This leaves a shortfall of 6.4 % to reach the EU's Kyoto commitment of an 8 % reduction in emissions in 2010 compared with base-year levels.

Savings from additional measures being planned would result in emission reductions of 6.8 %, still not sufficient to meet the target. The use of Kyoto mechanisms by various Member States would reduce emissions by a further 2.5 %, leading to total reductions of 9.3 %, sufficient to reach the EU-15 target. This would, however, rely on over-delivery by some Member States. All EU-10 project that existing domestic measures will be sufficient to meet their Kyoto targets in 2010, in one by using carbon sinks. Regarding other EEA countries, Iceland and the EU candidate countries Bulgaria and Romania are on track to achieving their Kyoto targets while Norway and Liechtenstein will, with existing domestic policies and measures, fall short of theirs.

Indicator assessment

For the EU-15, aggregate projections of total GHG emissions for 2010 based on existing ([1]) domestic policies and measures show a small fall to 1.6 % below base-year levels. This means that the current emission reduction of 1.7 % achieved by 2003 compared with the base-year level is projected to stabilise by 2010. This development, assuming only existing domestic policies and measures, leads to a shortfall of 6.4 % in meeting the EU's Kyoto commitment of an 8 % reduction in emissions in 2010 from base-year levels. The use of Kyoto mechanisms by Austria, Belgium, Denmark, Finland, Ireland, Italy, Luxembourg, the Netherlands and Spain, for which quantitative effects have been approved by the Commission in the EU emission trading scheme, would reduce the EU-15 gap by a further 2.5 %. This would lead to a shortfall of 3.9 % for the EU-15 with the combination of existing domestic measures and the use of Kyoto mechanisms. Sweden and the United Kingdom project that their existing domestic policies and measures will be sufficient to meet their burden-sharing targets. These Member States may even over-deliver on their targets. Emissions in Austria, Belgium, Denmark, Finland, France, Germany, Greece, Ireland, Italy, Luxembourg, the Netherlands, Portugal and Spain are all projected to be significantly above their commitments on the basis of their existing domestic measures. The relative gaps range from more than 30 % for Spain to about 1 % for Germany. By using the Kyoto mechanisms combined with existing domestic measures, Luxembourg would meet its target. Savings from additional policies and measures being planned by Member States would result in total emission reductions of about 6.8 % from 1990, still not sufficient to meet the shortfall for the EU-15 projected on existing domestic policies and measures.

Regarding the EU-10, all those with existing measures, except for Slovenia, have projections resulting in emissions in 2010 being lower than the Kyoto commitments. Slovenia's Kyoto target can be met by accounting for carbon sinks from LULUCF (land use, land use change and forestry).

Regarding other EEA countries, Iceland and the EU candidate countries Bulgaria and Romania will over-achieve their Kyoto targets while Norway and Liechtenstein will fall short with existing domestic policies and measures.

([1]) A 'with existing domestic measures' projection encompasses currently implemented and adopted policies and measures.

Projections of greenhouse gas emissions and removals

PART B

| Figure 1 | Relative gaps between GHG projections and 2010 targets, based on existing and additional domestic policies and measures, and changes by the use of Kyoto mechanisms |

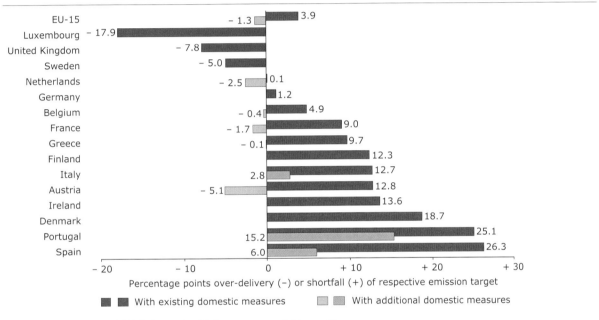

Note: Data source: EEA Data service (Ref: www.eea.eu.int/coreset).

| Figure 2 | Actual and projected EU-15 greenhouse gas emissions compared with Kyoto target for 2008–2012 |

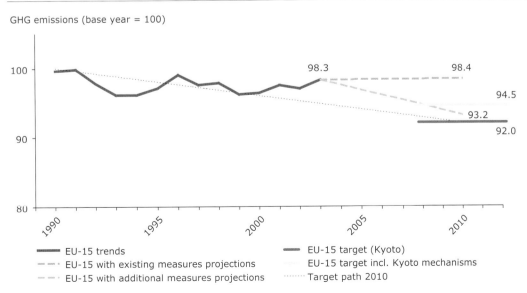

Note: Data source: EEA Data service (Ref: www.eea.eu.int/coreset).

The European environment | State and outlook 2005

Figure 3   Actual and projected greenhouse gas emissions aggregated for new Member States

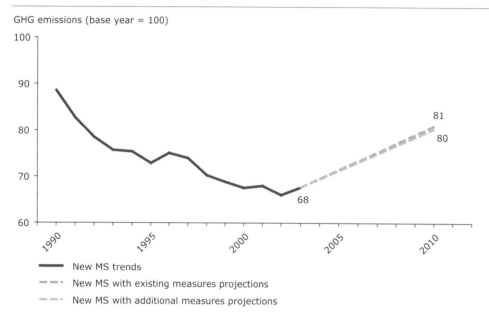

Note: Past GHG emissions and GHG projections include the eight new Member States which have Kyoto targets (not Cyprus and Malta).

Data source: (Ref: www.eea.eu.int/coreset).

Total GHG emissions from the combustion of fossil fuels in power plants and other sectors (e.g. households and services; industry) excluding the transport sector (60 % of total EU-15 GHG emissions) are projected to stabilise at 2003 level (or 3 % below 1990 level) by 2010 with existing measures and to decrease to 9 % below 1990 levels with additional measures.

Total GHG emissions from transport (21 % of total EU-15 GHG emissions) are projected to increase to 31 % above 1990 levels by 2010 with existing measures and to be 22 % above 1990 levels with additional measures.

Total GHG emissions from agriculture (10 % of total EU-15 GHG emissions) are projected to decrease to 13 % below 1990 levels by 2010 with existing measures and 15 % below 1990 levels with additional measures. The main reasons are decreasing cattle numbers and declining fertiliser and manure use.

Total GHG emissions from industrial processes (6 % of total EU-15 GHG emissions) are projected to be 4 % below base-year levels by 2010 with existing measures and 20 % below with additional measures.

GHG emissions from waste management (2 % of total EU-15 GHG emissions) are projected to decrease to 52 % below 1990 levels by 2010 with existing measures. The decline in biodegradable waste being landfilled and the growing share of CH_4 recovery from landfill sites are the main reasons for falling emissions.

Indicator definition

This indicator illustrates the projected trends in anthropogenic greenhouse gas emissions in relation to the EU and Member State targets, using existing policies and measures and/or additional policies and/or use of Kyoto mechanisms. Greenhouse gas emissions are presented by type of gas and weighted by their global warming potentials. The indicator also provides information on emissions by sectors: combustion of fossil fuels in power plants and other sectors (e.g. households and services; industry); transport; industrial processes; waste; agriculture and other (including solvents). All data are in million tonnes CO_2- equivalent.

Indicator rationale

There is growing evidence that emissions of greenhouse gases are causing global and European surface air temperatures to increase, resulting in climate change. The potential consequences at the global level include rising sea levels, increasing frequency and intensity of floods and droughts, changes in biota and food productivity and increases in diseases. Efforts to reduce or limit the effects of climate change are focused on limiting the emissions of all greenhouse gases.

This indicator supports the Commission's annual assessment of progress in reducing emissions in the EU and the individual Member States to achieve the Kyoto Protocol targets under the EU greenhouse gas monitoring mechanism (Council Decision 280/2004/EC concerning a mechanism for monitoring Community greenhouse gas emissions and for implementing the Kyoto Protocol).

Policy context

For the EU-15 Member States, the targets are those set out in Council Decision 2002/358EC in which Member States agreed that some countries be allowed to increase their emissions, within limits, provided these are offset by reductions in others. The EU-15 Kyoto Protocol target for 2008–2012 is a reduction of 8 % from 1990 levels for the basket of six greenhouse gases. For the EU-10 and the accession countries and other EEA member countries, the targets are included in the Kyoto Protocol. For an overview of the national Kyoto targets see the IMS website.

Indicator uncertainty

Uncertainties in the projections in GHG emissions have not been assessed. However, several countries carry out sensitivity analyses on their projections.

12 Global and European temperature

Key policy question

Will the increase in global average temperature stay within the EU policy target of not more than 2 °C above pre-industrial levels by 2100, and will the rate of increase in global average temperature stay within the proposed target of not more than 0.2 °C per decade?

Key message

The increase in global mean temperature observed over recent decades is unusual in terms of both magnitude and rate of change. The temperature increase up to 2004 was about 0.7 +/– 0.2 °C compared with pre-industrial levels, which is about one-third of the EU policy target of not more than 2 °C. According to the Intergovernmental Panel on Climate Change (IPCC), global mean temperature is likely to increase by 1.4–5.8 °C between 1990 and 2100, and thus the EU target might be exceeded between 2040 and 2070.

The current global rate of change is about 0.18 +/– 0.05 °C per decade, a value probably exceeding any 100-year average rate of warming during the past 1 000 years.

Indicator assessment

The earth in general and Europe in particular have experienced considerable temperature increases in the past 100 years (Figure 1), especially in the most recent decades.

Globally, the temperature increase up to 2004 was about 0.7 +/– 0.2 °C compared with pre-industrial levels, which means about one-third of the EU policy target for limiting global average warming to not more than 2 °C above pre-industrial levels. These changes are unusual in terms of both magnitude and rate of change (Figure 2). The 1990s was the warmest decade on record, and 1998 was the warmest year, followed by 2003, 2002, and 2004.

Global mean temperature is likely to increase by 1.4–5.8 °C between 1990 and 2100, assuming no climate change policies beyond the Kyoto protocol and taking the uncertainty in climate sensitivity into account. Considering this projected range, the EU target might be exceeded between 2040 and 2070.

The rate of global temperature increase is currently about 0.18 +/– 0.05 °C per decade, which is already close to the indicative target of 0.2 °C per decade. Under the range of scenarios assessed by the IPCC, the indicative proposed target of 0.2 °C per decade is likely to be exceeded in the next few decades.

Europe has warmed more than the global average with an increase of nearly 1 °C since 1900. The warmest year in Europe was 2000 and the next seven warmest years were all in the last 14 years. The temperature increase was larger in winter than in summer.

Indicator definition

The indicator shows trends in annual average global and European temperature and European winter/summer temperatures (all compared with the 1961–1990 average). The units are °C and °C per decade.

Indicator rationale

Surface air temperature gives one of the clearest signals of climate change, especially in recent decades. It has been measured for many decades or even centuries. There is mounting evidence that anthropogenic emissions of greenhouse gases are (mostly) responsible for the recently-observed rapid increases in average temperature. Natural factors such as volcanoes and solar activity could to a large extent explain the temperature variability up to middle of the 20th century, but can explain only a small part of the recent warming.

Figure 1   Global annual average temperature deviations, 1850–2004, compared with the 1961–1990 average (in °C)

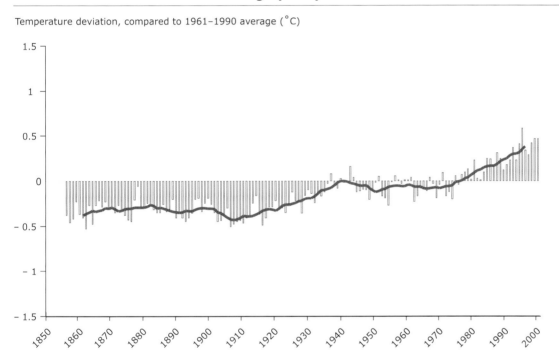

Note: Data source: KNMI, Climate Research Unit (CRU), http://www.cru.uea.ac.uk/cru/data file tavegl.dat (Ref: www.eea.eu.int/coreset).

Possible effects of climate change include rising sea levels, increasing frequency and severity of floods and droughts, changes in biota and food productivity and increase of infectious diseases. Trends and projections for the global annual average temperature can be related to indicative EU targets. However, temperature in Europe shows large differences from west (maritime) to east (continental), south (Mediterranean) to north (Arctic), and regional differences; winter/summer temperatures and cold/hot days illustrate temperature variations within a year. The rate and spatial distribution of temperature change is important, for example to determine the possibility of natural ecosystems adapting to climate change.

Policy context

The indicator can answer policy-relevant questions: will the global average temperature increase stay within the EU policy target (2 °C above pre-industrial levels)? Will the rate of global average temperature increase stay within the indicative proposed target of 0.2 °C increase per decade?

To avoid serious climate change impacts, the European Council, in its sixth environment action programme (6EAP, 2002), reaffirmed by the Environment Council and the European Council of March 2005, proposed that the global average temperature increase

Figure 2 Global average rate of temperature change (in °C per decade)

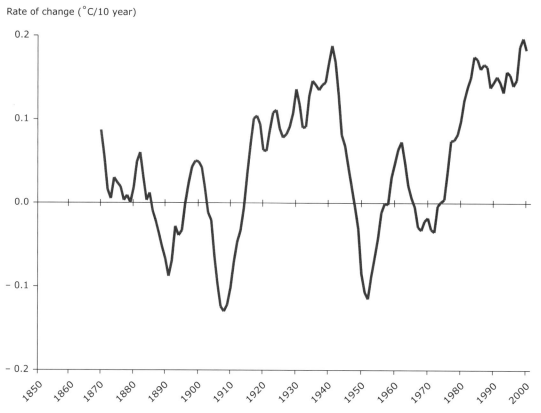

Note: Data source: KNMI, Climate Research Unit (CRU), http://www.cru.uea.ac.uk/cru/data file tavegl.dat. (Ref: www.eea.eu.int/coreset).

should be limited to not more than 2 °C above pre-industrial levels (about 1.3 °C above current global mean temperature). In addition, some studies have proposed a 'sustainable' target of limiting the rate of anthropogenic warming to 0.1 to 0.2 °C per decade.

The targets for both absolute temperature change (i.e. 2 °C) and rate of change (i.e. 0.1–0.2 °C per decade) were initially derived from the migration rates of selected plant species and the occurrence of past natural temperature changes. The EU target for global temperature increase (i.e. 2 °C) has recently been confirmed as a suitable target from both a scientific and a political perspective.

Indicator uncertainty

The observed increase in average air temperature, particularly during recent decades, is one of the clearest signals of global climate change.

Temperature has been measured over centuries. There is a generally agreed methodology with low uncertainty. Data sets used for the indicator have been checked and corrected for changing methodologies and location (rural in the past, now more urban). The uncertainty is larger for projected temperature changes, partly resulting from a lack of knowledge of parts of the climate system, including climate sensitivity

Global and European temperature

Figure 3 European annual, winter and summer temperature deviations (in °C, expressed as 10 year mean compared with the 1961–1990 average)

Note: Data source: KNMI, (http://climexp.knmi.nl) based on Climate Research Unit (CRU), file CruTemp2v. (Ref: www.eea.eu.int/coreset).

(temperature increase that results from doubling CO_2-concentrations) and seasonal temperature variability.

Temperature has been measured at many locations in Europe for many decades. The uncertainty has decreased over recent decades due to wider use of agreed methodologies and denser monitoring networks.

Annual values of global and European temperature are accurate to approximately +/– 0.05 °C (two standard errors) for the period since 1951. They were about four times as uncertain during the 1850s, with the accuracy improving gradually between 1860 and 1950 except for temporary deterioration during data-sparse, wartime intervals. New technologies, especially related to the use of remote sensing, will increase the coverage and reduce the uncertainty in temperature.

13 Atmospheric greenhouse gas concentrations

Key policy question

Will greenhouse gas (GHG) concentrations remain below 550 ppm CO_2-equivalent in the long term, the level needed to limit global temperature rise to 2 °C above pre-industrial levels ([1])?

Key message

The atmospheric concentration of carbon dioxide (CO_2), the main GHG, has increased by 34 % compared with pre-industrial levels as a result of human activities, with an accelerated rise since 1950. Other GHG concentrations have also risen as a result of human activities. The present concentrations of CO_2 and CH_4 have not been exceeded during the past 420 000 years and the present N_2O concentration during at least the past 1 000 years.

IPCC baseline projections show that GHG concentrations are likely to exceed the level of 550 ppm CO_2-equivalent in the next few decades (before 2050).

Indicator assessment

The concentration of GHGs in the atmosphere increased during the 20th century as a result of human activities, mostly related to the use of fossil fuels (e.g. for electric power generation), agricultural activities and land-use change (mainly deforestation), and continue to increase. The increase has been particularly rapid since 1950. Compared with the pre-industrial era (before 1750), concentrations of carbon dioxide (CO_2), methane (CH_4) and nitrous oxide (N_2O) have increased by 34 %, 153 %, and 17 %, respectively. The present concentrations of CO_2 (372 parts per million, ppm) and CH_4 (1772 part per billion, ppb) have not been exceeded during the past 420 000 years (for CO_2 probably not even during the past 20 million years); the present N_2O concentration (317 ppb) has not been exceeded during at least the past 1 000 years.

The IPCC showed various projected future GHG concentrations for the 21st century, varying due to a range of scenarios of socio-economic, technological and demographic developments. These scenarios assume no implementation of specific climate-driven policy measures. Under these scenarios, GHG concentrations are estimated to increase to 650–1 350 ppm CO_2-equivalent by 2100. It is very likely that fossil fuel burning will be the major cause of this increase in the 21st century.

The IPCC projections show that global atmospheric GHG concentrations are likely to exceed 550 ppm CO_2-equivalent in the next few decades (before 2050). If this level is exceeded, there is little chance that global temperature rise will stay below the EU target of not more than 2 degrees C above pre-industrial levels. Substantial global emission reductions are therefore necessary to meet this target.

Indicator definition

The indicator shows the measured trends and projections of GHG concentrations. GHGs that fall under the Kyoto Protocol (CO_2, CH_4, N_2O, HFCs, PFCs, and SF_6) are covered. The effect of GHG concentrations on the enhanced greenhouse effect is presented as CO_2-equivalent concentration. Global annual averages are considered. CO_2-equivalent concentrations are calculated from measured GHG concentrations (parts per million in CO_2-equivalent).

---

[1] Recent scientific insight shows that in order to have a high chance of meeting the EU policy target of limiting global temperature rise to 2 °C above pre-industrial levels, global GHG concentrations may need to be stabilised at much lower levels, e.g. 450 ppm CO_2-equivalent.

Atmospheric greenhouse gas concentrations PART B

Indicator rationale

The indicator shows the trend in GHG concentrations. It is the key indicator used for international negotiations for future (post-2012) emission reductions. Increase in GHG concentrations is considered to be one of the most important causes of global warming. The increase leads to enhanced radiative forcing and a more intense greenhouse effect, causing the global mean temperature of the earth's surface and lower atmosphere to rise.

Although most of the emissions occur in the northern hemisphere, the use of global average values is justified because the atmospheric lifetime of GHGs is long compared with the timescales of global atmospheric mixing. This leads to a rather uniform mixing around the globe. The indicator also expresses the relative importance of different gases for the enhanced greenhouse effect.

Enhanced GHG concentrations lead to radiative forcing and affect the earth's energy budget and climate system. To express instantaneous disturbance of the earth's radiation budget, both radiative forcing and CO_2-equivalent concentration can be used as an indicator. The CO_2-equivalent concentration is defined as the concentration of CO_2 that would cause the same

Figure 1   Measured and projected concentrations of 'Kyoto' greenhouse gases

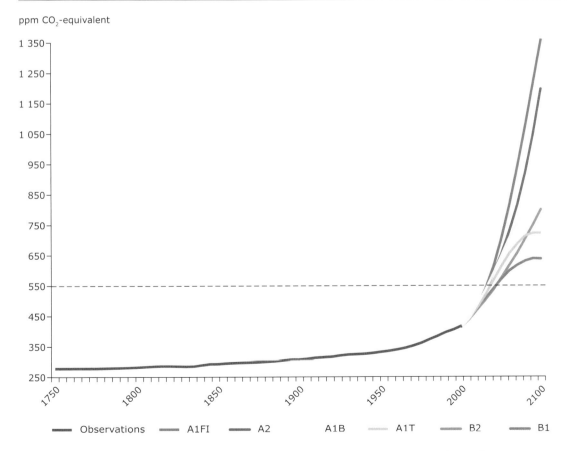

Note: Data source: SIO; ALE/GAGE/AGAGE; NOAA/CMDL; IPCC, 2001 (Ref: www.eea.eu.int/coreset).

amount of radiative forcing as the mixture of CO_2 and other GHGs. Here CO_2-equivalent concentrations rather than radiative forcings are presented, because they are more easily understandable by the general public. CO_2-equivalent concentrations can also easily be used to track progress towards the long-term EU climate objective to stabilise GHG concentrations at well below 550 ppm CO_2-equivalent. CFCs and HCFCs are not considered for this indicator, because the EU concentration stabilisation target applies only to the Kyoto GHGs. Increases in GHG concentrations are mostly from emissions from human activities, including the use of fossil fuels for power and heat generation, transport and households, and agriculture and industry.

Policy context

The indicator is aimed at supporting assessment of progress towards the long-term EU target to limit global temperature increase to below 2 °C above pre-industrial levels, and, derived from this, stabilisation of GHG concentrations at well below 550 ppm CO_2-equivalent (Decision No 1600/2002/EC of the European Parliament and of the Council of 22 July 2002, laying down the sixth Community environment action programme, confirmed by the Environment Council conclusions of March 2005).

The ultimate objective of the United Nations Framework Convention on Climate Change (UNFCCC) is to achieve *stabilization of GHG concentrations in the atmosphere at a level that would prevent dangerous anthropogenic interference with the climate system. Such a level should be achieved within a time-frame sufficient to allow ecosystems to adapt naturally to climate change, to ensure that food production is not threatened and to enable economic development to proceed in a sustainable manner*.

To reach the UNFCCC objective, the EU has specified more quantitative targets in its 6th environment action programme (6th EAP) which mentions a long-term EU climate change objective of limiting global temperature rise to a maximum of 2 °C compared with pre-industrial levels. This target was confirmed by the Environment Councils of 20 December 2004 and 22–23 March 2005. According to the Environment Council conclusions of December 2004, stabilisation of concentrations at well below 550 ppm CO_2-equivalent may be needed and global GHG emissions would have to peak within two decades, followed by substantial reductions in the order of at least 15 % and perhaps as much as 50 % by 2050 compared with 1990 levels.

Indicator uncertainty

Global average concentrations since approximately 1980 are determined by averaging measurements from several ground-station networks (SIO, NOAA/CMDL, ALE/GAGE/AGAGE), each consisting of several stations distributed across the globe. The use of global average values is justified because the time-scale at which sources and sinks change is long compared with that of global atmospheric mixing.

Absolute accuracies of global annual average concentrations are in the order of 1 % for CO_2, CH_4 and N_2O, and CFCs; for HFCs, PFCs, and SF_6, absolute accuracies can be up to 10–20 %. However, the year-to-year variations are much more accurate. Radiative forcing calculations have an absolute accuracy of 10 %; trends in radiative forcing are much more accurate.

The dominant sources of error for radiative forcing are the uncertainties in modelling radiative transfer in the earth's atmosphere and in the spectroscopic parameters of the molecules involved. Radiative forcing is calculated using parameterisations that relate the measured concentrations of GHGs to radiative forcing. The overall uncertainty in radiative forcing calculations (all species together) is estimated to be 10 %. Radiative forcing is also expressed as CO_2-equivalent concentration; both have the same uncertainty. The uncertainty in the trend in radiative forcing/CO_2-equivalent concentration is determined by the precision of the method rather than the absolute uncertainty discussed above. The uncertainty in the trend is therefore much less than 10 %, and is determined by the accuracy of concentration measurements (0.1 %).

It is important to note that global warming potentials are not used to calculate radiative forcing. They are used only to compare the time-integrated climate effects of emissions of different GHGs.

Uncertainties in model projections are related to uncertainties in the emission scenarios, the global climate models and the data and assumptions used.

Direct measurements have good comparability. Although methods for calculating radiative forcing and CO_2-equivalent are expected to improve further, any update of these methods will be applied to the complete dataset covering all years, so this will not affect the comparability of the indicator over time.

14 Land take

Key policy question

How much and in what proportions is agricultural, forest and other semi-natural and natural land being taken for urban and other artificial land development?

Key message

Land take by the expansion of artificial areas and related infrastructure is the main cause of the increase in the coverage of land at the European level. Agricultural zones and, to a lesser extent, forests and semi-natural and natural areas, are disappearing in favour of the development of artificial surfaces. This affects biodiversity since it decreases habitats, the living space of a number of species, and fragments the landscapes that support and connect them.

Indicator assessment

The largest land-cover category being taken by urban and other artificial land development (average for 23 European countries) is agriculture land. During 1990–2000, 48 % of all areas that changed to artificial surfaces were arable land or permanent crops. This process is particularly important in Denmark (80 %) and Germany (72 %). Pastures and mixed farmland are, on average, the next category being taken, representing 36 % of the total. However, in several countries or regions, these landscapes are the major source for land take (in a broad sense), for example in Ireland (80 %) and the Netherlands (60 %).

The proportion of forested and natural land taken for artificial development during the period is important in Portugal (35 %), Spain (31 %) and Greece (23 %).

Specific policy question: What are the drivers of uptake for urban and other artificial land development?

At the European level, housing, services and recreation make up half of the overall increase in urban and other artificial areas between 1990 and 2000. But the situation varies from countries with proportions of new land take for housing, services and recreation higher than 70 % (Luxembourg and Ireland) to countries such as Greece (16 %) and Poland (22 %) where urban development is due mainly to industrial/commercial activity.

Industrial/commercial sites is the next sector responsible for land take, with 31 % of the average European new land uptake during the period. However, this sector is taking the largest proportion of new uptake in Belgium (48 %), Greece (43 %) and Hungary (32 %).

Land take for mines, quarries and waste dumpsites was relatively important in countries with low artificial land take from 1990–2000 as well as in Poland (43 %) where mines are a key sector of the economy. At the European level, the percentage of the total new land take for mines, quarries and waste dumpsites is 14 %.

Land take for transport infrastructures (3.2 % of the total new artificial cover) is underestimated in surveys that are based on remote-sensing such as Corine land cover (CLC). Land take by linear features such as roads and railways is not included in the statistics, which focus only on area infrastructures (e.g. airports and harbours). Soil sealing and fragmentation by linear infrastructures therefore need to be observed by different means.

Specific policy question: Where have the more important artificial land uptakes occurred?

Land uptake by urban and other artificial development in the 23 European countries covered by Corine Land Cover 2000 amounted to 917 224 hectares in 10 years. It represents 0.3 % of the total territory of these countries. This may seem low, but spatial differences are very important and urban sprawl in many regions is very intense.

Considering the contribution of each country to new total urban and infrastructure sprawl in Europe, mean annual values range from 22 % (Germany) to 0.02 % (Latvia), with intermediate values in France (15 %), Spain (13.3 %) and Italy (9.1 %). Differences between countries are strongly related to their size and population density (Figure 3).

The pace of land take observed by comparing it with the initial extent of urban and other artificial areas in 1990 gives another picture (Figure 4). From this perspective, the average value in the 23 European countries covered by CLC2000 ranges up to an annual increase of 0.7 %. Urban development is fastest in Ireland (3.1 % increase in urban area per year), Portugal (2.8 %), Spain (1.9 %) and the Netherlands (1.6 %). However, this comparison reflects different initial conditions. For example, Ireland had a very small amount of urban area in 1990 and the Netherlands one of the largest in Europe. Urban sprawl in EU-10 is generally lower than in the EU-15 countries, in absolute and relative terms.

Indicator definition

Increase in the amount of agriculture, forest and other semi-natural and natural land taken by urban and other artificial land development. It includes areas sealed by construction and urban infrastructure as well as urban green areas and sport and leisure facilities. The main drivers of land take are grouped in processes resulting in the extension of:

Figure 1 Relative contribution of land-cover categories to uptake by urban and other artificial land development

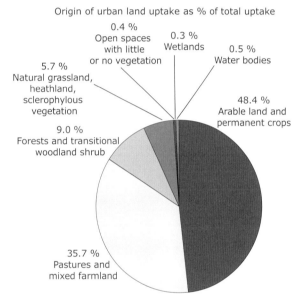

Note:

- housing, services and recreation,
- industrial and commercial sites,
- transport networks and infrastructures, and
- mines, quarries and waste dumpsites.

Figure 2 Land take by several types of human activity per year in 23 European countries, 1990–2000

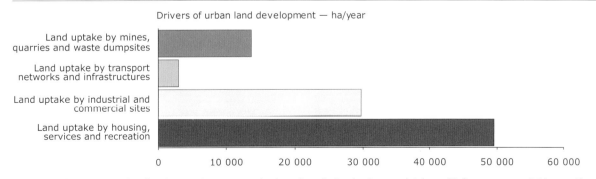

Note: Data source: Land and ecosystems accounts, based on Corine land cover database (Ref: www.eea.eu.int/coreset).

Indicator rationale

Land use by urban and related infrastructures has the highest impacts on the environment due to sealing of soil as well as disturbances resulting from transport, noise, resource use, waste dumping and pollution. Transport networks which connect cities add to the fragmentation and degradation of the natural landscape. The intensity and patterns of urban sprawl are the result of three main factors: economic development, demand for housing and extension of transport networks. Although subsidiarity rules assign land and urban planning responsibilities to national and regional levels, most European policies have a direct or indirect effect on urban development.

Figure 3 Mean annual urban land take as a percentage of total Europe-23 urban land take 1990–2000

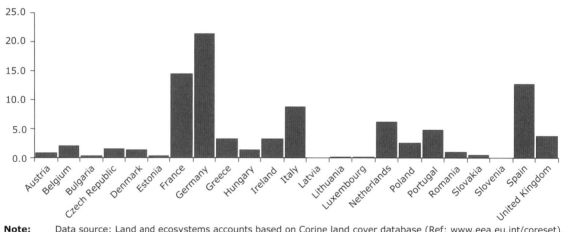

Note: Data source: Land and ecosystems accounts based on Corine land cover database (Ref: www.eea.eu.int/coreset).

Figure 4 Mean annual urban land take 1990–2000 as a percentage of 1990 artificial land

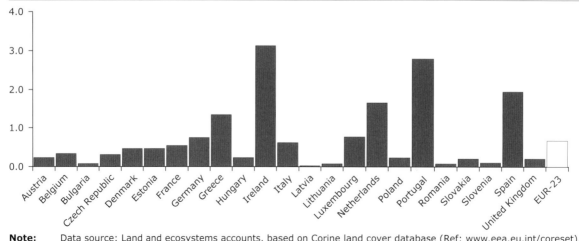

Note: Data source: Land and ecosystems accounts, based on Corine land cover database (Ref: www.eea.eu.int/coreset).

Built-up areas have been increasing steadily all over Europe for ten years, continuing the trend observed during the 1980s. The same is true for transport infrastructures, as a result of rising living standards, people living further from work, liberalisation of the EU internal market, globalisation of the economy, and more complex chains and networks of production. Increasing prosperity is increasing the demand for second homes. The growth in demand for land, both for building and for new transport infrastructure, is continuing.

Policy context

The main policy objective of this indicator is to measure the pressure from the development of urban and other artificial land on natural and managed landscapes that are necessary 'to protect and restore the functioning of natural systems and halt the loss of biodiversity' (included in the 6th environment action programme).

Important references can be found in the 6th Environment Action Programme (6EAP, COM (2001) 31) and the thematic documents related to it, such as the Commission Communication 'Towards a Thematic Strategy on the Urban Environment' (COM (2004) 60), the EU Strategy for Sustainable Development (COM (2001) 264, the new general regulation for the Structural Funds (Council Regulation EC no 1260/1999), the guidelines for INTERREG III (published on 23/05/2000 (OJ C 143)) and the ESDP Action programme and ESPON guidelines for 2001–2006.

There are no quantitative targets for land take for urban development at the European level, although different documents reflect the need for better planning of urban development and the extension of infrastructures.

Indicator uncertainty

Surfaces monitored with Corine land cover relate to the extension of urban systems that may include parcels not covered by construction, streets or other sealed surfaces. This is particularly the case for discontinuous urban fabric, which is considered as a whole. Monitoring the indicator with satellite images leads to the exclusion of small urban features in the countryside and most of the linear transport infrastructures, which are too narrow to be observed directly. Therefore, differences exist between CLC results and other statistics collected with different methodologies such as point or area sampling or farm surveys; this is often the case for agriculture and forest statistics. However, the trends are generally similar.

Geographical and time coverage at the EU level

All the EU-25 (except Sweden, Finland, Malta and Cyprus) as well as Bulgaria and Romania are covered with both '1990' and 2000 results. '1990' refers to the first experimental phase of CLC, which ran from 1986 up to 1995. 2000 is considered to be a reasonable characterisation (a few satellite images only being from 1999 or 2001, for cloud coverage reasons). Comparisons between countries therefore have to be done on the basis of annual mean values. The average number of years between two CLCs in each country can be seen in Table 1.

Representativeness of data at the national level

At the national level, there may be time differences between regions in large countries and these are documented in the CLC metadata.

Table 1   Average number of years between two CLCs per country

AT	BE	BG	CZ	DE	DK	EE	ES	FR	GR	HU	IE	IT	LT	LU	LV	NL	PL	PT	RO	SI	SK	UK
15	10	10	8	10	10	6	14	10	10	8	10	10	5	11	5	14	8	14	8	5	8	10

15 Progress in management of contaminated sites

Key policy question

How are the problems of contaminated sites being addressed (clean-up of historical contamination and prevention of new contamination)?

Key message

Several economic activities are still causing soil pollution in Europe, particularly those related to inadequate waste disposal and losses during industrial operations. In the coming years, the implementation of preventive measures introduced by the legislation already in place is expected to limit inputs of contaminants into the soil. As a consequence, most of the future management efforts will be concentrated on the clean-up of historical contamination. This is going to require large sums of public money which currently already accounts for an average of 25 % of total remediation expenditure.

Indicator assessment

The major localised sources of soil contamination in Europe derive from the inadequate disposal of waste, losses during industrial and commercial operations, and the oil industry (extraction and transport). However, the range of polluting activities and their importance may vary considerably from country to country. These variations may reflect different industrial and commercial structures, different classification systems or incomplete information.

A broad range of industrial and commercial activities have produced impacts on soil through the release of a broad variety of pollutants. The main contaminants causing soil contamination from local sources at industrial and commercial sites are reported to be heavy metals, mineral oil, polycyclic aromatic hydrocarbons (PAH), and chlorinated and aromatic hydrocarbons. Globally, these alone affect 90 % of the sites for which information on contaminants is available, while their relative contribution may vary greatly from country to country.

The implementation of existing legislative and regulatory frameworks (such as the Integrated Pollution Prevention and Control Directive and the Landfill Directive) should result in less new contamination of soil. However, large amounts of time and financial resources from the private and the public sector are still needed to deal with historical contamination. This is a tiered process, where the final steps (remediation) involve much larger resources than the first steps (site investigations).

In most of the countries for which data are available, site identification activities are generally far advanced, while detailed investigations and remediation activities are generally progressing slowly (Figure 1). However, progress in management may vary considerably from country to country.

The progress in each country (i.e. the numbers of sites treated in each management step) cannot be compared directly, due to different legal requirements and different degrees of industrialisation, and local conditions and approaches. For example, a large percentage of completed remediations compared with the estimated remediation needs in some countries could be interpreted as a well-advanced management process. However, surveys in these countries are also usually incomplete, which generally results in an underestimation of the problem.

Although most of the countries in Europe have legislative instruments which apply the 'polluter-pays' principle to the clean-up of contaminated sites, large sums of public money — on average 25 % of total costs — have to be provided to fund the necessary remediation activities. This is a common trend across Europe (Figure 2). Annual expenditures on the full clean-up process in the countries analysed in the period 1999–2002 varied from less than EUR 2 to EUR 35 per capita per year.

Although a considerable amount of money has already been spent on remediation, this is relatively little (up to 8 %) compared with the estimated total costs.

Progress in management of contaminated sites

Figure 1 Overview of progress in control and remediation of soil contamination by country

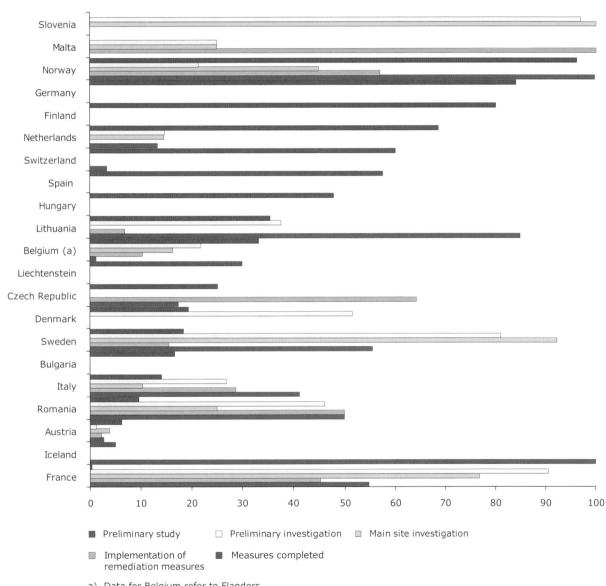

a) Data for Belgium refer to Flanders.

Note: Information on 'remediation completed' has not been included. Missing information indicates that no data have been reported for the particular country.

Data source: Eionet priority data flow; September 2003. 1999 and 2000 data: for EU countries and Liechtenstein: pilot Eionet data flow; January 2002; for accession countries: data request to new EEA member countries, February 2002 (Ref: www.eea.eu.int/coreset).

Core set of indicators | Terrestrial

Figure 2   Annual expenditure on contaminated site remediation by country

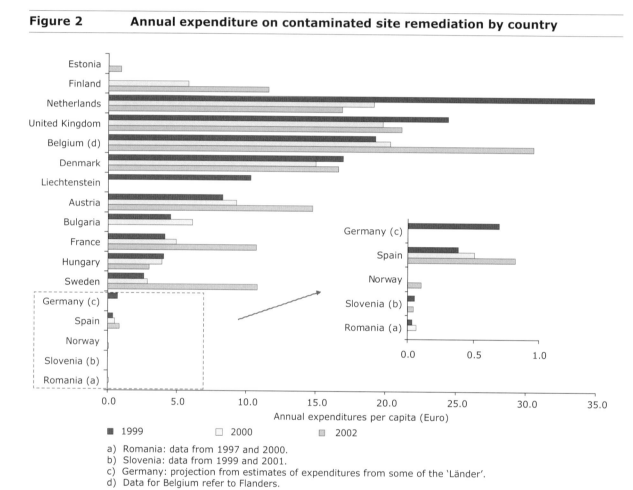

- 1999
- 2000
- 2002

a) Romania: data from 1997 and 2000.
b) Slovenia: data from 1999 and 2001.
c) Germany: projection from estimates of expenditures from some of the 'Länder'.
d) Data for Belgium refer to Flanders.

Note: Data source: (Ref: www.eea.eu.int/coreset).

Indicator definition

The term 'contaminated site' refers to a delimited area where the presence of soil contamination has been confirmed and the severity of possible impacts on ecosystems and human health are such that remediation is needed, specifically in relation to the current or planned use of the site. The remediation or clean-up of contaminated sites can result in full elimination or a reduction of these impacts.

The term 'potentially contaminated site' includes any site where soil contamination is suspected but not verified and investigations are needed to verify whether relevant impacts exist.

The management of contaminated sites is a tiered process, designed to ameliorate any adverse effects where impairment of the environment is suspected or has been proved, and to minimise any potential threats (to human health, water bodies, soil, habitats, foodstuffs, biodiversity, etc.). The management of a site starts with a basic survey and investigation, which may lead to remediation, after-care measures and brownfield redevelopment.

Indicator rationale

Emissions of dangerous substances from local sources may have far-reaching effects on the quality of soil and water, particularly groundwater, with important impacts on human and ecosystem health.

A number of economic activities causing soil pollution can be clearly identified across Europe. These relate, in particular, to losses during industrial operations and waste disposal from municipal and industrial sources. Management of contaminated sites aims at assessing the impacts of contamination by local sources and taking measures to satisfy environmental standards according to existing legal requirements.

The indicator tracks progress in the management of contaminated sites in Europe and related expenditures by the public and private sectors. It also shows the contributions of the main economic activities responsible for soil contamination and the major pollutants involved.

Policy context

The main policy objective of legislation aimed at protecting soil from contamination from local sources is to achieve a quality of the environment where the levels of contaminants do not give rise to significant impacts or risks to human health.

At the European level, remediation and prevention of soil contamination will be addressed by the forthcoming soil thematic strategy (STS). Existing EU legislation addresses the protection of water and sets standards for water quality, whereas no legal standards for soil quality exist or are likely to be established in the near future. Nevertheless, specific standards for soil quality and policy targets have been put in place in several EEA member countries. In general, legislation aims at preventing new contamination and setting targets for the remediation of sites where environmental standards have already been exceeded.

Indicator uncertainty

The information provided by this indicator has to be interpreted and presented with caution, due to uncertainties in methodology and problems of data comparability.

There are no common definitions of contaminated sites across Europe, which creates problems when comparing national data to produce European assessments. For this reason, the indicator focuses on the impacts of the contamination and progress in management, rather than on the extent of the problem (e.g. number of contaminated sites). Comparability of national data is expected to improve as common EU definitions are introduced in the context of the STS.

In reporting progress against a national baseline (number of sites expected) some countries may change their estimates in successive years. This may depend on the status of completion of national inventories (e.g. not all sites are included at the beginning of registration, but the number of sites may increase dramatically after more accurate screening; the reverse has also been observed due to changes in national legislation).

Moreover, cost estimates for remediation are difficult to obtain, especially from the private sector, and little information on quantities of contaminants is available.

Insufficiently clear methodology and data specifications may have resulted in countries interpreting requests for data in different ways and may therefore result in not fully comparable information. This is expected to improve in the future as better specifications and documentation of the methodology is provided.

Not all countries have been included in the calculations of the indicator (due to the unavailability of national data). The data available do not allow the evaluation of time trends. Most of the data integrates information from the whole country. However the process differs from country to country, depending on the degree of decentralisation. In general, data quality and representativeness increase with centralisation of the information (national registers).

16 Municipal waste generation

Key policy question

Are we reducing the generation of municipal waste?

Key message

The generation of municipal waste per capita in western European [1] countries continues to grow while remaining stable in central and eastern European [2] countries.

The EU target to reduce municipal waste generation to 300 kg/capita/year by 2000 was not achieved. No new targets have been set.

Indicator assessment

One of the targets set in the 5th environment action programme was to reduce the generation of municipal waste per capita per year to the average 1985 EU level of 300 kg by the year 2000 and then stabilise it at that level. The indicator (Figure 1) shows that the target is far from being reached. The target has not been repeated in 6th EAP.

The average amount of municipal waste generated per capita per year in many western European countries has reached more than 500 kg.

Municipal waste generation rates in central and eastern Europe are lower than in western European countries and generation is decreasing slightly. Whether this is due to different consumption patterns or underdeveloped municipal waste collection and disposal systems needs further clarification. Reporting systems also need further development.

Indicator definition

The indicator presents municipal waste generation, expressed in kg per person per year. Municipal waste refers to waste collected by or on behalf of municipalities; the main part originates from households, but waste from commerce and trade, office buildings, institutions and small businesses is also included.

Indicator rationale

Waste represents an enormous loss of resources in the form of both materials and energy. The amount of waste produced can be seen as an indicator of how efficient we are as a society, particularly in relation to our use of natural resources and waste treatment operations.

Municipal waste is currently the best indicator available for describing the general development of waste generation and treatment in European countries. This is because all countries collect data on municipal waste; data coverage for other waste, for example total waste or household waste, is more limited.

Municipal waste constitutes only around 15 % of total waste generated, but because of its complex character and its distribution among many waste generators, environmentally sound management of this waste is complicated. Municipal waste contains many materials for which recycling is environmentally beneficial.

Despite its limited share of total waste generation, the political focus on municipal waste is very high.

---

[1] Western European countries are the EU-15 countries + Norway and Iceland.
[2] Central and eastern European countries are the EU-10 + Romania and Bulgaria.

Municipal waste generation

Figure 1 Municipal waste generation in western European (WE) and central and eastern European (CEE) countries

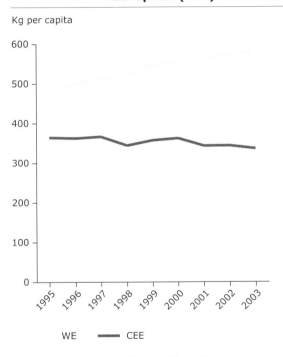

Note: Data source: Eurostat, World Bank (Ref: www.eea.eu.int/coreset).

Policy context

EU 6th environment action programme:

- Better resource efficiency and resource and waste management to bring about more sustainable production and consumption patterns, thereby decoupling the use of resources and the generation of waste from the rate of economic growth and aiming to ensure that the consumption of renewable and non-renewable resources does not exceed the carrying capacity of the environment.

- Achieving a significant overall reduction in the volumes of waste generated through waste prevention initiatives, better resource efficiency and a shift towards more sustainable production and consumption patterns.

- A significant reduction in the quantity of waste going to disposal and the volumes of hazardous waste produced while avoiding an increase in emissions to air, water and soil.

- Encouraging reuse. Preference should be given to recovery, and especially to recycling, of waste that are still generated.

EU waste strategy (Council Resolution of 7 May 1990 on waste policy):

- Where the production of waste is unavoidable, recycling and reuse of waste should be encouraged.

Communication from the Commission on the review of the Community strategy for waste management (COM(96) 399):

- There is considerable potential for reducing and recovering municipal waste in a more sustainable fashion, for which new targets need to be set.

This indicator is one of the structural indicators and is used for monitoring the Lisbon Strategy.

Target

The EU 5th EAP had a target of 300 kg household waste per capita per year, but no new targets have been set in the 6th EAP because of very little success with the 300 kg target. The target is therefore no longer relevant and is used here only for illustration purposes.

Indicator uncertainty

If no data on waste generation are available for a particular country and year, estimates are made by Eurostat to fill the gap, based on the linear best-fit method.

Core set of indicators | Waste

Table 1 Municipal waste generation in western European (WE) and central and eastern European (CEE) countries

Western Europe (municipal waste generation in kg per capita)

	1995	1996	1997	1998	1999	2000	2001	2002	2003
Austria	437	516	532	533	563	579	577	611	612
Belgium	443	440	474	470	475	483	461	461	446
Denmark	566	618	587	593	626	664	660	667	675
Finland	*413*	*410*	447	466	484	503	465	456	450
France	500	509	516	523	526	537	544	555	*560*
Germany	533	542	556	546	605	609	600	640	638
Greece	306	344	372	388	405	421	430	436	441
Ireland	513	*523*	*545*	554	*576*	598	700	695	*735*
Italy	451	452	463	466	492	502	510	519	*520*
Luxembourg	585	582	600	623	644	651	648	*653*	*658*
Netherlands	548	562	588	591	597	614	610	613	598
Portugal	391	404	410	428	432	447	462	454	461
Spain	*469*	*493*	*513*	526	570	587	590	587	*616*
Sweden	*379*	*397*	416	430	428	428	442	468	470
United Kingdom	433	510	531	541	569	576	590	599	610
Iceland	914	933	949	967	975	993	1 011	1 032	1 049
Norway	624	630	617	645	594	613	634	675	695
Western Europe	476	499	513	518	546	556	560	575	580

Central and eastern Europe (municipal waste generation in kg per capita)

	1995	1996	1997	1998	1999	2000	2001	2002	2003
Bulgaria	694	618	579	497	504	517	506	501	501
Cyprus	529	571	582	599	607	620	644	654	672
Czech Republic	*302*	310	318	293	327	334	274	279	280
Estonia	371	399	424	402	414	462	353	386	420
Hungary	465	474	494	492	491	454	452	457	*464*
Latvia	261	261	254	248	244	271	302	370	363
Lithuania	426	401	422	444	350	310	300	288	263
Malta	*331*	*342*	*352*	377	461	481	545	471	547
Poland	285	301	315	306	319	316	287	275	260
Romania	342	326	326	278	315	355	336	375	*357*
Slovak Republic	339	348	316	315	315	316	390	283	*319*
Slovenia	*596*	*590*	*589*	584	*549*	*513*	482	487	458
Central and eastern Europe	364	362	366	344	357	362	343	343	336

Note: Italics — estimates.

Data source: Eurostat, World Bank (Ref: www.eea.eu.int/coreset).

Because of different definitions of the concept 'municipal waste' and the fact that some countries have reported data on municipal waste and others on household waste, data are in general not comparable between member countries. Thus, Finland, Greece, Ireland, Norway, Portugal, Spain and Sweden do not include data on bulky waste as part of municipal waste, and very often not data on separately collected food and garden waste. Southern European countries in general include very few waste types under municipal waste, indicating that traditionally collected (bagged) waste is apparently the only big contributor to the total amount of municipal waste in these countries. The term, 'waste from household and commercial activities' is an attempt to identify common and comparable parts of municipal waste. This concept and further details on comparability were presented in EEA topic report No 3/2000.

17 Generation and recycling of packaging waste

Key policy question

Are we preventing the generation of packaging waste?

Key message

There is a general increase in per capita quantities of packaging being put on the market. This is not in line with the primary objective of the Directive on Packaging and Packaging Waste, which aims at preventing the production of packaging waste.

However, the EU target to recycle 25 % of packaging waste in 2001 has been significantly exceeded. In 2002 the recycling rate in the EU-15 was 54 %.

Indicator assessment

Only the United Kingdom, Denmark and Austria have reduced their per capita generation of packaging waste since 1997; in the remaining countries, the quantities have increased. However the 1997 data are less certain than those for later years, due to first-year problems of newly established data collection systems, which in turn may influence the apparent trends.

Between 1997 and 2002 the growth in packaging waste generation in the EU-15 almost followed the growth in GDP: generation increased by 10 % and GDP by 12.6 %.

There are large variations between Member States in the use of packaging per capita, ranging from

Figure 1   Packaging waste generation per capita and by country

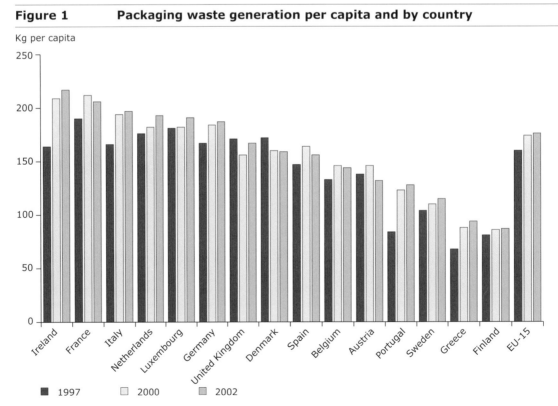

Note: Data source: DG Environment and the World Bank (Ref: www.eea.eu.int/coreset).

87 kg/capita in Finland to 217 kg/capita in Ireland (2002). The average 2002 figure for the EU-15 was 172 kg/capita. This variation can be partly explained by the fact that Member States have differing definitions of packaging and understanding of which types of packaging waste need to be reported to DG Environment. This illustrates the need to harmonise the methodology for reporting data in accordance with the directive on packaging and packaging waste.

The target of 25 % recycling of all packaging materials in 2001 was achieved by a good margin in virtually all countries. Seven Member States already comply with the overall recycling target for 2008, when not taking the 'new' material, wood, into account. The total EU-15 recycling rate increased from 45 % in 1997 to 54 % in 2002.

As with consumption of packaging per capita, the total recycling rate in Member States in 2002 varied greatly, from 33 % in Greece to 74 % in Germany.

To achieve these targets, several Member States have introduced producer responsibility and established packaging recycling companies. Other countries have improved their existing collection and recycling system.

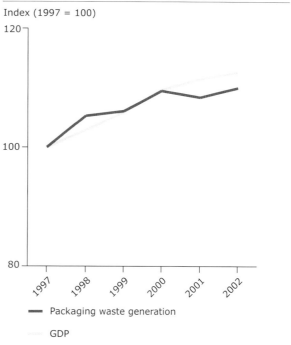

Figure 2   **Generation of packaging waste and GDP in the EU-15**

Note: Data source: DG Environment and Eurostat (Ref: www.eea.eu.int/coreset).

Indicator definition

The indicator is based on total packaging used in EU Member States expressed as kg per capita per year. The amount of packaging used is expected to equal the amount of packaging waste generated. This assumption is based on the short lifetime of packaging.

Packaging waste recycled as a share of packaging used in EU Member States is derived by dividing the quantity of packaging waste recycled by the total quantity of packaging waste generated and expressing this as a percentage.

Indicator rationale

Packaging uses a lot of resources, and typically has a short lifetime. There are environmental impacts from the extraction of resources, production of the packaging, collection of packaging waste and treatment or disposal of the waste.

Packaging waste is covered by specific EU regulations and there are specific targets for recycling and recovery. Information on the amounts of packaging waste generated therefore provides an indicator of the effectiveness of waste prevention policies.

Core set of indicators | Waste

Figure 3   Recycling of packaging waste by country, 2002

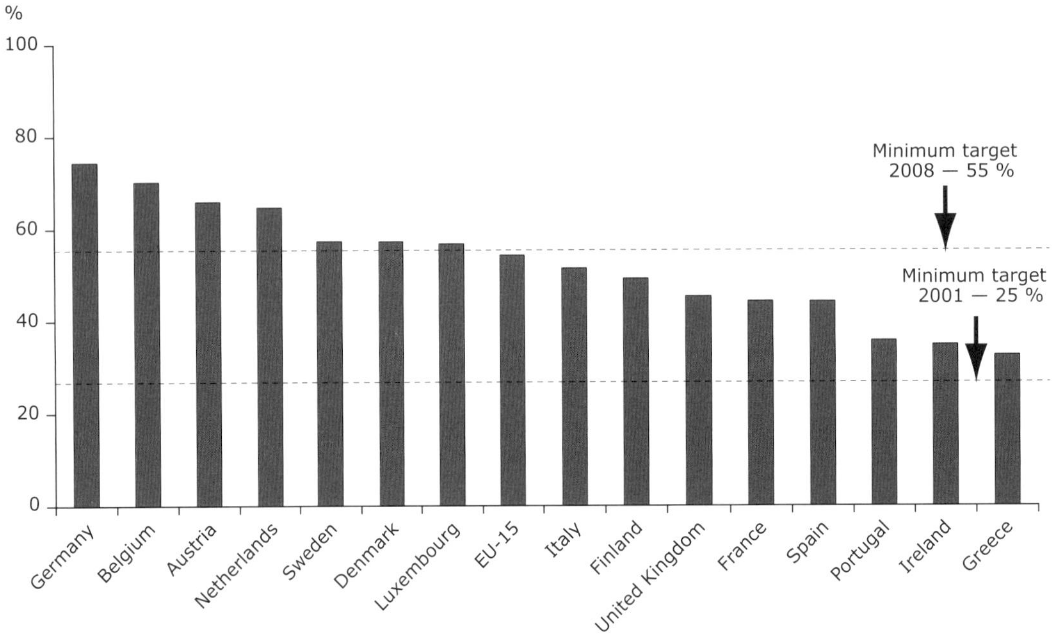

[Bar chart showing recycling percentages by country, from highest to lowest: Germany (~75%), Belgium (~70%), Austria (~66%), Netherlands (~65%), Sweden (~58%), Denmark (~58%), Luxembourg (~57%), EU-15 (~55%), Italy (~52%), Finland (~50%), United Kingdom (~46%), France (~45%), Spain (~45%), Portugal (~37%), Ireland (~36%), Greece (~34%). Dashed lines indicate Minimum target 2008 — 55 % and Minimum target 2001 — 25 %.]

Note: Data source: DG Environment (Ref: www.eea.eu.int/coreset).

Table 1   Packaging waste generation per capita and by country

	1997	1998	1999	2000	2001	2002
Ireland	164	184	187	209	212	217
France	190	199	205	212	208	206
Italy	166	188	193	194	195	197
Netherlands	176	161	164	182	186	193
Luxembourg	181	181	182	182	181	191
Germany	167	172	178	184	182	187
United Kingdom	171	175	157	156	158	167
Denmark	172	158	159	160	161	159
Spain	147	159	155	164	146	156
Belgium	133	140	145	146	138	144
Austria	138	140	141	146	137	132
Portugal	84	102	120	123	127	128
Sweden	104	108	110	110	114	115
Greece	68	76	81	88	92	94
Finland	81	82	86	86	88	87
EU-15	160	168	169	174	172	176

Note: Data source: DG Environment and the World Bank (see Figure 1) (Ref: www.eea.eu.int/coreset).

Table 2 — Targets of the packaging and packaging waste directive

By weight	Targets in 94/62/EC	Targets in 2004/12/EC
Overall recovery target	Min. 50 %, max. 65 %	Min. 60 %
Overall recycling target	Min. 25 %, max. 45 %	Min. 55 %, max. 80 %
Date to achieve targets	30 June 2001	31 December 2008

Policy context

Council Directive 94/62 of 15 December 1994 on packaging and packaging waste as amended by Directive 2004/12 of 11 February 2004 establishes targets for recycling and recovery of selected packaging materials.

The EU 6th environment action programme aims to achieve a significant overall reduction in the volumes of waste generated. This will be done through waste prevention initiatives, better resource efficiency, and a shift towards more sustainable production and consumption patterns. The 6th EAP also encourages reuse, recycling and recovery rather than disposal of waste that is still being generated.

Indicator uncertainty

The Commission decision of 3 February 1997 establishes the formats which Member States are to use in annual reporting on the directive on packaging and packaging waste. However, the decision does not define methods of estimating the quantities of packaging put on the market or calculating the recovery and recycling rates in enough detail to ensure full data comparability.

Due to the absence of harmonised methodology, national data on packaging waste are not always comparable. Some countries include all packaging waste in the figure for total packaging waste generation while others include only the total for the four obligatory packaging waste streams: glass, metal, plastics and paper.

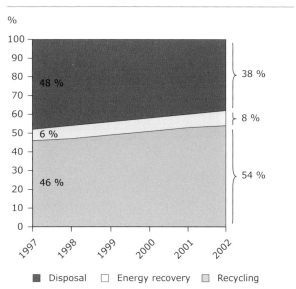

Figure 4 Treatment of packaging waste

Note: Data source: DG Environment (Ref: www.eea.eu.int/coreset).

18 Use of freshwater resources

Key policy question

Is the abstraction rate of water sustainable?

Key message

The water exploitation index (WEI) decreased in 17 EEA countries between 1990 and 2002, representing a considerable decrease in total water abstraction. But nearly half of Europe's population still lives in water-stressed countries.

Indicator assessment

The warning threshold for the water exploitation index (WEI), which distinguishes a non-stressed from a stressed region, is around 20 %. Severe water stress can occur where the WEI exceeds 40 %, indicating unsustainable water use.

Eight European countries can be considered water-stressed, i.e. Germany, England and Wales, Italy, Malta, Belgium, Spain, Bulgaria and Cyprus, representing 46 % of Europe's population. Only in Cyprus does the WEI exceed 40 %. However, it is necessary to take into account the high water abstraction for non-consumptive uses (cooling water) in Germany, England and Wales, Bulgaria and Belgium. Most of the water abstracted in the other four countries (Italy, Spain, Cyprus and Malta) is for consumptive uses (especially irrigation) and there is therefore higher pressure on water resources in these four countries.

The WEI decreased in 17 countries during the period 1990 to 2002, representing a considerable decrease in total water abstraction. Most of the decrease occurred in the EU-10, as a result of the decline in abstraction in most economic sectors. This trend was the result of institutional and economic changes. However, five countries (the Netherlands, the United Kingdom, Greece, Portugal, and Turkey) increased their WEI in the same period because of the increase in total water abstraction.

All economic sectors need water for their development. Agriculture, industry and most forms of energy production are not possible if water is not available. Navigation and a variety of recreational activities also depend on water. The most important uses, in terms of total abstraction, have been identified as urban (households and industry connected to the public water supply system), industry, agriculture and energy (cooling of power plants). The main water consumption sectors are irrigation, urban, and the manufacturing industry.

Southern European countries use the largest percentages of abstracted water for agriculture, generally accounting for more than two-thirds of total abstraction. Irrigation is the most significant use of water in the agriculture sector in these countries. Central and Nordic countries use the largest percentages of abstracted water for cooling in energy production, industrial production and public water supply.

The decrease in agricultural and industrial activities in the EU-10 and Romania and Bulgaria during the transition process led to decreases of about 70 % in water abstraction for agricultural and industrial uses in most of the countries. Agricultural activities reached their minima around the mid-1990s but more recently countries have been increasing their agricultural production.

Water use for agriculture, mainly irrigation, is on average four times higher per hectare of irrigated land in southern Europe than elsewhere. The water abstraction for irrigation in Turkey increased, and the increase in the area of irrigated land exacerbated the pressure on water resources; this trend is expected to continue with new irrigation projects.

Data show a decreasing trend in water use for public water supply in most countries. This trend is more pronounced in the EU-10 and Bulgaria and Romania, with a 30 % reduction during the 1990s. In most of these countries, the new economic conditions led to water supply companies increasing the price of water and installing water meters in houses. This resulted

Use of freshwater resources

Figure 1 Water exploitation index. Total water abstraction per year as a percentage of long-term freshwater resources in 1990 and 2002

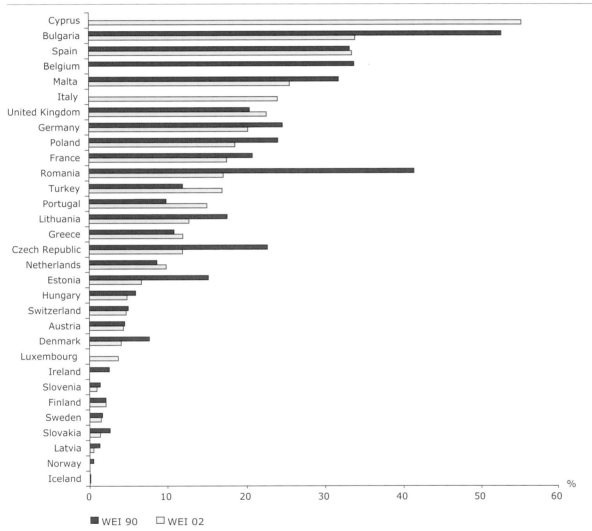

■ WEI 90 □ WEI 02

Note: 1990 = 1991 for Germany, France, Spain and Latvia;
1990 = 1992 for Hungary and Iceland;
2002 = 2001 for Germany, the Netherlands, Bulgaria and Turkey;
2002 = 2000 for Malta;
2002 = 1999 for Luxembourg, Finland and Austria;
2002 = 1998 for Italy and Portugal;
2002 = 1997 for Greece.

Belgium and Ireland 1994 data and Norway 1985 data.

Data source: EEA based on data from Eurostat data tables (Ref: www.eea.eu.int/coreset): renewable water resources (million m^3/year), LTAA and annual water abstraction by source and by sector (million m^3/year), total freshwater abstraction (surface and groundwater).

in people using less water. Industries connected to the public systems also reduced their industrial production and hence their water use. However the supply network in most of these countries is obsolete and losses in distribution systems require high abstraction volumes to maintain supply.

Water abstracted for cooling in energy production is considered a non-consumptive use and accounts for around 30 % of all water use in Europe. The western European countries and the central and northern countries of eastern Europe are the largest users of water for cooling; in particular more than half of water abstracted in Belgium, Germany and Estonia is used for this purpose.

Indicator definition

The water exploitation index (WEI) is the mean annual total abstraction of freshwater divided by the mean annual total renewable freshwater resource at the country level, expressed in percentage terms.

Indicator rationale

Monitoring the efficiency of water use by different economic sectors at the national, regional and local level is important for ensuring that rates of extraction are sustainable over the long term, an objective of the EU's sixth environment action programme (2001–2010).

Water abstraction as a percentage of the freshwater resource provides a good picture, at the national level, of the pressures on resources in a simple manner that is easy to understand, and shows trends over time. The indicator shows how total water abstraction puts pressure on water resources by identifying countries with high abstraction in relation to resources and therefore prone to water stress. Changes in the WEI help to analyse how changes in abstraction impact on freshwater resources by increasing pressure on them or making them more sustainable.

Policy context

Achieving the objective of the EU's sixth environment action programme (2001–2010), to ensure that rates of extraction from water resources are sustainable over the long term, requires monitoring of the efficiency of water use in different economic sectors at the national, regional and local level. The WEI is part of the set of water indicators of several international organisations such as UNEP, OECD, Eurostat and the Mediterranean Blue Plan. There is an international consensus on the use of this indicator.

There are no specific quantitative targets directly related to this indicator. However, the Water Framework Directive (2000/60/EC) requires countries to promote sustainable use based on long-term protection of available water resources and ensure a balance between abstraction and recharge of groundwater, with the aim of achieving good groundwater status by 2015.

Indicator uncertainty

Data at the national level cannot reflect water stress situations at the regional or local level. The indicator does not reflect the uneven spatial distribution of resources and may therefore mask regional or local risks of water stress.

Caution should be used when comparing countries, because of different definitions and procedures for estimating water use (for example, some include cooling water, others do not) and freshwater resources, in particular internal flows. Some sectoral abstractions, such as cooling water included in the industrial abstraction data, do not correspond to the specified uses.

Data need to be considered with reservation due to the lack of common European definitions and procedures for calculating water abstraction and freshwater resources. Current work is being carried out between Eurostat and EEA to standardise definitions and methodologies for data estimation.

Use of freshwater resources   PART B

Data are not available for all the countries considered, especially for 2000 and 2002, and the data series from 1990 are not complete. There are gaps in water use in some years and for some countries, particularly in the Nordic and the southern accession countries.

Accurate assessments that take climatic conditions into account would require the use of more disaggregated data at the spatial and geographical level.

Better indicators of the evolution of freshwater resources in each country are needed (for example by using information on trends in discharges at some representative gauging stations per country). If groundwater abstractions are considered separately from surface water abstractions, it would be necessary to have some indicators on the evolution of the groundwater resource (for example by using information on the head levels of selected piezometers per country). Better estimates of water abstraction could be developed by considering the uses involved in each economic sector.

The European environment | State and outlook 2005   327

19 Oxygen consuming substances in rivers

Key policy question

Is pollution of rivers by organic matter and ammonium decreasing?

Key message

Concentrations of organic matter and ammonium generally fell at 50 % of stations on European rivers during the 1990s, reflecting improvements in wastewater treatment. However, there were increasing trends at 10 % of the stations over the same period. Northern European rivers have the lowest concentrations of oxygen-consuming substances measured as biochemical oxygen demand (BOD) but concentrations are higher in rivers in some of the EU-10 Member States and accession countries where wastewater treatment is not so advanced. Ammonium concentrations in many rivers in EU Member States and accession countries are still far above background levels.

Indicator assessment

There has been a decrease in BOD and ammonium concentrations in the EU-15, reflecting implementation of the urban wastewater treatment directive and consequently an increase in the levels of treatment of wastewater. BOD and ammonium concentrations also declined in the EU-10 and accession countries, as a result partly of improved wastewater treatment but also of economic recession resulting in a decline in polluting manufacturing industries. However, levels of BOD and ammonium are higher in the EU-10 and accession countries in which wastewater treatment is still less advanced than in the EU-15. Ammonium concentrations in many rivers are considerably higher than the background concentrations of around 15 µg N/l.

The decline in the level of BOD is evident in nearly all countries for which data are available (Figure 2). The steepest declines are observed in the countries with the highest levels of BOD at the beginning of the 1990s (i.e.

the EU-10 and accession countries). However, some of these countries, such as Hungary, the Czech Republic and Bulgaria, although showing steep declines, still have the highest concentrations. There have also been dramatic decreases in the level of ammonium in some of the EU-10 and accession countries, such as Poland and Bulgaria (Figure 3). The EU-10 and accession countries have a wide range of median concentration values, with Poland and Bulgaria above 300 µg N/l, but Latvia and Estonia below 100 µg N/l. Levels are generally still highest in the eastern and lowest in the northern European countries.

Figure 1   **BOD and total ammonium concentrations in rivers between 1992 and 2002**

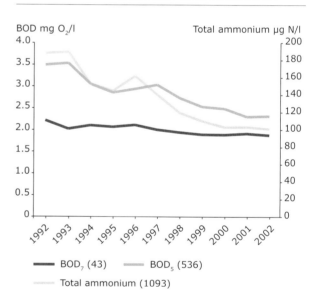

Note: BOD$_5$ data from Austria, Bulgaria, Czech Republic, Denmark, France, Hungary, Luxembourg, Slovak Republic and Slovenia; BOD$_7$ data from Estonia. Ammonium data from Austria, Bulgaria, Denmark, Estonia, Finland, France, Germany, Hungary, Latvia, Luxembourg, Poland, Slovak Republic, Slovenia, Sweden and the United Kingdom.

Number of river monitoring stations included in analysis noted in brackets.

Data source: EEA Data service (Ref: www.eea.eu.int/coreset).

Oxygen consuming substances in rivers

Figure 2 Trends in the concentration of BOD in rivers between 1992 and 2002 in different countries

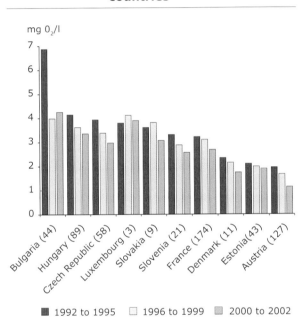

■ 1992 to 1995 □ 1996 to 1999 ■ 2000 to 2002

Note: BOD_5 data used for all countries except Estonia where BOD_7 data used.

Number of monitoring stations in brackets.

Data source: EEA Data service
(Ref: www.eea.eu.int/coreset).

Figure 3 Trends in the concentration of total ammonium in rivers between 1992 and 2002 in different countries

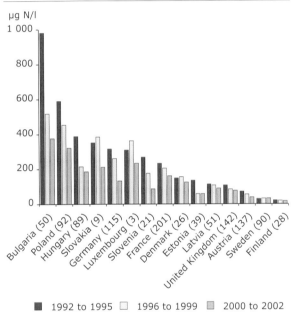

■ 1992 to 1995 □ 1996 to 1999 ■ 2000 to 2002

Note: Number of monitoring stations in brackets.

Data source: EEA Data service
(Ref: www.eea.eu.int/coreset).

In countries with a large proportion of its population connected to efficient sewage treatment plants, river concentrations of BOD and ammonia are low. Many of the EU-10 still have a lower proportion of their population connected to treatment plants (see indicator CSI 24), and when treatment is applied it is mainly primary or secondary. Concentrations in these countries are still high.

Indicator definition

The key indicator for the oxygenation status of water bodies is the biochemical oxygen demand (BOD) which is the demand for oxygen resulting from organisms in water that consume oxidisable organic matter. The indicator illustrates the current situation and trends regarding BOD and concentrations of ammonium (NH_4) in rivers. Annual average BOD after 5 or 7 days incubation (BOD_5/BOD_7) is expressed in mg O_2/l and annual average total ammonium concentrations in micrograms N/l. For all graphs, data are from representative river stations. Stations that have no designation of type are assumed to be representative and are included in the analysis. For Figures 1, 2 and 3, consistent time-series trends are calculated, using only stations that have recorded concentrations for each year included in the time-series; for Figures 2 and 3, consistent time-series are averaged for the three time periods 1992 to 1995, 1996 to 1999 and 2000 to 2002.

Core set of indicators | Water

Indicator rationale

Large quantities of organic matter (microbes and decaying organic waste) can result in reduced chemical and biological quality of river water, impaired biodiversity of aquatic communities, and microbiological contamination that can affect the quality of drinking and bathing water. Sources of organic matter are discharges from wastewater treatment plants, industrial effluents and agricultural run-off. Organic pollution leads to higher rates of metabolic processes that demand oxygen. This could result in the development of water zones without oxygen (anaerobic conditions). The transformation

Figure 4 Present concentration of BOD_5, BOD_7 (mg O_2/l) in rivers

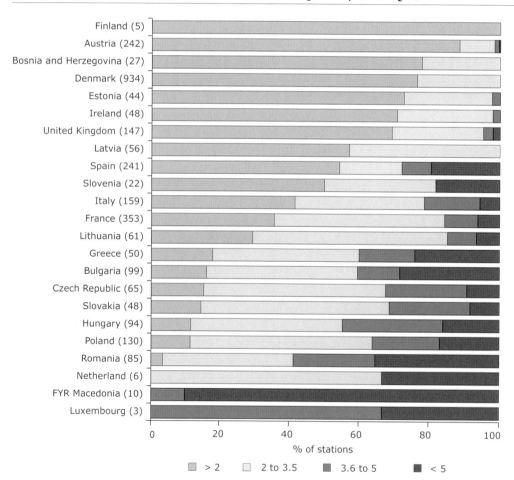

Note: BOD_5 data used for all countries except Estonia, Finland, Latvia and Lithuania where BOD_7 data used; The number of stations with annual means within each concentration band are calculated for the latest year for which data are available. The latest year is 2002 for all countries except the Netherlands (1998), Ireland (2000) and Romania (2001).

Number of river monitoring stations in brackets.

Data source: EEA Data service (Ref: www.eea.eu.int/coreset).

of nitrogen to reduced forms under anaerobic conditions in turn leads to increased concentrations of ammonium, which is toxic to aquatic life above certain concentrations, depending on water temperature, salinity and pH.

Policy context

The indicator is not related directly to a specific policy target but shows the efficiency of wastewater treatment (see CSI 24). The environmental quality of surface waters with respect to organic pollution and ammonium and the reduction of the loads and impacts of these pollutants are, however, objectives of several directives including: the Surface Water for Drinking Directive (75/440/EEC) which sets standards for BOD and ammonium content of drinking water, the Nitrates Directive (91/676/EEC) aimed at reducing nitrate and organic matter pollution from agricultural land, the Urban Waste Water Treatment Directive (91/271/EEC) aimed at reducing pollution from sewage treatment works and certain industries, the Integrated Pollution Prevention and Control Directive (96/61/EEC) aimed at controlling and preventing the pollution of water by industry, and the water framework directive which requires the achievement of good ecological status or good ecological potential of rivers across the EU by 2015.

Indicator uncertainty

The data sets for rivers include almost all countries in the EEA area, but the time coverage varies from country to country. The data set provides a general overview of concentration levels and trends of organic matter and ammonia in European rivers. Most countries measure organic matter as BOD over five days but a few countries measure BOD over seven days, which may introduce a small uncertainty in comparisons between countries.

20 Nutrients in freshwater

Key policy question

Are concentrations of nutrients in our freshwaters decreasing?

Key message

Concentrations of phosphorus in European inland surface waters generally decreased during the 1990s, reflecting the general improvement in wastewater treatment over this period. However, the decrease was not sufficient to halt eutrophication.

Nitrate concentrations in Europe's groundwaters have remained constant and are high in some regions, threatening drinking water abstractions. There was a small decrease in nitrate concentrations in some European rivers during the 1990s. The decrease was less than for phosphorus because of limited success with measures to reduce agricultural inputs of nitrate.

Indicator assessment

Concentrations of orthophosphate in European rivers have been decreasing steadily in general over the past 10 years. In the EU-15 this is because of the measures introduced by national and European legislation, in particular the urban waste water treatment directive which has increased levels of wastewater treatment with, in many cases, increased tertiary treatment that involves the removal of nutrients. There has also been an improvement in the level of wastewater treatment in the EU-10, though not to the same levels as in the EU-15. In addition, the transition recession in the economies of the EU-10 may have played a part in the decreasing phosphorus trends because of the closure of potentially-polluting industries and a decrease in agricultural production leading to less use of fertilisers. The economic recession in many of the EU-10 ended by the end of the 1990s. Since then many new industrial plants with better effluent treatment technologies have been opened. Fertiliser applications have also started to increase to some extent.

During the past few decades there has also been a gradual reduction in phosphorus concentrations in many European lakes. However the rate of decrease appears to have slowed or even stopped during the 1990s. As with rivers, discharges of urban wastewater have been a major source of pollution by phosphorus, but as purification has improved and many outlets have been diverted away from lakes, this source of pollution is gradually becoming less important. Agricultural sources of phosphorus, from animal manure and from diffuse pollution by erosion and leaching, are both important and need increased attention to achieve good status in lakes and rivers.

The improvements in some lakes have generally been relatively slow despite the pollution abatement measures taken. This is at least partly because of the slow recovery due to internal loading and because the ecosystems can be resistant to improvement and thereby remain in a bad state. Such problems may call for restoration measures, particularly in shallow lakes.

At the European level, there is some evidence of a small decrease in concentrations of nitrate in rivers. The decrease has been slower than for phosphorus because measures to reduce agricultural inputs of nitrate have not been implemented in a consistent way across EU countries and because of the probable time lags between reduction of agricultural nitrogen inputs and soil surpluses, and resulting reductions in surface and groundwater concentrations of nitrate. In terms of nitrate, 15 of the 25 countries with available information had a number of river stations where the drinking water directive guide concentration for nitrate of 25 mg NO_3/l was exceeded, and three of these countries had stations where the maximum allowable concentration of 50 mg NO_3/l was also exceeded. Countries with the highest agricultural land-use and highest population densities (such as Denmark, Germany, Hungary and the United Kingdom), generally had higher nitrate concentrations than those with the lowest (such as Estonia, Norway, Finland, and Sweden) reflecting the impact of emissions of nitrate from agriculture in the former and wastewater treatment works in the latter group of countries.

Nutrients in freshwater

PART B

Figure 1 Nitrate and phosphorus concentrations in European freshwater bodies

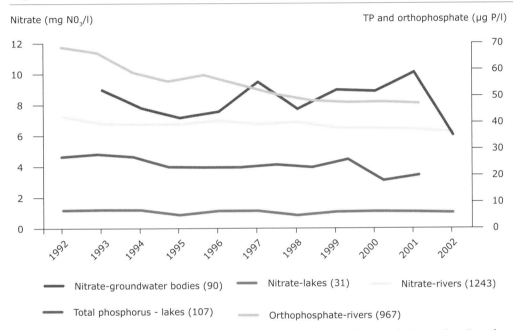

Note: Concentrations are expressed as annual median concentrations in groundwater, and median of annual average concentrations in rivers and lakes.

Numbers of groundwater bodies, lake and river monitoring stations in brackets.

Lakes: nitrate data from: Estonia, Finland, Germany, Hungary, Latvia and United Kingdom; total phosphorus data from Austria, Denmark, Estonia, Finland, Germany, Hungary, Ireland and Latvia.

Groundwater bodies: data from Austria, Belgium, Bulgaria, Denmark, Estonia, Finland, Germany, Lithuania, the Netherlands, Norway, Slovak Republic and Slovenia.

Rivers: data from Austria, Bulgaria, Denmark, Estonia, Finland, France, Germany, Hungary, Latvia, Lithuania, Poland, Slovenia, Sweden and the United Kingdom.

Data are from representative river and lake stations. Stations that have no designation of type are assumed to be representative and are included in the analysis.

Data source: EEA Data service (Ref: www.eea.eu.int/coreset).

Mean nitrate concentrations in groundwaters in Europe are above background levels (< 10 mg/l as NO_3) but do not exceed 50 mg/l as NO_3. At the European level, annual mean nitrate concentrations in groundwaters have remained relatively stable since the early 1990s but show different levels regionally. Due to a very low level of mean nitrate concentrations (< 2 mg/l as NO_3) in the Nordic countries, the European mean nitrate concentration shows an unbalanced view of nitrate distribution. The presentation above is therefore separated in the following sub-indicators into western, eastern and Nordic countries.

On average, groundwaters in western Europe have the highest nitrate concentration, due to the most intensive agricultural practices, twice as high as in eastern Europe, where agriculture is less intense. Groundwaters in Norway and Finland generally have low nitrate concentrations.

Agriculture is the largest contributor of nitrogen pollution to groundwater, and also to many surface water bodies, since nitrogen fertilisers and manure are used on arable crops to increase yields and productivity. In the EU, mineral fertilisers account for almost 50 % of nitrogen inputs into agricultural soils and manure for 40 % (other inputs are biological fixation and atmospheric deposition). Consumption of nitrogen fertiliser (mineral fertilisers and animal manure) increased until the late 1980s and then started to decline, but in recent years it has increased again in some EU countries. Consumption of nitrogen fertiliser per hectare of arable land is higher in the EU-15 than in the EU-10 and accession countries. Nitrogen from excess fertiliser percolates through the soil and is detectable as elevated nitrate levels under aerobic conditions and as elevated ammonium levels under anaerobic conditions. The rate of percolation is often slow and excess nitrogen levels may be the effects of pollution on the surface up to 40 years ago, depending on the hydrogeological conditions. There are also other sources of nitrate, including treated sewage effluents, which may also contribute to nitrate pollution in some rivers.

Indicator definition

Concentrations of orthophosphate and nitrate in rivers, total phosphorus and nitrate in lakes and nitrate in groundwater bodies. The indicator can be used to illustrate geographical variations in current nutrient concentrations and temporal trends.

The concentration of nitrate is expressed as mg nitrate (NO_3)/l, and orthophosphate and total phosphorus as µg P/l.

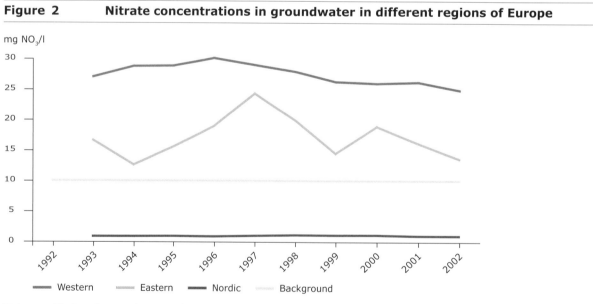

Figure 2 Nitrate concentrations in groundwater in different regions of Europe

Note: Western Europe: Austria, Belgium, Denmark, Germany, Netherlands; 27 GW-bodies.
Eastern Europe: Bulgaria, Estonia, Lithuania, Slovak Republic, Slovenia; 38 GW-bodies.
Nordic countries: Finland, Norway; 25 GW-bodies; Swedish data are not included due to a data gap.

The drinking water maximum admissible concentration (MAC) for nitrate of 50 mg NO_3/l is laid down in Council Directive 98/83/EC on the quality of water intended for human consumption.

Background concentrations of nitrate in groundwater (< 10 mg NO_3/l) are shown to aid the assessment of the significance of the nitrate concentrations (in association with the drinking water MAC).

Data source: EEA Data service (Ref: www.eea.eu.int/coreset).

Indicator rationale

Large inputs of nitrogen and phosphorus to water bodies from urban areas, industry and agricultural areas can lead to eutrophication. This causes ecological changes that can result in a loss of plant and animal species (reduction in ecological status) and have negative impacts on the use of water for human consumption and other purposes.

The environmental quality of surface waters with respect to eutrophication and nutrient concentrations is an objective of several directives: the water framework directive, the nitrate directive, the urban waste water treatment directive, the surface water directive and the freshwater fish directive. In future years, phosphorus concentrations in lakes will be highly relevant to work under the water framework directive.

Policy context

The indicator is not directly related to a specific policy target. The environmental quality of freshwaters with respect to eutrophication and nutrient concentrations is however an objective of several directives. These include: the Nitrates Directive (91/676/EEC) aimed at reducing nitrate pollution from agricultural land, the Urban Waste Water Treatment Directive (91/271/EEC) aimed at reducing pollution from sewage treatment works and certain industries, the Integrated Pollution Prevention and Control Directive (96/61/EEC) aimed at controlling and preventing pollution of water from industry, and the water framework directive which requires the achievement of good ecological status or good ecological potential of rivers across the EU by 2015. The water framework directive also requires the achievement of good groundwater status by 2015 and also the reversal of any significant and sustained upward trend in the concentration of any pollutant. In addition, the Drinking Water Directive (98/83/EC) sets the maximum allowable concentration for nitrate of 50 mg/l. It has been shown that drinking water in excess of the nitrate limit can result in adverse health effects, especially in infants less than two months of age. Groundwater is a very important source of drinking water in many countries and is often used untreated, particularly from private wells.

One key approach of the sixth environment action programme of the European Community 2001–2010 is to 'integrate environmental concerns into all relevant policy areas' which could result in a more intense consideration of applying agri-environmental measures to reduce nutrient pollution of the aquatic environment (e.g. in the common agricultural policy).

Indicator uncertainty

The data sets for groundwater and rivers include almost all EEA countries, but the time coverage varies from country to country. The coverage of lakes is less good. Countries are asked to provide data on rivers and lakes and on important groundwater bodies according to specified criteria. These rivers, lakes and groundwater bodies are expected to be able to provide a general overview, based on truly comparable data, of river, lake and groundwater quality at the European level.

Nitrate concentrations in groundwaters originate mainly from anthropogenic influence caused by agricultural land use. Concentrations in water are the effect of a multidimensional and time-related process which varies from groundwater body to groundwater body and is as yet less quantified. To evaluate the nitrate concentration in groundwater and its development, closely related parameters such as ammonium and dissolved oxygen have to be taken into account. However, there is a lack of data, especially for dissolved oxygen which provides information on the oxygen state of the water body (reducing or not).

21 Nutrients in transitional, coastal and marine waters

Key policy question

Are nutrient concentrations in our surface waters decreasing?

Key message

Phosphate concentrations in some coastal sea areas of the Baltic and North Seas have decreased over recent years, but they have remained stable in the Celtic Sea and increased in some Italian coastal areas. Nitrate concentrations have generally remained stable over recent years in the Baltic, North and Celtic Seas but have increased in some Italian coastal areas.

Indicator assessment

Nitrate

In the OSPAR (the North Sea, the English Channel and the Celtic Seas) and Helcom (the Baltic Sea bounded by the parallel of the Skaw in the Skagerrak at 57 °44.8'N) areas the available time-series show no clear trend in winter surface concentrations of nitrate. Both decreasing and increasing trends are observed at 3–4 % of the stations (Figure 1) which is certainly attributable to the temporal variability of nutrient loads resulting from varying run-offs.

In the Baltic Sea, winter surface nitrate concentrations are low, even in many coastal waters (the background concentration in the open Baltic Proper is around 65 µg/l). The higher concentrations observed in the Belt Sea and the Kattegat are due mainly to the mixing of Baltic waters with the more nutrient-rich North Sea and Skagerrak waters. The enhanced concentrations resulting from local *loading* are particularly noticeable in the coastal waters of Lithuania, the Gulf of Riga, the Gulf of Finland, the Gulf of Gdansk, the Pommeranian Bay and Swedish estuaries.

In the OSPAR area the nitrate concentrations are high (> 600 µg/l) due to land-based *loads* into the coastal waters of Belgium, the Netherlands, Germany, Denmark, and in a few UK and Irish estuaries.

Background concentrations in the open North Sea and Irish Sea are about 129 µg/l and 149 µg/l, respectively. In the Dutch coastal waters, an overall decrease of 10–20 % in winter nitrate concentrations has been observed. In the Mediterranean Sea, nitrate concentrations have increased at 24 %, and decreased at 5 % of the Italian coastal stations (Figure 1). The background concentration is low, i.e. 7 µg/l. Relatively low concentrations are observed in the Greek coastal waters, around Sardinia and the Calabrian Peninsula. Slightly higher concentrations are observed along the north-west and south-east Italian coasts. High concentrations are observed in most of the northern and western Adriatic Sea, as well as close to rivers and cities along the Italian west coast.

In the Black Sea, the background concentration of nitrate is very low, i.e. 1.4 µg/l. A slight decrease in nitrate concentration has been reported in the Romanian coastal waters, with a steady decline in the Turkish waters at the entrance to the Bosporus. An increased level of both nitrate and phosphate in Ukrainian waters during recent years is connected to high river run-offs.

Phosphate

In the Baltic and North Seas, phosphate concentrations have decreased at 25 % and 33 % of the coastal stations, respectively (Figure 1). In the Greater North Sea, the decline in phosphate concentrations is especially evident in the Dutch and Belgian coastal waters, which is probably due to reduced phosphate loads from the river Rhine. Decreases in phosphate concentrations have also been observed at some stations in the German, Norwegian and Swedish coastal waters, and in the open North Sea (more than 20 km from the coast). In the Baltic Sea area, decreases in phosphate concentrations were observed in the coastal waters of most countries, except Poland, as well as in the open waters.

In the Baltic Sea area, the winter surface phosphate concentration is very low in the Bothnian Bay compared with the background concentrations in the open Baltic Proper, and is potentially limiting primary production in the area. The concentration is slightly higher in the Gulf of Riga, the Gulf of Gdansk, in some Lithuanian, German and Danish coastal waters and in estuaries. Remedial measures have been taken in the catchment areas and a reduction in the use of fertilisers has occurred. However, recent research indicates that phosphate concentrations, for example in the open Baltic waters including the Kattegat, are strongly influenced by processes and transport within the water body due to variable oxygen regimes in the bottom water layer. The phosphate concentration is exceptionally high in the Gulf of Finland due to hypoxia and the up-welling of phosphate-rich bottom water in the late 1990s. In the North Sea, the English Channel and the Celtic Seas, phosphate concentrations in the coastal waters of Belgium, the Netherlands, Germany and Denmark are elevated compared to those of the open North Sea. The concentrations in the estuaries are generally high due to local loads.

In the Mediterranean Sea, phosphate concentrations have increased at 26 % and decreased at 8 % of the Italian coastal stations (Figure 1). Concentrations higher than the background value (i.e. about 1 µg/l) are observed in most coastal waters, and much higher concentrations are observed in hot spots along the east and west coasts of Italy.

In the open Black Sea, the background phosphate concentration is relatively high (about 9 µg/l) compared with the Mediterranean Sea and the background nitrogen value. This is probably due to the permanently anoxic conditions in the bottom waters of most of the

Figure 1   Summary of trends in winter nitrate and phosphate concentration, and N/P ratio in the coastal waters of the North Atlantic (mostly Celtic Seas), the Baltic Sea, the Mediterranean and the North Sea

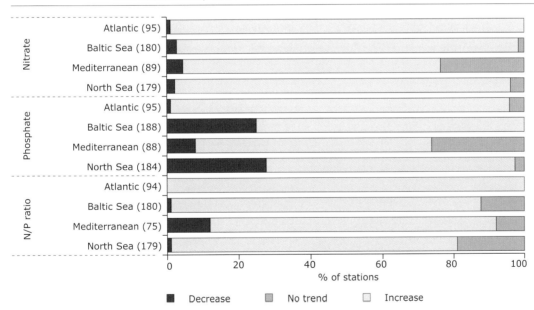

Note: Trend analyses are based on time-series 1985–2003 from each monitoring station having at least 3 years data in the period 1995–2003 and at least 5 years data in all. Number of stations in brackets.

Atlantic (incl. the Celtic Seas) data from: the United Kingdom, Ireland and ICES. Baltic Sea (incl. the Belt Sea and the Kattegat) data from: Denmark, Finland, Germany, Lithuania, Poland, Sweden and ICES. Mediterranean data from: Italy. North Sea (incl. the Channel and the Skagerrak) data from: Belgium, Denmark, Germany, the Netherlands, Norway, Sweden, the United Kingdom and ICES.

Data source: EEA Data service, data from OSPAR, Helcom, ICES and EEA member countries (www.eea.eu.int).

Black Sea, which prevent the phosphate from being bound in the sediments. The phosphate concentration along the Turkish coast is lower than in the open sea, while it is higher in the Romanian coastal waters influenced by the Danube River. In the Black Sea, a slow decline in the concentrations of phosphate has been reported in the Turkish waters at the entrance to the Bosporus.

N/P Ratio
In the Baltic Sea, the N/P ratio, based on winter surface nitrate and phosphate concentrations, is increasing in all areas (Figure 1) except the Polish coastal waters. The N/P ratio is high (> 32) in the Bothnian Bay, where it is likely that phosphorus limits the primary production of phytoplankton. However, the N/P ratio is low (< 8) to relatively low (< 16) in most of the open and coastal Baltic Sea area, indicating that nitrogen can be a potential growth-limiting factor.

In the Greater North Sea and Celtic Seas, high N/P ratios (> 16) are observed in the Belgian, Dutch, German and Danish coastal waters and estuaries, indicating potential phosphorus limitation, at least early in the growing season. In more open waters, the N/P ratio is generally below 16, indicating potential nitrogen limitation.

In the Mediterranean Sea, high N/P ratios (> 32) are found along the northern Adriatic coast and at hot spots along the Italian coasts and the north coast of Sardinia, indicating potential phosphorus limitation, at least during some periods of the growing season.

In the Black Sea, the N/P ratio is generally low, especially in the open sea and along the Turkish coast, indicating potential nitrogen limitation. High N/P ratios (> 32) are found only at a few Romanian coastal stations, indicating potential phosphorus limitation.

Indicator definition

The indicator illustrates overall trends in winter nitrate and phosphate concentration (microgram/l), and N/P ratio in the regional seas of Europe. The N/P ratio is based on molar concentrations. The winter period is January, February and March for stations east of longitude 15 degrees (Bornholm) in the Baltic Sea, and January and February for all other stations. The following sea areas are covered: the Baltic including the Belt Sea and the Kattegat; the North Sea — the OSPAR Greater North Sea including the Skagerrak and the Channel, but not the Kattegat; the Atlantic — the north-east Atlantic including the Celtic Seas, the Bay of Biscay and the Iberian coast; and the whole Mediterranean Sea.

Indicator rationale

Nitrogen and phosphorus enrichment can result in a chain of undesirable effects, starting from excessive growth of plankton algae that increases the amount of organic matter settling on the bottom. This may be enhanced by changes in the species composition and functioning of the pelagic food web (e.g. growth of small flagellates rather than larger diatoms), which leads to lower grazing by copepods and increased sedimentation. The consequent increase in oxygen consumption can, in areas with stratified water masses, lead to oxygen depletion, changes in community structure and death of the benthic fauna. Eutrophication can also increase the risk of algal blooms, some of them consisting of harmful species that cause the death of benthic fauna, wild and caged fish, and shellfish poisoning of humans. Increased growth and dominance of fast-growing filamentous macroalgae in shallow sheltered areas is another effect of nutrient overload which can change the coastal ecosystem, increase the risk of local oxygen depletion and reduce biodiversity and nurseries for fish.

The N/P ratio provides information on the potential nitrogen or phosphorus limitation of the primary phytoplankton production.

Policy context

Measures to reduce the adverse effects of excess anthropogenic inputs of nutrients and protect the marine environment are being taken as a result of various initiatives at all levels — global, European, national and regional conventions and Ministerial Conferences. There are a number of EU directives aimed at reducing the loads and impacts of nutrients, including the Nitrates Directive (91/676/EEC) aimed at reducing nitrate pollution from agricultural land; the Urban Waste Water Treatment Directive (91/271/EEC) aimed at reducing pollution from sewage treatment works and from certain industries; the Integrated Pollution Prevention and Control Directive (96/61/EEC) aimed at controlling and preventing pollution of water from industry; and the Water Framework Directive (2000/60/EC) which requires the achievement of good ecological status or good ecological potential of transitional and coastal waters across the EU by 2015. The European Commission is also developing a Thematic Strategy on the Protection and Conservation of the Marine Environment. Additional measures arise from international initiatives and policies including: the UN Global Programme of Action for the Protection of the Marine Environment from Land-based Activities; the Mediterranean Action Plan (MAP) 1975; the Helsinki Convention 1992 (Helcom); the OSPAR Convention 1998; and the Black Sea Environmental Programme (BSEP).

Targets

The most pertinent target with regard to concentrations of nutrients in water arises from the Water Framework Directive where one of the environmental objectives is to achieve good ecological status. This equates to water body type-specific nutrient concentrations/ranges that support the biological quality elements in a good state. As natural and background concentrations of nutrients vary between and within the regional seas, and between types of coastal water bodies, nutrient targets or thresholds for achieving good ecological status have to be determined locally.

Indicator uncertainty

The Mann-Kendall test for the detection of trends is a robust and accepted approach. Due to the multiple trend analyses, approximately 5 % of the tests conducted will turn out significant if in fact there is no trend. Data for this assessment are still scarce considering the large spatial and temporal variations inherent to the European transitional, coastal and marine waters. Long stretches of European coastal waters are not covered in the analysis due to lack of data. Trend analyses are consistent only for the North Sea and the Baltic Sea (data updated yearly within the OSPAR and Helcom conventions) and Italian coastal waters. Due to variations in freshwater discharge and the hydro-geographic variability of the coastal zone and internal cycling processes, trends in nutrient concentrations as such cannot be directly related to measures taken. For the same reasons the N/P ratio based on winter surface nutrient concentrations cannot be used directly to determine the degree of nutrient limitation of the primary phytoplankton production. Assessments based on N/P ratios can be regarded as describing only a potential nitrogen or phosphorus limitation to the marine plants.

Core set of indicators | Water

22 Bathing water quality

Key policy question

Is bathing water quality improving?

Key message

The quality of water at designated bathing beaches in Europe (coastal and inland) has improved throughout the 1990s and early 2000s. In 2003, 97 % of coastal bathing waters and 92 % of inland bathing waters complied with the mandatory standards.

Indicator assessment

The quality of EU bathing waters in terms of compliance with the mandatory standards laid down in the bathing waters directive has improved, but at a slower rate than initially envisaged. The original target of the 1975 directive was for Member States to comply with the mandatory standards by the end of 1985. In 2003, 97 % of coastal bathing waters and 92 % of inland bathing waters complied with these standards. Despite the significant improvement in bathing water quality since the adoption of the bathing water directive 25 years ago, 11 % of Europe's coastal bathing waters and 32 % of Europe's inland bathing beaches still did not meet (non-mandatory) guide values in 2003. The level of achievement of (non-mandatory) guide levels has been much lower than that for the mandatory standards. This is probably because the achievement of the guide levels would entail considerably more expenditure by Member States for sewage treatment works and the control of diffuse pollution sources.

Two countries (the Netherlands and Belgium) achieved 100 % compliance with mandatory standards in their coastal bathing waters in 2003 (Figure 2). The worst performance in terms of coastal waters and mandatory standards was found in Finland with 6.8 % non-compliant bathing waters in 2003. In contrast to its 100 % compliance with mandatory standards, only 15.4 % of Belgium's coastal bathing water met the guide levels, the lowest for the EU countries.

Three countries, Ireland, Greece and the United Kingdom, achieved 100 % compliance with mandatory standards in their inland bathing waters in 2003 (Figure 3). It should, however, be noted that these countries have designated the least number of inland bathing waters in the EU (9, 4 and 11, respectively) compared with Germany (1 572) and France (1 405) which have designated the highest number. Italy had the lowest compliance rate with mandatory standards (70.6 %) for its inland bathing waters in 2003.

Figure 1 Percentage compliance of EU coastal and inland bathing waters with mandatory standards of the bathing water directive, 1992 to 2003 for EU-15

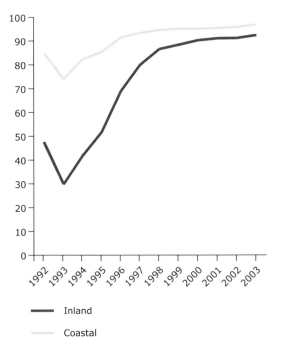

Note: 1992–1994, 12 EU Member States; 1995–1996, 14 EU Member States; 1997–2003, 15 EU Member States.

Data source: DG Environment from annual Member States' reports (Ref: www.eea.eu.int/coreset).

Bathing water quality

In 2003, the European Commission conducted infringement proceedings against nine of the EU-15 Member States (Belgium, Denmark, Germany, Spain, France, Ireland, the Netherlands, Portugal, and Sweden) for non-compliance with aspects of the bathing waters directive. Common reasons were non-compliance with standards and insufficient sampling. The Commission also noted that the number of inland UK bathing waters is low by comparison with most other Member States.

Indicator definition

The indicator describes the changes over time in the quality of designated bathing waters (inland and marine) in EU Member States in terms of compliance with standards for microbiological parameters (total coliforms and faecal coliforms) and physico-chemical parameters (mineral oils, surface-active substances and phenols) introduced by the EU Bathing Water Directive

Figure 2 Percentage of EU coastal bathing waters complying with mandatory standards and meeting guide levels of the bathing waters directive for the year 2003 by country

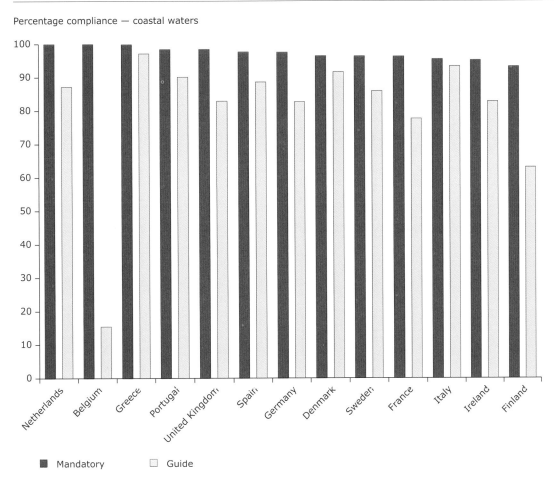

Note: Data source: DG Environment from annual Member States' reports (Ref: www.eea.eu.int/coreset).

Core set of indicators | Water

(76/160/EEC). The compliance status of individual Member States is presented for the last reported year. The indicator, based on the annual reports made by Member States to the European Commission, is expressed in terms of percentage of inland and marine bathing waters complying with the mandatory standards and guide levels for microbiological and physico-chemical parameters.

Indicator rationale

The Bathing Water Directive (76/160/EEC) was designed to protect the public from accidental and chronic pollution incidents which could cause illness from recreational water use. Examining compliance with the directive therefore indicates the status of bathing water quality in terms of public health and also

Figure 3 Percentage of EU inland bathing waters complying with mandatory standards and meeting guide levels of the bathing waters directive for the year 2003 by country

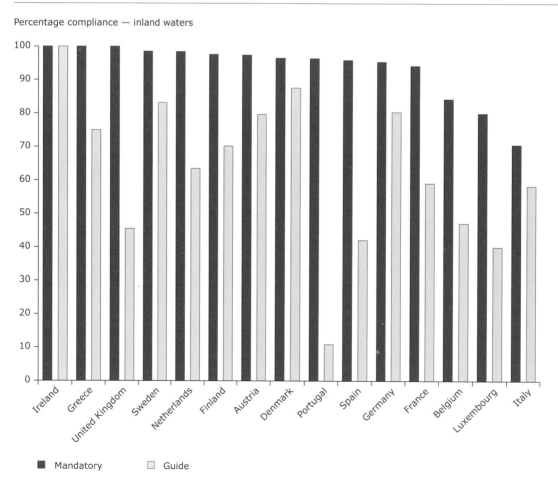

Note: Data source: DG Environment from annual Member States' reports (Ref: www.eea.eu.int/coreset).

the effectiveness of the directive. The bathing water directive is one of the oldest pieces of environmental legislation in Europe and data on compliance goes back to the 1970s. Under the directive, Member States are required to designate coastal and inland bathing waters and monitor the quality of the water throughout the bathing season.

Policy context and targets

Under the Bathing Water Directive (76/160/EEC) Member States are required to designate coastal and inland bathing waters and monitor the quality of the water throughout the bathing season. Bathing waters are designated where bathing is authorised by the competent authority and also where bathing is traditionally practised by a large number of bathers. The bathing season is then determined according to the period when there are the largest number of bathers (May to September in most European countries). The quality of the water has to be monitored fortnightly during the bathing season and also two weeks before. The sampling frequency may be reduced by a factor of two when samples taken in previous years show results better than the guide values and when no new factor likely to lower the quality of the water has appeared. Annex 1 of the directive lists a number of parameters to be monitored but the focus has been on bacteriological quality. The directive sets both minimum standards (mandatory) and optimum standards (guide). For compliance with the directive, 95 % of the samples must comply with the mandatory standards. To be classified as achieving guide values, 80 % of the samples must comply with the total and faecal coliform standards and 90 % with the standards for the other parameters. On 24 October 2002, the Commission adopted the proposal for a revised Directive of the European Parliament and of the Council concerning the Quality of Bathing Water (COM(2002)581). The draft directive proposes the use of only two bacteriological indicator parameters, but sets a higher health standard than the 1976/160 Directive. Based on international epidemiological research and the experience with implementing the current bathing water and water framework directives, the revised directive provides long-term quality assessment and management methods in order to reduce both monitoring frequency and monitoring costs.

Indicator uncertainty

There are differences in how countries have interpreted and implemented the directive, leading to differences in the representativeness of bathing waters included in terms of recreational water use.

During the life of the directive, the EU expanded from 12 countries in 1992 to 15 in 2003. The time-series is thus not consistent in terms of geographic coverage. The EU-10 Member States are expected to report on the quality of their bathing waters in 2005.

Human enteric viruses are the most likely pathogens responsible for waterborne diseases from recreational water use but detection methods are complex and costly for routine monitoring, and so the main parameters analysed for compliance with the directive are indicator organisms: total and faecal coliforms. Compliance with the mandatory standards and guide levels for these indicator organisms does not therefore guarantee that there is no risk to human health.

Core set of indicators | Water

23 Chlorophyll in transitional, coastal and marine waters

Key policy question

Is eutrophication in European surface waters decreasing?

Key message

There has been no general reduction in eutrophication (as measured by chlorophyll-a concentrations) in the Baltic Sea, the Greater North Sea or the coastal waters of Italy and Greece. Chlorophyll-a concentrations have increased in a few coastal areas and decreased in others.

Indicator assessment

No overall trend has been observed in summer surface chlorophyll-a concentrations, either in the open-sea areas of the Baltic Sea and the Greater North Sea, or the coastal waters of Italy and Greece in the Mediterranean Sea (Figure 1). The majority of the coastal stations in the three seas show no trend, however some stations show increasing or decreasing trends. For example, in the Baltic Sea, 11 % of the coastal stations show an increase in chlorophyll-a concentrations and 3 % a decrease. This lack of a clear general trend indicates that measures to reduce loads of nutrients have not yet succeeded in significantly reducing eutrophication.

In the Baltic Proper and the Gulf of Finland, high mean summer surface chlorophyll-a concentrations (> 2.8 µg/l) are found in open waters, probably due to summer blooms of cyano-bacteria, specific to the Baltic Sea. Concentrations > 4 µg/l are observed in estuaries and coastal waters influenced by rivers or cities in some Swedish, Estonian, Lithuanian, Polish and German coastal waters.

In the North Sea, high chlorophyll-a concentrations (> 5.8 µg/l) are observed in the Elbe estuary and Belgian, Dutch and Danish coastal waters influenced by river discharges. High concentrations are also observed in Liverpool Bay in the Irish Sea. In the open North Sea and Skagerrak, chlorophyll-a concentrations are generally low (< 1.4 µg/l).

In the Mediterranean Sea, 12 % of the stations in Italian coastal waters show a decrease in concentrations of chlorophyll-a, while 8 % show an increase (Figure 1). The lowest concentrations (< 0.35 µg/l) are observed around Sardinia and in southern Italian and Greek coastal waters. Higher concentrations (> 0.6 µg/l) are observed along the Italian east and west coasts and in the Greek Saronikos Bay. High concentrations (> 1.95 µg/l) are found in the northern Adriatic and along the Italian west coast from Naples to the north of Rome.

Very few chlorophyll-a data are available for the Black Sea. The available data show the highest level (> 1.7 µg/l) in the Ukrainian waters of the north-western Black Sea.

Indicator definition

The indicator illustrates trends in mean summer surface concentrations of chlorophyll-a in the regional seas of Europe. The concentration of chlorophyll-a is expressed as microgram/l in the uppermost 10 m of the water column during summer.

The summer period is:

- June to September for stations north of latitude 59 degrees in the Baltic Sea (Gulf of Bothnia and Gulf of Finland);

- May to September for all other stations.

The following sea areas are covered:

- Baltic: the Helcom area including the Belt Sea and the Kattegat;

- North Sea: the OSPAR Greater North Sea including the Skagerrak and the Channel, but not the Kattegat;

- Atlantic: the north-east Atlantic including the Celtic Seas, the Bay of Biscay and the Iberian coast;

- Mediterranean: the whole Mediterranean Sea.

Indicator rationale

The objective of the indicator is to demonstrate the effects of measures taken to reduce discharges of nitrogen and phosphate on coastal concentrations of phytoplankton expressed as chlorophyll-a. This is an indicator of eutrophication (See also CSI 21 Nutrients in transitional, coastal and marine waters).

The primary effect of eutrophication is excessive growth of plankton algae increasing the concentration of chlorophyll-a and the amount of organic matter settling to the bottom. The biomass of phytoplankton is most frequently measured as the concentration of chlorophyll-a in the euphotic part of the water column. Measurements of chlorophyll-a are included in most eutrophication monitoring programmes, and chlorophyll-a represents the biological eutrophication indicator with best geographical coverage at the European level.

The negative effects of excessive phytoplankton growth are 1) changes in species composition and functioning of the pelagic food web, 2) increased sedimentation, and 3) increase in oxygen consumption that may lead to oxygen depletion and the consequent changes in community structure or death of the benthic fauna.

Eutrophication can also promote harmful algal blooms that may cause discoloration of the water, foam formation, death of benthic fauna, wild and caged fish, or shellfish poisoning of humans. The shadowing effect of increased phytoplankton biomass will reduce the depth distribution of sea grasses and macroalgae. Secondary production of benthic fauna is most often food-limited and related to the input of phytoplankton settling at the bottom, which in turn is also related to the chlorophyll-a concentration.

Figure 1 **Trends in mean summer chlorophyll-a concentrations in coastal waters of the Baltic Sea, the Mediterranean (mainly Italian waters) and the Greater North Sea (mainly the eastern North Sea and the Skagerrak)**

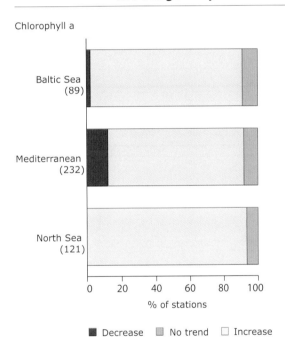

Note: Trend analyses are based on time series 1985–2003 from each monitoring station having at least three years of data in the period 1995–2003 and at least 5 years of data in all. Number of stations in brackets.

Baltic Sea (incl. the Belt Sea and the Kattegat) data from: Denmark, Finland, Lithuania, Sweden and the International Council for the Exploration of the Seas (ICES).

Mediterranean data from: Greece and Italy.

North Sea (incl. the Skagerrak) data from: Belgium, Denmark, Norway, Sweden, the United Kingdom and ICES.

Data source: EEA Data service, data from OSPAR, Helcom, ICES and EEA member countries (www.eea.eu.int).

Table 1   Number of coastal stations per country showing no trend, decreasing or increasing trend in summer surface concentrations of chlorophyll-a

Country	Chlorophyll			Number of stations
	Decrease	No trend	Increase	Total
Baltic Sea area				
Denmark	1	31	1	33
Finland	0	2	1	3
Lithuania	0	3	3	6
Open waters	0	23	1	24
Sweden	1	20	2	23
Mediterranean				
Greece	0	6	0	6
Italy	28	178	19	225
Open waters	0	1	0	1
North Sea area				
Belgium	0	12	3	15
Denmark	0	9	0	9
United Kingdom	0	3	0	3
Norway	0	20	0	20
Open waters	0	64	2	66
Sweden	0	5	3	8

Note: Trend analyses are based on time series 1985–2003 from each monitoring station having at least 3 years of data in the period 1995–2003 and at least 5 years of data in all (Ref: www.eea.eu.int/coreset).

Policy context

There are a number of EU directives aimed at reducing the loads and impacts of nutrients. These include: the Nitrates Directive (91/676/EEC) aimed at reducing nitrate pollution from agricultural land; the Urban Waste Water Treatment Directive (91/271/EEC) aimed at reducing pollution from sewage treatment works and certain industries; the Integrated Pollution Prevention and Control Directive (96/61/EEC) aimed at controlling and preventing pollution of water from industry; and the Water Framework Directive (2000/60/EC) which requires the achievement of good ecological status or good ecological potential of transitional and coastal waters across the EU by 2015. The European Commission is also developing the thematic strategy on the protection and conservation of the marine environment which will incorporate open marine waters and principal environmental threats, such as impact of eutrophication.

Measures also arise from a number of other international initiatives and policies including: the UN global programme of action for the protection of the marine environment against land-based activities; the Mediterranean action plan (MAP) 1975; the Helsinki Convention 1992 (Helcom) on the protection of the marine environment of the Baltic Sea area; OSPAR Convention 1998 for the protection of the marine environment of the North East Atlantic; and the Black Sea environmental programme (BSEP).

Targets

The most pertinent target with regard to concentrations of chlorophyll in water arises from the water framework directive where one of the environmental objectives is to achieve good ecological status. Good ecological status equates to water body type-specific chlorophyll concentrations/ranges that support the biological quality elements at a good status.

Type-specific chlorophyll concentrations/ranges do not necessarily relate to natural or background concentrations. Natural and background concentrations of chlorophyll vary between regional seas, from sub-area to sub-area within regional seas, and between types of coastal water bodies within a sub-area, depending on factors such as natural nutrient loads, water residence time and annual biological cycling. Chlorophyll targets or thresholds for achieving good ecological status therefore have to be determined locally.

Indicator uncertainty

Because of confounding factors such as variations in freshwater discharge, hydro-geographic variability of the coastal zone and internal nutrient cycling in water, biota and sediments, trends in chlorophyll-a concentrations are sometimes difficult to relate directly to, or demonstrate, nutrient reduction measures.

The Mann-Kendall test for the detection of trends used for statistical analysis of the data is a robust and accepted approach. Due to the multiple trend analyses, approximately 5 % of the conducted tests will turn out significant if in fact there is no trend.

Data for this assessment are still scarce considering the large spatial and temporal variations inherent in European transitional, coastal and marine waters. Long stretches of European coastal waters are not covered by the analysis due to lack of data. Trend analyses are only consistent for the eastern North Sea, the Baltic Sea area and Italian coastal waters.

24 Urban wastewater treatment

Key policy question

How effective are existing policies in reducing loading discharges of nutrients and organic matter?

Key message

Wastewater treatment in all parts of Europe has improved significantly since the 1980s, however the percentage of the population connected to wastewater treatment in southern and eastern Europe and the accession countries is relatively low.

Indicator assessment

Over the past twenty years, marked changes have occurred in the proportion of the population connected to wastewater treatment and in the technology involved. Implementation of the urban waste water treatment (UWWT) directive has largely accelerated this trend. Decreases in discharges in eastern Europe (EU-10) and the accession countries are due to economic recession resulting in a decline in polluting manufacturing industries.

Most of the population in the Nordic countries is connected to wastewater treatment plants with the highest levels of tertiary treatment, which efficiently removes nutrients (phosphorus or nitrogen or both) and organic matter. More than half of the wastewater in central European countries receives tertiary treatment. Only around half of the population in southern and eastern countries and the accession countries is currently connected to any wastewater treatment plants and 30 to 40 % to secondary or tertiary treatment. This is because policies to reduce eutrophication and improve bathing water quality were implemented earlier in the northern and central than in the southern and eastern countries and in the accession countries.

A comparison with indicators CSI 19 and CSI 20 shows that these changes in treatment have improved surface water quality, including bathing water quality, with a decrease in the concentrations of orthophosphates, total ammonium and organic matter over the past ten years. Member States have made considerable investments to achieve these improvements but most of them are however late in implementing the UWWT directive or have interpreted it differently and in ways that differ from the Commission's view.

The UWWT directive requires Member States to identify water bodies as sensitive areas, for example according to the risk of eutrophication. Wastewater treatment facilities with tertiary treatment had to be available in all agglomerations with a population equivalent greater than 10 000 discharging into a sensitive area by 31 December 1998. As shown in Figure 2, only two EU Member States, Denmark and Austria, were close to conforming to the directive's requirements in these terms. Germany and the Netherlands have designated their whole territory as a sensitive area, but are not in conformity with the goal of 75 % reduction of nitrogen.

For large cities with population equivalents greater than 150 000, Member States were required to provide more advanced (than secondary) treatment by 31 December 1998 when discharging into sensitive areas, and at least secondary treatment by 31 December 2000 for those discharging into 'normal' waters. However, on 1 January 2002, 158 of the 526 cities with population equivalents greater than 150 000 did not have a sufficient standard of treatment, and 25 agglomerations had no treatment at all, including Milan, Cork, Barcelona and Brighton. The situation has since improved, partly due to more comprehensive reporting to the Commission, partly to real improvements in treatment. Some of the cities made the necessary investment during 1999–2002, others plan to complete work soon.

An additional threat to the environment comes from the disposal of the sewage sludge produced in the treatment plants. The increase in the proportion of the population connected to wastewater treatment, as well as in the level of treatment, leads to an increase in the quantities of sewage sludge. This has to be disposed of, mainly by spreading on soils, to landfills or by incineration. These disposal routes can transfer

Figure 1 Changes in wastewater treatment in regions of Europe between the 1980s and the late 1990s

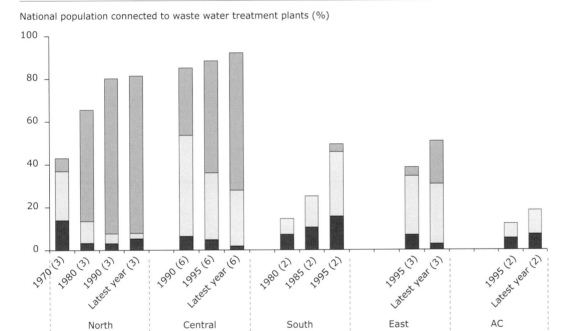

Note: Only countries with data from all periods included, the number of countries in parentheses.
Nordic: Norway, Sweden, Finland.
Central: Austria, Denmark, England and Wales, the Netherlands, Germany, Switzerland.
Southern: Greece, Spain.
East: Estonia, Hungary and Poland.
AC: Bulgaria and Turkey.

Data source: EEA Data service, based on Member States data reported to OECD/Eurostat, Joint questionnaire, 2002 (Ref: www.eea.eu.int/coreset).

pollution from water to soil or air and have to be taken into account in the respective policy implementation processes.

Indicator definition

The indicator tracks the success of policies to reduce pollution from wastewater by tracking the trends in the percentage of population connected to primary, secondary and tertiary wastewater treatment plants since the 1980s.

The level of conformity with the UWWTD is illustrated in terms of the percentage of the total load to sensitive area from large agglomerations and in terms of the levels of urban wastewater treatment in large cities in the EU (agglomerations > 150 000 p.e.).

Indicator rationale

Wastewater from households and industry represents a significant pressure on the water environment because of the loads of organic matter and nutrients as well as hazardous substances. With high levels of the population in EEA member countries living in urban

Core set of indicators | Water

Figure 2 Percentage of total load in sensitive area, and percentage of load in sensitive area by country, not conforming to the requirements of the urban waste water treatment directive, 2001

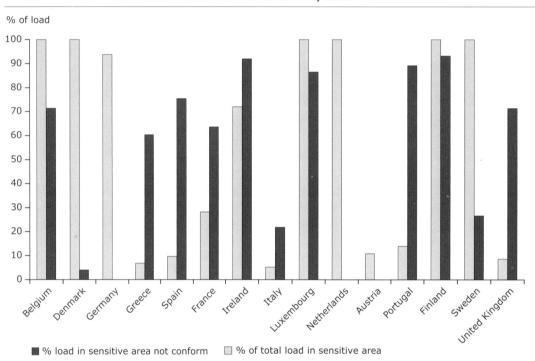

Note: For Sweden change in methodology between 1995 and 2000.

Data source: DG Environment, 2004 (Ref: www.eea.eu.int/coreset).

agglomerations, a significant fraction of wastewater is collected by sewers connected to public wastewater treatment plants. The level of treatment before discharge and the sensitivity of the receiving waters determine the scale of impacts on aquatic ecosystems. The types of treatments and conformity with the directive are seen as proxy indicators for the level of purification and the potential improvement of the water environment.

Primary (mechanical) treatment removes part of the suspended solids, while secondary (biological) treatment uses aerobic or anaerobic micro-organisms to decompose most of the organic matter and retain some of the nutrients (around 20–30 %). Tertiary (advanced) treatment removes the organic matter even more efficiently. It generally includes phosphorus retention and in some cases nitrogen removal. Primary treatment alone removes no ammonium whereas secondary (biological) treatment removes around 75 %.

Policy context and targets

The Urban Waste Water Treatment Directive (UWWTD; 91/271/EEC) aims to protect the environment from the adverse effects of urban wastewater discharges. It prescribes the level of treatment required before discharge and has to be fully implemented in the EU-15 by 2005 and in the EU-10 by 2008–2015. The directive requires Member States to provide all agglomerations of more than 2 000 population equivalent (p.e.) with collecting systems and all wastewaters collected to be provided with appropriate treatment by 2005.

Figure 3 Number of EU-15 agglomerations of more than 150 000 p.e. by treatment level, situation on 1 January 2002

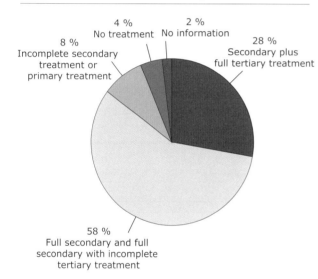

Note: Data source: DG Environment, 2004 (Ref: www.eea.eu.int/coreset).

Secondary treatment (i.e. biological treatment) must be provided for all agglomerations of more than 2 000 p.e. that discharge into fresh waters, while more advanced treatment (tertiary treatment) is required for discharges into sensitive areas. To help minimise pollution from various point sources, the integrated pollution prevention and control directive (IPPC), which came into force 1996, has a set of common rules on permission for industrial installations.

The achievements through the UWWTD and the IPPC directive have to be seen as an integrated part of objectives under the water framework directive (WFD) which aim at a good chemical and ecological status for all waters by 2015.

The European Commission reported on Member States implementation of the urban waste water treatment directive in 2002 and 2004 (http://europa.eu.int/comm/environment/water/water-urbanwaste/report/report.html and http://europa.eu.int/comm/environment/water/water-urbanwaste/report2/report.html).

Indicator uncertainty

For the assessment shown in Figure 1, countries have been grouped to show the relative contribution on a larger statistical basis and to overcome the incomplete nature of the data. The data and time trends are most complete for central Europe and the Nordic countries and least complete for the southern European and accession countries, with the exception of Estonia and Hungary.

Data gained from the UWWTD focuses on the performance of the treatment plant alone. But wastewater treatment systems could also include sewer networks with storm water overflows and storages which are complex and whose overall performance is difficult to assess. In addition to the treatments covered by the UWWTD there are other possible treatments, mostly industrial, but also independent treatments of smaller settlements outside urban agglomerations not included in UWWTD reporting. Compliance with the levels defined in the directive therefore does not guarantee that there is no pollution due to urban wastewater. To deal with the independent treatments, different methodologies for calculating connectivity have been applied, for example Sweden uses persons connected instead of person-equivalents ([1]).

([1]) For 1985 and 1995 loadings per person equivalents, for 2000 and 2002 loadings per person connected were used instead; Based on register studies on wastewater conditions in rural areas, the following assumption have been made (year 2000): everybody living in urban areas is connected to a treatment plant (MWWTP) Among people not living in urban areas, 192 000 persons are connected to MWWTP, 70 000 have no treatment at all and the remaining 1 163 000 have septic tanks. 60 % of the septic tanks have at least secondary treatment.

Core set of indicators | Agriculture

25 Gross nutrient balance

Key policy question

Is the environmental impact of agriculture improving?

Key message

The agricultural gross nutrient balance shows whether nutrient inputs and outputs per hectare farmland are in balance or not. A large positive nutrient balance (i.e. inputs are larger than the outputs) indicates a high risk of nutrient leaching and subsequent water pollution.

The gross nitrogen balance at the EU-15 level in 2000 was calculated to be 55 kg/ha, which is 16 % lower than the estimate for 1990, 66 kg/ha. It ranged from 37 kg/ha (Italy) to 226 kg/ha (the Netherlands). All national gross nitrogen balances showed a decline between 1990 and 2000, apart from Ireland (22 % increase) and Spain (47 % increase). The general decline in nitrogen balance surpluses is due to a small decrease in nitrogen input rates (by 1 %) and a significant increase in nitrogen output rates (by 10 %).

Indicator assessment

- The gross nutrient balance for nitrogen provides an indication of the risk of nutrient leaching by identifying agricultural areas that have very high nitrogen loadings. As the indicator integrates the most important agricultural parameters with regard to potential nitrogen surplus, it is currently the best-available approximation of agricultural pressures on water quality. High nutrient balances exert pressures on the environment in terms of an increased risk of leaching of nitrates to groundwater. The application of mineral and organic fertilisers can also lead to emissions to the atmosphere in the form of nitrous dioxide and ammonia, respectively.

- Gross nitrogen balances are particularly high (i.e. above 100 kg N per ha and year) in the Netherlands, Belgium, Luxembourg and Germany. They are particularly low in most Mediterranean countries which is linked to the overall lower livestock production in this part of Europe. It is currently not possible to provide gross nitrogen balance estimates for the EU-10 or the accession countries, since the relevant statistical data are under elaboration.

- National balances, however, can mask important regional differences in the gross nutrient balance that determine the actual risk of nitrogen leaching at the regional or local level. Individual Member States can thus have overall acceptable gross nitrogen balances at the national level but still experience significant nitrogen leaching in certain regions, for example in areas with high concentrations of livestock. There are a number of regions with particularly high livestock densities in the EU-15 (for example northern Italy, western France, north-eastern Spain and parts of the Benelux countries), which are likely to be regional hot spots for high gross nitrogen balances that lead to environmental pressures. Member States with high nitrogen balances are making efforts to reduce these pressures on the environment. They build on a range of different policy instruments, requiring considerable political effort to succeed given the significant social and economic consequences of reducing livestock production in the affected areas.

Indicator definition

The indicator estimates the potential surplus of nitrogen on agricultural land. This is done by calculating the balance between all nitrogen added to an agricultural system and all nitrogen removed from the system per hectare of agricultural land.

The inputs consist of the amount of nitrogen applied via mineral fertilisers and animal manure as well as nitrogen fixation by legumes, deposition from the air, and some other minor sources. Nitrogen output is that contained in the harvested crops, or grass and crops eaten by livestock. Escape of nitrogen to the atmosphere, e.g. as N_2O, is difficult to estimate and therefore not taken into account.

Figure 1 Gross nutrient balance at the national level

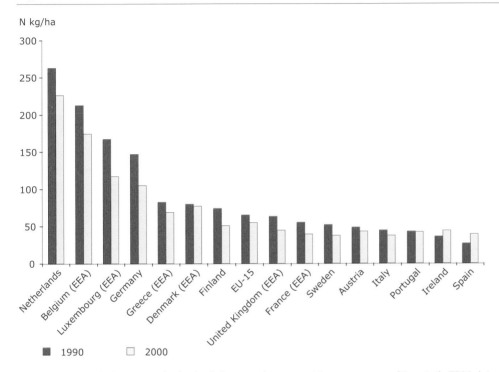

Note: EEA calculations on the basis of: harvested crops and forage crop area (Eurostat's ZPA1 data set or farm structure survey); livestock numbers (Eurostat's ZPA1 data set or farm structure survey); livestock excretion rates (OECD or averaged coefficients from Member States); fertiliser rates (EFMA); nitrogen fixation (OECD or averaged coefficients from Member States farm structure survey); atmospheric deposition (EMEP); yields (Eurostat's ZPA1 data set or average coefficients from Member States).

Data source: OECD website (http://webdomino1.oecd.org/comnet/agr/aeiquest.nsf) and EEA calculations.

Indicator rationale

Nutrient or mineral balances provide insight into links between agricultural nutrient use, changes in environmental quality, and the sustainable use of soil nutrient resources. A persistent surplus indicates potential environmental problems; a persistent deficit indicates potential agricultural sustainability problems. With respect to environmental impacts, however, the main determinant is the absolute size of the nutrient surplus/deficit linked to local farm nutrient management practices and agro-ecological conditions, such as soil type and weather patterns (rainfall, vegetation period, etc.).

The gross nutrient balance for nitrogen provides an indication of the risks of nutrient leaching by identifying agricultural areas that have very high nitrogen loadings. As the indicator integrates the most important agricultural parameters with regard to potential nitrogen surplus, it is currently the best available measure for the risk of nutrient leaching.

Policy context

The gross nitrogen balance is relevant to two EU directives: the Nitrates Directive (91/676/EC) and the Water Framework Directive (2000/60/EC). The nitrates directive has the general purpose of 'reducing water pollution caused or induced by nitrates from agricultural sources and prevent further such pollution' (Art. 1). A threshold nitrate concentration of 50 mg/l is set as the maximum permissible level, and the directive limits applications of livestock manure to land to 170 kg N/ha/yr. The water framework directive requires all inland and coastal waters to reach 'good status' by 2015. Good ecological status is defined in terms of the quality of the biological community, hydrological characteristics and chemical characteristics. The sixth environment action programme encourages the full implementation of both the nitrates and the water framework directives in order to achieve levels of water quality that do not give rise to unacceptable impacts on and risks to human health and the environment.

Indicator uncertainty

The approach used for calculating gross nutrient balance partly requires expert estimates of different physical relations for the country as a whole. However, in reality there may be large regional variations in some of these, and the regional figures should therefore be interpreted with care. Before comparing Member States, it should also be borne in mind that the calculations are based on a harmonised methodology, which may not in all cases reflect country-specific particularities. Moreover, the N-coefficients supplied by Member States also differ remarkably between countries, to an extent which is sometimes difficult to explain.

As a general rule, the data on inputs are estimated to be more accurate and reliable than those on outputs. Not only are the calculations on outputs mainly based on statistics at the national level extrapolated to the regional level, but the lack of (reliable) data on harvested fodder and grass adds a further element of uncertainty to the figures. As this uncertainty is carried through to the total N-balance, the same precautions should be taken before drawing conclusions from the results for the total balance. Nevertheless, the indicator is a good tool for identifying agricultural areas at risk of nutrient leaching.

Areas where data sets are not sufficiently developed include statistics on organic fertilisers, areas under cultivation for secondary crops, statistics for seeds and other planting material, and statistics for non-marketed production and residues.

Gross nutrient balance — PART B

Core set of indicators | Agriculture

26 Area under organic farming

Key policy question

What are environmentally relevant key trends in agricultural production systems?

Key message

The share of organic farming is increasing strongly and now stands at about 4 % of the agricultural area of the EU-15 and the EFTA countries. EU agri-environment programmes and consumer demand have been key factors for this strong increase. The share of organic land remains far below 1 % in most of the EU-10 Member States and the accession countries.

Indicator assessment

- The share of organic farming is far higher in northern and central European countries than in other parts of Europe — with the exception of Italy. Furthermore, there is considerable regional variation of this share within individual countries. In contrast, the share of organic farming is particularly low in most of the EU-10 and accession countries. The overall distribution seems to be influenced by the presence of consumer demand for organic products and government support in the form of agri-environment schemes and other measures.

- Recent literature reviews provide information on the environmental impacts of organic agriculture compared with conventional management systems but the results are not always unambiguous. The environmental benefits of organic farming are most clearly documented for biodiversity as well as for water and soil conservation. However, there is no clear evidence of reduced greenhouse gas emissions. Organic agriculture is likely to have a more positive environmental impact in areas with highly intensive agriculture than in areas with low-input farming systems. So far the regional uptake of organic farming is concentrated in extensive grassland regions where fewer changes are needed to convert to organic farming than in regions dominated by intensive, arable farming, where the benefits would be greater.

Indicator definition

Share of organic farming area (sum of current organically farmed areas and areas in process of conversion) as a proportion of total utilised agricultural area (UAA).

Organic agriculture can be defined as a production system which puts a high emphasis on environmental protection and animal welfare by reducing or eliminating the use of GMOs and synthetic chemical inputs such as fertilisers, pesticides and growth promoters/regulators. Instead, organic farmers promote the use of cultural and agro-ecosystem management practices for crop and livestock production. The legal framework for organic farming in the EU is defined by Council Regulation 2092/91 and amendments.

Indicator rationale

Organic farming is a system that has been explicitly developed to be environmentally sustainable, and is governed by clear, verifiable rules. It thus appears most suited for identifying environment-friendly farming practices compared with other types of farming that also take account of environmental requirements, such as integrated farming.

Farming is only considered to be organic at the EU level if it complies with Council Regulation (EEC) No 2092/91 (and amendments). In this framework, organic farming is differentiated from other approaches to agricultural production by the application of regulated standards (production rules), certification procedures (compulsory inspection schemes) and a specific labelling scheme, resulting in the existence of a specific market, partially isolated from non-organic foods.

Area under organic farming

Figure 1   Organic farming area in Europe

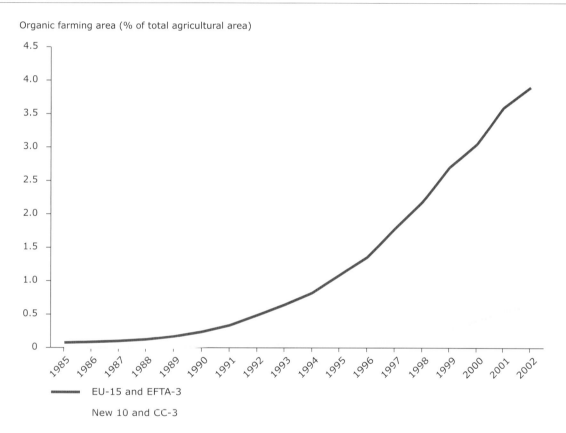

Note: Data source: Institute of Rural Sciences, University of Wales, Aberystwyth (Ref: www.eea.eu.int/coreset).

Policy context

Organic farming aims to establish environmentally sustainable agricultural production systems. Its legal framework is defined by Council Regulation 2092/91 and amendments. The adoption of organic farming methods by individual farmers is supported through agri-environment scheme payments and other rural development measures at the Member State level. In 2004 the EU Commission published a 'European Action Plan for Organic Food and Farming' (COM(2004) 415 final) to further promote this farming approach.

There are no specific EU targets for the share of organic farming area. However, a number of EU Member States have already set targets for area under organic farming, often 10–20 % in 2010.

Core set of indicators | Agriculture

Figure 2   Share of organic farming area in total utilised agricultural area

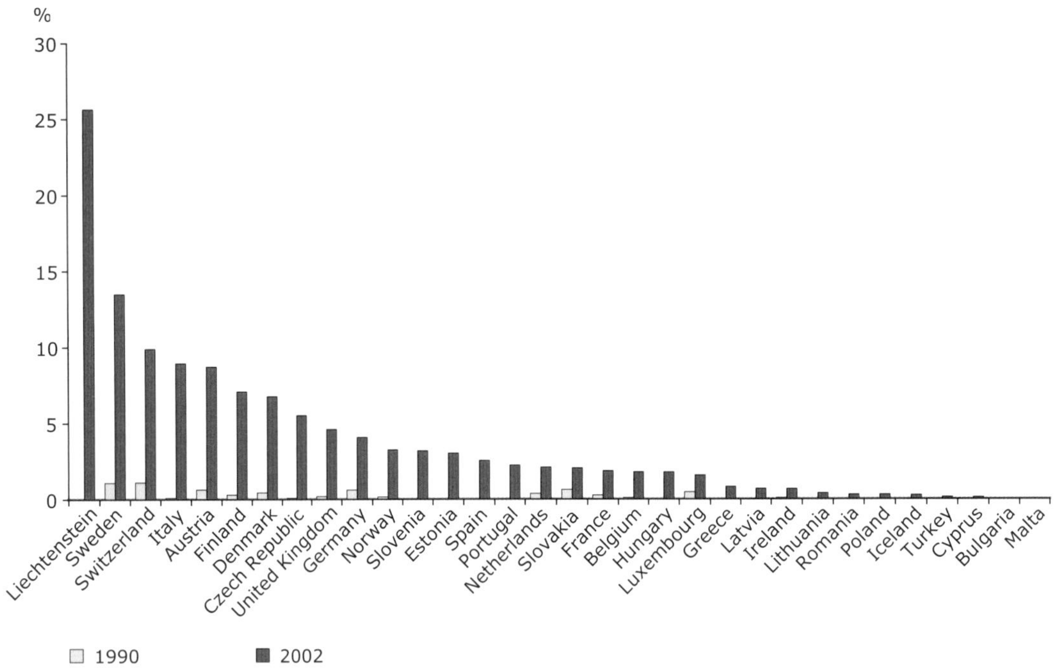

%
30
25
20
15
10
5
0

Liechtenstein, Sweden, Switzerland, Italy, Austria, Finland, Denmark, Czech Republic, United Kingdom, Germany, Norway, Slovenia, Estonia, Spain, Portugal, Netherlands, Slovakia, France, Belgium, Hungary, Luxembourg, Greece, Latvia, Ireland, Lithuania, Romania, Poland, Iceland, Turkey, Cyprus, Bulgaria, Malta

☐ 1990   ■ 2002

Note: Data source: Institute of Rural Sciences, University of Wales, Aberystwyth (Ref: www.eea.eu.int/coreset).

Table 1   Member States targets for area under organic farming

Member State	Name of programme	Target year	Target
EU	European action plan for organic food and farming (2004)	None	Sets out 21 key actions regarding the organic food market, public policy, standards and inspection
Austria	Aktionsprogramm Biologische Landwirtschaft 2003–2004	2006	At least 115 000 ha of arable land in 2006 (~ 8 % of arable land) *
Belgium	'Vlaams actieplan biologische landbouw' — Flemish Action Plan (2000–2003)	2010	10 % of farmland by 2010
Germany	'Bundesprogramm Ökologischer Landbau' (2000)	2010	20 % of farmland by 2010
Netherlands	'An organic market to conquer' (2001–2004)	2010	10 % of farmland by 2010
Sweden	Action plan (1999)	2005	20 % of farmland by 2005 10 % of all dairy cattle/beef cattle/lambs
United Kingdom	'Action Plan to develop organic food and farming in England — two years on' (2004)	2010	The United Kingdom produced share of the market for organic food products should be 70 % by 2010

* Austria has a higher share of grassland under organic production than of arable; hence the focus of the target on arable land.

Indicator uncertainty

The accuracy of data on organic farming varies somewhat between countries and includes provisional estimates. Nevertheless, available data are considered to be very representative and comparable ([1]). Some countries still have a rather low share of organic farming which limits the possibility of identifying trends at national level that may be not be significant from a European perspective.

A drawback of the data set used is that its maintenance depends on research funding and support from organic farming associations.

[1] Please note that the Swedish organic farming area includes a large share of farmland that is not certified according to Regulation 2092/91 but farmed in line with its specifications.

27 Final energy consumption by sector

Key policy question

Are we using less energy?

Key message

Final energy consumption in the EU-25 increased by about 8 % over the period 1990 to 2002. Transport has been the fastest-growing sector since 1990 and is now the largest consumer of final energy.

Indicator assessment

Final energy consumption in the EU-25 increased by about 8 % between 1990 and 2002, thus partly counteracting the reductions in the environmental impact of energy production achieved as a result of fuel-mix changes and technological improvements. Between 2001 and 2002, final energy consumption decreased by 1.4 percentage points, driven mainly by reductions in the household sector as a result of lower space heating requirements due to higher than average temperatures during 2002.

The structure of final energy consumption has undergone significant changes in recent years. Transport was the fastest-growing sector in the EU-25 between 1990 and 2002, with final energy consumption increasing by 24.3 %. Final energy consumption by services (including agriculture) and households grew by 10.2 % and 6.5 % respectively while final energy consumption in the industry sector fell by 7.7 % over the same period. These developments meant that by 2002, transport was the largest consumer of final energy, followed by industry, households and services.

Changes in the structure of final energy consumption were stimulated by the rapid growth of a wide range of service sectors and a shift to less energy-intensive manufacturing industries. The development of the internal market has resulted in increased freight transport as companies exploit the competitive advantages of different regions. Rising personal incomes have permitted higher standards of living, with resultant increases in the ownership of private cars and domestic appliances. Higher comfort levels, reflected in increased demand for space heating and cooling, have also contributed to higher final energy consumption.

There are significant differences in the pattern of final energy consumption between the pre-2004 EU-15 and EU-10 Member States. The EU-10 have seen falling final energy consumption mainly as a result of economic restructuring following the political changes of the early 1990s. However, with the economic recovery in these countries, final energy consumption since 2000 has increased slightly.

Indicator definition

Final energy consumption covers energy supplied to the final consumer for all energy uses. It is calculated as the sum of final energy consumption of all sectors. These are disaggregated to cover industry, transport, households, services and agriculture.

The indicator can be presented in relative or absolute terms. The relative contribution of a specific sector is measured by the ratio between the final energy consumption of that sector and total final energy consumption calculated for a calendar year. It is a useful indicator which highlights a country's sectoral needs in terms of final energy demand. Because sectoral shares depend on the country's economic circumstances, country comparisons of the shares are meaningless unless accompanied by a relevant measure of the importance of the sector in the economy. Because the focus is on the reduction of final energy consumption and not on the sectoral redistribution of such consumption, the trends in the absolute values (in thousand tonnes of oil equivalent) should be preferred as a more meaningful indicator of progress.

Indicator rationale

The trend in final energy consumption by sector provides a broad indication of progress in reducing energy consumption and associated environmental

Final energy consumption by sector — PART B

Figure 1 Final energy consumption by sector, EU-25

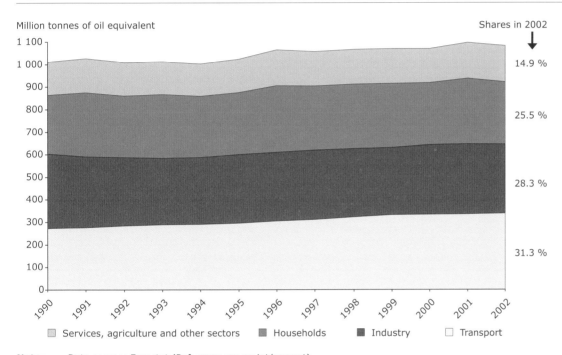

Note: Data source: Eurostat (Ref: www.eea.eu.int/coreset).

impacts by the different end-use sectors (transport, industry, services and households). It can be used to help monitor the success of key policies that attempt to influence energy consumption and energy efficiency.

Final energy consumption helps to estimate the scale of environmental impacts of energy use, such as air pollution, global warming and oil pollution. The type and extent of energy-related pressures on the environment depends both on the sources of energy (and how they are used) and on the total amount of energy consumed. One way of reducing energy-related pressures on the environment is thus to use less energy. This may result from reducing the energy consumption of energy-related activities (e.g. for warmth, personal mobility or freight transport), or by using energy in a more efficient way (thereby using less energy per unit of demand), or from a combination of the two.

Policy context

The reduction in final energy consumption should be seen in the context of reaching the target of an 8 % reduction in greenhouse gas emissions by 2008–2012 from 1990 levels for the EU-15 and individual targets for most EU-10, as agreed in 1997 under the Kyoto Protocol of the United Nations framework convention on climate change, and of enhancing the security of energy supply.

The Action Plan to Improve Energy Efficiency in the European Community (COM(2000)247 Final) outlines a wide range of policies and measures aimed at removing barriers to energy efficiency. It builds on the Communication (COM(98)246 Final) 'Energy efficiency in the European Community — towards a strategy for the rational use of energy' (supported by Council Resolution 98/C 394/01 on energy efficiency in the

PART B | Core set of indicators | Energy

Table 1 — Final energy consumption by country

	Final energy consumption (1000 TOE) 1990–2002								
	1990	1995	1996	1997	1998	1999	2000	2001	2002
EEA	1 108 173	1 116 435	1 168 855	1 156 256	1 164 531	1 169 296	1 174 172	1 198 205	1 187 846
EU-25	1 002 778	1 023 541	1 065 662	1 056 682	1 066 852	1 069 130	1 068 965	1 096 900	1 082 742
EU-15 pre-2004	858 290	895 951	933 514	926 098	942 069	947 238	950 282	972 694	959 928
EU-10	151 657	127 590	132 148	130 581	124 781	121 891	118 683	124 206	122 815
Austria	18 595	20 358	21 976	21 580	22 256	21 855	22 280	24 583	24 990
Belgium	31 277	34 489	36 383	36 529	37 092	36 931	36 922	37 211	35 816
Bulgaria	16 041	11 402	11 520	9 247	9 772	8 782	8 485	8 532	8 621
Cyprus	1 264	1 409	1 458	1 461	1 531	1 575	1 634	1 689	1 647
Czech Republic	36 678	25 405	25 612	25 566	24 323	23 167	24 114	24 131	23 829
Denmark	13 797	14 736	15 322	14 955	14 997	14 933	14 608	14 947	14 708
Estonia	6 002	2 648	2 895	2 962	2 609	2 355	2 362	2 516	2 586
Finland	21 634	22 227	22 478	23 484	24 172	24 637	24 555	24 739	25 489
France	135 709	141 243	148 621	145 654	150 829	150 719	151 624	158 652	152 686
Germany	227 142	222 342	230 895	226 131	224 450	219 934	213 270	215 174	210 485
Greece	14 534	15 811	16 870	17 257	18 159	18 157	18 508	19 112	19 497
Hungary	18 751	15 155	15 863	15 160	15 274	15 853	15 798	16 400	16 915
Iceland	1 602	1 660	1 726	1 753	1 819	1 953	2 057	2 071	2 152
Ireland	7 265	7 910	8 229	8 655	9 308	9 835	10 520	10 932	11 038
Italy	106 963	113 563	114 339	115 335	118 451	123 073	123 005	125 625	125 163
Latvia	3 046	2 845	3 118	2 930	2 688	2 755	2 913	3 642	3 620
Lithuania	9 423	4 097	3 931	3 930	4 340	3 954	3 639	3 778	3 902
Luxembourg	3 325	3 148	3 235	3 224	3 183	3 341	3 544	3 689	3 732
Malta	332	435	505	548	529	551	522	445	445
Netherlands	42 632	47 431	51 413	49 103	49 307	48 470	49 745	50 775	50 641
Norway	16 087	16 854	17 669	17 466	18 187	18 659	18 087	18 561	18 125
Poland	59 574	63 414	66 189	65 312	60 377	58 843	55 573	56 196	54 418
Portugal	11 208	13 042	13 863	14 550	15 421	15 982	16 937	18 069	18 342
Romania	33 251	25 187	30 410	27 702	25 012	21 611	22 436	22 742	23 247
Slovakia	13 219	8 242	8 218	8 242	8 838	8 486	7 605	10 883	10 864
Slovenia	3 368	3 940	4 359	4 470	4 272	4 352	4 523	4 526	4 589
Spain	56 647	63 536	65 259	67 986	71 750	74 378	79 411	83 221	85 379
Sweden	30 498	33 679	34 603	34 119	34 251	34 076	34 532	33 132	33 668
Turkey	31 245	37 791	41 868	43 409	42 891	49 162	54 142	49 399	52 958
United Kingdom	137 064	142 436	150 028	147 536	148 443	150 917	150 821	152 833	148 294

Note: TOE refers to tonnes of oil equivalent. No energy data for Liechtenstein available from Eurostat.

Data source: Eurostat (Ref: www.eea.eu.int/coreset).

European Community). It proposed an indicative EU target of reducing final energy intensity by 1 % per year above 'that which would have otherwise been attained during the period 1998–2010'.

The proposal for a Directive of the European Parliament and the Council on Energy End-use Efficiency and Energy Services (COM(2003) 739) aims at boosting the cost-effective and efficient use of energy in the EU by fostering energy-efficiency measures and promoting the market for energy services. It proposes that Member States adopt and meet mandatory targets of saving 1 % more of the energy previously used every year — this means 1 % of the average annual amount of energy distributed or sold to final customers the previous five years — through increased energy efficiency for a period of six years. In the sixth year, final energy consumption will then be 6 % lower than it would have been without the efficiency measures. The savings will have to be registered in the following sectors: households, agriculture, commerce and public, transport (excluding air and maritime transport), and industry (excluding energy-intensive industry).

The recent Green Paper on Energy Efficiency (COM(2005)265 final) states that overall, as much as 20 % of energy savings could be realised in a cost-effective way by 2020. It aims at identifying such cost-effective options and at opening a discussion on how to reach them.

Indicator uncertainty

Data have traditionally been compiled by Eurostat through the annual joint questionnaires (shared by Eurostat and the International Energy Agency), following a well-established and harmonised methodology. Data are transmitted to Eurostat electronically, using a common set of tables. Data are then treated to find inconsistencies and entered into the database. Estimations are not normally necessary since annual data are complete.

The sectoral breakdown of final energy consumption includes industry, transport, households, services, agriculture, fisheries and other sectors. The 'European energy and transport trends to 2030' produced for the DG for Energy and Transport of the European Commission aggregates agriculture, fisheries and other sectors together with the services sector, and projections are based on such aggregation. To be consistent with these projections, the core set indicator uses the same aggregation. The inclusion of agriculture and fisheries together with the services sector is however questionable given their divergent trends. Separate assessments are therefore made where appropriate.

A crude cross-country comparison of the relative sectoral distribution of final energy consumption (i.e. each sector's energy consumption as a percentage of the total for all sectors) is meaningless unless accompanied by some indications of the importance of the sector in the economy of the country. But even if the same sectors in two countries are equally important to the economy, the gross (primary) consumption of energy needed before it reaches the final user might draw from energy sources that pollute the environment in different ways. Thus, from an environmental point of view, the final energy consumption of a sector should be analysed in that broader context. Also, a decrease in the final energy consumption of one sector could result in increasing pressure on the environment if the net reduction in energy use in that sector results in a net increase in energy use in another sector or if there is a switch to more environmentally damaging energy sources.

28 Total energy intensity

Key policy question

Are we decoupling energy consumption from economic growth?

Key message

Economic growth is requiring less additional energy consumption, mainly as a result of structural changes in the economy. However, total energy consumption is still increasing.

Indicator assessment

Total energy consumption in the EU-25 grew at an average annual rate of just below 0.7 % over the period 1990 to 2002, while gross domestic product (GDP) grew at an estimated average annual rate of 2 %. As a result, total energy intensity in the EU-25 fell at an average rate of 1.3 % per year. Despite this relative decoupling of total energy consumption and economic growth, total energy consumption increased by 8.4 % over the period.

All the EU-25 countries except Portugal, Spain and Latvia experienced a decrease in total energy intensity between 1990 and 2002. The average annual decrease was 3.3 % in the EU-10 and 1 % in the pre-2004 EU-15 Member States. Despite this converging trend, total energy intensity in the EU-10 in 2002 was still significantly higher than in the EU-15 Member States.

Much of the reduction of total energy intensity was due to structural changes in the economy. These included a shift from industry towards services which are typically less energy-intensive, a shift within the industrial sector from energy-intensive industries towards higher value-added, less energy-intensive industries, and one-off changes in some Member States.

Trends in final energy consumption intensity by sector during 1990–2002 suggest that there have been substantial improvements in the energy intensity in the industry and services sectors. In contrast, the transport and households sectors show only limited decoupling of energy consumption from economic growth and population growth, respectively. The lack of improvement in final energy intensity in the household sector is influenced by rising living standards, leading to a larger number of households, lower occupancy levels and increased use of household appliances.

Indicator definition

Total energy intensity is the ratio between gross inland consumption of energy (or total energy consumption) and gross domestic product (GDP) calculated for a calendar year. It shows how much energy is consumed per unit of GDP.

Gross inland consumption of energy is calculated as the sum of gross inland consumption of the five sources of energy: solid fuels, oil, gas, nuclear and renewables. The GDP figures are taken at constant prices to avoid the impact of inflation, with 1995 as the base year.

Gross inland energy consumption is measured in thousand tonnes oil equivalent (ktoe) and GDP in million Euro at 1995 market prices. To make comparisons of trends across countries more meaningful, the indicator is presented as an index. An additional column is included to show the actual energy intensity in purchasing power standards for the latest available year.

Indicator rationale

The type and extent of energy-related pressures on the environment, such as air pollution and global warming, depends on the sources of energy and how and in what quantities they are used. One way of reducing energy-related pressures on the environment is to use less energy. This may result from reducing the demand for energy-related activities (e.g. for warmth, personal mobility or freight transport), or by using energy in a more efficient way (thereby using less energy per unit of demand), or a combination of the two.

The indicator identifies the extent, if any, of decoupling between energy consumption and economic growth.

Total energy intensity

Figure 1 Total energy intensity, EU-25

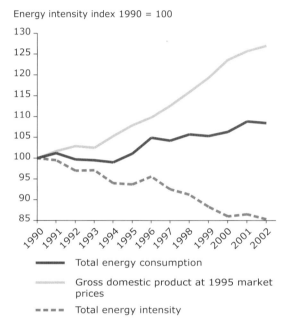

Energy intensity index 1990 = 100

— Total energy consumption
— Gross domestic product at 1995 market prices
- - - Total energy intensity

Note: Some estimates have been necessary in order to compute the EU-25 GDP index for 1990. Eurostat data was not available for a particular year for some EU-25 Member States. The European Commission's annual macroeconomic database (Ameco) was therefore used as an additional data source. GDP for the missing year is estimated on the basis of the annual growth rate from Ameco, the rate being applied to the latest available GDP from Eurostat. This method was used for the Czech Republic (1990–1994), Hungary (1990), Poland (1990–1994), Malta (1991–1998) and Germany (1990). For some other countries and particular years, however, GDP was not available from Eurosat or from Ameco. Few assumptions were made for the purpose of estimating the EU-25. For Estonia, GDP in 1990–1992 is assumed constant and takes the value observed in 1993. For Slovakia, GDP in 1990–1991 takes the value ffor 1992. For Malta, GDP in 1990 is assumed to be equal to GDP in 1991. These assumptions do not distort the trend observed for EU-25 GDP, since the latter three countries represent about 0.3–0.4 % of EU-25 GDP.

Data source: Eurostat and Ameco database, European Commission (Ref: www.eea.eu.int/coreset).

Relative decoupling occurs when energy consumption grows, but more slowly than gross domestic product. Absolute decoupling occurs when energy consumption is stable or falls while GDP grows. From an environmental point of view, however, overall impacts depend on the total amount of energy consumption and the fuels used to produce the energy.

The indicator does not show any of the underlying reasons that affect the trends. A reduction in total energy intensity can be the result of improvements in energy efficiency or changes in energy demand resulting from other factors including structural, societal, behavioural or technical change.

Policy context

Even though there is no target for total energy intensity, there are a number of EU directives, action plans and Community strategies that are directly or indirectly related to energy efficiency, e.g. the sixth environment action plan calls for the promotion of energy efficiency. Several energy and environment targets are also influenced by changes in energy intensity:

- The indicative target for final energy consumption intensity in the EU, set in the 1998 Communication 'Energy Efficiency in the European Community: Towards a Strategy for the Rational Use of Energy' (COM(98) 246 final), of 1 % per year improvement in the intensity of final energy consumption from 1998 'over and above that which would otherwise be attained'.

- The EU and EU-10 targets under the Kyoto Protocol of the United Nations Framework Convention on Climate Change (UNFCCC) to reduce greenhouse gas emissions.

- The EU Indicative Combined Heat and Power Target Set in the Community Strategy on Cogeneration to Promote Combined Heat and Power (COM(97) 514 final), for an 18 % share of CHP electricity production in total gross electricity production by 2010.

Core set of indicators | Energy

Table 1 — Total energy intensity by country

	Total energy intensity 1995–2002 (1995 = 100)								Annual average change 1995–2002	Energy intensity in 2002 (TOE per million GDP in PPS)
	1995	1996	1997	1998	1999	2000	2001	2002		
EEA	100.0	102.0	98.6	96.9	93.7	91.5	91.9	90.6	– 1.4 %	177
EU-25	100.0	102.0	98.8	97.3	94.2	91.8	92.4	91.0	– 1.3 %	174
EU-15 pre-2004	100.0	102.0	99.0	98.2	95.6	93.5	94.0	92.7	– 1.1 %	167
EU-10	100.0	99.9	93.6	87.3	81.2	77.1	77.5	75.5	– 3.9 %	249
Austria	100.0	103.5	101.6	99.2	95.7	92.1	100.2	98.2	– 0.3 %	148
Belgium	100.0	105.7	104.4	104.3	102.3	99.0	95.6	89.5	– 1.6 %	207
Bulgaria	100.0	109.4	102.8	96.8	85.4	81.7	81.8	76.6	– 3.7 %	392
Cyprus	100.0	105.5	100.7	107.5	100.4	100.5	97.7	96.1	– 0.6 %	194
Czech Republic	100.0	98.7	100.0	97.7	89.7	91.8	91.4	90.0	– 1.5 %	282
Denmark	100.0	110.0	99.7	95.8	90.0	85.1	85.9	83.6	– 2.5 %	144
Estonia	100.0	101.5	90.4	81.4	76.1	66.1	69.3	62.9	– 6.4 %	371
Finland	100.0	104.0	102.9	99.4	95.0	89.5	90.8	93.6	– 0.9 %	282
France	100.0	104.3	99.9	99.6	96.4	95.7	96.4	95.3	– 0.7 %	180
Germany	100.0	102.7	100.3	98.1	94.4	92.3	94.2	92.4	– 1.1 %	178
Greece	100.0	102.8	99.9	101.5	97.8	98.2	97.0	96.2	– 0.5 %	165
Hungary	100.0	100.9	94.6	89.4	86.7	81.1	79.5	77.6	– 3.6 %	204
Iceland	100.0	109.6	109.1	110.3	121.3	120.6	122.3	124.2	3.1 %	473
Ireland	100.0	98.3	92.9	90.7	86.5	80.7	79.5	76.6	– 3.7 %	138
Italy	100.0	98.8	98.2	99.5	99.2	97.1	95.6	95.7	– 0.6 %	132
Latvia	100.0	92.6	79.7	74.5	84.6	76.1	82.2	75.4	– 4.0 %	218
Lithuania	100.0	102.1	89.8	93.6	80.9	71.1	75.7	75.2	– 4.0 %	280
Luxembourg	100.0	98.7	89.8	82.1	80.0	77.4	79.1	81.5	– 2.9 %	199
Malta	100.0	106.1	106.9	108.6	103.8	94.7	84.9	82.8	– 2.7 %	135
Netherlands	100.0	100.9	95.7	91.6	87.4	85.9	86.8	87.0	– 2.0 %	188
Norway	100.0	93.1	93.2	94.8	97.2	92.2	92.6	89.3	– 1.6 %	184
Poland	100.0	101.1	91.2	82.0	75.5	70.2	69.6	67.6	– 5.4 %	241
Portugal	100.0	96.3	98.3	100.8	104.3	101.8	102.7	107.3	1.0 %	155
Romania	100.0	103.2	99.1	94.0	85.3	87.5	82.2	76.2	– 3.8 %	272
Slovakia	100.0	90.8	91.2	86.1	84.2	82.5	88.9	85.7	– 2.2 %	319
Slovenia	100.0	101.2	97.8	93.6	87.6	84.8	87.4	86.2	– 2.1 %	217
Spain	100.0	96.3	97.4	97.8	99.3	99.3	99.3	100.1	0.0 %	154
Sweden	100.0	101.1	96.2	93.6	89.7	81.0	86.2	84.5	– 2.4 %	238
Turkey	100.0	101.6	99.5	98.3	101.3	102.8	103.2	100.0	0.0 %	193
United Kingdom	100.0	101.8	96.2	96.5	93.2	90.4	88.9	85.3	– 2.2 %	154

Note: The year for the reference index value is 1995 because GDP for 1990 was not available for all countries. The last column shows the energy intensity measured in purchasing power standards. These are currency conversion rates that convert to a common currency and equalise the purchasing power of different currencies. They eliminate differences in price levels between countries, allowing meaningful volume comparisons of GDP. They are an optimal unit for benchmarking country performance in a particular year. TOE refers to tonnes of oil equivalent. No energy data for Liechtenstein available from Eurostat.

Data source: Eurostat (Ref: www.eea.eu.int/coreset).

Total energy intensity

- The EU Directive 2004/8/EC on the promotion of cogeneration based on useful heat demand in the internal energy market. The purpose of this Directive is to increase energy efficiency and improve security of supply by creating a framework for the promotion and development of high-efficiency cogeneration of heat and power based on useful heat demand and primary energy savings in the internal energy market.

- The proposed Directive on Energy End-use Efficiency and Energy Services (COM(2003) 739 final), sets targets for Member States to save 1 % per year of all energy supplied between 2006 and 2012 compared with current supply.

Indicator uncertainty

Data have traditionally been compiled by Eurostat through the annual joint questionnaires (shared by Eurostat and the International Energy Agency) following a well-established and harmonised methodology. Data are transmitted to Eurostat electronically, using a common set of tables. Data are then treated to find inconsistencies and entered into the database. Estimations are not normally necessary since annual data are complete.

There is no estimate of GDP for the EU-25 in 1990, needed to compute the EU-25 GDP index in 1990, available from Eurostat. Eurostat data was not available for a particular year for some EU-25 Member States. The European Commission's annual macroeconomic database (Ameco) has been used to estimate GDP for the missing years and countries by applying annual growth rates from Ameco to the latest available GDP data from Eurostat. This method was used for the Czech Republic (1990–1994), Hungary (1990), Poland (1990–1994), Malta (1991–1998) and Germany (1990). In some cases, however, GDP was not available from Eurostat or from Ameco. With the sole aim of having an estimate for the EU-25, the following assumptions were made: for Estonia, GDP in 1990–1992 is assumed constant and takes the value observed in 1993; for Slovakia, GDP in 1990–1991 takes the value for 1992; for Malta, GDP in 1990 is assumed to be equal to GDP in 1991. These assumptions are consistent with the trend observed for the EU-25, since the latter three countries represent about 0.3–0.4 % of EU-25 GDP. 1995 was chosen as the base year for the indices in the country table in order to avoid estimations.

The intensity of energy consumption is relative to changes in real GDP. Cross-country comparisons of energy intensity based on real GDP are relevant for trends but not for comparing energy intensity levels in specific years and specific countries. This is why the core set indicator is expressed as an index. In order to compare the energy intensity between countries for a specific year, an additional column is shown with the energy intensities in purchasing power standards.

Energy intensity is not sufficient to measure the environmental impact of energy use and production. Even when two countries have the same energy intensity or show the same trend over time there could be important environmental differences between them. The link to environmental pressures has to be made on the basis of the absolute amounts of the different fuels used to produce that energy. Energy intensity should therefore always be put in the broader context of the actual fuel mix used to generate the energy.

29 Total energy consumption by fuel

Key policy question

Are we switching to less polluting fuels to meet our energy consumption?

Key message

Fossil fuels continue to dominate total energy consumption, but environmental pressures have been limited by switching from coal and lignite to relatively clean natural gas.

Indicator assessment

The share of fossil fuels such as coal, lignite, oil and natural gas in total energy consumption declined only slightly between 1990 and 2002, to reach 79 %. Their use has considerable impact on the environment and is the main cause of greenhouse gas emissions. However, changes to the fossil fuel mix have benefited the environment, with the share of coal and lignite declining continuously and being replaced by relatively cleaner natural gas, which now has a 23 % share.

Most of the switch between fossil fuels was in the power generation sector. In the pre-2004 EU-15 Member States this was supported by implementation of environmental legislation and liberalisation of electricity markets, which stimulated the use of combined-cycle gas plants due to their high efficiency, low capital cost and low gas prices in the early 1990s, and by the expansion of the trans-EU gas network. Fuel mix changes in the EU-10 were induced by the process of economic transformation, which led to changes in fuel prices and taxation and removal of energy subsidies, and policies to privatise and restructure the energy sector.

Renewable energy, which typically has lower environmental impacts than fossil fuels, has seen rapid growth in absolute terms, but from a low starting point. Despite increased support at the EU and national level, its contribution to total energy consumption remains low at almost 6 %. The share of nuclear power has grown slowly to reach almost 15 % of total energy consumption in 2002. While nuclear power produces little pollution under normal operations there is a risk of accidental radioactive releases, and highly radioactive waste are accumulating for which no generally acceptable disposal route has yet been established.

Overall, the changes in the fuel mix of total energy consumption contributed to reducing emissions of greenhouse gases and acidifying substances. Rising total energy consumption, however, counteracted some of the environmental benefits of the fuel switch. Total energy consumption in the EU-25 increased by 8.4 % over the period 1990–2002 although it decreased slightly between 2001 and 2002 due to higher than average temperatures and a slowing of GDP growth.

Indicator definition

Total energy consumption or gross inland energy consumption represents the quantity of energy necessary to satisfy the inland consumption of a country. It is calculated as the sum of gross inland consumption of energy from solid fuels, oil, gas, nuclear and renewable sources. The relative contribution of a specific fuel is measured by the ratio between the energy consumption originating from that specific fuel and the total gross inland energy consumption calculated for a calendar year.

Energy consumption is measured in thousand tonnes of oil equivalent (ktoe). The share of each fuel in total energy consumption is presented in the form of a percentage.

Indicator rationale

Total energy consumption is a driving force indicator providing an indication of the environmental pressures caused by energy production and consumption. It is disaggregated by fuel source as the environmental impact of each fuel is very specific.

Figure 1   Total energy consumption by fuel in the EU-25

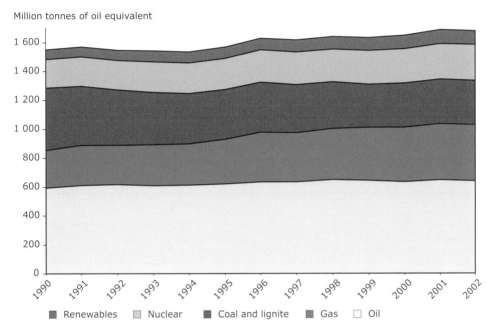

Million tonnes of oil equivalent

■ Renewables   □ Nuclear   ■ Coal and lignite   ■ Gas   □ Oil

Note: Data source: Eurostat (Ref: www.eea.eu.int/coreset).

The consumption of fossil fuels (such as crude oil, oil products, hard coal, lignite and natural and derived gas) provides a proxy indicator of resource depletion, CO_2 and other greenhouse gas emissions and air pollution (e.g. SO_2 and NO_x). The degree of the environmental impact depends on the relative share of different fossil fuels and the extent to which pollution abatement measures are used. Natural gas, for instance, has approximately 40 % less carbon per unit of energy than coal and 25 % less carbon than oil and contains only marginal quantities of sulphur.

The level of nuclear energy consumption provides an indication of the trends in the amount of nuclear waste generated and the risks associated with radioactive leaks and accidents. Increasing consumption of nuclear energy at the expense of fossil fuels would on the other hand contribute to reductions in CO_2 emissions.

Renewable energy consumption measures the contribution from technologies that are more environmentally benign, as they produce no (or very little) net CO_2 and usually significantly lower levels of other pollutants. Renewable energy can, however, have impacts on landscapes and ecosystems. The incineration of municipal waste uses both renewable and non-renewable material and may also generate local air pollution. However, emissions from the incineration of waste are subject to stringent regulations including tight controls on quantities of cadmium, mercury and other such substances. Similarly, the inclusion of both large and small-scale hydropower provides only a broad indicator of environmentally benign energy supply. While small-scale hydro schemes generally have little environmental impact, large-scale hydro can have major adverse impacts (flooding, impact on ecosystems, water levels, requirements for population resettlement).

Core set of indicators | Energy

Table 1   Total energy consumption by fuel (%)

	Total energy consumption by fuel (%) in 2002							
	Coal and lignite	Oil	Gas	Nuclear	Renewables	Industrial waste	Imports-exports of electricity	Total energy consumption (1 000 TOE)
EEA	18.5	37.6	23.1	13.8	6.8	0.2	0.0	1 843 310
EU-25	18.2	38.0	23.1	14.8	5.7	0.2	0.1	1 684 042
EU-15 pre-2004	14.7	39.9	23.6	15.6	5.8	0.2	0.3	1 482 081
EU-10	43.5	23.8	19.5	8.8	5.0	0.3	– 1.0	201 961
Austria	12.3	41.5	21.4	0.0	24.0	0.6	0.2	30 909
Belgium	12.7	35.5	25.4	23.2	1.6	0.4	1.2	52 570
Bulgaria	35.6	23.4	11.6	27.9	4.4	0.0	– 2.9	18 720
Cyprus	1.5	96.7	0.0	0.0	1.9	0.0	0.0	2 420
Czech Republic	49.9	19.9	18.9	11.1	2.2	0.3	– 2.4	40 991
Denmark	21.1	44.1	23.3	0.0	12.3	0.0	– 0.9	19 821
Estonia	57.2	21.5	12.0	0.0	10.5	0.0	– 1.2	4 963
Finland	18.5	28.9	10.5	16.4	22.2	0.6	2.9	35 136
France	5.2	34.7	14.1	42.4	6.1	0.0	– 2.5	265 537
Germany	24.9	37.1	22.0	12.4	3.1	0.4	0.3	343 671
Greece	31.4	57.0	6.1	0.0	4.7	0.0	0.8	29 736
Hungary	14.1	24.8	42.2	14.0	3.5	0.0	1.4	25 633
Iceland	2.9	24.3	0.0	0.0	72.8	0.0	0.0	3 382
Ireland	17.0	56.6	24.3	0.0	1.9	0.0	0.3	15 139
Italy	7.9	50.9	33.2	0.0	5.3	0.2	2.5	173 550
Latvia	2.4	27.2	30.8	0.0	34.8	0.0	4.8	4 189
Lithuania	1.7	29.4	25.3	42.1	8.0	0.0	– 6.4	8 671
Luxembourg	2.3	62.4	26.5	0.0	1.4	0.0	7.4	3 979
Malta	0.0	100.0	0.0	0.0	0.0	0.0	0.0	823
Netherlands	10.7	37.9	45.8	1.3	2.2	0.3	1.8	78 195
Norway	3.1	29.0	23.4	0.0	47.7	0.0	– 3.2	26 278
Poland	61.7	22.4	11.4	0.0	4.7	0.6	– 0.7	88 837
Portugal	13.4	61.4	10.5	0.0	14.0	0.0	0.6	25 966
Romania	22.0	26.7	37.2	4.0	10.5	0.3	– 0.7	35 753
Slovakia	22.9	18.4	31.6	24.9	3.9	0.3	– 1.9	18 570
Slovenia	22.8	35.5	11.3	20.8	11.0	0.0	– 1.4	6 864
Spain	16.7	50.5	14.4	12.5	5.6	0.0	0.4	130 063
Sweden	5.5	30.7	1.6	34.2	27.1	0.1	0.9	51 435
Turkey	26.3	40.8	19.6	0.0	12.9	0.0	0.4	75 135
United Kingdom	15.8	34.7	37.9	10.0	1.2	0.0	0.3	226 374

Note: TOE refers to tonnes of oil equivalent. No energy data for Liechtenstein available from Eurostat.

Data source: Eurostat (Ref: www.eea.eu.int/coreset).

Policy context

Total energy consumption disaggregated by fuel type provides an indication of the extent of environmental pressure caused (or at risk of being caused) by energy production and consumption. The relative shares of fossil fuels, nuclear power and renewable energies together with the total amount of energy consumption are valuable in determining the overall environmental burden of energy consumption in the EU. Trends in the shares of these fuels will be one of the major determinants of whether the EU meets its target for reduction of greenhouse gas emissions agreed under the Kyoto Protocol.

There are two targets indirectly related to this indicator: 1) The EU target of an 8 % reduction in greenhouse gas emissions by 2008–2012 from 1990 levels, as agreed in 1997 under the Kyoto Protocol of the United Nations Framework Convention on Climate Change (UNFCCC); and 2) The White Paper for a Community Strategy and Action Plan (COM(97) 599 final) which provides a framework for action by Member States to develop renewable energy and sets an indicative target of increasing the share of renewable energy in total energy consumption in the pre-2004 EU-15 to 12 % by 2010.

Indicator uncertainty

Data have traditionally been compiled by Eurostat through the annual joint questionnaires (shared by Eurostat and the International Energy Agency) following a well-established and harmonised methodology. Data are transmitted to Eurostat electronically, using a common set of tables. Data are then treated to find inconsistencies and entered into the database. Estimations are not normally necessary since annual data are complete.

The share of energy consumption for a particular fuel could decrease even though the actual amount of energy used from that fuel grows. Similarly, its share could increase despite a possible reduction in the total consumption of energy from that fuel. What makes a share for a particular fuel increase or decrease depends on the change in its energy consumption relative to the total consumption of energy.

From an environmental point of view, however, the relative contribution of each fuel has to be put in the wider context. Absolute (as opposed to relative) volumes of energy consumption for each fuel are the key to understanding the environmental pressures. These depend on the total amount of energy consumption as well as on the fuel mix used and the extent to which pollution abatement technologies are used.

Total energy consumption may not accurately represent the energy needs of a country (in terms of final energy demand). Fuel switching may in some cases have a significant effect in changing total energy consumption even though there is no change in (final) energy demand. This is because different fuels and different technologies convert primary energy into useful energy with different efficiency rates.

30 Renewable energy consumption

Key policy question

Are we switching to renewable energy sources to meet our energy consumption?

Key message

The share of renewable energies in total energy consumption increased over the period 1990–2002, but still remains at a low level. Significant further growth will be needed to meet the EU indicative target of a 12 % share by 2010.

Indicator assessment

The contribution of renewable energy sources to total energy consumption increased between 1990 and 2001 in the EU-25, but fell slightly in 2002 due to lower production of hydroelectricity (as a result of low rainfall) to reach 5.7 %. This is still substantially short of the indicative target set in the White Paper on Renewable Energy (COM(97) 599 final) to derive 12 % of EU total energy consumption from renewable sources by 2010 (at present, the 12 % aim applies only to the pre-2004 EU-15 Member States).

Between 1990 and 2002, the fastest-growing renewable energy source was wind with an average increase of 38 % per year, followed by solar energy. The increase in the use of wind to produce electricity was accounted for mainly by strong growth in Denmark, Germany and Spain, encouraged by support policies for the development of wind power. However, as wind and solar energy started from a very low level, they accounted for only 3.2 % and 0.5 % of total renewable energy consumption in 2002. Geothermal energy contributed 4.0 % of total renewable energy in 2002. The main sources of renewable energy were biomass and waste, and hydropower, accounting for 65.6 % and 26.7 % of total renewables, respectively.

A number of environmental concerns and a lack of suitable sites mean that large-scale hydropower is unlikely to contribute to significant future increases in renewable energy in the EU-25. Growth will therefore need to come from other sources such as wind, biomass, solar energy and small-scale hydropower. Expanding the use of biomass for energy purposes needs to take account of conflicting land-use for agricultural and forestry areas, and in particular nature conservation requirements.

Indicator definition

The share of renewable energy consumption is the ratio between gross inland consumption of energy from renewable sources and total gross inland energy consumption calculated for a calendar year, expressed as a percentage. Both renewable energy and total energy consumption are measured in thousand tonnes of oil equivalent (ktoe).

Renewable energy sources are defined as renewable non-fossil sources: wind, solar energy, geothermal energy, wave, tidal energy, hydropower, biomass, landfill gas, sewage treatment plant gas and biogases.

Indicator rationale

The share of energy consumption from renewable energy provides a broad indication of progress towards reducing the environmental impact of energy consumption, although its overall impact has to be seen within the context of total energy consumption, the total fuel mix, potential impacts on biodiversity and the extent to which pollution abatement equipment is fitted.

Renewable energy sources are generally considered environmentally benign, with very low net emissions of CO_2 per unit of energy produced, even allowing for the emissions associated with the construction of the plant. Emissions of other pollutants are also often lower for renewable than for fossil fuel energy production. The exception is municipal and solid waste (MSW) incineration which, because of the cost associated with separation, usually involves the combustion of some mixed waste including materials contaminated

Figure 1 Contribution of renewable energy sources to total energy consumption, EU-25

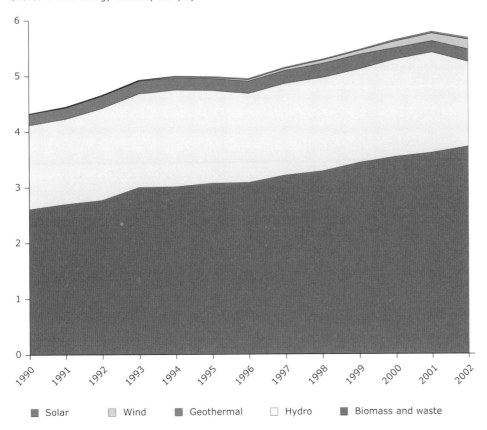

Note: Data source: Eurostat (Ref: www.eea.eu.int/coreset).

with heavy metals. However, emissions from MSW incineration are subject to stringent regulations including tight controls on quantities of cadmium, mercury, and other such substances.

Most renewable (and non-renewable) energy sources have some impact on landscapes, noise and ecosystems, although many of these can be minimised through careful site selection. Large hydropower schemes in particular can have adverse impacts including flooding, disruption of ecosystems and hydrology, and socio-economic impacts if resettlement is required. Some solar photovoltaic schemes require relatively large quantities of heavy metals in their construction and geothermal energy can release pollutant gases carried by its hot fluid if not properly controlled. Some types of biomass and biofuel crops also have considerable land, water and agricultural input requirements such as fertilisers and pesticides.

Core set of indicators | Energy

Table 1 Share of renewable energy in total energy consumption (%)

	Share of renewable energy in total energy consumption (%) 1990–2002								
	1990	1995	1996	1997	1998	1999	2000	2001	2002
EEA	5.4	6.1	6.1	6.3	6.5	6.7	6.8	6.8	6.8
EU-25	4.3	5.0	4.9	5.2	5.3	5.5	5.6	5.8	5.7
EU-15 pre-2004	4.9	5.3	5.3	5.5	5.6	5.6	5.8	5.9	5.8
EU-10	1.4	3.1	2.9	3.0	3.4	4.1	4.3	4.7	5.0
Austria	20.3	22.0	20.6	21.1	20.8	22.4	22.7	23.6	24.0
Belgium	1.4	1.4	1.3	1.2	1.3	1.3	1.3	1.4	1.6
Bulgaria	0.6	1.6	2.0	2.3	3.4	3.5	4.2	3.6	4.4
Cyprus	0.3	2.1	2.0	2.0	1.9	1.9	1.8	1.8	1.9
Czech Republic	0.3	1.5	1.4	1.6	1.6	2.0	1.6	1.8	2.2
Denmark	6.7	7.6	7.2	8.3	8.7	9.6	10.7	11.1	12.3
Estonia	4.7	9.1	10.4	10.7	9.7	10.4	11.0	10.6	10.5
Finland	19.2	21.3	19.8	20.6	21.8	22.1	24.0	22.7	22.2
France	7.0	7.6	7.2	6.9	6.8	7.0	6.8	6.8	6.1
Germany	1.6	1.9	1.9	2.2	2.4	2.6	2.9	2.8	3.1
Greece	5.0	5.3	5.4	5.2	4.9	5.4	5.0	4.6	4.7
Hungary	0.1	0.1	0.1	0.1	0.1	1.5	1.7	1.6	3.5
Iceland	65.8	64.9	65.5	66.8	67.6	71.3	71.4	73.2	72.8
Ireland	1.6	2.0	1.6	1.6	2.0	1.9	1.8	1.8	1.9
Italy	4.2	4.8	5.2	5.3	5.4	5.8	5.2	5.5	5.3
Latvia	9.4	6.8	4.5	7.6	11.4	30.1	28.8	35.0	34.8
Lithuania	0.2	0.4	0.3	0.3	6.5	7.9	9.0	8.3	8.0
Luxembourg	1.3	1.4	1.2	1.4	1.6	1.3	1.5	1.3	1.4
Malta	0.0	0.0	0.0	0.0	0.0	0.0	0.0	0.0	0.0
Netherlands	1.1	1.2	1.6	1.8	1.9	2.1	2.1	2.1	2.2
Norway	53.1	48.9	43.3	43.7	44.0	44.8	51.0	44.1	47.7
Poland	1.6	4.0	3.6	3.7	4.0	4.0	4.2	4.5	4.7
Portugal	15.9	13.3	16.1	14.7	13.6	11.1	12.9	15.7	14.0
Romania	4.2	6.2	12.9	11.2	11.8	12.5	10.9	9.3	10.5
Slovakia	1.6	3.0	2.8	2.6	2.7	2.8	3.0	4.1	3.9
Slovenia	4.6	8.9	9.4	7.7	8.3	8.8	11.6	11.5	11.0
Spain	7.0	5.5	7.0	6.4	6.3	5.2	5.8	6.5	5.6
Sweden	24.9	26.1	23.6	27.6	28.2	27.8	31.6	28.8	27.1
Turkey	18.5	17.4	16.6	15.8	15.9	15.1	13.1	13.1	12.9
United Kingdom	0.5	0.9	0.8	0.9	1.0	1.1	1.1	1.1	1.2

Note: Data source: Eurostat. No energy data for Liechtenstein available from Eurostat (Ref: www.eea.eu.int/coreset).

Policy context

Energy use (both energy production and final consumption) is the biggest contributor to greenhouse gas emissions in the EU. The energy-related share of these emissions increased from 79 % in 1990 to 82 % in 2002. Increased market penetration of renewable energy will help to reach the EU commitment under the Kyoto Protocol of the United Nations Framework Convention on climate change. The overall Kyoto target for the pre-2004 EU-15 Member States requires a 8 % reduction in emissions of greenhouse gases by 2008–2012 from 1990 levels, while most new Member States have individual targets under the Kyoto Protocol.

The main target for the indicator is defined in the White Paper for a Community Strategy and Action Plan (COM(97) 599 final), which provides a framework for action by Member States to develop renewable energy and sets an indicative target to increase the share of renewable energy in total energy consumption (GIEC) in the EU-15 to 12 % by 2010.

The biofuels directive (2003/30/EC) aims at promoting the use of biofuels to replace diesel and petrol in transport and sets an indicative target of a 5.75 % share of biofuels by 2010.

The Renewable Electricity Directive (2001/77/EC) sets an indicative target of 21 % of gross electricity consumption to be produced from renewable energy sources in the EU-25 by 2010.

Indicator uncertainty

Data have traditionally been compiled by Eurostat through the annual joint questionnaires, shared by Eurostat and the International Energy Agency, following a well-established and harmonised methodology. Methodological information on the annual joint questionnaires and data compilation can be found in Eurostat's website for metadata on energy statistics.

Biomass and waste, as defined by Eurostat, cover organic, non-fossil material of biological origin, which may be used for heat production or electricity generation. They comprise wood and wood waste, biogas, municipal solid waste (MSW) and biofuels. MSW comprises biodegradable and non-biodegradable waste produced by different sectors. Non-biodegradable municipal and solid waste are not considered to be renewable, but current data availability does not allow the non-biodegradable content of waste to be identified separately, except for industry.

The indicator measures the relative consumption of energy from renewable sources in total energy consumption for a particular country. The share of renewable energy could increase even if the actual energy consumption from renewable sources falls. Similarly, the share could fall despite an increase in energy consumption from renewable sources. CO_2 emissions depend not on the share of renewables but on the total amount of energy consumed from fossil sources. Therefore, from an environmental point of view, attaining the 2010 target for the share of renewable energy does not necessarily imply that CO_2 emissions from energy consumption will fall.

31 Renewable electricity

Key policy question

Are we switching to renewable energy sources to meet our electricity consumption?

Key message

The share of renewable energy in EU electricity consumption grew slightly over the period 1990–2001 but decreased in 2002 due to lower production from hydropower. Significant further growth will be needed to meet the EU indicative target of a 21 % share by 2010.

Indicator assessment

Renewable energy makes an important contribution to meeting electricity consumption with a share of 12.7 % in 2002. However, this share has not increased significantly since 1990 (12.2 %) despite growth in absolute terms. Total renewable electricity production grew by 32.3 % over the period 1990 to 2002, but this was only slightly faster than the growth in gross electricity consumption. Compared with 2001, the share of renewables in gross electricity consumption in 2002 declined by 1.5 percentage points due to lower production from hydropower, as a result of lower rainfall. Substantial growth is needed to meet the EU-25 indicative target of 21 % by 2010 set in Directive 2001/77/EC.

There are significant differences in the share of renewables between the EU-25 Member States. These reflect differences in the policies chosen by each country to support the development of renewable energy and the availability of natural resources.

Among the EU-25 in 2002, Austria had the largest share of renewable electricity in gross electricity consumption including large hydropower, and the third largest share excluding large hydropower. Denmark and Finland have the largest shares of renewable electricity in gross electricity consumption when large hydropower is excluded. Finland's high share is due mainly to electricity production from biomass, while Denmark's renewable electricity is produced by wind power and, to a much lesser extent, biomass and waste. In both these countries, government policies have been in place to encourage the growth of these technologies. In absolute terms, Germany has the largest production of renewable electricity excluding large hydropower, mainly from wind and biomass.

While large hydropower dominates renewable electricity production in most Member States, it is unlikely to increase significantly in the future in the EU-25 as a whole, due to environmental concerns and a lack of suitable sites. Other renewable energy sources, such as wind, biomass, solar and small-scale hydropower will therefore have to grow substantially if the 2010 target is to be met.

Indicator definition

The share of renewable electricity is the ratio between the electricity produced from renewable energy sources and gross national electricity consumption calculated for a calendar year, expressed as a percentage. It measures the contribution of electricity produced from renewable energy sources to the national electricity consumption.

As well as being one of the EEA's core set indicators, it is also one of the *structural indicators* used to underpin the European Commission's analysis in its annual Spring report to the European Council. The methodologies are identical for both indicators.

Renewable energy sources are defined as renewable non-fossil energy sources: wind, solar, geothermal, wave, tidal, hydropower, biomass, landfill gas, sewage treatment plant gas and biogases.

Electricity produced from renewable energy sources comprises the electricity generation from hydroplants (excluding that produced as a result of pumping storage systems), wind, solar energy, geothermal energy and electricity from biomass/waste. Electricity from biomass/waste comprises electricity generated from wood/wood waste and the burning of other solid waste of a renewable nature (straw, black liquor), municipal solid

Renewable electricity

PART B

Figure 1 Share of renewable electricity in gross electricity consumption in the EU-25 in 2002

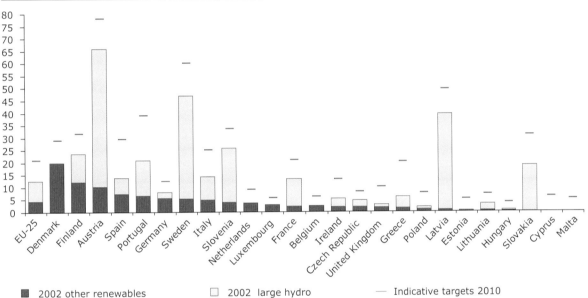

■ 2002 other renewables   □ 2002 large hydro   — Indicative targets 2010

Note: The Renewable Electricity Directive (2001/77/EC) defines renewable electricity as the share of electricity produced from renewable energy sources in gross electricity consumption. The latter includes imports and exports of electricity. The electricity generated from hydropower storage systems is included in gross electricity consumption but it is not input as a renewable source of energy. Large hydropower plants have a capacity of more than 10 MW.

Data source: Eurostat.

waste incineration, biogas (including landfill, sewage, farm gas) and liquid biofuels.

Gross national electricity consumption comprises total gross national electricity generation from all fuels (including autoproduction), plus electricity imports, minus exports.

Indicator rationale

The share of electricity consumption from renewable energy sources provides a broad indication of progress towards reducing the environmental impact of electricity consumption, although its overall impact has to be seen within the context of total electricity consumption, the total fuel mix, potential impacts on biodiversity and the extent to which pollution abatement equipment is fitted.

Renewable electricity is generally considered environmentally benign, with very low net emissions of CO_2 per unit of electricity produced, even allowing for emissions associated with the construction of the electricity production facilities. Emissions of other pollutants are also generally lower for renewable electricity production than for electricity produced from fossil fuels. The exception to this is the incineration of municipal and solid waste (MSW), which, because of the high costs of separation, usually involves the combustion of some mixed waste including materials contaminated with heavy metals. Emissions to atmosphere from MSW incineration are subject to stringent regulations including tight controls on emissions of cadmium, mercury, and other such substances.

The exploitation of renewable energy sources usually has some negative impact on landscapes, habitats

Core set of indicators | Energy

Table 1   **Share of renewable electricity in gross electricity consumption in the EU-25 (includes 2010 indicative targets)**

	Share of renewable electricity in gross electricity consumption (%) 1990–2002 and 2010 indicative targets									
	1990	1995	1996	1997	1998	1999	2000	2001	2002	2010 targets
EEA	17.1	17.5	16.6	17.2	17.7	17.5	18.2	17.8	17.0	–
EU-25	12.2	12.7	12.4	12.8	13.1	13.1	13.7	14.2	12.7	21.0
EU-15 pre-2004	13.4	13.7	13.4	13.8	14.1	14.0	14.7	15.2	13.5	22.1
EU-10	4.2	5.4	4.8	5.0	5.7	5.5	5.4	5.6	5.6	–
Austria	65.4	70.6	63.9	67.2	67.9	71.9	72.0	67.3	66.0	78.1
Belgium	1.1	1.2	1.1	1.0	1.1	1.4	1.5	1.6	2.3	6.0
Bulgaria	4.1	4.2	6.4	7.0	8.1	7.7	7.4	4.7	6.0	–
Cyprus	0.0	0.0	0.0	0.0	0.0	0.0	0.0	0.0	0.0	6.0
Czech Republic	2.3	3.9	3.5	3.5	3.2	3.8	3.6	4.0	4.6	8.0
Denmark	2.4	5.8	6.3	8.8	11.7	13.3	16.4	17.4	19.9	29.0
Estonia	0.0	0.0	0.1	0.1	0.2	0.2	0.2	0.2	0.5	5.1
Finland	24.4	27.6	25.5	25.3	27.4	26.3	28.5	25.7	23.7	31.5
France	14.6	17.7	15.2	14.8	14.3	16.4	15.0	16.4	13.4	21.0
Germany	4.3	4.7	4.7	4.3	4.9	5.5	6.8	6.2	8.1	12.5
Greece	5.0	8.4	10.0	8.6	7.9	10.0	7.7	5.1	6.0	20.1
Hungary	0.5	0.7	0.8	0.8	0.7	1.1	0.7	0.8	0.7	3.6
Iceland	99.9	99.8	99.9	99.9	99.9	99.9	99.9	100.0	99.9	–
Ireland	4.8	4.1	4.0	3.8	5.5	5.0	4.9	4.2	5.4	13.2
Italy	13.9	14.9	16.5	16.0	15.6	16.9	16.0	16.8	14.3	25.0
Latvia	43.9	47.1	29.3	46.7	68.2	45.5	47.7	46.1	39.3	49.3
Lithuania	2.5	3.3	2.8	2.6	3.6	3.8	3.4	3.0	3.2	7.0
Luxembourg	2.1	2.2	1.7	2.0	2.5	2.5	2.9	1.5	2.8	5.7
Malta	0.0	0.0	0.0	0.0	0.0	0.0	0.0	0.0	0.0	5.0
Netherlands	1.4	2.1	2.8	3.5	3.8	3.4	3.9	4.0	3.6	9.0
Norway	114.6	104.6	91.4	95.3	96.2	100.7	112.2	96.2	107.2	–
Poland	1.4	1.6	1.7	1.8	2.1	1.9	1.7	2.0	2.0	7.5
Portugal	34.5	27.5	44.3	38.3	36.1	20.5	29.4	34.2	20.8	39.0
Romania	23.0	28.0	25.3	30.5	35.0	36.7	28.8	28.4	30.8	–
Slovakia	6.4	17.9	14.9	14.5	15.5	16.3	16.9	17.4	18.6	31.0
Slovenia	25.8	29.5	33.0	26.9	29.2	31.6	31.4	30.4	25.9	33.6
Spain	17.2	14.3	23.5	19.7	19.0	12.8	15.7	21.2	13.8	29.4
Sweden	51.4	48.2	36.8	49.1	52.4	50.6	55.4	54.1	46.9	60.0
Turkey	40.9	41.9	43.0	38.1	37.3	29.5	24.3	19.1	25.6	–
United Kingdom	1.7	2.0	1.6	1.9	2.4	2.7	2.7	2.5	2.9	10.0

Note: Almost all electricity generated in Iceland and Norway comes from renewable energy sources. The renewable electricity share in Norway is above 100 % in some years because a part of the (renewable) electricity generated domestically is exported to other countries. The share of renewable electricity in Germany in 1990 refers to West Germany only. National indicative targets for the share of renewable electricity in 2010 are taken from Directive 2001/77/EC. Notes to their 2010 indicative targets are made by Italy, Luxemburg, Austria, Portugal, Finland and Sweden in the directive; Austria and Sweden note that reaching the target is dependent upon climatic factors affecting hydropower production, with Sweden considering 52 % a more realistic figure if long-range models on hydrologic and climatic conditions were applied. No energy data for Liechtenstein available from Eurostat.

Data source: Eurostat (Ref: www.eea.eu.int/coreset).

and ecosystems, although many of the impacts can be minimised through careful site selection. Large hydropower schemes in particular can have adverse impacts including flooding, disruption of ecosystems and hydrology, and socio-economic impacts if resettlement is required. Some solar photovoltaic schemes require relatively large quantities of heavy metals in their construction, and geothermal energy can release pollutant gases carried by hot fluids if not properly controlled. Wind turbines can have visual and noise impacts on the areas in which they are sited. Some types of biomass crops have considerable land, water and agricultural input requirements such as fertilisers and pesticides.

Policy context

The original EU Directive on the promotion of electricity from renewable energy sources in the internal electricity market (2001/77/EC) sets an indicative target of 22.1 % of gross EU-15 electricity consumption from renewable sources by 2010. It requires Member States to set and meet national indicative targets consistent with the directive and national Kyoto Protocol commitments. For the EU-10 Member States, national indicative targets are included in the accession treaty: the 22.1 % target set initially for the EU-15 for 2010 becomes 21 % for the EU-25.

The power sector is responsible for a significant share of European greenhouse gas emissions and increased market penetration of renewable electricity would therefore help to reach the EU commitment under the Kyoto Protocol. The overall Kyoto target for the pre-2004 EU-15 Member States requires an 8 % reduction in emissions of greenhouse gases by 2008–2012 from 1990 levels, while most EU-10 Member States have individual targets under the Kyoto Protocol.

Indicator uncertainty

Data have traditionally been compiled by Eurostat through the annual joint questionnaires, shared by Eurostat and the International Energy Agency, following a well-established and harmonised methodology. Methodological information on the annual joint questionnaires and data compilation can be found on Eurostat's website for metadata on energy statistics.

The Renewable Electricity Directive (2001/77/EC) defines the share of renewable electricity as the percentage of electricity produced from renewable energy sources in gross electricity consumption. The numerator includes all electricity generated from renewable sources, most of which is for domestic use. The denominator contains all electricity consumed in a country, thus including imports and excluding exports of electricity. Therefore, the share of renewable electricity can be higher than 100 % in a country if all electricity is produced from renewable sources and some of the over-generated renewable electricity is exported to a neighboring country.

Biomass and waste, as defined by Eurostat, cover organic, non-fossil material of biological origin, which may be used for heat production and electricity generation. They comprise wood and wood waste, biogas, municipal solid waste (MSW) and biofuels. MSW comprises biodegradable and non-biodegradable waste produced by different sectors. Non-biodegradable municipal and solid waste are not considered to be renewable, but current data availability does not allow the non-biodegradable content of waste to be identified separately, except for industry.

The electricity produced as a result from hydropower storage systems (i.e. that needed electricity to be filled) is not classified as a renewable source of energy in terms of electricity production, but is part of the gross electricity consumption in a country.

The share of renewable electricity could increase even if the actual electricity produced from renewable sources falls. Similarly, the share could fall despite an increase in electricity generation from renewable sources. Therefore, from an environmental point of view, attaining the 2010 target for the share of renewable electricity does not necessarily imply that carbon dioxide emissions from electricity generation will fall.

32 Status of marine fish stocks

Key policy question

Is the use of commercial fish stocks sustainable?

Key message

Many commercial fish stocks in European waters remain non-assessed. Of the assessed commercial stocks in the north-east Atlantic, 22 to 53 % are outside safe biological limits (SBL). Of the assessed stocks in the Baltic Sea, the West Ireland Sea and the Irish Sea, 22, 29 and 53 %, respectively, remain outside SBL. In the Mediterranean, the percentage of stocks outside SBL range from 10 to 20 %.

Indicator assessment

Many commercial fish stocks in European waters remain non-assessed. In the north-east Atlantic, the percentage of non-assessed stocks of economic importance range from a minimum of 20 % (North Sea) to a maximum of 71 % (West Ireland) which is an increase from 13 % and 59 % respectively in the previous assessment in 2002. The Baltic Sea also shows a high percentage of non-assessed stocks at 67 % compared with the previous 56 %. In the Mediterranean region, the percentage is much higher with an average of 80 %, and a range from 65 % in the Aegean Sea to 83 % in the Adriatic (the previous highest value was 90 % in the South Alboran Sea).

Of the assessed commercial stocks in the north-east Atlantic, 22 to 53 % are outside safe biological limits (SBL). This is an improvement compared with the last record of 33-60 %. Of the assessed stocks in the Baltic and West Ireland Seas, 22 and 29 %, respectively, are over-fished (33 % in the past) while 53 % of stocks in the Irish Sea remain outside SBL (past record held by West of Scotland at 60 %). In the Mediterranean the percentage of stocks outside SBL range from 10 to 20 %, with the Aegean and the Cretan Sea being in the worst condition.

Examination of 'safe' stocks in the north-east Atlantic shows a slight decline ranging between 0 and 33 %; these values correspond to the West Ireland and North Sea, respectively. The last assessment of 2002 showed a range of 5 to 33 % for the Celtic Sea/Western Channel and the Arctic, respectively. In the Mediterranean, the range extends from 0 % (Cretan Sea) to 11 % (Sardinia) compared with a minimum of 0 % (S. Alboran and Cretan Seas) and a maximum of 15 % (Aegean Sea) in 2002.

When examining the European stocks more closely, the following conclusions can be drawn:

- The recovery of herring stocks appears to continue.

- Almost all round fish stocks have declined and are currently not sustainable.

- Pelagic and industrial species remain in better condition but still need to be subject to reduced fishing rates.

- In the Mediterranean region, only two demersal and two small pelagic stocks are monitored by the General Fisheries Commission for the Mediterranean (GFCM), with a limited spatial coverage. Demersal stocks remain outside safe biological limits. Many assessments that cover wider areas are based on preliminary results. Small pelagic stocks in the same area exhibit large-scale fluctuations but are not fully exploited anywhere, except for anchovy and pilchard in the Southern Alboran and Cretan Seas.

- According to the latest assessment by the International Commission for the Conservation of Atlantic Tunas (ICCAT) a strong recruitment of swordfish over recent years has rendered the exploitation of the stock sustainable. Concern still remains about the over-exploitation of bluefin tuna. Uncertainties of stock assessment and lack of documented reporting (including EU Member States) still hinder management of these highly migratory species. Bluefin tuna catches continue to exceed the sustainable rate and, despite ICCAT recommendations for both the Atlantic and the Mediterranean, no measures (despite reductions in total allowable catches) have been enforced.

Status of marine fish stocks

Map 1 Status of commercial fish stocks in European Seas, 2003–2004

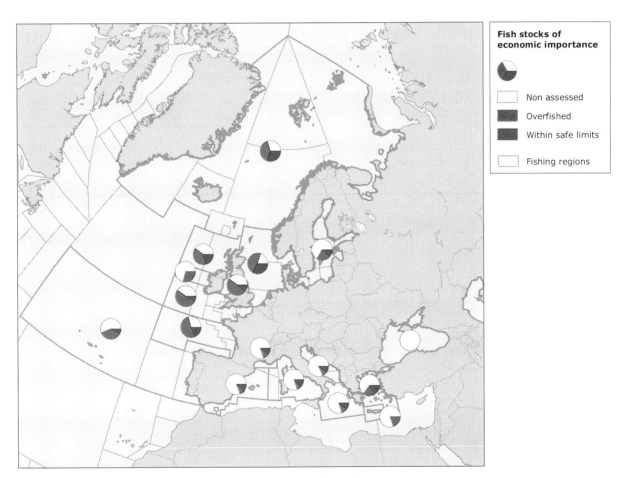

Note: Data source: GFCM, ICCAT, ICES (Ref: www.eea.eu.int/coreset).

Indicator definition

The indicator tracks the ratio of the number of over-fished stocks to the total number of commercial stocks per fishing area in European seas. The indicator also contains information on: 1) number of commercial, exploited and over-fished stocks by sea area and 2) the state of commercial stocks (over-fished stocks per area), safe stocks, stocks for which an assessment has not been carried out, and stocks of non-commercial importance in the particular area.

Landings and spawning stock biomass are given in thousand tonnes, recruitment in million tonnes; fishing mortality is expressed as the proportion of a stock that is removed by fishing activities in a year.

Indicator rationale

EU policies, and in particular the common fisheries policy (CFP), aim for sustainable fishing over a long period through appropriate management of fisheries

Core set of indicators | Fisheries

Figure 1   State of commercial fish stocks in the Mediterranean Sea up to 2004

	1	2	3	4	5	6	7	8	9	10	11	12	13	14	15	16	17	18	19	20	21	22	23	24	25	26	27	28	29	30
Anchovy	4		2			4	4	1	1	1	1	1	1	1	1	1	2	4	1	1	1	1	1							
Black Sea whiting																														
Blue whiting																														
Bogue																						1								
Breams			1																			1								
Flat fish																														
Greater forkbread																														
Gurnads																														
Grey mullet																														
Hake	4					n	4	3	1	3	1	1	1	1	1	1	1	1	1	1		1	1							
Horse mackerel			n																			1								
Mackerel																														
Megrim																														
Pilchard	4		n			4	4	1	1	1	1	1	1	1	1	1	4	1	1	1		1	1							
Poor cod																														
Red mullet	4		n		n	4	1	1	3	3	4	1	1	1	1	1	1	1	1	1		1	1							
Sea bass																														
Sardinella																														
Sole																														
Sprat																														
Bluefin tuna																														
Swordfish	4	4	4	4	4	4	4	4	4	4	4	4	4	4	4	4	4	4	4	4	4	4	4	4	4	4	4	4		

Note: 1. Northern Alboran, 2. Alboran Island Sea, 3. Southern Alboran Sea, 4. Algeria, 5. Balearic Island, 6. Northern Spain, 7. Gulf of Lions, 8. Corsica Island, 9. Ligurian and North Tyrrhenian Sea, 10. South and Central Tyrrhenian Sea, 11. Sardinia, 12. Northern Tunisia, 13. Gulf of Hammamet, 14. Gulf of Gabes, 15. Malta Island, 16. South of Sicily, 19. Western Ionian Sea, 20. Eastern Ionian Sea, 21. Libya, 17. Northern Adriatic, 18. Southern Adriatic Sea, 22. Aegean Sea, 23. Crete Island, 24. South of Turkey, 25. Cyprus Island, 26. Egypt, 27. Levant, 28. Marmara Sea, 29. Black Sea, 30. Azov Sea.
Colour coding:
Blue = within safe biological limits;
Red = outside safe biological limits;
Grey = no assessment;
1, 2, 3, 4 in the cells refer to the year of assessment, i.e. 2001 (in 2002 report), 2002, 2003 and 2004 respectively;
n = new assessment.
Data source: GFCM, ICCAT (Ref: www.eea.eu.int/coreset).

within a healthy ecosystem, while offering stable economic and social conditions for all involved in the activity. An indication of the sustainability of fisheries in a particular area is the ratio of the number of over-fished stocks (those that are outside safe biological limits) to the total number of commercial stocks (for which an assessment of status has been carried out). A high value of this ratio identifies areas under heavy pressure from fishing.

In general, a stock becomes over-fished when mortality from fishing and other causes exceeds recruitment and growth. A fairly reliable picture of stock development can be derived by comparing trends over time in recruitment, spawning stock biomass, landings and fish mortality. Hence not only the quantity of fish taken from the sea is important, but also their species and size, and the techniques used to catch them.

Policy context

The sustainable exploitation of fish stocks is regulated through the EU Common Fisheries Policy (OJ C 158 27.06.1980). Regulatory arrangements, identifying harvesting levels based on the CFP, the precautionary principle and multiannual fisheries plans, were set through the Cardiff European Council (COM (2000) 803). Total Allowable Catches (TAC) and quotas for the stocks in the north-east Atlantic and the Baltic Sea are set annually by the Fisheries Council. In the Mediterranean Sea, where no TAC have been set except for the highly migratory tuna and swordfish, fisheries management is achieved by means of closed areas and seasons to keep fishing effort under control and make exploitation patterns more rational. The General Fisheries Council for the Mediterranean (GFCM) attempts to harmonise the process.

The latest action plan on fisheries management as part of the CFP reform was presented to the Fisheries Council in October 2002, and Council Regulation (EC) No 2371/2002 of 20 December 2002 on the conservation and sustainable exploitation of fisheries resources under the common fisheries policy is now in force.

A new set of regulations has since been adopted on specific issues.

Indicator uncertainty

All international fisheries organisations use the same principles to determine the state of the stocks, and ICES has fine-tuned the methodology used. However, decisions are based on safety margins usually set at 30 % above safe limits which in turn bears a degree of uncertainty since estimates of fishing mortality (F) and spawning stock biomass (SSB) are themselves uncertain; the decision on the reference points is then a task for managers, not scientists.

Species and spatial coverage for the Mediterranean is limited. No reference points have been defined for the Mediterranean stocks. The detailed stock assessments for the north-east Atlantic and Baltic are obtained through the International Council for the Exploration of the Seas (ICES). In the Mediterranean, stock assessments are carried out by the General Fisheries Council for the Mediterranean (GFCM) and, in the absence of complete or independent information on fishing intensity or fishing mortality, are based mainly on landings. Stock assessment is thus based mainly on analysis of landing trends, biomass surveys, and analysis of commercial catch per unit effort (CPUE) data.

Data sets are fragmented both temporally and spatially. Monitoring activities are based on scientific surveys rather than commercial catches, resulting in low values of SSB estimates and thus biased exploitation patterns. In the Mediterranean, fisheries management is considered to be at an early stage compared with the north-east Atlantic. Catch and effort statistics are not considered to be fully reliable and much effort is directed at estimation of corrective factors.

Different approaches are being used in the Mediterranean and the north-east Atlantic to determine whether a stock is outside safe biological limits.

33 Aquaculture production

Key policy question

Is the current level of aquaculture sustainable?

Key message

European aquaculture production has continued to increase rapidly during the past 10 years due to expansion in the marine sector in the EU and EFTA countries. This represents a rise in pressure on adjacent water bodies and associated ecosystems, resulting mainly from nutrient release from aquaculture facilities. The precise level of local impact will vary according to production scale and techniques as well as the hydrodynamics and chemical characteristics of the region.

Indicator assessment

A significant increase in total European aquaculture production has been observed in the past 10 years. However it has not been uniform across countries or production systems. Only the mariculture sector has experienced a significant increase, while brackish water production has increased at a much slower rate and the levels of freshwater production have declined. Europe's fish farms fall into two distinct groups: the fish farms in western Europe grow high-value species such as salmon and rainbow trout, frequently for export, whereas lower-value species such as carp are cultivated in central and eastern Europe, mainly for local consumption.

The biggest European aquaculture producers are found in the EU and EFTA region. Norway has the highest production with more than 500 000 tonnes in 2001, followed by Spain, France, Italy and the United Kingdom. These five countries account for 75.5 % of all aquaculture production in 34 European countries. Turkey's production of 67 000 tonnes represents the highest production in the EU accession countries and Balkan region. The country ranking in 2001 in terms of production was very similar to that in 2000.

Norway is the dominant aquaculture producer with about 90 % of its production being Atlantic salmon. It is noteworthy that in 2001, farming of this single species in Norway exceeded the combined total of all production species from all EU accession countries and Balkan countries. Spain is the next biggest producer with production dominated by blue mussel, followed by France, with production dominated by the Pacific cupped oyster (*Crassostrea gigas*). Turkish production consists mainly of trout, sea bream and sea bass.

The major part of the increase in aquaculture production has been in marine salmon culture in northwest Europe, and to a lesser extent trout culture (throughout western Europe and Turkey), seabass and seabream cage culture (mainly Greece and Turkey), and mussel and clam cultivation (throughout western Europe), which, however, exhibits a downward trend since 1999. In contrast, inland aquaculture of carp (mainly common and silver carp) has declined significantly throughout eastern and central Europe (EU accession countries and Balkan countries) due partly to political and economic changes in eastern Europe. As in the case of production per country, no significant changes have been observed in production by major species since the last assessment (2000).

Different types of aquaculture generate very different pressures on the environment, the main ones being discharges of nutrients, antibiotics and fungicides. The main environmental pressures are associated with intensive finfish production, mainly salmonids in marine, brackish and freshwaters, and seabass and seabream in the marine environment — sectors which have experienced the highest growth rate in recent years. The pressures associated with the cultivation of bivalve molluscs are generally considered to be less severe than those from intensive finfish cultivation. Pond aquaculture of carp in inland waters usually requires less intensive feeding, and in most cases a greater proportion of the nutrients discharged are assimilated locally. Chemicals, particularly formalin and malachite green, are used in freshwater farms to control fungal and bacterial diseases. In marine farms, antibiotics are used for disease control but the amounts used have been reduced drastically in recent years following the introduction of vaccines. In general, significant improvements in the efficiency of feed and nutrient utilisation as well as environmental management have served to partially mitigate the associated increase in environmental pressure.

The environmental pressures exerted by aquaculture are not uniform. The level of local impact will vary according to production scale and techniques as well as the hydrodynamics and chemical characteristics of the region.

Of the EU countries, Spain, France and the Netherlands, and of the accession countries, Turkey, have the greatest marine aquaculture production in relation to coastline length. Aquaculture production intensity as measured per unit coastline length has reached an average of around 8 tonnes per km of coastline in EU and EFTA countries compared with 2 tonnes per km in the EU accession countries and Balkan region. The pressure is likely to continue to increase as the production of new species such as cod, halibut and turbot becomes more reliable.

Marine finfish culture (mainly Atlantic salmon) is making a significant contribution to nutrient loads in coastal waters, particularly in the case of countries with relatively small total nutrient discharges to coastal waters. For example in Norway (Norwegian and North Sea coasts), phosphorus discharge from mariculture appear to exceed the total from other sources. In general, the pressure from nutrients from the intensive cultivation of marine and brackish water is becoming significant in the context of total nutrient loadings to coastal environments. However the published data on total nutrient loadings to coastal waters remains poor in quality and inconsistent in coverage; the conclusions should therefore be treated with caution.

Indicator definition

The indicator quantifies the development of European aquaculture production by major sea area and country as well as the contribution of aquaculture discharges

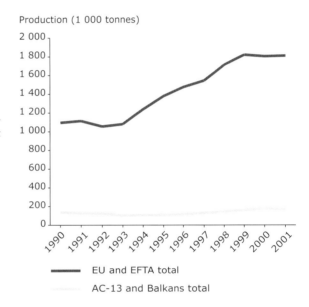

Figure 1 Annual aquaculture production by major area (EU and EFTA, and EU accession countries and Balkans), 1990–2001

Note: Aquaculture production includes all environments, i.e. marine, brackish and freshwater.

EU and EFTA: Austria, Belgium, Denmark, Finland, France, Germany, Greece, Ireland, Italy, the Netherlands, Portugal, Spain, Sweden, the United Kingdom, Iceland, Norway and Switzerland; EU accession countries and Balkans: Albania, Bulgaria, Czech Republic, Croatia, Estonia, FYR Macedonia, Hungary, Latvia, Lithuania, Poland, Romania, Yugoslavia, Slovak Republic, Slovenia, Cyprus, Malta and Turkey.

Luxembourg, Liechtenstein and Bosnia-Herzegovina are not included due to either no aquaculture production or lack of data.

Data source: UN Food and Agriculture Organization (FAO) Fishstat Plus (Ref: www.eea.eu.int/coreset).

Figure 2 Annual production of major commercial aquaculture species groups, 1990–2001

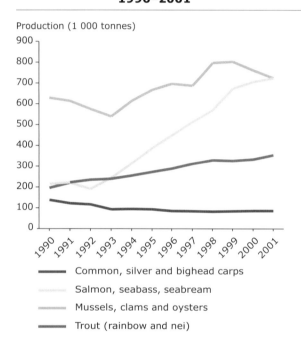

Common, silver and bighead carps
Salmon, seabass, seabream
Mussels, clams and oysters
Trout (rainbow and nei)

Note: Includes all countries and production environments for which data are available.

nei = not elsewhere indicated; trout (rainbow and nei) includes all species of trout.

Data source: FAO Fishstat Plus (Ref: www.eea.eu.int/coreset).

of the pressures of aquaculture on the marine environment. It is a simple and readily-available indicator but, as a stand-alone indicator, its meaning and relevance are limited because of widely varying production practices and local conditions. It needs to be integrated with other indicators related to production practices (such as total nutrient production or total chemical discharge) to generate a more specific indicator of pressure. Coupled with information on the assimilative capacity of different habitats, such an indicator would allow estimation of impact and ultimately the proportion of the carrying capacity of the surrounding environment used and the limits to expansion.

Policy context

Until recently there was no general policy for European aquaculture, although the Environmental Impact Assessment (EIA) Directive (85/337/EEC and amendment 97/11/EEC) requires specific farms to undergo EIA and the water framework directive requires all farms to meet environmental objectives for good ecological and chemical status of surface waters by 2015. There are few national policies specifically addressing the diffuse and cumulative impacts of the sector as a whole on aquatic systems, or the need to limit total production in line with the assimilative capacity of the environment. However, limits on feed inputs in some countries, such as Finland, effectively limit production.

The new reformed common fisheries policy (CFP) aims to improve the management of the sector. In September 2002, the Commission presented a communication on 'a strategy for the sustainable development of European aquaculture' to the Council and to the European Parliament. The main aim of the strategy is the maintenance of competitiveness, productivity and sustainability of the European aquaculture sector. The strategy has three main objectives: 1) to create secure employment; 2) to provide safe and good quality fisheries products and promote animal health and welfare standards; and 3) to ensure an environmentally sound industry.

of nutrients relative to the total discharges of nutrients into coastal zones.

Production is measured in thousand tonnes, while marine aquaculture production relative to coastline length is given in tonnes/km.

Indicator rationale

The indicator tracks aquaculture production and nutrient discharges and thereby provides a measure

Aquaculture production

Map 1 Marine aquaculture production relative to coastline length

Aquaculture production relative to coastline length

Tonnes/km
- < 5
- 5–10
- 10–20
- 20–30
- 30–40
- No data
- Non EU

Note: Only marine and brackish waters production.

Average production density values for countries with a coastline and with coastline data available. Based on latest year for which there are data, i.e. 2001 for all countries except Bulgaria (2000), Estonia (1995) and Poland (1993).

Data source: FAO Fishstat Plus and World Resources Institute (Ref: www.eea.eu.int/coreset).

Indicator uncertainty

The weakness of the indicator relates to the validity of the relationship between production and pressure. Production acts as a useful, coarse indicator of pressure but variations in culture species, production systems and management approaches mean that the relationship between production and pressure is non-uniform.

34 Fishing fleet capacity

Key policy question

Is the size and capacity of the European fishing fleet being reduced?

Key message

The size of the EU fishing fleet is following a downward trend, with reductions of 19 % in power and 11 % in tonnage in the period 1989–2003, and 15 % in numbers in the period 1989–2002. Similarly, the combined fleet of Estonia, Cyprus, Lithuania, Latvia, Malta, Poland and Slovenia decreased its tonnage by 50 % over the period 1992–1995. However, the EFTA fleet increased in terms of power (by 12 %; 1997–2002) and tonnage (by 34 %; 1989–2003) despite a drop in numbers by 40 % (1989–2002).

Indicator assessment

Power and tonnage are the main factors that determine the capacity of a fleet and thus approximate to the pressure on the fish stocks. Excess power is considered to be one of the major factors that lead to over-fishing.

Currently, the total power of the fishing fleet amounts to 7 122 145 kW in the EU-15 (2003) and 2 503 580 kW in EFTA (2002). Data for Estonia, Cyprus, Lithuania, Latvia, Malta, Poland, Slovenia, Bulgaria and Romania are not available. Over the past 15 years the EU fleet capacity in terms of power has been gradually decreasing, but the power of the EFTA fleet increased at a considerable rate of almost 13 % over the period 1997–2002. Norway, Italy, Spain, France and the United Kingdom retain the largest power in their fleets, which accounted for almost 70 % of the total fleet in 2003.

In 2003, the fishing fleet tonnage (GRT) consisted of 1 922 912 tonnes in the EU-15 and 579 097 tonnes in the EFTA countries. The last recorded census for Estonia, Cyprus, Lithuania, Latvia, Malta, Poland and Slovenia, in 1995, reported 543 631 tonnes. In the period 1989–2003, the EU fleet was gradually reduced in tonnage by approximately 10 %; at the same time the EFTA fleet experienced an almost 30 % increase (Figure 3). The fleets of Estonia, Cyprus, Lithuania, Latvia, Malta, Poland and Slovenia faced a dramatic decrease of 50 %, and those of Bulgaria and Romania 70 %, due to the restructuring of the economies of the new EEA member countries; there are no data available on fleet tonnage in these countries beyond 1995. Currently, Spain, Norway,

Figure 1   Changes in European fishing fleet capacity: 1989–2003

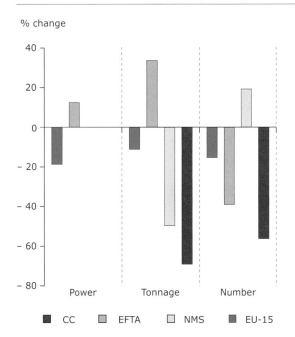

Note: Power changes refer to 1989–2003 for the EU-15 and 1997–2002 for EFTA.
Tonnage changes refer to 1989–2003 for the EU and EFTA; 1992–1995 for NMS and CC countries (see legend).
Number changes refer to 1989–2002 for the EU and EFTA; 1992–2001 for NMS; and 1992–1995 for CC countries.

Legend: Countries have been grouped into the following categories:
EU-15 (Austria, Belgium, Denmark, Germany, Greece, Spain, France, Ireland, Italy, Luxembourg, the Netherlands, Portugal, Finland, Sweden, the United Kingdom);
EFTA (Iceland and Norway);
New Member States (Estonia, Cyprus, Lithuania, Latvia, Malta, Poland and Slovenia);
Candidate countries (Bulgaria and Romania).

Data source: DG Fisheries, Eurostat, UN Food and Agriculture Organization (FAO).

Figure 2 European fishing fleet capacity: number of vessels

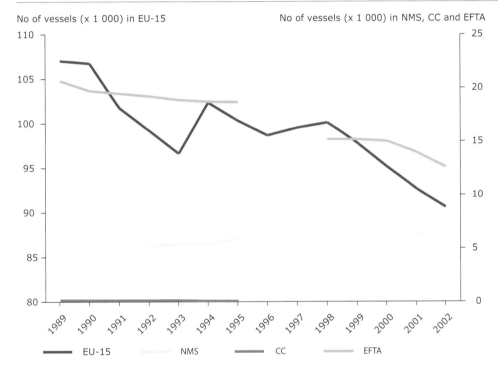

Note: Data availability: Number of vessels 1989–2002 for EU-15; 1989–1992 and 1998–2002 for EFTA; 1989–1995 and 2001 for NMS (see legend); 1992–1995 and 2001 for Bulgaria and Romania.

Legend: Countries have been grouped into the categories as in Figure 1.

Data source: DG Fisheries, Eurostat, FAO (Ref: www.eea.eu.int/coreset).

the United Kingdom, France, Italy and the Netherlands retain the fleets of largest tonnage, accounting for almost 70 % of the total fleet in 2003.

In 2002 there were 90 595 fishing vessels in the EU-15 and 12 589 in the EFTA countries. According to DG Fisheries, the fleets of Estonia, Cyprus, Lithuania, Latvia, Malta, Poland and Slovenia amounted to approximately 6 200 vessels in 2001. Both EU and EFTA fleets have been gradually reduced in size over the past 15 years, whereas the fleet of Estonia, Cyprus, Lithuania, Latvia, Malta, Poland and Slovenia has increased gradually over the past 10 years (Figure 2). It is noteworthy that the peak value observed in 1994 was due to the introduction of new countries, namely Finland and Sweden, into the registry. Greece, Italy, Spain, Norway and Portugal retain the largest number of vessels, accounting for almost 70 % of the total fleet in 2003. In the case of Greece and Portugal, a comparison of the number of vessels with the fleet capacity indicates that these two fleets consist mainly of small vessels.

Despite the overall drop in size and capacity (power and tonnage) experienced by the EU fleet in the past 15 years, no visible improvement in the condition of the fish stocks has been observed. According to DG Fisheries *One of the most fundamental and enduring problems of the common fisheries policy has been the chronic overcapacity of the EU fleet. Conservation measures have persistently been undermined by fishing activities at levels well beyond the level of pressure that the available fish stocks*

Core set of indicators | Fisheries

Figure 3 European fishing fleet capacity: tonnage

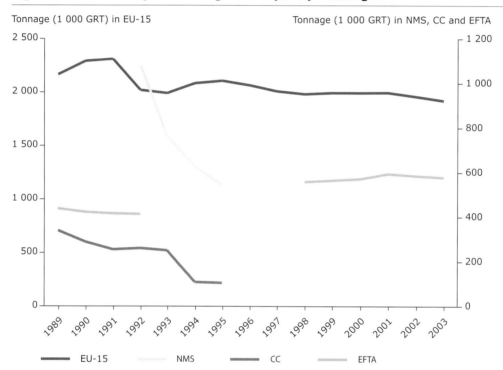

Note: Data availability: 1989–2003 for EU-15; 1989–1992 and 1998–2003 for EFTA; 1992–1995 for NMS (see legend); 1989–1995 for CC countries.

Legend: Countries have been grouped into the categories as in Figure 1.

Data source: DG Fisheries, Eurostat, FAO (Ref: www.eea.eu.int/coreset).

could safely withstand. As new technology makes fishing vessels ever more efficient, the capacity of the fleet should be reduced to maintain a balance between fishing capacity and the quantities of fish that can safely be taken out of the sea by fishing. The multiannual guidance plans (MAGPs) have proved inadequate and have been replaced by a simpler scheme in the reformed common fisheries policy (January 2003).

Indicator definition

The indicator is a measure of the size and capacity of the fishing fleet, which in turn is assumed to approximate to the pressure on marine fish resources and the environment.

The size of the European fishing fleet is presented as the number of vessels, the capacity as the total engine power in kW, and the total tonnage in tonnes.

Indicator rationale

Fishing capacity, defined in terms of tonnage and engine power and sometimes number of vessels, is one of the key factors that determine the fishing mortality caused by the fleet. In simple terms, excess capacity leads to over-fishing and increased environmental pressure which undermines the principle of sustainable use. As new technology makes fishing vessels ever more efficient, the size and capacity of the fleet should be reduced to maintain a balance between fishing pressure and the quantities of fish available. Four multiannual guidance plans (MAGPs) were established to achieve sustainability by setting, for each coastal Member State, maximum levels of fishing capacity by types of vessel. However, MAGPs failed to meet expectations and proved cumbersome to manage. MAGP IV, which ended in December 2002, has therefore been replaced by a simpler scheme. Under the new scheme the fleet capacity will be reduced gradually, i.e. the introduction of new capacity into the fleet without public aid must be compensated by the withdrawal of at least an equivalent capacity, also without public aid.

Policy context

EU policies aim to achieve sustainable fishing over a long period within a sound ecosystem through appropriate management of fisheries, while offering stable economic and social conditions for all those involved in the fishing activity.

Sustainable exploitation of the fish stocks is ensured through the EU Common Fishery Policy (OJ C 158 27.06.1980).

Within the four MAGPs, an effort has been made to achieve a sustainable balance between the fleet and available resources. Commission Regulation (EC) No 2091/98 of 30 September 1998 dealt with the segmentation of the Community fishing fleet and fishing effort in relation to the multiannual guidance programmes, and Council regulation (EC) 2792/1999 laid down the detailed rules and arrangements regarding Community structural assistance in the fisheries sector, mainly through the structural funds and the financial instrument for fisheries such as the financial instrument for fisheries guidance (FIFG).

According to the reformed common fisheries policy, MAGPs failed to meet expectations and proved cumbersome to manage. Subsidies for construction/modernisation and running costs have undermined the efforts made, also with public aid, to eliminate overcapacity by helping the introduction of new vessels into the fleet. MAGP IV, which ended in December 2002, has been replaced by a simpler scheme under the reform of the CFP (Council Regulation (EC) No 2371/2002 on the Conservation and Sustainable Exploitation of Fisheries Resources under the common fisheries policy).

Targets

No specific target exists. However, the aim under the reformed CFP is to reduce the size and capacity of the fishing fleet to achieve sustainable fishing.

Indicator uncertainty

Data sets are fragmented both temporally and spatially. Data for Estonia, Cyprus, Lithuania, Latvia, Malta, Poland, Slovenia, Bulgaria and Romania are only covered by FAO, apart from a not very accurate assessment of the number of vessels reported by DG Fisheries for 2001. Data for EFTA are covered by Eurostat. Data for the EU-15 come from Eurostat and DG Fisheries. Data on power for Estonia, Cyprus, Lithuania, Latvia, Malta, Poland, Slovenia, Bulgaria and Romania are lacking, and in the case of tonnage and number of vessels they exist for the majority of these countries but only for a limited period, 1992–1995.

Restructuring the fleet and reducing its capacity do not necessarily lead to reduction in fishing pressure as advances in technology and design allow new vessels to exert more fishing pressure than older vessels of equivalent tonnage and power.

35 Passenger transport demand

Key policy question

Is passenger transport demand being decoupled from economic growth?

Key message

Growth in the volume of passenger transport has nearly paralleled that in GDP. Transport growth was marginally lower than GDP growth between 1997 and 2001, but once again exceeded it in 2002. Decoupling between transport demand and GDP over the period has been less than 0.5 % per year compared with transport growth of 2.1 % per year, and decoupling has not been achieved each year.

Indicator assessment

Over the past decade, passenger transport demand has grown steadily in the EEA countries as a whole, thereby making it increasingly difficult to stabilise or reduce the environmental impacts of transport. Most countries saw growth every year, but there are a few exceptions, notably Germany, where demand has remained almost stable since 1999. Transport demand per capita has also grown, and by 2002 had reached more than 10 000 km in the countries for which data are available.

The main underlying factor is the growth in incomes coupled with a tendency to spend more or less the same share of disposable income on transport. Additional income therefore means additional travel budget, which allows more frequent, faster, farther and more luxurious travelling. The average daily distance travelled by EU-15 citizens increased from 32 km in 1991 to 37 km in 1999, the fastest-growing modes of transport being private car and aviation.

Overall growth in passenger transport demand has been very similar to that of GDP. Transport growth was marginally lower than GDP growth between 1997 and 2001, but once again exceeded it in 2002. From 1997, decoupling between transport demand and GDP growth was less than 0.5 % per year compared with transport growth of 2.1 % per year.

One explanatory factor for the slight decoupling is a greater instability in fuel prices from 1997 onwards, which may have reduced the tendency to invest in additional cars. The 'fuel price protests' in 2000, albeit primarily by hauliers, illustrated the reaction of road users to higher prices. This is also consistent with the higher growth in 2002, because fuel prices by then had once again come down. But increasing congestion in some cities has also been put forward as an explanatory factor.

EU-wide data on travel purposes are not available. However, based on national mobility surveys, 40 % of passenger transport demand in the 1990s was for leisure. Tourism is an important travel motive, and most of the trips attributed to tourism are long-distance ones. The importance of tourism for air traffic is highlighted by the presence of the tourist destinations Palma de Mallorca, Tenerife and Malaga in the top 20 airports that handle most passengers.

The stated objective of the common transport policy of maintaining the 1998 modal shares is not currently being met. The share of car transport is stable at around 72 % while air transport is growing and bus plus rail is declining steadily. In absolute numbers, bus and rail are roughly maintaining their respective markets, while all growth is in road and in particular air transport.

Increasing wealth among citizens give more people the option to buy a car and use the added flexibility that it provides. Only in dense urban centres and for longer distances can public transport compete in terms of travel time.

Aviation saw a small drop in market share following the 11 September 2001 terrorist attacks on the World Trade Centre and the Pentagon, the subsequent wars and the SARS epidemic. This led to increased consolidation of the airline industry but also provided opportunities for low-cost airlines, which are rapidly gaining market share. Thus the relative cost of air travel has dropped, further fuelling the recent growth in air travel.

Passenger transport demand PART B

Indicator definition

To measure decoupling of passenger demand from economic growth, the volume of passenger transport relative to GDP (i.e. the intensity) is calculated. Separate trends for the two components of intensity are shown for the EU-25. Relative decoupling occurs when passenger transport demand grows at a rate below that of GDP. Absolute decoupling occurs when passenger transport demand falls while GDP rises or remains constant.

The unit is the passenger-kilometre (passenger-km), which represents one passenger travelling a distance of one kilometre. It is based on passenger transport by car, bus, coach and train. Estimates of passenger transport by air are, where available (EU-15), included in total inland passenger transport. All data are based on movements within the national territory, regardless of the nationality of the vehicle.

Passenger transport demand and real GDP are shown as an index (1995 = 100). The ratio of the former to the latter is indexed on the previous year (i.e. annual decoupling/intensity changes) in order to be able to observe changes in the annual intensity of passenger transport demand relative to economic growth.

The indicator can also be presented as the share of transport by passenger car in total inland transport (i.e. modal split share for passenger transport). Eurostat is currently working on methods for the calculation and territorial attribution of performance data for air transport which, if included, would have a significant impact on the passenger modal shares. When Eurostat's results become available, the core set indicator will be reviewed and the modal split shares shown.

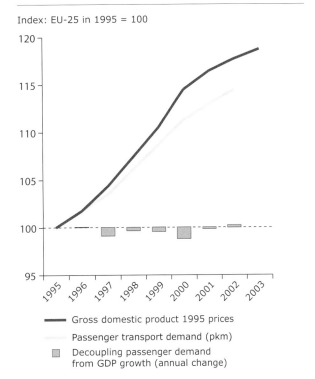

Figure 1 Trend in passenger transport demand and GDP

Index: EU-25 in 1995 = 100

— Gross domestic product 1995 prices
— Passenger transport demand (pkm)
▪ Decoupling passenger demand from GDP growth (annual change)

Note: If the decoupling indicator (vertical bars) is above 100 transport demand is outpacing GDP growth (i.e. positive bar = no decoupling) whereas a value below 100 is means transport demand growing less rapidly than GDP (i.e. negative bar = decoupling). The EU-25 index for passenger transport demand does not include Malta, Cyprus, Estonia, Latvia, and Lithuania because of lack of a complete time series in these countries. Decoupling for passenger demand also excludes the GDP of these 5 countries, together representing about 0.3–0.4 % of EU-25 GDP. See also indicator definition.

Data source: Eurostat and DG Energy and Transport, European Commission (Ref: www.eea.eu.int/coreset).

Indicator rationale

Transport is one of the main sources of greenhouse gases and also gives rise to significant air pollution, which can seriously damage human health and ecosystems. The indicator helps to understand developments in the passenger transport sector (transport's 'magnitude'), which in turn explains observed trends in the impact of transport on the environment.

The relevance of the modal split policy to the environmental impact of passenger transport arises from differences in the environmental performance

The European environment | State and outlook 2005

Core set of indicators | Transport

Table 1 Trend in the annual intensity of passenger transport demand

Trends in passenger transport demand (passenger/km for car, train and buses/coaches); Index 1995 = 100

	1995	1996	1997	1998	1999	2000	2001	2002
EEA	100	102	103	106	108	110	112	113
EU-25	100	102	103	106	108	110	112	113
EU-15 pre-2004	100	102	103	105	108	110	112	113
EU-10	n.a.	n.a.	n.a.	n.a.	n.a.	n.a.	n.a.	n.a.
Belgium	100	101	102	105	108	108	110	112
Denmark	100	103	105	107	110	110	109	111
Germany	100	100	100	101	104	102	104	105
Greece	100	104	108	113	119	125	131	137
Spain	100	104	107	112	118	121	124	133
France	100	102	104	107	110	110	114	115
Ireland	100	107	115	120	129	138	144	152
Italy	100	102	104	107	107	116	115	115
Luxembourg	100	102	104	105	105	107	109	111
Netherlands	100	101	104	105	107	108	108	110
Austria	100	100	99	101	102	103	103	104
Portugal	100	105	112	118	126	131	134	140
Finland	100	101	103	105	108	109	111	113
Sweden	100	101	101	102	105	106	108	111
United Kingdom	100	102	103	104	104	105	106	108
Cyprus	n.a.	n.a.	n.a.	n.a.	n.a.	n.a.	n.a.	n.a.
Czech Republic	100	102	102	102	105	108	109	110
Estonia	100	n.a.	n.a.	n.a.	n.a.	n.a.	n.a.	n.a.
Hungary	100	100	101	102	104	106	106	108
Latvia	n.a.	n.a.	n.a.	n.a.	n.a.	n.a.	n.a.	n.a.
Lithuania	100	n.a.	n.a.	n.a.	n.a.	n.a.	n.a.	123
Malta	n.a.	n.a.	n.a.	n.a.	n.a.	n.a.	n.a.	n.a.
Poland	100	102	108	114	115	120	123	127
Slovenia	100	108	104	95	92	92	90	85
Slovakia	100	98	95	94	97	106	105	108
Island	100	105	111	118	122	124	125	127
Norway	100	104	104	106	107	108	110	112
Bulgaria	n.a.	n.a.	n.a.	n.a.	n.a.	n.a.	n.a.	n.a.
Romania	n.a.	n.a.	n.a.	n.a.	n.a.	n.a.	n.a.	n.a.
Turkey	100	107	n.a.	n.a.	121	n.a.	n.a.	n.a.

Note: Total passenger transport demand data including air are not available for all countries and years. To guarantee a fairer comparison of trends, the index shown in the table does not include air transport demand. The aggregate EU-25 excludes Cyprus, Estonia, Latvia, Lithuania, Malta, because of lack of available passenger demand data since 1995.

Data source: Passenger demand data used in the structural indicators (February 2005), Eurostat (Ref: www.eea.eu.int/coreset).

(resource consumption, emissions of greenhouse gases, pollutants and noise, land consumption, accidents, etc.) of different transport modes. These differences are becoming smaller on a passenger-km basis, which makes it increasingly difficult to determine the direct and future overall environmental effects of modal shifts. The total environmental effect of modal shifts can in fact only be determined on a case-by-case basis, where local circumstances and specific local environmental effects can be taken into account (e.g. transport in urban areas or over long distances).

Policy context

The objective of decoupling was first defined in the transport and environment integration strategy that was adopted by the Council of Ministers in Helsinki (1999). The objective of decoupling is also mentioned in the sustainable development strategy, adopted by the European Council in Gothenburg, in order to reduce congestion and other negative side-effects of transport. The Council reaffirmed the objective of decoupling in the review of the integration strategy in 2001 and 2002.

Decoupling of economic growth and transport demand is mentioned in the sixth environment action programme as a key action in order to deal with climate change and alleviate the health impacts of transport in urban areas.

Shifting transport from road to rail is an important strategic element in the EU transport policy. The objective was first formulated in the sustainable development strategy (SDS). In the review of the transport and environment integration strategy in 2001 and 2002, the Council states that the modal split should remain stable for at least the next ten years, even with further traffic growth.

Modal shift is central and the Commission proposes measures aimed at modal shift in the white paper on the common transport policy (CTP) 'European Transport Policy for 2010: Time to Decide'. The target is to decouple transport growth significantly from growth in GDP in order to reduce congestion and other negative side effects of transport. Another target is to bring about a shift in transport use from road to rail, water and public passenger transport so that the share of road transport in 2010 is no greater than in 1998.

Indicator uncertainty

All data should be based on movements within the national territory, regardless of the nationality of the vehicle. However, data collection methodology is not harmonised at the EU level and the coverage is incomplete.

In relation to air transport, Eurostat does not currently collect data on transport performance within the national territory of the countries where this performance takes place, as would be required by the 'national territory principle'. Eurostat is working on methods for the calculation and territorial attribution of performance data for air transport. Until such data become available, the EU-25 aggregate for the core set indicator will include estimates of air transport demand from the European Commission's DG for Energy and Transport. The same estimates are not available for individual countries and for the same years.

Loading of the vehicle is a factor which plays a key role in assessing whether or not there is decoupling of passenger transport demand from GDP growth. Load factors for car passenger transport (i.e. the average number of passengers per car) are not mandatory variables in the data on passenger transport performance collected through the Eurostat/ECMT/UNECE common questionnaire on transport statistics. Since load factors are not always available, a sound assessment of passenger transport trends becomes very difficult. One could not, for instance, properly determine what share of the observed passenger-km trend results from changes in the average number of passengers per vehicle. For a complete picture of transport demand and the related environmental problems it would therefore be valuable to complement the data on the number of passenger-km with data on vehicle-km.

36 Freight transport demand

Key policy question

Is freight transport demand being decoupled from economic growth?

Key message

Freight transport volume has grown rapidly and has generally been strongly coupled with growth in GDP. Consequently the objective of decoupling GDP and transport growth has not been achieved. Closer inspection reveals great regional differences, with growth faster than GDP in the EU-15 and slower than GDP in the EU-10 Member States. This is mainly a result of the economic restructuring in the EU-10 Member States over the past decade.

Indicator assessment

Freight transport demand has grown significantly since 1992, thereby making it increasingly difficult to limit the environmental impacts of transport. But underlying the almost parallel growth with GDP is a more complex picture. Freight transport demand has grown significantly faster than GDP in the EU-15 whereas the picture for the EU-10 is the opposite.

For the EU-15, the main explanation is that the internal market is leading to some relocation of production processes, causing additional growth in transport demand over and above the steady growth in GDP. For the EU-10, the main reason is the large shift in production away from traditional relatively heavy low-value industry towards higher-value production and services. This, coupled with strong economic growth, means that freight transport growth is not keeping up with GDP growth. Both effects are temporary, but the data do not contain any indication that real decoupling is taking place.

The share of alternative modes (rail and inland waterways) in freight transport has declined during the past decade. As a result, the objective outlined in the common transport policy (CTP) of stabilising the shares of rail, inland waterways, short-sea shipping and oil pipelines, and shifting the balance from 2010 onwards, will not be achieved unless there is a strong reversal of the current trend.

This development can be explained by looking at the type of goods transported. This plays an important role in choice of mode. Perishable and high-value goods require fast and reliable transportation — road transport is often the fastest and most reliable form available, providing much flexibility with pickup and delivery points. Agricultural products and manufactured goods are some of the most important goods transported throughout Europe. Their shares in tonne-km are also rising.

Because the transport system allows it, modern production prefers 'just-in-time' delivery of goods. Transport speed and flexibility are therefore of great importance. Despite congestion, road transport is often faster and more flexible than rail or water transport. In addition, as a result of spatial planning and infrastructure development, many destinations can only be reached by road, and combined transport is used only to a limited extent. Furthermore, the road sector is liberalised to a great extent, while the inland waterway and rail sectors have only relatively recently been opened up to broad competition. Finally the average tonne of goods carried by road travels about 110 km, a distance over which rail or inland waterways are less efficient because road transport is needed to and from the points of loading. Moreover, in using multi-modal transport for such short distances, valuable time is lost due to lack of standardisation of loading units and convenient and fast connections between inland waterways and rail. For short-sea shipping, the average tonne of goods is carried more than 1 430 km. Here, time is less of an issue. The low price of shipping is probably of overriding importance.

Indicator definition

To measure decoupling of freight transport demand from economic growth, the volume of freight transport relative to GDP (i.e. the intensity) is calculated. Separate trends for its two components are shown for the EU-25. Relative decoupling occurs when freight transport demand grows at a rate below that of GDP. Absolute decoupling occurs when freight transport demand falls and GDP continues to rise or remains constant. If demand and GDP both fall, they remain coupled.

The unit is the tonne-kilometre (tonne-km), which represents the movement of one tonne over a distance of one kilometre. It includes transport by road, rail and inland waterways. Rail and inland waterways transport are based on movements within national territory, regardless of the nationality of the vehicle or vessel. Road transport is based on all movements of vehicles registered in the reporting country.

Freight transport demand and GDP are shown as an index (1995=100). The ratio of the former to the latter is indexed on the previous year (i.e. annual decoupling/ intensity changes) in order to be able to observe changes in the annual intensity of freight transport demand relative to economic growth.

The indicator can also be presented as the share of road in total inland transport (i.e. modal split for freight transport). Eurostat is currently working on methods regarding the calculation and territorial attribution of performance data for maritime transport which, if included, would have a significant impact on the modal shares. When Eurostat's results become available, the core set indicator will be reviewed and the modal shares shown.

Indicator rationale

Transport is one of the main sources of greenhouse gas emissions and also gives rise to significant air pollution, which can seriously damage human health and ecosystems. Reducing demand would therefore reduce the environmental burden of freight transport. Decoupling freight transport from GDP growth is only indirectly linked to environmental impact.

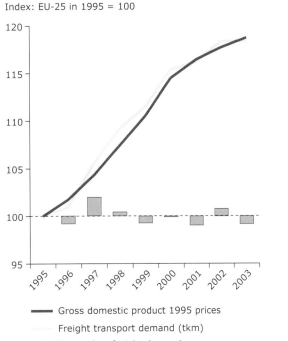

Figure 1 Trends in freight transport demand and GDP

Index: EU-25 in 1995 = 100

— Gross domestic product 1995 prices
— Freight transport demand (tkm)
▨ Decoupling freight demand from GDP growth (annual change)

Note: The decoupling indicator is calculated as the ratio of freight transport demand to GDP measured in 1995 market prices. The bars depict the intensity of transport demand in the current year in relation to the intensity in the previous year. An index above 100 results from transport demand outpacing GDP growth (i.e. positive bar = no decoupling) whereas an index below 100 is explained by transport demand growing less rapidly than GDP (i.e. negative bar = decoupling). See also indicator definition.

Data source: Eurostat
(Ref: www.eea.eu.int/coreset).

The relevance of the modal split policy for the environmental impact of freight transport arises from the differences in environmental performance (resource consumption, greenhouse gas emissions, pollutant and noise emissions, land consumption, accidents etc.) of different transport modes. These differences are becoming smaller on a tonne-km basis, which makes

Core set of indicators | Transport

Table 1   Trends in the annual intensity of freight transport demand

Trends in freight transport demand (tonne/km for road, rail and inland waterways); index 1995 = 100

	1995	1996	1997	1998	1999	2000	2001	2002	2003
EEA	100	102	106	109	111	114	115	117	118
EU-25	100	101	106	109	112	115	116	118	118
EU-15 pre-2004	100	102	105	110	113	117	118	120	119
EU-10	100	98	106	106	104	106	105	109	115
Belgium	100	93	97	93	87	112	115	116	112
Denmark	100	95	96	96	103	107	99	100	103
Germany	100	99	103	106	111	114	115	114	115
Greece	100	120	136	155	161	162	162	163	164
Spain	100	100	108	121	129	142	153	174	181
France	100	101	104	108	114	115	114	113	111
Ireland	100	113	123	142	176	209	211	241	263
Italy	100	106	106	112	108	112	113	115	105
Luxembourg	100	69	84	93	115	136	152	157	164
Netherlands	100	102	109	116	122	119	118	116	109
Austria	100	104	107	113	123	130	136	140	141
Portugal	100	120	130	131	136	139	154	153	144
Finland	100	100	105	113	117	125	119	123	121
Sweden	100	102	106	103	102	109	105	109	111
United Kingdom	100	104	106	108	106	105	105	105	106
Cyprus	100	103	105	108	110	114	118	122	130
Czech Republic	100	97	114	97	99	101	103	110	115
Estonia	100	113	146	183	209	223	245	261	298
Hungary	100	99	103	120	115	119	116	119	118
Latvia	100	126	149	148	141	156	169	183	214
Lithuania	100	99	111	112	126	135	129	165	185
Malta	100	103	106	109	113	116	116	116	116
Poland	100	104	110	109	105	106	103	103	107
Slovenia	100	95	106	104	110	128	131	121	125
Slovakia	100	71	70	74	72	65	62	62	66
Island	100	103	109	112	121	127	130	132	139
Norway	100	123	138	143	144	147	146	147	156
Bulgaria	100	88	86	73	61	31	33	35	38
Romania	100	102	102	78	66	73	81	94	104
Turkey	100	120	123	133	132	142	131	131	133

Note: Data source: Freight demand data used in the structural indicators (February 2005), Eurostat (Ref: www.eea.eu.int/coreset).

it increasingly difficult to determine the direct and future overall environmental effects of modal shifts. The differences in performance within specific modes can also be substantial, for example old versus new trains. The total environmental effects of modal shifts can only be determined on a case-by-case basis, where local circumstances and specific local environmental effects can be taken into account (e.g. transport in urban areas or through sensitive areas). The magnitude of the environmental effects of modal shifts may be limited, since modal shift is only an option for small market segments. Opportunities for modal shift depend, for example, on the type of goods carried — e.g. perishable goods or bulk goods — and the specific transport requirements of these goods.

Policy context

The EU has set itself the objective of reducing the link between economic growth and freight transport demand ('decoupling') in order to achieve more sustainable transport. Reducing the link between transport growth and GDP is a central theme in EU transport policy for reducing the negative impacts of transport.

The objective of decoupling freight transport demand from GDP was first mentioned in the transport and environment integration strategy adopted by the Council of Ministers in Helsinki (1999). This named the expected growth in transport demand as an area where urgent action was needed. In the sustainable development strategy adopted by the European Council in Gothenburg, the objective of decoupling is set in order to reduce congestion and other negative side-effects of transport. In the review of the integration strategy in 2001 and 2002, the Council reaffirmed the objective of reducing the link between the growth of transport and GDP.

In the sixth environment action programme, decoupling of economic growth and transport demand is named as one of the key objectives in order to deal with climate change and alleviate the health impacts of transport in urban areas.

Shifting freight from road to water and rail is an important strategic element in the EU transport policy. The objective was first formulated in the sustainable development strategy (SDS). In the review of the transport and environment integration strategy in 2001 and 2002, the Council stated that the modal split should remain stable for at least the next ten years, even with further traffic growth.

In the white paper on the common transport policy (CTP) 'European Transport Policy for 2010: Time to Decide', the Commission proposes a number of measures aimed at modal shift. The target is to decouple transport growth significantly from growth in GDP in order to reduce congestion and the other negative side effects of transport. A second target is to stabilise the shares of rail, inland waterways, shortsea-shipping and oil pipelines at 1998 level and bring about a shift in transport use from road to rail, water and public passenger transport from 2010 onwards.

Indicator uncertainty

Total inland freight transport demand excludes maritime transport because of methodological problems related to the allocation of international maritime transport to specific countries. Thus, the effect of globalisation (production being moved from Europe, for example to China) does not have a measurable impact on the indicator in spite of having large real consequences for total freight transport demand.

Load factors for road freight transport are not mandatory and are collected only in the framework of Council Regulation (EC) No 1172/98. Even for the countries that measure such variables, data have been reported to Eurostat only since 1999. Assessment of the loading of vehicles was not foreseen by the Regulation. Loading is a factor which plays a key role in assessing whether or not there is decoupling of freight transport demand from economic activity.

37 Use of cleaner and alternative fuels

Key policy question

Is the EU making satisfactory progress towards using cleaner and alternative fuels?

Key message

- Many Member States have introduced incentives to promote the use of low and zero-sulphur fuels ahead of the mandatory deadlines (a maximum of 50 ppm 'low' in 2005 and a maximum of 10 ppm 'zero' in 2009). The combined penetration increased from around 20 to almost 50 % between 2002 and 2003, but this is still some way off the 2005 target of 100 %.

- The penetration of biofuels and other alternative fuels is low. The share of biofuels in the EU-25 is less than 0.4 %, still far off the 2 % target set for 2005. However, following the adoption of the Biofuels Directive in 2003, national initiatives are rapidly changing the situation.

Indicator assessment

A reduction in the sulphur content of petrol and diesel fuels is expected to have a significant impact on exhaust emissions as it will enable the introduction of more sophisticated after-treatment systems. In view of the 2005 (50 ppm) and 2009 (10 ppm) mandates, many Member States have introduced incentives to promote these fuels. However, the capacity of refineries to supply the fuels affects the time it takes for them to penetrate the market.

In 2003, the combined share of low and zero-sulphur petrol and diesel in the EU-15 was 49 % and 45 % respectively, with a nearly equal split between low and zero-sulphur fuels. Compared with the 2002 figures of around 20 %, these fuels have seen significant growth. If this continues at the same pace, both the 2005 and the 2009 targets are within reach. Many countries have abandoned the sale of regular (350 ppm sulphur) petrol and diesel fuel. In particular, Germany leads the way by being the only country offering only zero-sulphur fuel. At the other end of the scale, four countries (France, Italy, Portugal and Spain) do not yet offer low or zero-sulphur fuels in their markets.

Assessment of the market penetration of biofuels is hampered by incomplete data sets, as not all countries have yet set up reporting for this. Based on the available data, the share of biofuels in the EU-25 in 2002 was still low, accounting for 0.34 % of all petrol and diesel sold for transport purposes (reported biofuels consumption as a percentage of total gasoline and diesel consumption). This share has more than doubled over the past eight years; however more effort is needed to reach the 2 % and 5.75 % objectives by the end of 2005 and 2010 respectively. France and Germany have the highest shares of biofuels sold in their markets.

Indicator definition

The use of cleaner and alternative fuels is measured using two different indicators:

1) The share of regular, low and zero-sulphur fuels in total fuel consumption for road transport. Fuels with less than 50 parts of sulphur per million (ppm) are often referred to as low-sulphur and those with less than 10 ppm as zero-sulphur.

2) The percentage of final energy consumption of biofuels for transport in the total combined final energy consumption of gasoline, diesel and biofuels for transport.

Petrol and diesel fuels are measured in millions of litres and presented as shares of regular, < 50 ppm sulphur and < 10 ppm sulphur.

Final energy consumption of biofuels, diesel and gasoline for transport are measured in Terajoules of net calorific value (NCV) and the share of biofuels is presented as a percentage of the sum of all three fuels.

Use of cleaner and alternative fuels   PART B

Figure 1 Low and zero-sulphur fuel use (%), EU-15

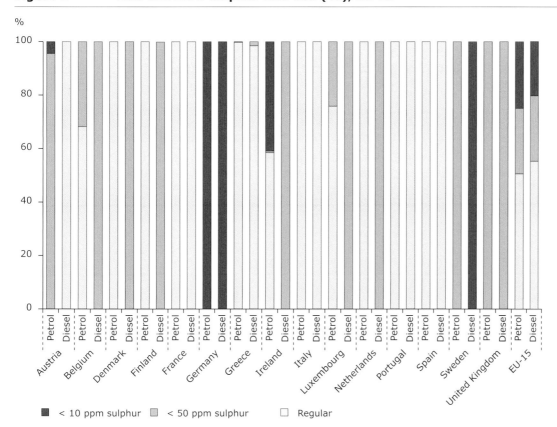

Note: Data source: European Commission, 2005. Quality of petrol and diesel fuel used for road transport in the European Union: Second annual report (reporting year 2003). Report from the European Commission (COM (2005) 69 final) (Ref: www.eea.eu.int/coreset).

Indicator rationale

EU legislation has set requirements for the sulphur content of road transport fuels and the minimum share of biofuels in total road transport fuel consumption. The indicator has been selected to follow these policy requirements by monitoring the progress achieved.

The promotion of low and zero-sulphur fuels will enable a further decrease in emissions of pollutants from road vehicles, while the promotion of biofuels is essential for reducing greenhouse gas and especially CO_2 emissions.

Policy context

EU legislation requires a reduction of the sulphur content of road transport fuels to 50 mg/kg (low-sulphur) by 2005 and a further reduction to below 10 mg/kg (zero-sulphur) by 2009. It also suggests that EU road transport fuel consumption should have a 2 % share of biofuels by 2005 and 5.75 % by 2010.

Indicator uncertainty

The data are collected on an annual basis by the European Commission and can thus be considered reliable and accurate. The requirement for data

Figure 2 **Share of biofuels in transport fuels (%)**

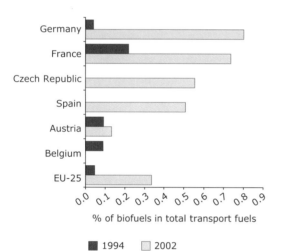

Note: The biofuels directive aims at promoting the use of biofuels for transport to replace diesel or petrol. The primary objective is to increase the consumption of biofuels, as opposed to its production, which may or not be exported to other countries. The share of biofuels should reach 2 % by 2005 and 5.75 % by 2010. The denominator includes all EU-25 countries with consumption of diesel and gasoline. The numerator refers to the final energy consumption of biofuels in the transport sector. By 2002, only a few EU countries had consumption of biofuels or were reporting consumption of biofuels to Eurostat. A progressively larger number of EU countries are expected to report biofuels consumption to Eurostat when data become available for 2003, the year of entry into force of the directive.

Data source: Eurostat
(Ref: www.eea.eu.int/coreset).

collection for low and zero-sulphur fuels and biofuels is mandatory and thus the results are harmonised at the EU level.

Data on the share of low and zero-sulphur fuels are currently available only for the EU-15 and for three years (2001, 2002 and 2003), resulting from their reporting obligations. Data on biofuels are currently available for eight of the EU-25 countries (data for Italy and Denmark available, but reported as zero); however it is very likely that these countries represent the vast majority of biofuel consumption for transport purposes in the time-frame indicated.

Use of cleaner and alternative fuels — PART B

Table 1 Final energy consumption in the transport sector

	1994						2002					
	Final energy consumption in terajoules (net calorific value)			Fuel shares in final energy consumption (%)			Final energy consumption in terajoules (net calorific value)			Fuel shares in final energy consumption (%)		
	Motor spirit (gasoline)	Gas/diesel oil	Biofuels	Motor spirit (gasoline)	Gas/diesel oil	Biofuels	Motor spirit (gasoline)	Gas/diesel oil	Biofuels	Motor spirit (gasoline)	Gas/diesel oil	Biofuels
EU-25	5 541 712	4 864 585	4 896	53.2	46.7	0.05	5 242 160	6 635 686	40 052	44.0	55.7	0.34
EU-15	5 105 540	4 574 576	4 896	52.7	47.2	0.05	4 791 160	6 192 212	38 964	43.5	56.2	0.35
EU-10	436 172	290 009	0	60.1	39.9	0.0	451 000	443 473	1 088	50.4	49.5	0.12
Belgium	125 004	178 591	272	41.1	58.8	0.09	91 960	244 452	0	27.3	72.7	0.00
Czech Republic	69 256	50 591	0	57.8	42.2	0.0	84 876	110 445	1 088	43.2	56.2	0.55
Denmark	81 048	71 995	0	53.0	47.0	0.0	84 216	78 509	0	51.8	48.2	0.0
Germany	1 301 344	983 687	952	56.9	43.0	0.04	1 187 516	1 127 380	18 700	50.9	48.3	0.80
Estonia	12 540	6 683		65.2	34.8	0.0	13 464	13 790		49.4	50.6	0.0
Greece	116 424	83 669		58.2	41.8	0.0	153 692	97 079		61.3	38.7	0.0
Spain	403 040	511 830	0	44.1	55.9	0.0	361 636	881 363	6 358	28.9	70.5	0.51
France	660 352	934 576	3 502	41.3	58.5	0.22	570 196	1 256 818	13 566	31.0	68.3	0.74
Ireland	43 340	34 940		55.4	44.6	0.0	69 784	80 074		46.6	53.4	0.0
Italy	721 952	622 487	0	53.7	46.3	0.0	703 692	831 237	0	45.8	54.2	0.0
Cyprus	7 920	11 040		41.8	58.2	0.0	10 076	14 382		41.2	58.8	0.0
Latvia	18 700	11 125		62.7	37.3	0.0	14 960	18 950		44.1	55.9	0.0
Lithuania	18 568	14 678		55.9	44.1	0.0	15 796	25 676		38.1	61.9	0.0
Luxembourg	23 980	24 746		49.2	50.8	0.0	24 464	48 307		33.6	66.4	0.0
Hungary	63 492	33 502		65.5	34.5	0.0	58 740	74 617		44.0	56.0	0.0
Malta	3 740	4 484		45.5	54.5	0.0	2 244	4 991		31.0	69.0	0.0
Netherlands	172 128	187 178		47.9	52.1	0.0	183 656	256 507		41.7	58.3	0.0
Austria	101 684	82 612	170	55.1	44.8	0.09	91 036	165 393	340	35.5	64.4	0.13
Poland	187 044	111 926		62.6	37.4	0.0	185 548	119 117		60.9	39.1	0.0
Portugal	81 532	88 196		48.0	52.0	0.0	91 036	173 642		34.4	65.6	0.0
Slovenia	33 704	14 890		69.4	30.6	0.0	33 792	22 631		59.9	40.1	0.0
Slovakia	21 208	31 091		40.6	59.4	0.0	31 504	38 874		44.8	55.2	0.0
Finland	84 128	69 457		54.8	45.2	0.0	80 520	84 938		48.7	51.3	0.0
Sweden	183 216	88 365		67.5	32.5	0.0	180 048	110 826		61.9	38.1	0.0
United Kingdom	1 006 368	612 250		62.2	37.8	0.0	917 708	755 690		54.8	45.2	0.0
Iceland	6 072	2 496		70.9	29.1	0.0	6 424	2 242		74.1	25.9	0.0
Norway	73 744	72 798		50.3	49.7	0.0	72 336	87 011		45.4	54.6	0.0
Bulgaria	43 428	21 573		66.8	33.2	0.0	26 884	35 955		42.8	57.2	0.0
Romania	51 568	66 538		43.7	56.3	0.0	76 648	89 845		46.0	54.0	0.0
Turkey	174 856	228 293		43.4	56.6	0.0	137 280	262 514		34.3	65.7	0.0

Note: By 2002, only a few EU countries had consumption of biofuels or were reporting consumption of biofuels to Eurostat. A progressively larger number of EU countries are expected to report biofuels consumption to Eurostat when data become available for 2003, the year of entry into force of the directive.

Data source: Eurostat (Ref: www.eea.eu.int/coreset).

Country analysis

Country analysis

Setting the scene — main results	408
EEA scorecard	412
Thematic assessment	
Greenhouse gas emissions	414
Total energy consumption	416
Renewable electricity	418
Emissions of acidifying substances	420
Emissions of ozone precursors	422
Freight transport demand	424
Area under organic farming	426
Municipal waste generation	428
Use of freshwater resources	430
Country perspectives	432
Austria	434
Belgium	436
Bulgaria	438
Cyprus	440
Czech Republic	442
Denmark	444
Estonia	446
Finland	448
France	450
Germany	452
Greece	454
Hungary	456
Iceland	458
Ireland	460
Italy	462
Latvia	464
Liechtenstein	466
Lithuania	468
Luxembourg	470
Malta	472
The Netherlands	474
Norway	476
Poland	478
Portugal	480
Romania	482
Slovak Republic	484
Slovenia	486
Spain	488
Sweden	490
Switzerland	492
Turkey	494
United Kingdom	496
Methodology and main decision points	498
Ranking graphs	508

PART C

Setting the scene — main results

This part of the report focuses on country level analysis and the relative environmental performance of the EEA member countries. For this purpose a *scorecard* is presented which uses indicators from the EEA core set of indicators (CSI) ([1]). The use of the CSI ensures that the data underpinning the scorecard is the best European environmental information available at this time.

The scorecard allows the reader to make policy-relevant and informative comparisons between countries and amongst issues against performance benchmarks ([2]). The scorecard covers the following key environmental issues: climate change, air pollution, waste and water, and some of the main sectors that impact on the environment: energy, transport and agriculture. Most of the indicators are also of relevance to additional issues such as sustainable consumption, sustainable use of resources, and human health.

Scorecards are based on limited sets of indicators and can therefore never be exhaustive in their coverage of environmental issues. Although the scorecard presented here should not be seen as a definitive overall analysis of environmental performance it does show patterns of results between the EEA member countries. Some of the variations in results come from natural climatic and geographic conditions as well as from historical, social, economic and environmental factors.

The grouping of the countries used in the scorecard reflects some of the patterns in performance, as well as some of the underlying factors. This grouping is an aid to understanding and no overall ranking is implied by the order in which countries or groups are listed in the scorecard.

- The group of countries with the most consistent pattern across all the indicators are the EU-10 Member States and the accession countries, Bulgaria and Romania. These countries still have economies with relatively high energy and emission intensities, but all 1990 air-related targets are within reach. A number of the countries in this group have high freight transport intensities, both per capita and per unit of GDP. It is also in this group that one finds the lowest generation of municipal waste (either because waste generation is lower, or because it is collected less systematically). Only Slovenia does not fit the overall pattern but resembles Belgium or Norway more closely, although it is more emission-intensive.

- Another group of countries in which all air targets are within reach includes the western European countries that have recently restructured their industries and/or have long experience of environmental policies. This group comprises Germany, the United Kingdom, the Netherlands, France and Sweden. These countries have economies that are generally less energy and emission intensive, but their energy use per capita is far higher than in the previous group.

- Three western countries (Portugal, Spain and Ireland) which have rapidly developing economies, have difficulties in reaching any of the environmental targets included in the scorecard, and generally have emission-intensive economies.

- Another group of countries (Luxembourg, Slovenia, Belgium, Norway, Austria, Italy, Denmark, Finland and Greece) have difficulties reaching either their Kyoto (burden-sharing) targets or their ozone precursor emission targets. Although there is considerable variation within this group, a common feature of all these countries is the relatively high emission of ozone precursors. Energy use and greenhouse gas

---

([1]) More information about how the scorecard was constructed and main decision points can be found in the last section of this part of the report.
([2]) Information on performance benchmarks and targets can be found on the core set of indicators website www.eea.eu.int/coreset.

emissions per capita are also on the high side. All of these may be related to the high transport intensities found in these countries.

- The remaining six countries cannot be properly compared with others as there are currently gaps in the data provided to the EEA, or they do not have targets for the selected indicators. A number of these have recently joined EEA and Eionet and data exchange procedures in these cases are still under development.

Understanding the diversity between countries and how this can systematically impact environmental policy implementation is very relevant to decision-making at European level. Once this diversity is taken into account, it can also be very instructive to examine differences in performance and how they relate to the different types of responses adopted at the country level. The timing of policy action and the level of ambition in setting national targets and objectives can also substantially influence apparent performance.

The importance of the scorecard lies in its relevance to policy-makers, who need to understand changes in the environment and how these result directly or indirectly from the implementation of policy. The aim is to provide a deeper understanding of country conditions, behaviour and response to environmental problems which can begin to explain some of the differences in performance, and perhaps highlight areas for future work.

The scorecard encourages such critical assessment by presenting an array of policy relevant environmental indicators across a broad selection of themes together on one page. The scorecard also facilitates communication, and the aim is to support shared policy learning between the EU, member countries and other actors from which lessons and good practices can be derived.

In making these assessments attention has been paid to adhere to the principles of fairness and acceptability to the countries being compared, and the methodology has been developed in conjunction with the countries themselves. In particular the following underlying principles are used throughout:

- In recognition of the inherited environmental legacy of a country, both geographical and political, the scorecard focuses equally on the present status and on progress made over time. Thus a country that is a poor performer at the moment has the chance to show that it has made progress, while a country that has excellent status at present is not unfairly targeted for slowing its rate of improvement.

- The diversity in economic and social starting conditions that exists within the wide EEA area is recognised and all country comparisons are therefore made on per capita environmental performance unless the political targets for an environmental indicator are specifically set in relation to GDP, in which case both comparisons are shown.

During the development of the scorecard (a multiple-year effort with countries and experts), the need to avoid the pitfalls of oversimplification, irrelevance for policy-making and lack of credibility or legitimacy with the countries, was recognised. The scorecard represents a cautious and transparent first step in an ongoing process with EEA countries which will continue to evolve over the coming years. In particular, the balance between ease of communication and relevance for supporting decision-making will be continuously reviewed.

This part of the report has three components:
(i) **Thematic assessment:** an assessment of each of the indicators which explores trends across countries;
(ii) **Country analysis:** an analysis for each country for each of the nine indicators presented in the scorecard; and (iii) **Methodology:** a final section describing how the scorecard was developed and the main decision points.

Country analysis

> **Reading the scorecard**
>
> - The nine indicators used in the scorecard are a subset of the EEA core set of indicators (CSI).
> - The country scorecard uses two symbols. The arrow symbols are used for progress indicators (usually covering the period 1992–2002), while the solid colours represent the status indicators (based on 2002 or 2003 data).
> - The country scorecard also uses two colour schemes. Green and red are used when the values being compared relate directly to agreed policy targets. Alternatively, and in most cases where there are no hard or useable policy targets or where the data do not allow country comparisons to be made against the actual targets, three shades of blue are used to show comparisons against average European performance (light blue indicates the top relative performers).
> - Overall the lighter the colour or tone the better the performance.
> - For the green and red scale target indicators, countries which are not on track to meet policy targets (assessed on a linear progression to target) are shown in red. When countries are on track they are marked green. Countries substantially above the target line are marked in light green indicating that even with the uncertainties in the data there is a high confidence of them being on track. Countries closer to the target line are marked in dark green. More detailed information on targets can be found on the CSI website (www.eea.eu.int/coreset).
> - For the blue scale status indicators, country comparisons are based on absolute values in the current year, such that the top 25 % of the range of results and the bottom 25 % are clearly marked. In several cases, notably organic farming, the majority of countries falls into the lowest 25 % of the range.
> - Outliers which significantly distort the distribution of the data are excluded (this is the case for municipal waste generation, share of organic farming and emissions of ozone precursors).
> - For the blue scale progress indicators, countries are ranked according to the difference in their performance to the average progress in the EU-25 over the past ten years. Again, the top 25 % ([3]) of the range and the bottom 25 % are identified.

The scorecard is complemented by the country analysis section that has been prepared in partnership between the countries and the EEA. This section provides an opportunity for countries to raise issues specific to their situation and to bring in relevant information that serves to balance the scorecard results

The methodology section is designed to help in the reading and interpretation of the scorecard and outlines some of the key decision points encountered during the development of the scorecard. A more technical methodology for practitioners wishing to create scorecards is published separately.

([3]) For further information please refer to the box in the section 'Methodology and main decision points'.

Setting the scene — main results PART C

PART C | Country analysis

EEA scorecard

In the EEA scorecard of relative environmental performance presented below countries are grouped by roughly similar patterns of results as well as socio-economic and geographical factors. The scorecard presents results for nine indicators from the EEA core set of indicators, for a combination of *progress* over time (usually ten years 1992–2002, boxes with arrows) and *status* for the latest year available (2002/2003, solid colour boxes). The scorecard also uses two colour schemes: red/green when the values being compared relate directly to agreed policy targets; and three shades of blue to show comparisons against average European performance (light blue indicates the top relative performers). For more information see the section *Methodology and main decision points*.

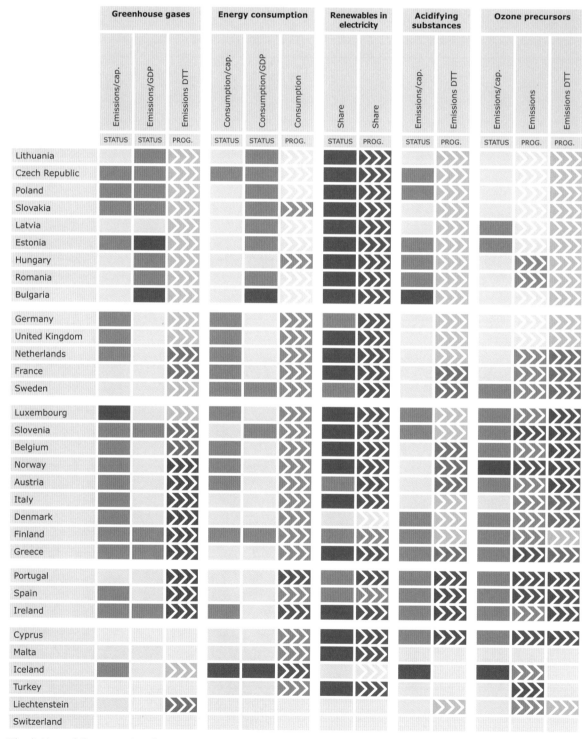

The listing of the countries does not represent an overall ranking

EEA scorecard

PART C

STATUS (2002/2003)
- Top 25 % of indicator values
- Middle 50 % of indicator values
- Lowest 25 % of indicator values

PROGRESS (1992–2002/2003)
- Top 25 % of indicator values
- Middle 50 % of indicator values
- Lowest 25 % of indicator values

Distance to target (DTT)
- On track to meet the target
- Within ± 5 percentage points of the target line
- Not on track to meet the target

Overall, the lighter the colour or tone the better the performance.

The listing of the countries does not represent an overall ranking

The European environment | State and outlook 2005

PART C | Country analysis | Thematic assessment

Greenhouse gas emissions

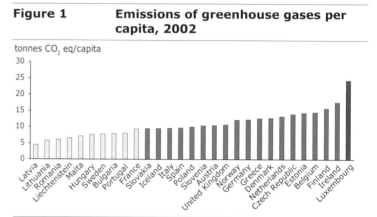

Figure 1 — Emissions of greenhouse gases per capita, 2002

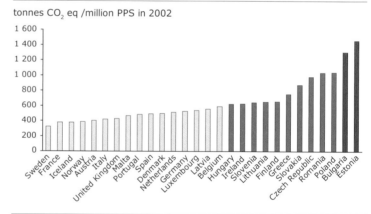

Figure 2 — Emissions of greenhouse gases per unit of GDP, 2002

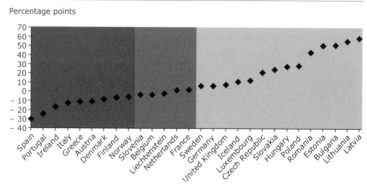

Figure 3 — Distance to Kyoto target, linear target path, 2002

Notes: For technical explanations and caveats on country base-years and targets see CSI 10 and CSI 11 in the EEA indicator management system (www.eea.eu.int/coreset). The distance to target calculation is explained in the text. The comparison per unit of national income is done on the basis of gross domestic product (GDP) expressed in purchasing power parities (PPP). Hence the unit is not euro but purchasing power standard (PPS).

Sources: GHG emissions: EEA; GDP: Eurostat; Population: the World Bank.

Greenhouse gas emissions

Combating climate change and minimising its potential consequences are key objectives of the UN Framework Convention on Climate Change (UNFCCC) and a high priority for the EU. Achieving this aim would require substantial (15 to 50 %) reductions in global greenhouse gas emissions. As a first step, the parties to the UNFCCC in 1997 adopted the Kyoto Protocol, which requires developed countries to reduce emissions of the six greenhouse gases to 5.2 % below their levels in a given base-year (1990 in most cases) by the period 2008–2012. Individual country targets are found on the CSI website (www.eea.eu.int/coreset). The 'burden sharing targets' of the EU-15 Member States reflect the economic development of countries, but also the ambition of governments at the time the targets were agreed (1998).

In 2003, aggregate greenhouse gas emissions in the EU-15 Member States were 1.7 % below the base-year level, and emissions increased by 1.3 % between 2002 and 2003 (excluding land-use changes and carbon sequestration in forests). After more than half the time period between 1990 and the first commitment period (2008–2012) under the Kyoto Protocol, the reduction by 2003 was less than a quarter of that needed to reach the EU-15 target.

Progress in reducing greenhouse gas emissions is illustrated by comparison with an assumed linear target path to the Kyoto Protocol target (Figure 3). 2002 data were used since these were available for all countries (^{1}). For some countries (Austria, Belgium, Denmark, Ireland, Luxembourg and the Netherlands) the expected use of the so-called Kyoto mechanisms, which allow the achievement of national targets by taking measures in other countries, is taken into account. The recently-launched European Union greenhouse gas emission trading scheme is not taken into account.

Austria, Denmark, Finland, Greece, Ireland, Italy, Portugal and Spain are not on track (more than five percentage points below the target path) and are therefore marked in red in the scorecard. Of the countries that are close to the target path, Belgium and Slovenia are slightly below the target line, while France and the Netherlands are slightly above. Given the inherent uncertainties in the data it is impossible to say with certainty whether or not these countries are on track, and they are therefore marked dark green.

The countries that are not on track do not stand out in terms of high emissions per capita or per unit of national income (GDP). The extreme ends of the bar graphs (the low side of the per capita graph and the high side of the per GDP graph) are occupied by the EU-10 and the accession countries. These are on track for reaching their Kyoto targets mainly as a result of the economic transition. However, due to data gaps, the uncertainty in the EU-10 data is higher than for the EU-15 countries.

Lowest emission intensity in 2002:
Latvia per capita (4.5 tonnes CO_2-equivalent/capita; Figure 1) and **Sweden** by GDP (320 tonnes CO_2-equivalent/million PPS GDP; Figure 2)

Highest emission intensity in 2002:
Luxembourg per capita (24 tonnes CO_2-equivalent/capita; Figure 1) and **Estonia** by GDP (1 460 tonnes CO_2-equivalent/million PPS GDP; Figure 2)

Factor difference:
x5 (per capita) and x5 (per unit of GDP)

(^{1}) At the time of writing, 2003 data in EEA indicator format were available only for the EU-15 countries. Ranking this 2003 data within the EU-15 group per capita or per unit of GDP does not result in significant changes compared with 2002. The relative distance-to-target indicator is however not possible to reproduce as new data on the use of the Kyoto mechanisms is still being submitted.

Country analysis | Thematic assessment

Total energy consumption

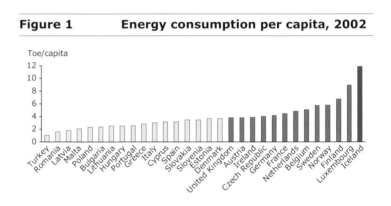

Figure 1   Energy consumption per capita, 2002

Figure 2   Energy consumption per unit of GDP, 2002

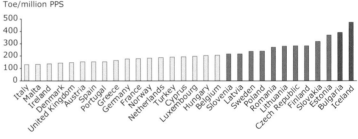

Figure 3   Energy consumption changes, 1992–2002, compared with the EU-25 average

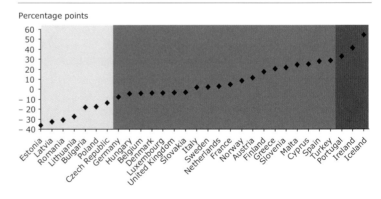

Notes: For technical specifications see CSI 28 in the EEA indicator management system (www.eea.eu.int/coreset). All indicators are calculated on the basis of total, or gross inland, consumption. The comparison per unit of national income is done on the basis of gross domestic product (GDP) expressed in purchasing power parities (PPP). Hence the unit is not euro but purchasing power standard (PPS).

Sources: Energy and GDP: Eurostat; Population: the World Bank.

Total energy consumption

The largest contributor to greenhouse gas emissions in the EU is energy consumption, where emissions originate both during the transformation of one energy form to another and during final consumption. There are no direct targets for total energy consumption (which includes the energy losses during transformation, sectoral energy-use, and energy production). Total energy consumption increased by 11 % between 1992 and 2002 in the EU-15 and decreased by 5 % in the EU-10.

Drilling down into the data using the country scorecard, it can be seen that there is a wide variation in energy intensity among countries. Iceland stands out because of energy intensive industries and abundance of geothermal energy, which is used for heating and electricity production; the latter is used mainly by the aluminium industry, which has chosen Iceland as a production location because of the availability of cheap and 'green' electricity. Since the efficiency of producing electricity from geothermal energy is very low (an estimated 90 % of the energy content is lost), this pushes up the total consumption figure. The high consumption in Iceland, however, does not have the same environmental consequences as energy consumption in other countries. A number of the EU-10 Member States and accession countries also have high energy consumption per unit of GDP. The longer and colder winters under a continental climate contribute to relatively higher consumption in a number of these countries. However, besides that, the countries face the challenge of improving their efficiency by replacing old power plants and industrial installations, by improving maintenance, and through insulating and installing heating controls in buildings. At the other end of the scale are the countries with low energy intensity, such as Austria, Denmark, Ireland, Italy and Malta falling into the lowest 25 % of the distribution (Figure 2).

The Nordic countries Finland, Norway and Sweden, because of their geographical location, rank high on per capita consumption, with Finland and Sweden also ranking high on the energy intensity per GDP scale. The EU-10 Member States and the accession countries have relatively low per capita energy consumption compared with the EU-15 Member States.

Progress in reducing total energy consumption is compared with the EU-25 average. Largely due to economic transition the best progress compared with this average was in the EU-10 Member States and the EU accession countries (Figure 3). The least progress was seen in Iceland, Ireland and Portugal.

Lowest energy consumption in 2002:
Turkey per capita (1.1 toe/capita; Figure 1) and **Italy** by GDP (132 toe/million PPS GDP; Figure 2)

Highest energy consumption in 2002:
Iceland, but without the same environmental consequences. Next in row: **Luxembourg** per capita (9 toe/capita; Figure 1) and **Bulgaria** by GDP (392 toe/million PPS GDP; Figure 2)

Factor difference:
x8 (per capita) and x3 (per unit of GDP)

PART C   Country analysis | Thematic assessment

Renewable electricity

| Figure 1 | Share of electricity from renewables other than large hydro in electricity consumption, 2002 |

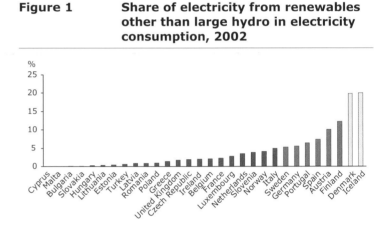

| Figure 2 | Change, 1992–2002, in the share of electricity from renewables other than hydro in electricity consumption, compared with the EU-25 average |

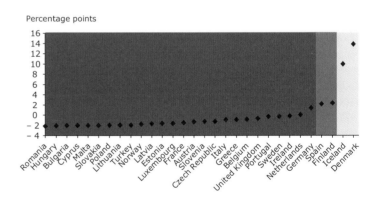

Notes: For technical specifications and definitions of the variables see CSI 31 in the EEA indicator management system (www.eea.eu.int/coreset). Note that the comparison over time is on the basis of renewables other than all hydro, as the data set available for large hydro is poor prior to 2002.
Source: Eurostat.

Share of renewables in electricity (other than large hydro)

Increased market penetration of renewable energy will help EEA member countries to reach their commitments under the UNFCCC Kyoto Protocol. The White Paper for renewables (COM(97) 599 final) included an indicative target of increasing the share of renewable energy in total energy consumption in the EU-15 to 12 % by 2010. Four years later the EU Directive on the promotion of electricity from renewable energy sources in the internal electricity market (2001/77/EC) set an indicative target of 22.1 % of gross EU-15 electricity consumption to come from renewable sources, including large hydro, by 2010. It required Member States to set and meet annual national indicative targets consistent with the directive. For the EU-10 Member States, national indicative targets are included in the Accession Treaty: the 22.1 % target set initially for the EU-15 for 2010 has become 21 % for the EU-25.

Countries with a high share of all renewables in electricity generation are Norway and Iceland (both close to 100 %), Austria (66 %), Sweden (47 %) and Latvia (39 %). The distance to the national targets for the share of renewable energy in electricity consumption is included in CSI 31. Countries with a more than ten percentage point distance to the indicative target include Austria, Greece, Italy, Portugal, Slovakia, Spain, and Sweden.

As can be seen from Figure 1 of CSI 31 (see Part B of this report), much of the available renewable electricity in Europe comes from existing large hydropower plants (> 10 MW). As hydropower production depends to a large extent on rainfall in a specific year, country comparison in a certain year can be flawed. In addition, with the exception of a few countries, the growth potential of large hydropower plants in Europe is limited, due in part to a lack of suitable sites, environmental concerns and the water framework directive. Hence, to focus attention on environmental protection it makes sense to compare trends in the use of renewable electricity excluding the share produced by large hydropower plants. Unfortunately, data prior to 2002 do not allow hydropower plants to be differentiated by size for all countries and therefore for the progress indicator, which uses 1992 data, hydro of all sizes is excluded in the present analysis — this is a temporary measure until improved data become available.

Cyprus and Malta are alone among the 31 EEA member countries in having no renewable electricity at all. They are closely followed by Bulgaria, Estonia, Hungary, Latvia, Lithuania, Romania, the Slovak Republic and Turkey which all had less than a 1 % share of renewables (other than large hydropower plants) in total electricity in 2002. The capacity in Turkey has increased in more recent years. After Iceland, with its geothermal energy sources, Denmark leads in Europe with a share of 19.9 % of electricity from renewables other than large hydropower plants, largely wind energy (Figure 1).

Denmark's lead is reflected in the progress made between 1992 and 2002 when there was a 16 percentage point increase in the share of renewables other than hydropower plants. During the same period Iceland increased by 12 percentage points, while Finland, Germany and Spain also showed a substantial increase (between 3.4 and 4.4 percentage points; Figure 2).

With the exception of Spain, all countries that are more than 10 percentage points away from the renewable electricity target (see Figure 1 in CSI 31), have shown a lower than average progress (Figure 2) in using renewables other than hydro. This implies that countries that are not meeting their renewable electricity targets are not investing sufficiently in these renewables.

Highest share:
Iceland and **Denmark**
20 % of electricity from renewables other than large hydro in 2002, see text above and under total energy consumption

Lowest share:
Cyprus and **Malta** no renewable electricity in 2002

Emissions of acidifying substances

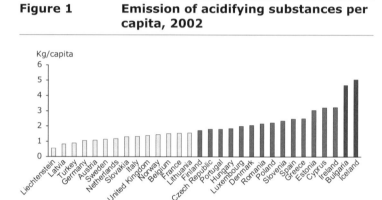

Figure 1 Emission of acidifying substances per capita, 2002

Note: SO_2, NH_3 and NO_x (expressed as NO_2); The factors are NO_x 0.021, SO_2 0.031 and NH_3 0.058. Results are expressed in acidification equivalents.

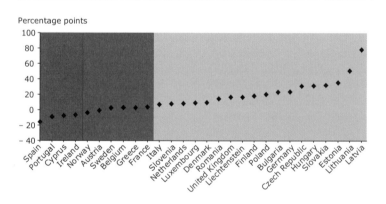

Figure 2 Emission of acidifying substances: distance to NECD targets, linear target path, 2002

Notes: For technical specifications see CSI 01 in the EEA indicator management system (www.eea.eu.int/coreset).
Sources: Emissions: EEA; GDP: Eurostat.

Emissions of acidifying substances

Emissions of acidifying substances into the atmosphere result in deposition that can damage ecosystems, buildings and materials. Emissions of the key acidifying gases, nitrogen oxides (NO_x), sulphur dioxide (SO_2) and ammonia (NH_3), are covered by the EU National Emission Ceilings Directive (NECD) (2001/81/EC) and the Gothenburg protocol under the United Nations Convention on Long-range Transboundary Air Pollution (CLRTAP) (UNECE 1999). The NECD generally involves slightly stricter emission reduction targets than the Gothenburg Protocol for EU-15 Member States for the period 1990–2010. The Gothenburg Protocol entered into force in May 2005.

Emissions of acidifying substances in Europe have been reduced substantially (by 44 % in the EEA member countries excluding Malta, between 1990 and 2002), mainly due to the increased use of pollution-abatement equipment, e.g. flue gas desulphurisation, together with the use of low-sulphur fuels in power plants

Countries have individual targets for each of the acidifying substances to be reached in 2010. These targets have been established with the aim of reducing the exceedance of deposition above a critical load. A critical load is the highest deposition that will not cause long-term harmful effects on ecosystems. In Scandinavia where soils have a low buffering capacity, critical loads are low. The opposite holds for the Mediterranean countries. Country targets reflect the gap between the 1990 emissions and these critical loads taking into account long-distance transport of pollutants. Of course the agreed targets are also the result of a political negotiation. For this indicator, both targets and emissions have been recalculated in 'acidification equivalents' to allow for aggregation. Progress towards the targets is measured as the distance to an assumed linear target line.

There is almost an order of magnitude variation between the countries emitting most and least acidifying substances per capita in 2002. Latvia, Germany and Austria are among the lowest per capita emitters, together with Liechtenstein and Turkey.

Iceland is the highest, which has a lot to do with a very small population; a large part of their sulphur emissions stem from their fishing fleet. Others on the relatively high side are Bulgaria, Cyprus, Estonia, and Ireland, all of which emitted more than 3 kg of acidifying substances per capita in 2002 (Figure 1).

Cyprus, Ireland, Portugal and Spain are not on track to meet their 2010 targets. Norway and Austria are close to being on track and the EU as a whole is well on track, due to the good performance of Germany, the United Kingdom and the EU-10 Member States (Figure 2).

Lowest emission intensity in 2002:
Liechtenstein
0.56 kg acidifying emissions/capita

Highest emission intensity in 2002:
Iceland 5.06 kg acidifying emissions/capita (but see text above), next in row
Bulgaria 4.67 kg acidifying emissions/capita

Factor difference:
x9

PART C Country analysis | Thematic assessment

Emissions of ozone precursors

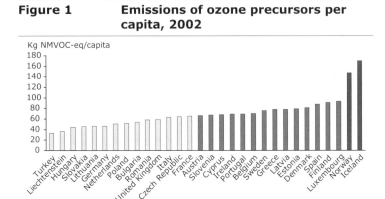

Figure 1 Emissions of ozone precursors per capita, 2002

Note: Includes NO_x, NMVOCs, CO and CH_4.

Figure 2 Change, 1990–2002, in emissions of ozone precursors, compared with the EU-25 average

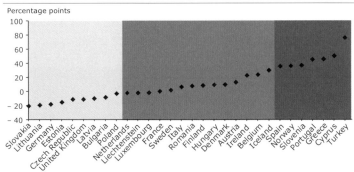

Note: Includes NO_x, NMVOCs, CO and CH_4.

Figure 3 Emissions of ozone precursors, distance to NECD targets, linear path, 2002

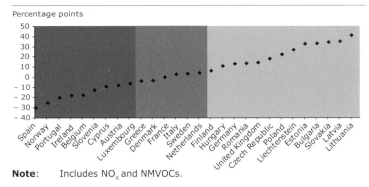

Note: Includes NO_x and NMVOCs.

Notes: For technical specifications and aggregation methodology see CSI 02 in the EEA indicator management system (www.eea.eu.int/coreset). The distance-to-target calculation is explained in the text. Quartile groupings in the progress column and graph are calculated excluding data on Turkey.

Sources: Emissions: EEA; GDP: Eurostat.

Emissions of ozone precursors

High concentrations of ground-level ozone adversely affect the human respiratory system and are harmful to crops and forests. The key ozone-forming precursor gases are NO_x, NMVOCs, CO and CH_4. Emissions of NMVOCs and NO_x are covered by the EU National Emission Ceilings Directive (NECD) (2001/81/EC) and the Gothenburg Protocol of the UNECE Convention on Long-range Transboundary Air Pollution. Each country has its own target for each of these two substances. There are no EU emission targets set for either CO or CH_4. However, there are several directives and protocols that indirectly affect emissions of these two substances. For example, CO is covered by the second daughter directive under the Air Quality Directive, and CH_4 is included in the basket of six greenhouse gases under the Kyoto Protocol.

Ozone formation depends on climatic conditions and is mainly a warm-weather phenomenon. The southern European countries are therefore more predisposed to ozone formation. This means that there is not a level playing field in terms of the relation between ozone precursor emissions and ozone formation, and poor performance in emission reduction in southern countries will probably lead to greater levels of impact, for example on human health, than similar poor performance in more northern countries.

This indicator includes the status (current emissions per capita) and the progress (emission changes 1990–2002) for these four precursors. The distance-to-target comparison is made only for NO_x and NMVOCs, for which there are NECD targets. The distance to target is measured to an assumed linear target path towards each country target. Both targets and emissions have been recalculated in NMVOC equivalents.

The EU-15 as a whole is on track to reach the targets for emission reduction in 2010. However, at country level only four countries are clearly on track (Finland, Germany, Liechtenstein and the United Kingdom). Five additional countries (Denmark, France, Greece, Italy and Sweden) are within + or – 5 percentage points of the linear target line (2002) and are also likely, although less certain, to meet the target. On the other hand, Austria, Belgium, Ireland, Luxembourg, Norway, Portugal and Spain are not on track to meet their NECD targets (and neither are Cyprus and Slovenia in the EU-25, and Norway). The EU-10 Member States have substantially decreased their emissions over the past decade, led by the Slovak Republic with a 57 % reduction (Figure 3).

Estonia, Germany, Lithuania and the Slovak Republic have shown the greatest overall reduction in the basket of four tropospheric ozone-forming gases over the ten years between 1990 and 2002, with over a 50 % reduction in emissions in each case.

Lowest emission intensity in 2002:
Turkey 33 kg NMVOC equivalent/capita

Highest emission intensity in 2002:
Iceland 170 kg NMVOC equivalent/capita and **Norway** 147 kg NMVOC equivalent/capita

Factor difference:
x5

PART C

Country analysis | Thematic assessment

Freight transport demand

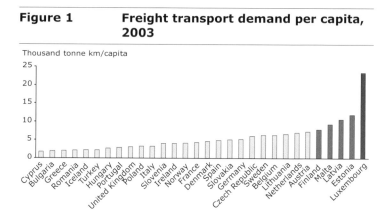

Figure 1 Freight transport demand per capita, 2003

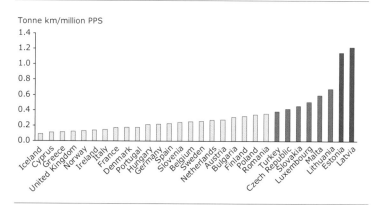

Figure 2 Freight transport demand per unit of GDP, 2003

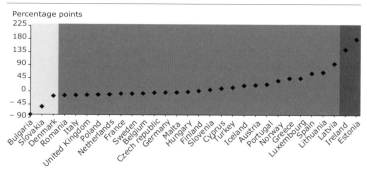

Figure 3 Change, 1995–2003, in freight transport demand, compared with the EU-25 average

Notes: For technical specifications see CSI 36 in the EEA indicator management system (www.eea.eu.int/coreset). The comparison per unit of national income is done on the basis of gross domestic product (GDP) expressed in purchasing power parities (PPP). Hence the unit is not euro but purchasing power standard (PPS).

Sources: Freight transport: Eurostat; GDP: Eurostat, Population: the World Bank.

Freight transport demand

Transport is a key problem area for Europe's environmental policy. Freight and passenger transport demand have both grown substantially since 1995, thereby making it increasingly difficult to limit the environmental consequences of transport.

The EU has set itself the objective of decoupling freight transport demand from GDP. This objective was first mentioned in the transport and environment integration strategy which was adopted by the Council of Ministers in Helsinki in 1999. In the Sustainable development strategy adopted by the European Council in Gothenburg, the objective of decoupling is described as a possible tool to help reduce congestion and other negative side-effects of transport. In the review of the transport and environment integration strategy in 2001 and 2002, the Council reaffirmed the objective of reducing the link between the growth of transport and GDP. In the EU's sixth environment action programme, decoupling of economic growth and transport demand is described as one of the key objectives in order to deal with climate change and alleviate the health impacts of transport in urban areas.

While some success in decoupling passenger transport from economic growth has been achieved, this is not echoed for freight transport, which increased by 18 % between 1995 and 2003. The passenger transport demand indicator showing modal share (CSI 35) is currently under revision to include air transport, and therefore cannot be included in the current EEA country scorecard (see the indicator selection table in the methodology section). On the basis of the present definitions, Europe is not showing signs of progress to the modal split target in freight transport (in the period 1998–2002 the share of road freight increased by 4 %).

From the country scorecard it may be noted that the three Baltic countries consistently end up at the high end of transport demand, both per capita and per unit of GDP, as well as in the relative growth of freight transport. Part of these increases is related to their geographical location on the transport route between Russia, Ukraine and western Europe. Another part of the explanation is low wages, giving these countries a comparable advantage for the location of international transport companies.

Care must therefore be taken when interpreting the scorecard to take the geographical location of countries into consideration, for example so that island states are not compared with the central European and Baltic transit states. Island states are often characterised by high freight transport efficiencies resulting from smaller distances and fewer centres of high population that need to be served by freight transport. From the scorecard it can also be seen that countries which are peninsulas also tend to have generally lower levels of freight transport, possibly for reasons similar to those of islands.

Because this indicator defines freight transport by country of registration rather than actual traffic on the road network of a particular country, many of the apparent trends seen, including decoupling, may instead refer to the relative competitive advantage of a country compared to its neighbours in attracting international transport companies.

Lowest freight transport demand in 2003 were in two island states:
Cyprus per capita (2 ktonne km/capita; Figure 1) and **Iceland** by GDP (90 tonne km/PPS GDP; Figure 2)

Lowest freight transport demand in states other than island states or 'peninsulas' were Hungary per capita (3 ktonne km/capita; Figure 1) and **France** by GDP (175 tonne km/PPS GDP; Figure 2)

Highest freight transport demand in 2003:
Luxembourg per capita (23 ktonne km/capita; Figure 1) and **Latvia** by GDP (1220 tonne km/PPS GDP; Figure 2)

Highest freight transport demand in an island state: **Malta** 9 ktonne km/capita Figure 1, and 600 tonne km/PPS GDP; Figure 2

Factor difference (non-island states): x8 (per capita) and x7 (per GDP)

PART C | Country analysis | Thematic assessment

Area under organic farming

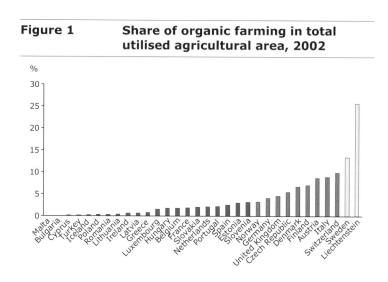

Figure 1 Share of organic farming in total utilised agricultural area, 2002

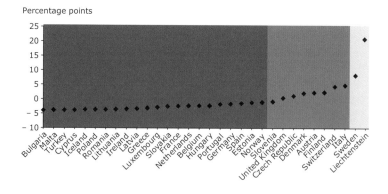

Figure 2 Change, 1992–2002, in the share of organic farming in total utilised agricultural area compared with the EU-25 average

Notes: For technical specifications and definition of organic agriculture see CSI 26 in the EEA indicator management system (www.eea.eu.int/coreset). Percentages refer to the area of certified and policy-supported organic land and land being converted to organic agriculture to total utilised agricultural area. The quartile partition in country groups is made excluding Liechtenstein data.

Sources: Institute of Rural Sciences, University of Wales, Aberystwyth

Area under (and share of) organic farming

Organic farming aims to be a more environmentally sustainable form of agricultural production. Its legal framework is defined by Council Regulation 2092/91 and amendments. The adoption of organic farming methods by individual farmers is supported through agri-environment scheme payments and other rural development measures at the Member State level. In 2004 the EU Commission published the European Action Plan for Organic Food and Farming (COM(2004) 415 final) to further promote this farming system.

This indicator tracks progress in that direction. However, it is only one element in the overall impact of agriculture on the environment. Due to economic transition and other factors, inputs of fertilisers and pesticides in many of the EU-10 Member States, the accession countries and Turkey are still low. Some of the farmed land in many of these countries does not receive fertiliser or pesticides and could be considered similar to organically farmed land from an environmental perspective. However such land is not classified as organic and therefore is not recorded in this dataset. In some cases the barriers to certification still to be overcome are structural, i.e. the organisational infrastructure to support the certification process is lacking. This indicator cannot be used to draw conclusions about the environmental impact of agriculture in different countries.

The share of organic farming has increased, but overall farming is organic on only 4 % of the total utilised agricultural area in the EU-15, 1.4 % in the new EU-10 and only 0.4 % in the remaining EEA countries.

As there is no specific EU target for the share of organic farming area, apart from a generic aim of increasing the share of organic farming, the country comparisons are undertaken on the basis of comparison with the EU-25 average increase. Some EU Member States have national targets for the area under organic farming, often aiming to reach 10–20 % in 2010. Despite progress in the share of organic farming at the European level, its share in many countries is still low. Only eight countries have 5 % or more of their utilised agricultural land under organic farming (Austria, the Czech Republic, Denmark, Finland, Italy, Liechtenstein, Sweden and Switzerland). Eleven countries (Bulgaria, Cyprus, Greece, Iceland, Ireland, Latvia, Lithuania, Malta, Poland, Romania and Turkey) have less than 1 %.

Austria, the Czech Republic, Denmark, Finland, Italy, Liechtenstein, Sweden, Switzerland, and the United Kingdom all show progress equal to or better than the EU-25 average.

Highest share in 2002: Liechtenstein
25.6 % certified organic land, but note that Liechtenstein has only 3 840 ha of utilised agricultural area. **Sweden** on the other hand has 3.8 million ha of which 13 % ([1]) is classified organic land

Lowest share in 2002: Malta
No organic farming reported in its 13 000 ha of utilised agricultural land
followed by **Bulgaria** with 0.01 % of certified organic land of 5.5 million ha of utilised agricultural area

---

([1]) Swedish organic farming area includes a large share of farmland that is not certified according to Regulation 2092/91, but farmed in line with the regulation's specifications.

PART C | Country analysis | Thematic assessment

Municipal waste generation

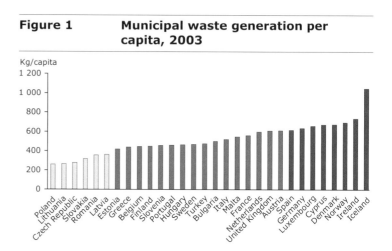

Figure 1 — Municipal waste generation per capita, 2003

Figure 2 — Change, 1995–2003, in municipal waste generation compared with the EU-25 average

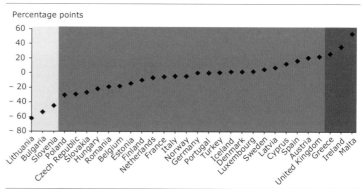

Figure 3 — Municipal waste generation distance to the '300 kg/capita target', 2003

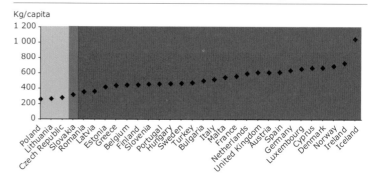

Notes: For technical specifications see CSI 16 in the EEA indicator management system (www.eea.eu.int/coreset). Definitions for municipal waste change from country to country which makes comparisons on a detailed level impossible (see text). In the status column and graph partitioning in quartile groups is done excluding the Icelandic data.

Sources: Eurostat and EEA.

Municipal waste generation

The amount of waste collected by or on behalf of municipalities (municipal waste) continues to increase and, although there are no recent updated targets for municipal waste, this trend is moving in the wrong direction, in Europe.

Key components for improving resource efficiency to ensure that society's consumption of renewable and non-renewable resources does not exceed the carrying capacity of the environment are: better waste management; reducing the volumes of waste generated; and encouraging the reuse, recovery and recycling of waste that are still generated.

In general, Europe is performing poorly in this area with a vast majority of countries experiencing a growth in the collection of waste in the period 1995–2003. The EU's fifth environment action programme (1993–2000) contained a target of 300 kg household waste per capita (expressed as returning to the 1985 level by 2000). There was very little success in meeting this target, and although it was not included in the sixth environment action programme and no new targets have since been set, the 300 kg per capita target is still a useful benchmark. Thus, it is this target which is used for comparing country performance.

However, waste collection systems vary between countries. For example including or excluding bulky waste, waste from rural areas, from small businesses, garden waste and some other categories in municipal waste means that the absolute country figures are difficult to compare. Trends over time in the generation of waste can be obscured by changes in the methods for collection. Different national definitions and methods for gathering the statistics further decrease comparability.

More importantly, however, the progress that several countries are making in stopping the illegal dumping of waste shows up on this indicator as a higher level of waste generation, where in fact such action is clearly environmentally beneficial. This indicator therefore becomes functional only once an optimal level of waste collection is attained, after which trends over time become meaningful. Caution should therefore be applied in the use of this indicator to gauge progress for some countries, in particular the candidate and accession states.

Nevertheless, for most other countries the general picture arising from this indicator provides some insights into both economic development and waste management. However, because of definitional issues caution should be exercised in using the data for comparisons.

The country scorecard shows that the Czech Republic, Latvia, Lithuania, Poland, Romania and the Slovak Republic are close to the old 300 kg per capita benchmark. Cyprus, Denmark, Iceland, Ireland, Luxembourg and Norway produced 650 kg or more municipal waste per person in 2003, more than twice the old target.

Lowest amount of municipal waste in 2003:
Poland 260 kg/capita

Highest amount of municipal waste in 2003:
Iceland 1049 kg/capita including a relatively high share of company waste ([1]) and **Ireland** 735 kg/capita

Factor difference:
x3

([1]) In Iceland, all collected waste is registered as 'municipal waste' but this has recently been changing, as municipal waste and company waste are increasingly collected separately which makes a better assessment of the different waste streams possible. These readjusted figures show that the value of 1 049 kg/year municipal waste generated per capita as cited above is likely to represent 559 kg/capita company waste and 490 kg/year of household waste. Similar considerations may also apply to other countries.

Use of freshwater resources

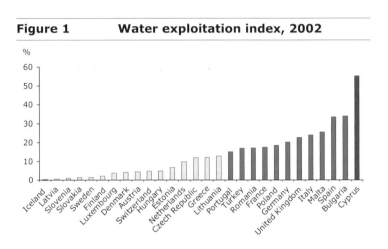

Figure 1   Water exploitation index, 2002

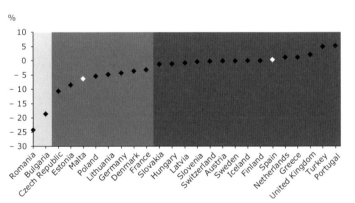

Figure 2   Change, 1990–2002, in the water exploitation index

Note: Malta and Spain were water stressed in 2002 (as were Cyprus and Italy, but progress for these two countries could not be calculated). Bulgaria, Germany and the United Kingdom all have large power plant cooling water abstractions. However, these abstractions are not considered to contribute to water stress.

Notes: The water exploitation index is the mean annual total water abstraction divided by the mean annual total renewable freshwater resource. For technical specifications see CSI 18 in the EEA indicator management system (www.eea.eu.int/coreset). For the United Kingdom data are only available for England and Wales.

Sources: Eurostat, EEA, national sources.

Use of freshwater resources (water exploitation index)

Water abstraction decreased in 17 EEA countries between 1990 and 2002. But several countries in southern Europe still experience water stress, abstracting a considerable part of available water.

The EU's sixth environment action programme aims at ensuring that rates of abstraction from water resources are sustainable over the long term, which implies improvement of the efficiency of water use in different economic sectors. There are no specific quantitative targets directly related to this indicator, as climatic conditions and land-use differ strongly across Europe. However, the Water Framework Directive (2000/60/EC) requires countries to promote sustainable use based on long-term protection of available water resources and ensure a balance between abstraction and recharge of groundwater, with the aim of achieving good groundwater status by 2015.

While the use of water for cooling (in electricity production and energy) is of little concern with regard to the available amounts of water, other uses do raise concerns. Countries with a high water stress, that is a high ratio of abstraction against water resources, have problems with groundwater table lowering, degradation of natural wetlands and salt water intrusion into coastal aquifers. The latter happens particularly in the Mediterranean region. The main water-consuming sectors are irrigation, urban use, and manufacturing industry. Water use for irrigation has been growing in many countries. Traditionally much of the irrigation in Europe consisted of gravity-fed systems where surface water is transported through small channels. In an increasing number of regions in the north and the south, sprinkler irrigation withdrawing groundwater is becoming common practice. This increases water consumption and contributes to the impacts mentioned above.

The decrease in agricultural and industrial activity in the EU-10 and Romania and Bulgaria during the transition process led to decreases of about 70 % in water abstraction for agricultural and industrial uses in most of the countries. Agricultural activities reached their minima around the mid-1990s, but more recently countries have been increasing production.

These developments can be summarised in terms of the water exploitation index, which is defined as the mean annual total abstraction of freshwater divided by the mean annual total renewable freshwater resource at the country level. Resources include inflow from upstream countries. To enable comparison, it is expressed as a percentage. The warning threshold for the index, which distinguishes a non-stressed from a stressed region, is around 20 %. Severe water stress can occur where the index exceeds 40 %, indicating unsustainable water use.

Excluding countries that abstract a lot of water for cooling purposes, there are four countries that can be considered as water-stressed (Italy, Malta, Spain and Cyprus). Only Cyprus has a water exploitation index exceeding 40 %. In water-stressed countries such as Cyprus there is increasing reliance on the desalination of seawater to meet rising needs. Five countries (the Netherlands, Greece, the United Kingdom, Turkey and Portugal) increased their water exploitation index in the period 1990–2002 because of increases in total water abstraction.

Least problem country in 2002:
Iceland Share of available water that is abstracted: 0.1 %

Problem country in 2002:
Cyprus Share of available water that is abstracted: 55.3 %

Factor difference:
x500

Country perspectives

The following country-by-country analyses were prepared in partnership with the countries to provide additional country level perspectives on the scorecard analyses. The selection of indicators included in the scorecard is the responsibility of the EEA and does not necessarily reflect the priorities of the countries. To allow a deeper understanding of the issues at country level, some figures included here are from national sources and so may not be fully comparable with data compiled by Eurostat, the EEA or other international bodies. The EEA takes full responsibility for the final result.

In an increasingly diverse European Union and set of EEA member countries, performance across a range of environmental issues and policies will be strongly influenced by the different starting positions and legacies in the countries such as geographic and climatic pre-conditions and socio-economic legacies and governance.

When put alongside the scorecard, these country analyses serve to flag up different issues. Together they help to open up a debate about country and European environmental performance which we hope will support shared policy learning.

For example, one of the key questions that arises from this assessment is: 'What does good environmental performance actually mean for a specific country'. This is highlighted by the way in which for some countries resource limitation is not a relevant issue, for example energy and water in Iceland. Understanding then how such a country should be assessed in a European context therefore becomes a key question and the scorecard serves an important function in helping to raise such questions.

Thus the aim of the country analysis is to complement the scorecard by providing a deeper understanding of country conditions, behaviour and response to environmental problems.

Each country analysis is introduced by an overarching summary for each country, followed by an assessment of each of the nine indicators included in the scorecard. Since the objective here is to connect the bottom-up country reality with the top down European assessment, the texts were prepared in partnership with the countries.

For ease of reference, the country level analyses are presented in alphabetical order. When reading the analyses, reference should be made to each country's line in the scorecard (which is also presented on each country page) and to the ranking graphs for each country (figures grouped at the end of this part of the report).

In addition Table 1 is provided to illustrate the socio-economic diversity of Europe. The table presents the EEA member countries in groups allowing the different sizes of the economies in these countries to be contrasted. This is an additional reference to help in understanding the analysis.

More detailed information at country level can be found in the latest national environmental reports catalogued in the SERIS data base http://countries.eea.eu.int/SERIS. For further information please contact the relevant national focal point. Contact details can be found on http://org.eea.eu.int/organisation/nfp-eionet_group.html.

Country perspectives   PART C

Table 1   Population and GDP for EEA member countries 2003

Country	Population (1 000s)	GDP (million EUR)	
Germany	82 551	2 072 162	Over 70 million inhabitants
Turkey	70 712	166 092	
France	59 725	1 407 304	
United Kingdom	59 280	1 078 600	
Italy	57 646	944 770	
Spain	41 101	582 408	Over 30 million inhabitants
Poland	38 195	141 807	
Romania	22 200	29 598	
Netherlands	16 215	385 436	
Greece	10 680	120 249	
Belgium	10 348	249 185	10 million inhabitants
Czech Republic	10 202	49 084	
Portugal	10 191	100 758	
Hungary	10 120	46 002	
Sweden	8 956	232 716	
Austria	8 059	217 399	
Bulgaria	7 824	10 959	
Switzerland	7 344	275 660	
Denmark	5 387	162 099	5 million inhabitants
Slovakia	5 381	20 066	
Finland	5 210	131 784	
Norway	4 560	141 203	
Ireland	3 947	94 404	
Lithuania	3 454	7 473	
Latvia	2 321	6 007	
Slovenia	1 964	20 548	
Estonia	1 350	4 515	
Cyprus	770	9 132	
Luxembourg	448	20 823	Under 1 million inhabitants
Malta	399	3 095	
Iceland	286	7 088	
Liechtenstein	33		

Note: These figures, from the EEA data service, are those used as the normalising variables in the scorecard.

Sources: GDP 1995 constant prices: Eurostat; Population: the World Bank.

Austria

Austria is one of the leading countries in organic farming and renewables. Despite a generally high level of environmental protection and eco-efficiency, it has problems meeting agreed reduction targets for greenhouse gas and NO_x emissions. Pressure on the environment is increasing due to a continuing rise of passenger and freight transport, causing various environmental problems, including air emissions, noise and fragmentation of ecosystems and landscapes.

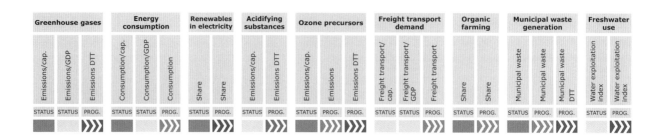

Greenhouse gas emissions

Because of rising emissions, in particular from the road transport and energy production sectors, total 2002 greenhouse gas (GHG) emissions were 8.5 % above 1990 levels. Austria needs to reduce emissions if its Kyoto Protocol target of 13 % below 1990 levels is to be met by the 2008–2012 commitment period. In particular, GHG emissions from transport have shown an increase of 82 % between 1990 and 2003 (though a considerable amount is due to 'fuel tourism' caused by low fuel taxes and thus prices in Austria). In 2005, the Austrian government launched a broad consultation process re-evaluating the national climate strategy with a view to taking appropriate measures.

Energy consumption

Total energy consumption increased by almost 5.6 % between 2000 and 2002. Fossil fuel sources contribute approximately 77 % of the total energy demand. Additional measures to abate increasing energy consumption are needed.

Renewable electricity

Production of renewable energy is increasing, mainly as a result of the *Ökostromgesetz* (target: additional 4 % of renewable energy in electricity production in 2008). Austria generates 67 % of its electricity from hydropower (including large hydropower plants) and 0.5 % from other renewable energy sources, based on data from 2002. However, the use of renewable energy sources (biofuels directive) in the transport sector, which is expected to increase substantially until the year 2008, is also relevant even though it is not included in this specific indicator.

Emissions of acidifying substances

Compliance with the ceilings of the EU national emission ceilings (NEC) directive for SO_2 and NH_3 will be achieved; high reductions, in particular in SO_2, have already been accomplished. The major challenge is NO_x, in particular emissions from the transport sector. Some reasons for higher NO_x emissions are: 'fuel tourism' (fuel prices in Austria are lower than in some neighbouring countries); high share of transit traffic and diesel vehicles (with higher specific emissions than petrol cars); higher real-life emissions of vehicles than during test cycles. Additional ambitious measures in main source sectors are currently being discussed.

Emissions of ozone precursors

Compliance with the ceiling of the NEC directive for NMVOC is likely to be achieved. For NO_x see comments above under 'acidifying substances'.

Population: 8 059 000
Size: 83 860 km²
GDP: 217 399 million euro

Country perspective

Freight transport demand

Pressure on the environment is increasing due to a continuing rise of both passenger and freight transport, causing various environmental problems, including air emissions as well as noise and fragmentation of ecosystems and landscapes.

Freight transport has grown steadily in recent years, in particular since the second half of the nineties. Freight transport demand has doubled between 1980 and 2002; roughly two-thirds of this transport is road transport. Increasing contribution of fuel tourism and transit traffic have further contributed to increasing emissions, as shown in the national inventory. A recent study indicates that up to one-third of the motor fuels sold in Austria is not consumed in Austria. With the implementation of the biofuels directive into national law a slowdown in the growth of national GHG emissions by the Austrian transport sector is expected.

Share of organic farming

The share of organic farming area in Austria increased once more from 8.8 % in 2002 to 9.6 % in 2003. This progress is particularly due to a considerable increase in organic arable farming. The total area of certified and supported organic farms (according to IACS, the integrated assessment and controlling system) amounts to 326 703 ha as of 2003.

Municipal waste

The high figures are not only the result of the generated waste quantities but also of a comprehensive waste collection system which covers all households in Austria, and of high waste quantities collected from communal services. Nearly half of the increase between 1995 and 2001 is caused by an improvement in data collection. During the same period, the percentage for recycling and composting of municipal waste increased from 51 % to 63 %, while the share of waste going to landfills dropped from 46 % to 33 %.

Use of freshwater resources

Between 3 505 and 3 850 million m³ water are abstracted from freshwater sources every year. From 1985 to 2002 the situation was more or less stable regarding total water abstraction. Variability from year to year can mainly be explained by variations in hydropower generation which are compensated by electricity production from caloric power plants, thus leading to variations in water abstraction used for cooling. However, since 1985, water abstraction for industrial production purposes has decreased from 43 % to 32 % of the total amount, whereas water abstraction for cooling purposes for production of electricity has increased from 32 % to 48 %. Irrigation plays a negligible role in Austria: less than 2 % of the water abstracted is used for irrigation purposes. The present per capita abstraction of freshwater is about 470 m³ (including cooling-water used for the production of electricity). Since important efforts have already been taken in the past to increase efficiency and to reduce losses, no major further reductions can be expected in the near future.

For more information please contact the relevant national focal point. Contact details can be found on:
http://org.eea.eu.int/organisation/nfp-Eionet_group.html

Belgium

Belgium has a high population density and is the main crossroads of western Europe. This leads to considerable pressure on the environment and the land. Belgium is performing moderately across most indicators. However, the country presents good results for municipal waste generation and is having some success with tackling freight transport. An area for particular attention is ozone pollution.

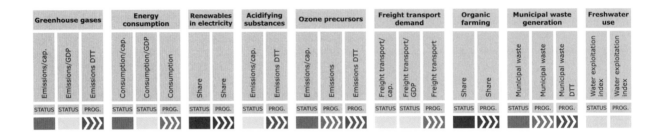

Greenhouse gas emissions

Total emissions in 2002 were at the same level as in 1990, but Belgium seems to be on track to reach the Kyoto target (7.5 % below the 1990 level). The regions showed different results for 1990–2002 GHG emissions: 3.1 % increase in Flanders; 7.3 % reduction in the Walloon region; and 9.6 % increase in the Brussels-Capital region. In Flanders GHG emission trends in the period 1990–2002 showed a 26 % increase from transport but reductions from industry (– 11 %) and agriculture (– 10 %). The Walloon region showed increases from the domestic sector (15 %), the tertiary sector (8.7 %) and the transport sector (3.6 %, with a dramatic increase of 376 % of chlorofluorocarbon (CFC) emissions) and a strong decline from the industry sector. Measures to reduce emissions include voluntary agreements with energy-intensive industries, performance standards for residents, and support for combined heat and power generation in Flanders. Internal and external measures (investing in the World Bank Community Development Carbon Fund) are being introduced in the Walloon and Brussels-Capital regions.

Energy consumption

Flanders reports improvements in energy intensity due to efforts in the industry and energy sectors since 1998. The Walloon region reports an increase in final energy consumption by 7.6 % (1990–2002). Final energy consumption in the Brussels-Capital region increased by 18 % between 1990 and 2003, to 2.16 million tonne oil equivalent in 2003. Flanders aims to reduce household energy use by 7.5 % in 2010 compared with 1999, despite an increase of 37 % from 1990 to 2002, through several measures to promote the rational use of energy. The Walloon *Plan pour la maîtrise durable de l'énergie* shows that total energy demand could be reduced by 9 % in 2010, compared with 1990, with detailed targets for various sectors. The key sectors in the Brussels-Capital region are housing, tertiary and transport, and the most important energy carriers are natural gas, oil and electricity.

Renewable electricity

There was an almost threefold increase in 2002 but the total share only reached 2 %. The share of renewable energy in electricity production in Flanders is growing (0.75 % in 2003). The use of the organic fraction of household waste will contribute to reaching targets. In the Walloon region, the share reached 2.3 % in 2003: hydroelectricity fell in 2003 due to unfavourable climatic conditions, and wind energy is growing rapidly but is less than 2 % of the total. The Brussels-Capital region applies 'green certificates' covering renewable energy production in the two other regions to boost demand.

Emissions of acidifying substances

Belgium seems to be on track to meet the NECD targets. Emissions in Flanders fell by 41 % (1990–2003), but deposition of acidifying substances is higher than the critical loads in 53 % of the nature area. New measures for the different industrial sub-sectors and lower

Population: 10 348 000
Size: 30 528 km²
GDP: 249 185 million euro

emissions from agriculture should enable the targets to be met. Emissions have also fallen in the Walloon and Brussels-Capital regions.

Emissions of ozone precursors

If no extra measures are taken, Belgium will not reach the target. In Flanders volatile organic compound (VOC) emissions decreased by 43 % during the period 1990–2003, NO_x emissions by 12 %. The Walloon plan prioritises the reduction of VOC emissions. Progress has already been made through the use of catalytic converters and the reduction of solvents in paints. Emissions of VOCs and NO_x fell by 25 % in the Brussels-Capital region (1990–2003).

Freight transport demand

Freight transport demand is still growing. In Flanders it has increased by 30 % (1995–2000), but has stabilised since 2000. The use of waterways is increasing (46 % in 1990–2003). Total transport demand in the Walloon region keeps increasing; freight transport by 17 % (1995–2000). Road transport represents up to 85 % of freight transport. Total road traffic in the Brussels-Capital region increased by 15 % (1990–2003) (small reduction in 2003).

Share of organic farming

The area under organic farming stabilised at around 1.7 % of the total agricultural area in 2004. Organic farming in Flanders covered only 0.5 % of the total agricultural area (2004) but new subsidies have recently been endorsed. In Wallonia 2.7 % utilised agricultural area is under organic farming (2004) and the number of farms converting to organic farming is still increasing. The Brussels-Capital region, although very urbanised, is developing a 'green network' of public spaces, including some nature reserves and parks. Some areas of this 'green network' are managed in a differentiated way, e.g. extensive gardening and protection of threatened species.

Municipal waste

There has been good progress in slowing the growth of municipal waste. Household waste generation in Flanders is decreasing: stabilisation in 2001, 0.2 % reduction in 2002 and 3.4 % in 2003. Waste generation per capita has fallen since 2001. 70 % of household waste is collected separately, most of this is reused, composted or recycled. Wallonia has seen a slow but irregular decline in municipal waste since 1997. A large proportion is recovered: in 2003 more than half went to material reclamation plants and less than 20 % to landfill. The amount of municipal waste collected in the Brussels-Capital region was stable between 1999 and 2002. Raw municipal waste fell by 9.4 % (1996–2002). Selective collection of waste for recycling of packaging increased by 42.9 % and of other types of paper and paperboards by 50.1 %.

Use of freshwater resources

Total use of water (excluding cooling-water) in Flanders decreased by 14 % (1991–2002). Industry use decreased by almost 40 % in the period 1996–2000. Water availability in Flanders is low and two-thirds is imported from Wallonia. Among the lowest in Europe the Walloon region uses 105 litres per person per day for domestic needs. This is due to increasing water prices, the use of more efficient equipment and increasing use of rainwater. In 2004, the Brussels region used 113 litres per person per day for domestic needs. 61 % of water in the Brussels-Capital region is used by households, 25 % by the tertiary sector and 11 % by fire control and other public services, including network losses.

For more information please contact the relevant national focal point. Contact details can be found on:
http://org.eea.eu.int/organisation/nfp-Eionet_group.html

Bulgaria

Bulgaria is performing well across the indicators. Emissions of ozone precursors have already exceeded the 2010 targets. Furthermore, most progress indicators show relatively good trends. Bulgaria is maintaining its vigilance to ensure that environmental trends do not deteriorate as economic and social changes occur, especially in the lead up to EU accession.

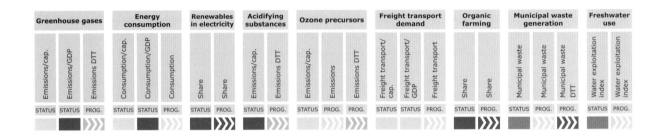

Greenhouse gas emissions

Greenhouse gas emissions substantially decreased in Bulgaria in the period 1988–2002. In 2002 Bulgaria reduced GHG emissions by 79 million tonnes (minus 56 %) compared to the base-year 1988. The main reasons for this reduction were:

- governmental policies for transition to the market economy, restructuring of industry, privatisation and liberalisation;
- energy policy towards liberalisation of the energy market and removal of subsidies;
- decrease of the population; and
- decrease of the GDP.

The result is that Bulgaria can confidently expect to comply with the emission target for the first commitment period.

Energy consumption

Bulgaria has implemented the EU directives on the liberalisation of the electricity market, which will substantially influence the way the Bulgarian government can manage environmental emissions from the energy sector and will induce a shift from 'command-and-control' type of policy instruments to market-based instruments, such as the Emissions Trading Scheme and green certificates.

Renewable electricity

For 2003 renewable energy sources provided 7.75 % of total electricity production. For renewable electricity production, the Renewable Electricity Directive (Council Directive 2001/77/EC on the Promotion of Electricity from Renewable Energy Sources in the Internal Electricity Market) is the most important piece of legislation. The adopted energy law in Bulgaria incorporates the measures listed in the Directive and provides the framework for green certificates. In 2004 the government started with the preparations of the implementation of the green certificates scheme in Bulgaria. The system is planned to become operational in 2006.

Emissions of acidifying substances

According to the CLRTAP and its protocols, national emissions of acidifying substances should, by 2010, be reduced by 57 % for SO_2, 26 % for NO_X and 25 % for NH_3 compared to the base-year 1990. In 2003 the national SO_2 emissions were reduced by 52 %, NO_X by 42 % and NH_3 by 64 % compared to the base-year. This means the targets for 2010 are in reach.

Emissions of ozone precursors

The national non-methane volatile organic compounds (NMVOC) emissions have to be reduced by 15 % by 2010 compared to the base-year 1990. The decreasing trend of NMVOC emissions shows a reduction of 45 % in 2003, which means the target for 2010 has already

Population: 7 824 000
Size: 110 910 km²
GDP: 10 959 million euro

been met. The reduction of NO_x in 2003 by 42 % compared to the base-year also means that the target for 2010 has already been fulfilled.

Freight transport demand

The changes in the Bulgarian economy after 1989 have led to a significant initial drop in the transport of goods. After 1997 the general tendency has gradually reversed. However, processes taking place across the different types of transport modes are diverse. Land transport trends in Bulgaria are similar to those in the European Union. Road transport continues to grow, showing a serious increase with respect to the amount of transported goods and work done (in tonne km). At the same time its main competitor, rail transport, shows a continuous drop in the amount of transported goods and work done. Marine transport remains a key means of long-distance transportation due to its advantages for transporting goods of significant size, while river transport has up to now been of little importance in the Bulgarian economy.

Share of organic farming

The development of organic farming in Bulgaria has been relatively slow, but is now increasing rapidly. Bulgaria has favourable soil and climatic conditions for organic farming and is already developing a reputation as a high-quality producer of speciality organic products (e.g. essential oils and medicinal herbs) with good export potential. There are also good opportunities for the development of a domestic market for organic products, including fresh and processed fruits and vegetables. At the end of 2004 the area under organic farming was approximately 11 771 ha, representing 0.2 % of the used agricultural area. 11 259 ha have passed the transition period while the remaining 512 ha are already in a process of transition. In 2005, a national strategy and an action plan for the development of organic farming were prepared, covering the period 2006–2013.

Municipal waste

Unlike in the rest of Europe, household waste production is on the decrease. The majority of collected municipal waste is landfilled. 0.4 % of this waste was designated for recovery operations in 2002, but further development of infrastructure will be needed to come within reach of the national priorities for decreasing landfill of biodegradable waste.

Use of freshwater resources

The total quantity of used freshwater in Bulgaria for the period of 1999–2003 is relatively stable and ranges from 6 820 to 6 918 millions m³. In 2000–2001 this quantity decreased from 6 130 to 5 833 million m³. The quantity of freshwater from surface water sources used in 2003 was 6 450 million m³, while the amount of used freshwater from groundwater sources was 467 million m³. The main consumers of freshwater are industry and agriculture. The share for drinking water consumption decreased twofold between 1999 and 2003, due to higher water prices and measures taken to decrease water losses in the water supply systems.

For more information please contact the relevant national focal point. Contact details can be found on:
http://org.eea.eu.int/organisation/nfp-Eionet_group.html

Cyprus

Cyprus has a large potential for renewable energy and organic farming and is implementing plans and measures to promote progress. Improvement is needed in air emissions and waste. While the 1992–2002 period studied is not representative, progress has been made following EU accession. The legislative changes and the necessary measures and programmes now in full implementation are expected to significantly improve the country's environmental performance.

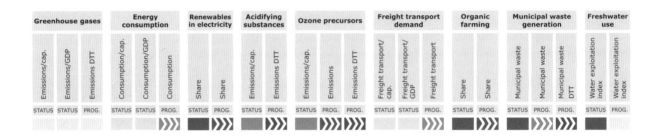

Greenhouse gas emissions

Cyprus has no quantified targets yet, however as a contracting party to the Kyoto Protocol it is expected that it will have to commit to important reductions in the rates of its greenhouse gas emissions. In response to this, a strategic plan for the reduction in the rate of increase of greenhouse gas emissions is being gradually implemented. This concentrates on the promotion of renewable energy sources, the use of natural gas, energy conservation and efficiency, transport management, changes in agriculture and industry and waste management. Additionally, a greenhouse gas emissions trading system has been prepared and approved by the EU. These measures are expected to reduce greenhouse gas emissions.

Energy consumption

The energy intensity in Cyprus is above average but progress is being made in energy efficiency and conservation, although there is still considerable room for improvement. Measures have been taken towards this, including the implementation of an action plan (2002–2010), the adoption of a new law on the promotion and utilization of renewable energy sources and energy conservation, and a new grants scheme for the promotion of renewable energy sources and the rational use of energy in all sectors.

Renewable electricity

Although Cyprus is one of the leading countries in the use of solar water heating systems, where solar energy constitutes approximately 2 % of the country's primary energy consumption, the share of renewables in electricity is still very small. Based on new measures that have been implemented, the target is to increase the share of renewables to 6 % by the year 2010 through the utilization of the wind, hydro and biomass potential, together with the use of photovoltaics and co-generation.

Emissions of acidifying substances

Emissions of acidifying substances are still high in Cyprus and there has been a 5 % increase in emissions in 2002 compared to 1990 levels. Legislation has been enacted for the control of air quality and atmospheric pollution together with integrated pollution prevention and control, therefore progress is expected in the near future.

Emissions of ozone precursors

Although emissions of ozone precursors in Cyprus are relatively low compared to other European countries, they have been increasing and the emission levels for 2002 were 14.3 % above 1990 levels. However, legislation has been enacted for the control of air quality and atmospheric pollution together with integrated pollution prevention and control, and therefore progress is expected in the near future.

Cyprus

Country perspective

Population: 770 000
Size: 9 250 km²
GDP: 9 132 million euro

Freight transport demand

Freight transport has increased by 29.7 % in 2003, compared to 1990 levels. The per capita freight transport demand is very low compared to other European countries. Shipping covers most of the transport demand, while road transport is relatively low due to the small size of the island and the short travel distances.

Share of organic farming

The share of organic farming in Cyprus is still relatively low and in 2002 amounted to less than 1 % of the total utilised agricultural area of the country. However, relevant legislation has been enacted, while measures for the development of organic agriculture have been included as part of the strategic development plan for agriculture (2004–2006). These measures have been designed to encourage organic farming and contribute to the creation of an economically viable sector by providing support for the conversion from conventional to organic agriculture. They include economic incentives and educational and awareness programmes for farmers and consumers. Such measures are expected to increase the share of organic farming in the future.

Municipal waste

In 2002 Cyprus was producing approximately 0.654 tonnes of municipal waste per person per year, meaning that waste production is on the increase with the volume of waste generated in 2002 being 29.2 % above 1995 levels. A waste management strategy has been drafted which includes objectives for the reduction, re-use and recycling of waste, as well as treatment and disposal. Additionally, a programme has been established to encourage the recycling of packaging waste in various municipalities. These measures are expected to reduce the volume of waste generated and promote reuse and recycling.

Use of freshwater resources

Although water abstraction per capita is comparatively low, Cyprus is not an efficient user of water for irrigation and water abstraction per irrigated area is very high. The government water policy is currently focused on the exploitation of other non-conventional water sources, such as recycled water. Additionally, measures are being taken to reduce irrigation water demand, including financial incentives and educational and awareness programmes.

For more information please contact the relevant national focal point. Contact details can be found on:
http://org.eea.eu.int/organisation/nfp-Eionet_group.html

Czech Republic

The Czech Republic shows both good progress and performance across the indicators and is on track to maintain and improve the quality of its environment in the future. Fast economic growth is now expected following EU accession and therefore issues which may gain in importance include emissions of acidifying substances, energy intensity, greenhouse gas emissions and freight transport intensity.

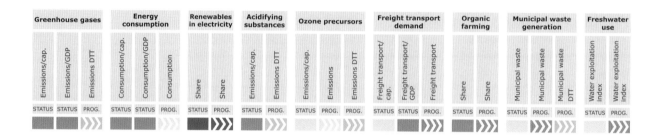

Greenhouse gas emissions

The total emissions of greenhouse gases decreased by 25 % between 1990 and 2003 and reached 143.4 million tonnes CO_2-equivalents in the year 2003. This implies that the Kyoto Protocol targets (to decrease by 8 % compared with 1990) are being fulfilled. Although the trend of GHG emissions is quite favourable, the absolute values are still a problem — however the emissions of greenhouse gases per capita in the Czech Republic are significantly above the average of the EU-15.

Energy consumption

The final energy consumption in 2004 was 1 099.3 PJ which is significantly lower than during the first half of the nineties due to the transformation of industry and the implementation of lower energy intensity technologies. However, energy consumption is now slowly increasing by approximately 1 % per year. Energy intensity is 60 % higher than the EU-25 average and still remains an issue. The implementation of the national programme for economical energy management and use of renewable and secondary energy resources is expected to improve the situation. This defines sustainable objectives for energy conservation and the use of renewable and secondary energy sources. The second phase of this programme will run from 2006 to 2009.

Renewable electricity

In 2004 the share of electricity generation from renewable sources was 4 % of the gross electricity consumption and 2.9 % of primary energy sources. Even though this has increased slightly from 2003, it remains too low to achieve the national indicative target of 8 % of gross electricity consumption from renewable sources by 2010. Following EU Directive 2001/77/EC to promote renewable electricity, the recently adopted Czech Act No. 180/2005 Coll. will significantly help to meet this target. Through guaranteed feed-in prices, this Act guarantees the recovery of investments in renewable energy sources (the lack of financial security for investors has been the main barrier to a more widespread use of renewable energy sources so far).

Emissions of acidifying substances

Since 1990 a dramatic decrease in emissions has taken place: almost 90 % for SO_2; 40 % for NO_x; and about 50 % for NH_3. The current and future expected decreasing trends of these emissions are politically supported by the implementation of two national programmes for the reduction of emissions (reduction in general and from especially large combustion sources). This situation should enable the national emission ceilings set for SO_2 and NH_3 to be complied with by 2010. However, there may be problems achieving the emission ceiling for NO_x by 2010 due to the extrapolation of the presently stagnating or even slightly increasing NO_x emission trends.

Population: 10 202 000
Size: 78 870 km²
GDP: 49 084 million euro

Country perspective

Emissions of ozone precursors

Because of its geographic location, the Czech Republic belongs to a middle vulnerable region with respect to ozone formation. Since 1990 a considerable decrease in emissions of all ozone precursors has taken place. The two national programmes mentioned above for the reduction of emissions into the atmosphere will contribute to further reductions.

Freight transport demand

From 1995 until recently, the trend in freight transport has followed GDP. In the period 1995–1998 there were significant variations in actual freight transport demand due to changes in the economy. In the period 1998–2001 there was a constant value of 36 tonne km/1 000 CZK (approximately 1 tonne km/euro). From 2002 to 2003 freight traffic performance increased more than GDP, but in 2004 it decreased in value, indicating decoupling for the first time.

Share of organic farming

Organic farming started in the early 1990s. In 2004, there was a total of 836 organic farmers and 263 299 ha of agricultural land managed organically, i.e. 6.16 % of the total agricultural area. Organic farming is mostly applied by farms in mountainous and sub-mountainous regions on permanent grassland. About 90 % of the organic area is grassland, the share of arable land amounted to 7.5 %, perennial crops to 0.4 % and the rest are other areas. The most important animals in organic husbandry are beef cattle. The largest expansion of land used for organic farming occurred in the years 1998–2001, particularly in connection with the renewal of government payments to organic farmers in 1998. Since 2004, organic farming has been supported within the framework of new agri-environmental schemes in the Czech horizontal rural development plan.

Municipal waste

The production of municipal waste increased from 1995 to 2002 and reached 4.6 million tonnes in 2002. Since 2002 municipal waste production has slightly decreased (between 2003 and 2004 minus 4.4 million tonnes). Landfilling remains the most usual form of disposal: with 67 % in 2004. 11.7 % of municipal waste was recycled in 2003; the amount of recycled and recovered municipal waste is increasing. Incineration has increased significantly since 1999, moving from less than 0.5 % of overall waste volume to almost 10 % in 2003.

Use of freshwater resources

In 2004 total withdrawals from surface waters amounted to 1 626.1 m³ and from groundwaters 401.9 m³. The public supply system accounted for 24 % of total withdrawals from surface waters and 86 % from groundwaters (2003–2004). The energy industry made up the largest share of withdrawals from surface waters (54 % in 2004). Agriculture is not an important consumer of water, accounting for 4 % of surface water withdrawals in 2003 (2 % for groundwater). The overall decreasing trend in withdrawals of surface waters (1990–2001) has not continued, however, the decrease for 1990–2003 was 41.3 %. In 2002 and 2003, there was an increase of withdrawals of 19.6 % seen in all categories of water use. Power engineering and heat generation contributed the largest share of the increase. Preliminary data for 2004 indicates that the trend has reversed and withdrawals from surface waters have dropped by 4.1 % since 2003 (by 4.5 % for groundwater).

For more information please contact the relevant national focal point. Contact details can be found on:
http://org.eea.eu.int/organisation/nfp-Eionet_group.html

Denmark

Denmark is a European leader in the areas of renewable energy, water use and organic farming. However, per capita it is still performing less well in some areas such as municipal waste generation and, whilst managing to keep close to target, also air emissions.

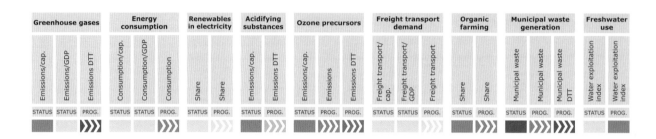

Greenhouse gas emissions

In 2002 Denmark's greenhouse gas emissions were 1 % below base-year levels. If the base-year is adjusted for electricity trade in 1990, GHG emissions were, in 2002, 9 % below base-year level with a distance-to-target indicator of + 3.5 percentage points. Main factors for decreasing emissions with regard to 2001 were decreases in fossil fuel combustion in households and industry, and emission decreases from agricultural soils. Between 1990 and 2002, large emission decreases from agricultural soils and from households counterbalanced increasing road transport emissions. Projections based on current measures are below target for 2010. Denmark will not achieve the Kyoto target based on these projections without additional measures. Denmark intends to achieve the Kyoto target through use of the Kyoto mechanisms and additional cost-effective domestic measures.

Energy consumption

In the period 1990–2003, the Danish gross energy consumption (adjusted for fuel consumption linked to foreign trade in electricity and climate variations in relation to a normal weather year) increased by 1 %, while GDP for the same period has increased by 30 %. In the same period efficiency measures have led to a reduction of adjusted CO_2 emissions of 14.9 %. The total national electricity consumption has increased by 14.2 % in the same period.

Renewable electricity

The share of renewable energy has gradually increased over the period. In 2003 renewable sources, including waste, constituted 24.8 % of national electricity consumption. Wind energy accounted for the majority of the consumption (15.8 %), but biomass has also increased in importance, contributing 4 % in 2003.

Emissions of acidifying substances

In 1990 the relative contribution in acid equivalents was almost equal for the three gases. In 2003 the most important acidification factor in Denmark was ammonia nitrogen and the relative contributions for SO_2, NO_x and NH_3 were 9 %, 40 % and 51 % respectively. From 1980 to 2003, SO_2 emissions decreased by 93 %. This large reduction is mainly due to the installation of desulphurisation plants and the use of fuels with lower sulphur content in public power and district heating plants. From 1985 to 2003, NO_x emissions decreased by 32 %, the contribution from public power and district heating plants decreased in this period by 47 % due to installation of low NO_x burners and de-nitrifying units. Almost all Danish emissions of NH_3 result from agricultural activities. The total ammonia emissions decreased by 32 % from 1985 to 2003. This is due to a vigorous national environmental policy during the last twenty years.

Population: 5 387 000
Size: 43 090 km²
GDP: 162 099 million euro

Emissions of ozone precursors

The emissions of non-methane volatile organic compounds (NMVOC) in Denmark originate from many different sources and can be divided into two main groups: incomplete combustion, and evaporation. Road transportation vehicles are still the main contributor to emissions from incomplete combustion, even though the emissions have declined since the introduction of catalytic converters in 1990. Evaporative emissions mainly originate from the use of solvents. Emissions from energy industries increased during the nineties because of increasing use of stationary gas engines, which have much higher emissions of NMVOC than conventional boilers. Total anthropogenic emissions decreased by 39 % between 1985 and 2003, mainly due to an increasing use of catalytic converters in cars and reduced emissions from the use of solvents. For NO_x, refer to the text under acidifying gases.

Freight transport demand

Transport trends in Denmark have not really moved in a sustainable direction. Inland freight transport may show weak signs of decoupling, while a shift towards rail transport is not taking place. For passenger transport there are signs of some decoupling. The Great Belt Link (inaugurated in 1998) has also meant a significant reduction in domestic air travel, boosting intercity rail (as well as road transport). The new Copenhagen Metro has led to increased urban rail use.

Share of organic farming

The relatively high percentage of area under organic farming in Denmark is due to a combination of a market-driven demand for organic dairy products and public support for conversion to organic cash crop production in order to reduce environmental impacts from farming. The large increase in organic areas in the late 1990s was followed by stagnation after 2000 due to falling market prices of organic food.

Municipal waste

In line with other European countries, household waste production has increased in the last 10 years, but has stabilised in the last few years. The majority of this waste is being incinerated (60 % in 2003), 31 % is being recycled and 6 % is being landfilled. The national targets are 33 % recycling, 60 % incineration and 7 % landfilled by 2008.

Use of freshwater resources

Drinking water abstraction in Denmark is based on groundwater resources. Sustainable use has been encouraged by the introduction of water taxes and promotion of awareness regarding wastage in the distribution systems. Use of water for irrigation depends strongly on weather conditions. There has, however, been a general decrease, especially following the introduction of tax on abstraction permits.

A recent important finding concerning the link between water and agriculture is that, since the implementation of the first action plan on the aquatic environment in 1987, the total discharges of nitrogen to coastal waters have fallen significantly. The decrease in nitrogen is attributable to a considerable reduction in both leaching from agricultural land and in wastewater discharges. With corrections for variations in run-off, the reduction in the marine nitrogen load is calculated at approximately 40 %.

For more information please contact the relevant national focal point. Contact details can be found on:
http://org.eea.eu.int/organisation/nfp-Eionet_group.html

Estonia

Estonia is showing good progress in several areas: achievement of air emissions targets, total energy consumption, use of freshwater resources and municipal waste generation. However, it is performing poorly in the area of freight transport, and has a low per capita performance for air emissions, as well as low performance per GDP for energy consumption and greenhouse gas emissions.

Greenhouse gas emissions

From 1990 to 2002, GHG emissions decreased by 55 % mainly due to economic restructuring and political measures. In April 2004 the government approved the national programme for the reduction of greenhouse gas emissions (2003–2012). In 2002 GHG emissions per capita were among the highest in Europe. Estonia's Kyoto target is to reduce these by 8 % by 2008–2012 against 1990. Energy related activities are the most significant contributors to GHG emissions. Oil shale is the main indigenous fuel with low net caloric value (8.5–9 MJ/kg), high ash (45–50 %) and sulphur (1.4–1.8 %) content.

Energy consumption

Primary energy use has stabilised after a downtrend at the beginning of the 1990s, reaching 200 000 TJ/a. In line with the objectives of the national development plan, the consumption level equals that of 2003 (201 892 TJ). Energy consumption per capita is relatively high due to climatic conditions and low population densities, somewhat similar to Nordic neighbours. Energy intensity per unit of GDP is similar to other (industrialised) EU-10 Member States.

Renewable electricity

The national indicative target (12 % of gross national energy consumption by 2010) has already been achieved due to the relatively high use of wood and wood waste for heat production, followed by biomass. In 2002 the share of renewable electricity was 0.2 %. In 2003 it stayed below 1 % despite a 2.7 increase from 7 GWh in 2002 to 19 GWh in 2003, mostly of wind and small hydro. In December 2004 the Estonian Parliament approved the long-term public fuel and energy sector development plan until 2015. This sets the main objectives of the energy sector and establishes the country's main political interests in the energy field. It foresees a 1.5 % share of renewable electricity sources in 2005 and a national target of 5.1 % for 2010. This will need investment in equipment for the production of renewable energy of 130–190 million euro.

Emissions of acidifying substances

Due to the decline in energy production, SO_2 emissions have decreased by about 63 % (1990 to 2003). The power industry is the greatest polluter (85.5 % of emissions). Under the EU accession treaty there is a transitional period for the rate of desulphurisation of combustion plants using oil shale. Estonia's objectives here are:

- to reduce total SO_2 emissions from stationary and mobile sources below 100 000 tonnes per year by 2010;
- to reduce SO_2 emissions from oil shale power plants below 25 000 tonnes per year by 2012; and
- to fix the maximum sulphur content in ship fuel at 1.5 % (in the development plan for transport 2004–2013).

Population: 1 350 000
Size: 45 100 km²
GDP: 4 515 million euro

Emissions of ozone precursors

From 1990 to 2003 volatile organic compound (VOC) emissions decreased by 42.8 %. Emissions of non-methane volatile organic compounds (NMVOCs) from transport decreased by 69 %. This is compared to the decrease in petrol and diesel consumption by 45 % and 36 % respectively. Emissions from non-industrial fuel combustion (households, agriculture, business and public sectors) have grown by 38.8 % due to an increase of wood and wood waste combustion. Between 1990 and 2003 emissions of CO decreased (~ 67 %) mainly due to a decrease in vehicle fuel use and recently by an increased use of diesel cars. In 2003 the largest CO polluters were small combustion facilities using solid fuel and household stoves (61 %), and transport (33 %). Transport is the largest source of nitrogen oxides pollution (57.8 %). Reduced NO_X emissions from mobile sources (1990–2003) are mainly due to the same factors as for volatile organic compounds (VOCs) and CO. The increase in the number of cars with catalytic converters has also played a role.

Freight transport demand

There has been a general growth in the number of vehicles over the past decade which is still growing since people prefer their own car to public transport. 69 % of cars and 70 % of lorries are older than 10 years. Although official statistics do not include transit traffic flows in freight transport demand, this also plays an important role.

Share of organic farming

In total there were around 4 000 ha of controlled agricultural land in 1999 (0.4 % of agricultural land in production). In 2000 there were at least 238 farmers with about 10 000 ha, who had applied for the state label 'MAHEMÄRK'. The marketing of organic products is weak and consumers have difficulties finding organic products in the shops. The most common marketing methods are on-farm sales, selling to hospitals, schools, kindergartens and local shops.

Taking into account the present agricultural situation and existing developments, there is great potential for the rapid development of the sector. It is estimated that there will be a 50 to 100 % annual increase of production over the next few years.

Municipal waste

Municipal waste constitutes around 4 % of total waste generation. About 60 % of mixed municipal waste is from households, the rest is from institutions and enterprises. Between 1999 and 2003, per capita municipal waste generation averaged 410 kg. In the same period the percentage of municipal waste sent to landfill significantly decreased, comprising only 67 % of the total amount of municipal waste generated in 2003. This was mainly due to the implementation of the packaging act and the packaging excise duty act, promoting the recovery of packaging and packaging waste, of alcoholic and non-alcoholic beverages and the extension of the separate collection of waste and sorting of municipal waste by fractions in the new Tallinn waste sorting plant.

Use of freshwater resources

In 2004 the abstraction of water was about 1.7 km³ (1.4 km³ surface water, of which 1.3 km³ was cooling-water and approximately 0.31 km³ groundwater, of which 0.26 km³ was mining water). This was almost half of that in 1991. Economic changes, the decrease of production, reorganisation of technology and increased water taxes (prices) has given rise to more sustainable water use in industry and households. Since 1996, the abstraction of water has remained about constant. A slight increase occurred in 2003 and 2004 due to increased consumption of water by power stations, mines and fish farms. However, the Water Exploitation Index is still in the low stress range of 10–20. Estonia's long-term average annual run-off amounts to over 11 km³ per year (more than 8 000 m³ of water per person per year).

For more information please contact the relevant national focal point. Contact details can be found on:
http://org.eea.eu.int/organisation/nfp-Eionet_group.html

Finland

Finland is performing well in renewable electricity for which it is ranked third, and is in the top ten for organic farming and municipal waste. However, the country shows medium performance across the other indicators, ranking in the bottom few countries for per capita performance in four out of six cases. In general Finland shows more positive trends for progress indicators, but in the areas of waste, water use and GHG emissions, Finland still has to improve to meet its targets.

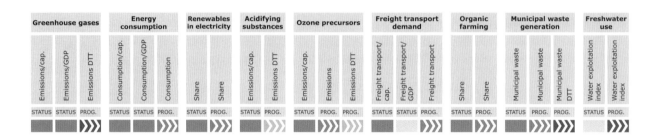

Greenhouse gas emissions

In 2002 total greenhouse gas emissions exceeded the 1990 level by 6.8 %. Finland's target is to keep its greenhouse gas emissions in the first commitment period at the 1990 level. Greenhouse gas emissions in Finland depend on many issues: prevailing economic situation, energy intensive industries; volumes of hydro-power produced; imports of energy and renewable sources; and climatic conditions. In the energy sector CO_2 emissions increase between 1990 and 2002 by 15 %. According to the latest United Nations Framework Convention on Climate Change report (April 2005), greenhouse gas emissions in Finland were 70.4 Tg CO_2 in 1990 and 77.2 Tg CO_2-equivalent in 2002. From these figures Finland's greenhouse gas emissions in 2002 exceeded the 1990 level by 9.7 % percent.

Energy consumption

The energy intensity per capita is rather high in Finland. This is due among other things to the climate, long distances to be covered, transport, and energy-intensive industry. Energy conservation has been practiced for a long time, however, and for more than twenty years the aim has been to produce as much electricity as possible from combined heat and power plants (CHP), where Finland ranks among the international top countries. Other means include for example voluntary energy conservation agreements and audits, and promotion of sustainable consumer behaviour

Renewable electricity

The share of renewable energy in electricity production has been increasing substantially and the action plan for renewable energy sources (2003–2006) aims at an increase of 30 % by 2010 compared to 2001. The proposed actions in the plan include the development and commercialisation of new technology and financial steering instruments such as energy taxation, investment aid and subsidies for the production chain of forest chip wood, and the general improvement of the competitive strength of renewable energy sources.

Emissions of acidifying substances

Emissions have decreased since 1990 with SO_X showing the largest decrease to a level well below the EU national emission ceilings (NEC) directive ceiling. Air emissions depend on climatological conditions, export/import of electricity, availability of hydropower, and many other factors that cause variation in annual emissions.

Emissions of ozone precursors

Ground level ozone concentrations are mainly low in Finland and the few occurrences of elevated levels are due to long-range transport of emissions. The emissions of the ozone precursors used in the indicator have been decreasing and the existing emission reduction targets are within reach.

Finland

Country perspective

Population: 5 210 000
Size: 338 150 km²
GDP: 131 784 million euro

Freight transport demand

The fact that Finland is large and sparsely populated sets the scene for transport demand. Even though rail freight transport has maintained its market share relatively well in Finland (around 25 % of all freight transport kilometres), road transport has continuously increased and waterway transport has diminished. Nevertheless, shipping plays an important role for Finnish foreign-trade transportation. Around 85 % of the Finnish foreign trade is carried by ships. To develop short-sea-shipping is a main goal but the inland waterways are not well suited to freight transport.

Share of organic farming

The share of organic farming has increased rapidly and this increase is expected to continue. Also the average size of organic farms has grown. Many factors, including the development of subsidies, will affect the rate of increase in organic farming.

Municipal waste

Trends in municipal waste quantities for the last few years have seen a decrease in total quantities generated, both in terms of tonnes/annum and per capita. There has been an 11 % decrease in total quantities generated between 2000 and 2003. For the same time period, gross domestic product (GDP) grew by an average of 6 %. In Finland, roughly 450 kg of municipal waste per capita are generated every year, which is some 100 kg less than the European average. The majority of this waste is landfilled (60 %); while 29 % is recovered as material and 9 % as energy.

Use of freshwater resources

Water abstraction per capita in Finland is above average but the abstracted amount of water is a very small proportion of the available water resources. Regarding the agricultural use of water, the irrigated area is very small and the amount of water used for irrigation is only about 2 % of the total abstracted amount.

For more information please contact the relevant national focal point. Contact details can be found on:
http://org.eea.eu.int/organisation/nfp-Eionet_group.html

France

France is near the median position among European countries for most of the selected indicators. Being a large country, France has to face a wide variety of environmental issues. The development of electricity production from nuclear power has contributed to low greenhouse gas emissions. Trends suggest that policy targets in the scorecard, except municipal waste production, are well on the way to being reached.

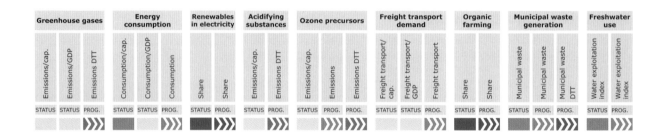

Greenhouse gas emissions

Under the Kyoto Protocol, France is committed to stabilising its emissions of greenhouse gases by 2008–2012 compared to the level of 1990. Over the period 1990–2003, France managed to stabilise its emissions, in particular due to the importance of the nuclear electricity production. However, the emissions from the transport sector, which represent the largest share of the total emissions (27 % of the national total in 2002) continue growing (1.7 % a year between 1990 and 2002). Emissions from the residential and tertiary sectors are growing too.

Energy consumption

Energy consumption increased by approximately 1.6 % a year between 1990 and 2000, mainly due to the demand from transport and residential-tertiary sectors. The pace of growth has slowed since 2000 (+ 0.6 %).

Renewable electricity

With one of the largest forest areas in western Europe, the second higest potential for wind energy, and high geothermal and hydro-energy production, France is rich in renewable energy resources. Of currently exploited energy, 50 % comes from wood biomass, 30 % from hydropower and 12 % from waste incineration. Other renewable sources are still little developed. Approximately 13 % of the primary electricity production is of renewable origin.

Emissions of acidifying substances

Acidifying emissions fell by almost 30 % between 1990 and 2003. Only ammonia emissions from agricultural activities were maintained at practically the same level over the period.

Emissions of ozone precursors

Ozone precursor emissions have almost halved since 1990. Nevertheless, the presence of ozone remains one of the most frequent causes of poor urban air quality, especially in the Mediterranean regions and Rhône-Alps where the average number of sunshine days a year is high.

Freight transport demand

Inland freight transport grew faster than the GDP between 1995 and 1999. Since then it has decreased appreciably to 6 % in 2003 under the level of 1995 per unit of GDP.

France PART C

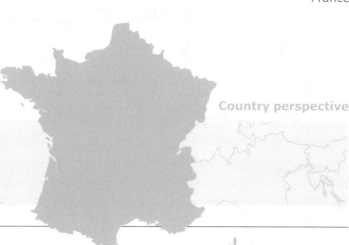

Country perspective

Population: 59 725 000
Size: 551 500 km²
GDP: 1 407 304 million euro

Share of organic farming

With less than 2 % of the utilised agricultural area, organic farming constitutes a relatively small proportion of agricultural production.

Municipal waste

Calculated per capita, the quantity and trends of municipal waste are similar to the EU-15 average. For waste management France is characterised by a higher share of incineration (34 % against 19 % for EU-15), and a lower share of waste reuse and dumping.

Use of freshwater resources

Available water reserves are sufficient on a national scale. However, locally, certain rivers or aquifers face water stress in summer periods and are then made subject to use restrictions.

For more information please contact the relevant national focal point. Contact details can be found on:
http://org.eea.eu.int/organisation/nfp-Eionet_group.html

Germany

One of the most densely populated countries in Europe, Germany shows above-average progress in reducing per capita emissions and municipal waste and has relatively high levels of eco-efficiency but its per capita performance could be improved. Progressive waste management policies are in place but there is still room to improve performance. Legislation and ecological reforms encourage energy savings and development of profitable renewable energy.

Greenhouse gas emissions

Total greenhouse gas emissions (in GHG eq.) were reduced by almost 19 % compared to the base-year. Reasons for this trend include: an improvement in energy efficiency, a switch in the types of fuel used (to gaseous and liquid fuels instead of solid fuels) and a replacement of 'old technologies' following German reunification in the early 1990s; new policies and measures resulting from climate protection programmes; and a decoupling of economic growth and energy consumption which relates to greenhouse gas emissions (especially carbon dioxide). This trend is also influenced by changing personal energy consumption habits and the implementation of eco-taxes. Reaching the burden-sharing reduction target (– 21 %) does not seem possible solely by using domestic actions. This will result in additional emission reductions by using the flexible mechanisms of the Kyoto Protocol and the EU emission trading scheme.

Energy consumption

As a highly industrialised country, energy consumption per capita is rather high compared to other European countries and energy intensity per GDP better reflects the energy efficiency in the energy and industry sector. In recent years energy consumption has decoupled from economic growth. Energy conversion efficiency has improved as highly efficient units replace older generating plants. Although additional measures to improve efficiency and reduce energy consumption in the industry sector have led to lower energy consumption per production unit, progress has slowed significantly in recent years.

Renewable electricity

The production of renewable energy in Germany is increasing, accounting for 3 % of the primary energy supply in 2002. The contribution of renewables to electricity production more than doubled between1992 and 2002. Electricity production by renewable sources is promoted by the German renewable energy sources act from 2000, which gives priority to feeding electricity from renewable energies into the grid at fixed price levels. It has triggered a boom in the construction of wind and biomass installations. Further measures to promote the use of renewable energy include the Act on the Further Development of the Ecological Tax Reform (2003), which provides incentives for saving energy and improving energy efficiency.

Emissions of acidifying substances

Emissions of acidifying substances have been reduced by more than two-thirds. This was mainly influenced by reductions in SO_2 emissions (– 88 % compared with 1990) and NO_x (– 48 %). To a large extent both compounds come from energy related processes so that the basic drivers for these trends could be summarised as: fuel switching; improved energy efficiency; replacement of 'old technologies'; and implementation of abatement technologies (e.g. DENOX, DESOX, car catalysts). More than 90 % of NH_3 emissions come from agricultural activities. After German reunification, there

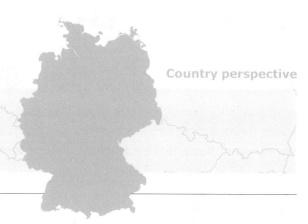

Population: 82 551 000
Size: 357 030 km²
GDP: 2 072 162 million euro

was a reduction in livestock numbers in former eastern Germany. This meant that emissions were reduced by almost 20 % up to 1994, but that since then emissions have more or less stabilised. Germany is working on additional policies and measures for further reducing acidifying substances (especially NO_x and NH_3).

Emissions of ozone precursors

Emissions of ozone precursors were reduced by approximately 54 % in the period 1990–2002. This trend is based on similar developments for all substances included (2002 compared to 1990: NO_x – 48 %, NMVOC – 58 %, CO – 62 % and CH_4 – 42 %). Germany has reached one of the best European emission ratios per GDP. The emission reductions are based on activities to improve energy efficiency, fuel switching, economic reconstruction, implementation of new technologies in the road transport sector, the replacement of natural gas distribution systems in the former GDR and the gasoline distribution systems in total. Similarly, livestock reduction in the early 1990s led to emission reductions for CH_4. Germany is working on further measures to reduce emissions from solvent use which dominate the NMVOCs.

Freight transport demand

Freight transport in Germany has been increasing but during the past three years development has levelled off and is in line with economic development. The increase mainly relates to road transport while the share of rail and inland shipping has decreased.

Share of organic farming

The share of organic farming in Germany was just 1 % in 1992 but increased to 4.1 % in 2002. The number of organic farms also rose from below 1 % to 4 % in the same time period. The market share for organically produced food amounted to 2.3 % (= 3 billion euro) in 2002. The yearly increase in share of organic farming and number of farms reached 22 % in 2000 and has been declining since then (2001: 15.6 %, 2002: 6 %). In a European comparison of the share of organic farming, Germany with 4.1 %, is in middle-place. However, comparing the actual area used for organic farming, Germany ranks in third place.

Municipal waste

In Germany the amount of municipal waste collected from households and small- and medium-sized enterprises has been relatively stable during recent years and has even decreased slightly. This has been influenced by an advanced waste management policy. Elements of this policy include the phasing-out of landfilling for municipal waste, and introducing both the concept of total recycling of municipal waste by 2020 (Strategy 2020), and mandatory take-back obligations for packaging waste. These elements have encouraged both recycling and avoidance of waste.

Use of freshwater resources

In Germany, the private and economic sectors use only 22 % of naturally available water. The volume of water consumption in the main sectors is continually decreasing. In fact, the demand for private households is 15 % below the 1991 level; water abstraction for cooling purposes in thermal power stations also decreased 15 % from the 1991 level, and the abstraction for manufacturing and quarrying industries decreased about 29 % over the same period. Water use for agricultural irrigation is not significant and has stabilised at a low level.

For more information please contact the relevant national focal point. Contact details can be found on:
http://org.eea.eu.int/organisation/nfp-Eionet_group.html

Greece

Progress has been made integrating environment into sectoral and economic policies to reduce environmental pressures. The most important environmental issues in Greece are: land use, waste management and water resources management. The relatively non-degraded natural environment has a rich biodiversity, a large variety of habitats, high-quality bathing waters and coastal areas and relatively good air quality.

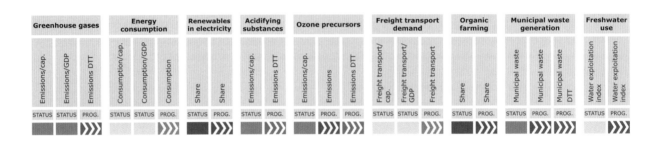

Greenhouse gas emissions

Greenhouse gas emissions increased steadily during the last decade, the most important gases being CO_2 and CH_4. The production and use of energy, as well as waste disposal and agriculture are the primary sources of emissions. Recent projections indicate that with a consistent implementation of its 2003 plan Greece will come close to meeting its target. An upcoming evaluation of the effectiveness plan will indicate if, and to what extent, Kyoto mechanisms need to be used.

Energy consumption

Per capita demand for primary energy in Greece is lower than the EU average. This high energy intensity presents opportunities to reduce the energy demand through rational use of energy resources and the promotion of energy-saving technologies. Up to now, the Greek energy sector has been dependent on conventional fuels, contributing significantly to the release of atmospheric pollutants. More specifically, in the electricity production sector, the choice to exploit domestic lignite resources as an appropriate response to the energy crisis of the 1970s, needs reconsidering in the light of network integration, market liberalisation and environmental protection. The total operational electrical capacity of natural gas plants will be increased by 52 % by 2010, of hydropower plants by 18 % and of renewables by at least 100 % while the capacity of lignite plants will be decreased by 3 %.

Renewable electricity

Renewable energy sources contributed 4.7 % of total energy demand in 2002 (5 % in 2003). Two-thirds of the total production comes in the form of heat from biomass and active solar systems, and the remaining third comes from hydropower plants and wind. It must be noted that electricity production from large hydro is largely affected by weather conditions (rainfall) and the availability of water in the reservoirs. The share of electricity from RES to total electricity consumption was 6 % for 2002, which was below the EU-15 average of 13.5 %. Due to high rainfall it was 9.6 % in 2003, almost half of the target set by the RES-E directive of 20.1 % by 2010.

Emissions of acidifying substances

Emissions of air pollutants increased following GDP growth with the exception of NO_x and SO_2. The reform and diversification of the energy sector offer:

- rational use and conservation of energy in the building sector;
- measures for the transport sector;
- measures for industry; and
- institutional and organisational measures.

Greece

Country perspective

Population: 10 680 000
Size: 131 960 km²
GDP: 120 249 million euro

Emissions of ozone precursors

Despite the partial decoupling of air pollutants from economic growth recorded during the last few years, considerable efforts are underway to ensure a permanent downward trend and to meet the targets set within the EU framework, particularly for NO_x and non-methane volatile organic compound (NMVOC) emissions. Between 1990 and 2002 emissions increased and were above the level that would be needed to meet 2010 NECD targets. Focus for these actions is the energy sector, responsible for the largest part of air quality degradation.

Freight transport demand

Following trends recorded throughout the EU during the last decade, the demand for transport services in Greece is rapidly growing. The main reason is change in the pattern of production and consumption. However, when comparing transport demand per capita and GDP, Greece is ranked among the best performing countries.

Share of organic farming

Organic farming in Greece started in 1992 with the inception of the Common Agricultural Policy (CAP) reform. The percentage of land dedicated to organic farming compared to total cultivated area has increased impressively over the last years reaching 1.41 % in 2004. Olive and citrus tree plantations were the dominant early organic cultivations, but during the last decade increased consumer demand and CAP incentives have given rise to a greater variety of crops such as arable and vineyards. One additional area of significant increase has been organic livestock production.

Municipal waste

Economic development, intense urbanisation and changes in consumption patterns have resulted in an increase in solid waste generation. The quantity of municipal waste generated increased 42.5 % from 1995 to 2002. Initiatives by local municipalities to reduce packaging waste, and the extensive involvement of private companies mainly in paper packaging recycling, are examples of Greece's practical approach to improving the waste management situation. Inappropriate waste disposal and management practices still persist, leading to the degradation of surface and groundwaters, air pollution and forest fires. However, significant progress has been made in the management of hazardous waste, and sludge and electricity production from biomass gases and waste has increased from 1 GWh in 1999 to 126 GWh in 2002.

Use of freshwater resources

The problems of water management mainly concern issues of quantity and not of quality. The uneven distribution of water resources and rainfall creates water availability problems. Agriculture is the most significant water consumer and demand for irrigation has doubled in the last twenty years. Irrigation is of paramount importance for agriculture productivity in Greece where water deficiencies in arid and semi-arid areas can severely curtail crop yields. Irrigation accounts for over 80 % of total water abstractions. Between 1992 and 2002, water abstraction for agricultural use was reduced by about 2.5 %. It is estimated that over the next years, further reductions will be achieved. These will arise from the implementation of new CAP and EU regulations, modernisation and renovation of irrigation networks, application of new technologies for irrigation, and the training of farmers in good agricultural practices. Significant progress has been made in wastewater management and approximately 70 % of the national population was serviced by wastewater treatment plants in 2004. For the 2004 bathing season, 99.9 % of Greek coasts complied with national requirements, while 97.6 % of coasts met EU requirements.

For more information please contact the relevant national focal point. Contact details can be found on:
http://org.eea.eu.int/organisation/nfp-Eionet_group.html

Hungary

Hungary shows average performance across the scorecard, and although less eco-efficient than the EU-15 Member States, is more so than many other of the EU-10. Emissions of acidifying substances have decoupled, while waste generation remains coupled to household consumption. A priority area for future consideration is how to develop markets for renewables.

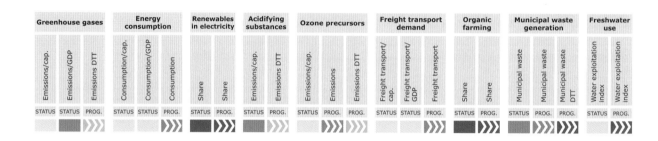

Greenhouse gas emissions

In the past decade, GDP has been decoupled significantly from greenhouse gas emissions as a result of economic recession and profound restructuring. Emissions trends and intensities are closely correlated with greenhouse gas emissions from the use of fossil energy sources (coal and hydrocarbons) and reflect the decarbonisation of the economy very well. By 2002, carbon-dioxide emissions decreased by almost one-third compared to the period between 1985 and 1987 (from 120 million tonnes in the basis period between 1985 and 1987 to 79 million tonnes in 2003).

Energy consumption

Before 1989, energy consumption per capita was several times higher than that of average western European consumption (as a consequence of energy-wasting industry). In 1997, however, energy use was only 79.5 % of the 1985 level, and it decreased by a further 15 % in 2002. Natural gas has strengthened its position among energy sources, contributing approximately 40 % in 2002. The decline of coal use continued from 29 % in 1980 to 12.5 % in 2002. Nuclear-generated electricity has provided 10–12 % of the country's energy for the last 15 years. Only one third of the total primary energy use comes from domestic production. Since 1990, import dependence of energy supply has been increasing slightly.

Renewable electricity

Over 16 % of the renewable energy source potential is used with the currently available technical solutions. The most widely used resources are water power (56 %) and biomass (19 %), while wind and solar energies are the least used (0.4 % and 1.7 % respectively). Nevertheless, the share of renewable energy sources in electricity generation is low at about 0.3 % which implies that further efforts are needed to meet EU requirements.

Emissions of acidifying substances

Total emissions of acidifying gases decreased to 60 % by the end of the 1990s, indicating a strengthening (although saturating) decoupling from GDP. These patterns are reflected in trends in emissions per capita. Emissions of acidifying substances per unit GDP show the most striking decoupling, which may be explained by the joint effects of restructuring and environmental policy measures in the 1990s. It is still not clear whether or not the brief halt in decoupling, or slight increase, is the beginning of a new trend.

Emissions of ozone precursors

CO, non-methane volatile organic compound (NMVOC), CH_4 and NO_x emissions expressed in TOFP equivalent exceeded the GDP index between 1992 and 1999, while signs of weak decoupling emerged in 2000.

Population: 10 120 000
Size: 93 030 km²
GDP: 46 002 million euro

Country perspective

Freight transport demand

The share of railways and waterways has decreased since the mid-1990s. Road freight transport grew evenly in the second half of the 1990s and it has increased to over 50 % in the last couple of years. General trends in air freight transport show slow growth, but despite slight fluctuations in the past couple of years; its share within total performance of freight transport is still insignificant (well below 0.5 %). In contrast to road freight transport, road passenger transport has not changed in the course of the last decade.

Share of organic farming

Since 1996, the share of controlled green farming areas (where farming activities are regularly supervised) has been growing dynamically, while agricultural area has decreased radically. Although the proportion of such controlled areas within all agricultural areas was below 0.2 % in 1996, by 2002 this proportion was almost 1.9 % (an almost tenfold increase). The volume of organic animal husbandry has so far remained at a low level. Only 10 % of Hungarian organic production is sold on domestic markets, while the rest is sold primarily on the markets of Switzerland and the European Union.

Municipal waste

The quantity of waste collected from the population increased by 30 % between 1992 and 2002 and exceeded the rate of GDP growth over the whole period. This trend is traceable to economic restructuring and the spread of more up-to-date and cleaner technologies. In Budapest the share of plastics in municipal waste grew drastically after 1996 and by the end of the decade was more than three times the 1990 level. The degradable organic material content of municipal waste has grown. The shares of paper, glass, metal and textile waste have decreased. Since 1993, the quantity of waste collected from the population has grown at the same (or greater) rate than the increase of average household consumption, so even any relative decoupling seen is insignificant.

Use of freshwater resources

In the early 1990s, the total annual water abstraction fell significantly as a result of declining industrial production and restructuring, the reduction in household consumption and shrinking size of irrigated areas. With the drastic fall of actual water use (over 50 %) and despite a minimal rise in usable water resources, the intensity of water use decreased significantly in absolute terms. In 1999, it was 8.5 %, which is considered low by international standards. Agriculture uses some 18 % of water abstraction (for fishponds and to a lesser extent irrigation). In 1997, overall irrigated agricultural area fell to 82 000 ha, only 2 % of the cultivated land, which was followed by an increase to over 200 000 ha in the early 1990s. In 2001, the area of irrigated land was around 105 000 ha.

For more information please contact the relevant national focal point. Contact details can be found on:
http://org.eea.eu.int/organisation/nfp-Eionet_group.html

Iceland

Iceland's environmental situation and problems differ from those of other European countries. Iceland is sparsely populated, depending primarily on natural resources and their efficient use for its economy. From an Icelandic perspective, a different set of indicators for analysing its environmental performance would be more suitable, focusing on the management of fish stocks, renewable energy sources and wilderness.

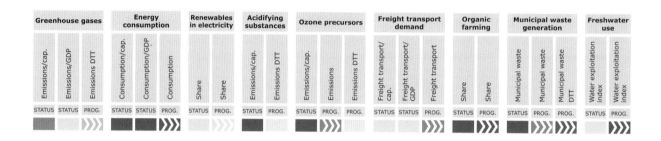

Greenhouse gas emissions

Emissions of greenhouse gases were 3.5 million tonnes in 2003 — an increase of 8 % since 1990. As the Icelandic population is approximately 300 000, this means 11 600 kg per capita, which is below average. The greenhouse gas emissions profile for Iceland is in many regards unusual. Firstly, emissions from the generation of electricity and from spatial heating are essentially non-existent since they are generated from renewable non-emitting energy sources. Secondly, more than 80 % of emissions from energy use come from transportation and fishing vessels. Finally, individual sources of emissions from industrial processes have significant proportional impacts on emissions at the national level. Iceland's obligations according to the Kyoto Protocol are therefore twofold; emissions should not increase more than 10 % based on the level of emissions in 1990, and 1 600 000 tonnes of CO_2 emissions from industrial processes, falling under 'single projects', should be exempt from the 10 % target. Taking this into consideration, emissions falling under the 10 % Kyoto target have decreased by 6 % since 1990.

Energy consumption

In 2002, primary energy consumption amounted to 500 GJ per capita (11.9 toe/capita), which ranks among the highest in the world. There are a number of reasons for this, in particular the high proportion of electricity used in power intensive industry, a relatively high amount of electricity production from geothermal energy, and substantial energy consumption for fishing and transportation. In addition much energy is used for space heating using abundant geothermal hot water.

Renewable electricity

Iceland is a top performing country when it comes to use of electricity from renewable energy sources. The main part of the energy needed for spatial heating also comes from renewable energy, with geothermal energy meeting 86 % of the spatial heating requirements in Iceland. Today geothermal energy and hydropower account for more than 70 % of the country's primary energy consumption.

Emissions of acidifying substances

These emissions become a problem when they are high per area, not necessarily per capita. In Iceland there are about three inhabitants/km² which is by far the lowest population density in Europe. These indicators are high if calculated on a per capita basis but are low in total amounts and very low if calculated per km² in an area grid system. It should be noted that the data used include considerable emissions from geothermal sources, which are no longer included in the United Nations Framework Convention on Climate Change reporting. When excluding geothermal sources, the fishing fleet accounts for more than one third of the emissions. This indicator is therefore not suitable for evaluating environmental deterioration and conditions in Iceland. In fact, acidification is simply not a problem in Iceland.

Iceland

Country perspective

Population: 286 000
Size: 103 000 km²
GDP: 7 088 million euro

Emission of ozone precursors

The same applies for ozone precursors as for acidifying substances; emissions of ozone precursors are not a problem in Iceland. Fishing ships are the largest contributors, releasing up to half of the TOFP.

Freight transport demand

Iceland scores highest for the amount of freight transport per unit of GDP, but comparing Iceland with other countries here is difficult. As approximately half of the population lives in Reykjavik and the communities surrounding the capital, the transport distances are limited. Moreover, marine transport is not included.

Share of organic farming

The conditions for agriculture in Iceland do not give scope for a high score for organic farming. Cultivated land in Iceland is 1 290 km² or approximately only 1.3 % of the total land area; 1.2 % is cultivated grass fields and 0.1 % horticulture, fodder and grain fields. Meadows and pastures cover 17 700 km². Most of the vegetated land, covering up to 80 000 km², is open, non-fertilised rangeland used for grazing purposes where the sheep roam free during the summer months.

Municipal waste

Waste management in Iceland in general has improved radically since the 1990s with rapidly increasing recycling and recovery figures, a ban on open pit-burning and fewer and bigger landfills that are operated in an environmentally sound way. The figures on municipal waste that Iceland have published have been revised recently taking into account that household waste and industrial waste are collected together in many of the municipalities in Iceland. This waste collection system makes it almost impossible to estimate the quantity of household waste in the total waste collected with an acceptable grade of certainty. Therefore, all collected waste is registered as 'municipal waste'. However, this has been changing, especially in the most densely populated areas, as municipal waste and idustrial waste are increasingly collected separately, making a better assessment of the different waste streams possible. Thus, it has become clear that municipal waste generated per capita is somewhere around 490 kg/year and not 1 030 kg/year as previously estimated.

Use of freshwater resources

Iceland has the highest renewable freshwater availability per capita in Europe. Heavy rainfall (an average of 2 000 mm per year) and the fact that Iceland is the most sparsely populated country in Europe, means that there is abundant water per inhabitant and the majority of the population has access to plentiful supplies of freshwater. Most of the water, over 95 % of the public water supply, is untreated groundwater originating from springs, boreholes and wells. Water stress related to abstraction is not known and water abstraction is sustainable.

For more information please contact the relevant national focal point. Contact details can be found on:
http://org.eea.eu.int/organisation/nfp-Eionet_group.html

Ireland

Meeting international commitments on air emissions and waste management are priority areas. Positive signals include recent modest improvements in greenhouse gas emissions and some acidifying gases and long-term major reductions in serious pollution in rivers and urban air. Increasing awareness of the environment and willingness to act is shown by the plastic bag levy success story, increased recycling and high compliance with the new smoking-ban in bars and restaurants.

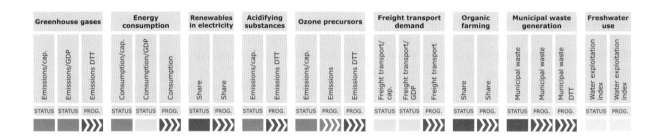

Greenhouse gas emissions

Annual emissions of greenhouse gases were down for a second consecutive year in 2003. Emissions were 25 % above the 1990 level in 2003 (Ireland's Kyoto Protocol target is to limit emissions to 13 % above the 1990 level in the period 2008–2012).

Although GHG emission levels in 2003 were 4 % lower than their peak 2001 level, the total exceeded 1990 levels by 25 %. While the downward trend from 2001 to 2003 is welcome, Ireland still faces a significant challenge meeting its Kyoto Protocol target.

Energy consumption

Ireland's energy consumption by GDP is low compared with most other EEA member countries. For per capita energy consumption, the figure is influenced by a higher dependency on energy for heating purposes than warmer countries. Total energy consumption is also influenced by the unprecedented period of economic growth in Ireland since the mid-1990s.

Historically poor building codes have been repeatedly improved, the new directive on the energy performance of buildings will help improve energy efficiency. Despite the continued heavy reliance on carbon-intensive fuels, there were some gains from energy efficiency and fuel switching as some new electricity suppliers entered the market in 2002 and 2003.

Renewable electricity

The significant increase in total primary energy requirement in the period 1998–2001 hides the large growth of renewable energy, rising by over 71 % between 1990 and 2002. More recently the rate of growth of electricity from renewable sources has increased substantially, particularly for wind. In 2004 the amount of electricity generated by wind increased by 69 % over 2002 levels. This rapid increase in electricity generated by renewables is set to continue over the coming years with wind being the biggest growth area.

Emissions of acidifying substances

Existing emission levels of acidifying gases are such that compliance with the EU national emission ceilings (NEC) directive represents a major challenge. Ammonia and volatile organic compounds (VOCs) have not previously been subject to abatement strategies, and progress on reducing total emissions of SO_2 and NO_X has been slower than in most other European countries. Reductions equal to 75 % and 100 % of those needed have been achieved for SO_2 and ammonia respectively. NO_X emission reductions have been more difficult, only showing decreases since 2000. Progress towards the NO_X ceiling of 65 kt will mainly depend on reductions in road traffic emissions and on the wide application of expensive control technologies (e.g. selective catalytic reduction) in industry and for electricity generation. The high level of ammonia emissions (mostly from agriculture) account for a major part of the per capita acidifying emissions. At 116 kt in 2003, the emissions of ammonia had decreased to the level of the 2010 ceiling.

Country perspective

Population: 3 947 000
Size: 70 270 km²
GDP: 94 404 million euro

Emissions of ozone precursors

A reduction in ozone precursor volatile organic compound (VOC) emissions of 58 % was achieved between 1990 and 2003. However, NO_x emissions only began to show a decrease in 2000, with the result that the total in 2003 remained marginally higher than the 1990 baseline value of 115 kt.

In Ireland ozone levels are influenced by transboundary sources and generally remain below the effects thresholds for human health and vegetation as set down in the 2002 ozone directive.

Freight transport demand

Ireland is an island nation of approximately 480 km in length and 240 km in width. The majority of Irish industry operates a 'just-in-time delivery' system and the products produced are generally of a low volume/high value. These factors contribute to low tonne/km transport efficiency.

Share of organic farming

Organic food represents approximately 1 % of the entire food market. A number of new instruments will have a positive effect and lead to an increase in organic farming. The rural environment protection scheme now provides additional payments to participating farmers to convert to organic farming methods. Grant assistance towards the development of the organic sector is also available. The recent decoupling of farm payments and food production gives farmers a positive incentive to diversify into organic farming.

Municipal waste

Due to lack of harmonisation of definitions and methods at EU level, Irish municipal waste statistics are over-stated because household waste in Ireland includes commercial waste and other waste which is similar to household waste.

The generation of municipal waste including household waste has increased significantly since the mid-1990s. The principal factors driving the increase are the economic boom and significant population increase. However, municipal waste recycling increased to 727 000 tonnes in 2003, representing a recycling rate of 28 %. Significant progress has been made in meeting the national target of 35 % municipal waste recycling by 2013.

Since the 1996 Waste Management Act, waste management in Ireland has been transformed. The number of landfills has decreased from over 100 unlined and unregulated dumps to 34 authorised municipal waste sites that operate to modern EU standards. Recycling has increased visibly with a 26 % reduction in the proportion of waste being sent to landfill. An ongoing challenge is the continued rise in absolute quantities of waste (up 10 % in 2003). Also, the successful clamp-down on large-scale illegal dumps has resulted in new threats including fly-tipping and backyard burning of rubbish. Considerable challenges remain, including a continuing deficit of infrastructure and the requirement to decouple waste production from economic growth.

Use of freshwater resources

Due to present climatic conditions, water abstraction is not a significant environmental issue in Ireland. Ireland's water quality overall remains of a high standard. Serious pollution in rivers and streams has been reduced to just 0.6 % of river channel, its lowest level since the early 1990s. Eutrophication of rivers, lakes and tidal waters continues to be the main threat to surface waters with agricultural run-off and municipal discharges being the key contributors.

For more information please contact the relevant national focal point. Contact details can be found on:
http://org.eea.eu.int/organisation/nfp-Eionet_group.html

Italy

Italy shows relatively good environmental performance and average progress across the scorecard indicators. Reducing GHG emissions to be on track with Kyoto targets is posing a challenge. In common with other southern European countries, priorities for Italy include improving the efficiency of its irrigation system in order to reduce water stress from agriculture.

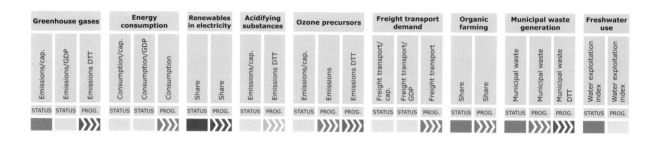

Greenhouse gas emissions

In the context of the convention on climate change and of the Kyoto Protocol, Italy has undertaken the commitment to reduce overall national emissions by 6.5 % with respect to the base-year by 2008–2012. However, the total emissions in 2002, in CO_2-equivalent terms, although fairly constant with respect to 2001, are 7 % higher than the base-year, and thus far from the fixed target. The emission trend is closely related to energy consumption

Energy consumption

A recent survey at European level has shown that the Italian energy system is characterised by a good performance in terms of energy intensity and the ratio of final to total energy consumption and a changing energy supply pattern, involving the increased use of natural gas, renewable energy, cogeneration and a recent increase in solid fuel consumption.

Renewable electricity

Recent data shows that production of renewable energy contributes only 5.9 % of the total energy produced, although showing an increase at national level of about 47 % in the period 1991–2003. Renewable sources primarily include hydroelectricity, biogas, wood and wind energy.

Emissions of acidifying substances

As a whole, emissions of acidifying substances are diminishing and they are nearing the European targets. Although close to the desired target, emissions of ammonia show a slight increase due to the transport sector.

Emissions of ozone precursors

After a slight increase of 6 % between 1980 and 1992, emissions of non-methane volatile organic compounds (NMVOCs) dropped by 37.6 % between 1992 and 2002, close to the European targets.

Freight transport demand

The freight transport intensity by GDP and per capita has increased in recent years, featuring an upward trend until 1995, after which it started fluctuating. Freight transport by road has continuously increased to 67.7 % of the total transport demand when distances in excess of 50 km are considered.

Share of organic farming

After a decade of continuous growth (1990–2001), organic farming indicators (utilised agricultural areas and farms) show that the sector is now consolidated

Italy

Country perspective

Population: 57 646 000
Size: 301 340 km²
GDP: 944 770 million euro

and mature. However, the sector underwent a slight drop in the last two years, mainly due to a widespread move away from organic farming in many southern regions as a consequence of a general delay in the implementation of the EC regulation.

Municipal waste

In 2003, municipal waste generation per capita was 524 kg. However, the growth rate of municipal waste generation has been decreasing since 2000. Thanks to the increase of separated collection and bio-mechanical treatments, the amount of municipal waste disposed of in landfill is decreasing, reaching 63 % in 2002.

Use of freshwater resources

The updated assessment of freshwater abstraction remains a priority in the management of water resources in Italy. The main water consuming sector is agriculture (irrigation), and the main source of water for this purpose remains groundwater, especially in southern Italy which suffers from water scarcity.

Groundwater bodies are then affected by imbalances in the recharge regime and salt intrusion along the coastline. Lack of updated information makes any estimate of this critical issue unreliable. However, since the reduction of water stress in agriculture is a priority in the national water policy, relevant actions have been put in place: district authorities for the management and cost recovery of the integrated water cycle (abstraction-treatment-distribution-wastewater treatment and reuse) have been established covering more than 80 % of the national territory; a yearly report is submitted to a national committee (report to the Parliament) on freshwater uses, network development, water uses, tariff, wastewater treatment and water reuse. In 2004, a ministry decree was enforced for treated wastewater reuse for irrigation and industrial reuse, including financial support. A national information system on water quality and quantity (including sectoral uses) taking into account the reporting requirements of all water directives was implemented in 2003 for a regular reliable assessment on water quality and water uses. By the end of 2005, all regional authorities will implement a regional water protection plan to comply with the environmental objectives and sustainable water use as required by EU sectoral legislation, including Directive 2000/60/CE.

For more information please contact the relevant national focal point. Contact details can be found on:
http://org.eea.eu.int/organisation/nfp-Eionet_group.html

Latvia

Latvia scores highest in Europe in reduction of air emissions, but is performing poorly in freight transport and municipal waste. Areas for attention include improving renewable electricity production and organic farming. Like other European countries, Latvia risks increasing waste volumes, but ongoing activities are tackling this. Actions are also needed to improve urban air quality where some seasonal exceedances of PM_{10} and NO_2 are observed.

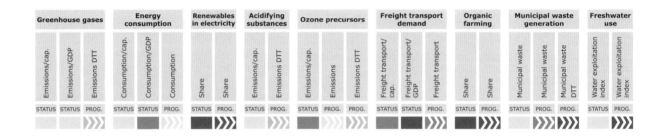

Greenhouse gas emissions

Significant decreases of GHG emissions since 1990 have put Latvia in first place in Europe. However, it is difficult to say if this is a good or a bad trend, since the decrease was the result of economic decline and re-structuring of industry. The Kyoto Protocol requires Latvia to make an 8 % reduction of GHG emissions by 2008–2012, feasible from present trends. Projections indicate that GHG emissions could even be reduced 50 % below target. However, with higher economic growth, carbon dioxide emissions would increase especially from increased mobility combined with low energy efficiency in the energy, industry and household sectors. Measures must therefore be taken to improve energy efficiency and promote the use of local renewable energy resources. The Latvian climate change mitigation programme 2005–2010 sets out principles for participation in flexible mechanisms, including GHG emissions trading.

Energy consumption

Many actions have been taken to promote energy efficiency within the state investment programme. Support is planned for heat energy production enterprises which use heavy fuel oil for heat energy production. Fuel conversion projects in enterprises to comply with Latvian and EU laws limiting sulphur content in particular types of liquid fuel, will also improve energy efficiency. A decrease in energy intensity since 1997 can mainly be explained by the growing GDP.

Renewable electricity

Under EU law, Latvia has undertaken to increase the quantity of electricity generated by renewable resources in 2010 to 49.3 % of the country's domestic electricity consumption. Assuming current electricity consumption and an average flow rate in the Daugava river, 44–45 % would be within reach. However, installed capacities for renewable electricity generation are being used fully. It is now planned to emphasise the development of co-generation. Latvia has a higher consumption of electricity by GDP than the developed European countries. This proportion is now decreasing, indicating increasing improvements in the effective use of electricity by businesses.

Emissions of acidifying substances

EMEP calculations show that less than 5 % of pollution in the country was generated in Latvia, the rest coming from other countries. The proportion of Latvian generated nitrogen (oxidized and reduced) deposition is on average 7.5 % of the total nitrogen deposits in Latvia.

Emissions of ozone precursors

Emissions of ozone precursors have been stable in recent years and relevant changes are not expected in the near future. Since 1990, emissions of nitrogen oxides decreased by 27 % and of non-methane volatile organic compounds (NMVOCs) by 29 %. These decreases have been brought about by the use of lower sulphur content

Population: 2 321 000
Size: 301 340 km²
GDP: 6 007 million euro

Country perspective

fuels, use of car catalytic converters and improvements in treatment plants. Concentrations of ozone precursors (nitrogen dioxide, benzene) and ground-level ozone concentrations in ambient air have been monitored since 1998. The highest hourly and maximum daily 8-hour mean concentrations of ozone were recorded in Riga. The target values, information and alert thresholds were not exceeded.

Freight transport demand

Since 1995 freight turnover by rail and roads has more than doubled. Rail provides 72 % of total freight turnover mainly related to transit freight traffic. However, domestic freight is mostly carried by road transport. Since 1995 transport energy use increased by 58 %. In 2002 road transport fuel consumption constituted 87 % of total transport energy use. Compared to 2001, gasoline consumption has slightly decreased, gas consumption has remained stable, while diesel fuel consumption has increased. The EU biofuels directive requires these fuels to be 2 % of the fuel market in 2005 (or 20 000 tonnes) and 5.75 % by 2010. This is planned to be reached using local raw materials. Several laws and regulations have been passed which fix fuel quality requirements. As a result, the proportions of fuel used, the quality of fuel, as well as emissions to air (especially lead and sulphur emissions) have significantly changed.

Share of organic farming

The national target is to increase organic farming to 3 % of total agricultural area by 2006 and increase the share of organic products sold in the country to 3 %. Support for the development of organic farming is foreseen in the rural development plan. Although state support (e.g. subsidies for organic farming) was available since 2001, no farmers applied for this programme under the frame of SAPARD. From 1998 to 2003 the area of organic farming increased from 0.6 % to 1 % of total agricultural area. The largest part (48 %) covers medium-sized farms (20–100 ha). The amount and assortment of organic products is insufficient and does not match demand. Processing of organic products is poorly developed so only unprocessed or primary processed products are sold. Although the share of organic farming in agriculture is very small, a rapid increase of organic area and products can be expected in coming years.

Municipal waste

Although waste management systems have improved and public awareness has increased, waste production continues to rise. Rising consumption is increasing the use of household packaging waste. More attention is being given to waste processing, reuse and recycling, and a better understanding of resource flows and waste arisings and how to influence them. Used packaging that contains economically worthwhile and reusable materials contributes to 20–30 % of non-hazardous waste. Adopted changes in the calculation of the natural resource tax encourage the management of packaging waste. Recovery of materials from waste, including the export of materials extracted from waste outside Latvia, is still limited by a poorly developed market.

Use of freshwater resources

The total amount of water abstraction is decreasing. This trend is explained by the stabilisation of industrial activity and the installation of water meters, which motivate water savings. Metered water constituted 69 % of the total amount of withdrawn water in 2003. Losses constitute approximately 9–13 % of total water taken and can include water which a consumer has received but not paid for. A decrease of losses has been encouraged by modernisation and reconstruction of water management in small towns supported by the state investment programme.

For more information please contact the relevant national focal point. Contact details can be found on:
http://org.eea.eu.int/organisation/nfp-Eionet_group.html

Liechtenstein

A small and centrally located country, Lichtenstein is subject to pressures from increasing non-domestic freight transport, especially in the Rhine valley. There are plans to increase domestic energy from renewable sources up to 10 % of total energy requirements. The population has grown rapidly, and although waste generation is increasing in parallel, GHG emissions and other air pollutants are decreasing.

Greenhouse gas emissions

Despite an increase in population of over 20 % in the past 13 years, emissions of greenhouse gases have decreased substantially due to the implementation of technical measures in combustion plants, motor vehicles and in industry. Liechtenstein aims to reduce emissions of greenhouse gases by 8 % by 2008–2012, compared to the base-year 1990, and will therefore go beyond its commitments under the Kyoto Protocol.

Energy consumption

Total energy consumption increased by around 20 % during the past 10 years to approximately 1.2 million MWh/per year. This more or less corresponds to the population growth during this period. Half of total energy consumption is taken up by the heating requirements and management of buildings and plants, transport uses 30 % and 20 % is used in commercial and industrial production.

Renewable electricity

Liechtenstein imports most of the energy for its requirements. Energy produced in Liechtenstein makes up 7 % of total energy consumption and consists entirely of renewable energy (25 % wood, 75 % hydropower). The priorities for the coming years are: increasing the share of renewable energy to 10 % of total energy consumption by using more domestic biomass; increasing the use of solar energy through thermal solar plants; increasing the amount of electricity gained through solar energy by using photovoltaic technology; investing in combined heat and power plants in major projects and increasing the use of wood as an energy source.

Emissions of acidifying substances

Liechtenstein has radically reduced NO_x emissions by 40 % over the last 10 years. This was mainly due to technical measures such as the introduction of catalytic converters for motor vehicles and the implementation of technologies with low NO_x emissions for heating. Since the end of the 1980s, SO_2 emissions have also radically decreased due to the use of natural gas and of products with low SO_2 emissions in combustion.

Emissions of ozone precursors

Liechtenstein is ranked in the top third for ozone precursors. However, exceedances of the long-term objective for the protection of human health still occur frequently during the summer months. Measures undertaken up to now in the transport, combustion, industry and commercial sectors have not been sufficient to reduce the harmful ozone concentrations on a long-term basis. In addition, Liechtenstein is heavily affected by increasing transit traffic in the Rhine valley (north-south axis).

Liechtenstein

Country perspective

Population: 33 000
Size: 160 km²

Freight transport demand

The number of transport vehicles has increased in the last five years by 5 %. In Liechtenstein, 32 % of freight transport is domestic, 63 % is transport with destination and origin in Switzerland and Austria and about 5 % is other transit transport.

Share of organic farming

Liechtenstein has the highest percentage of organic farms in Europe. The organically farmed area is continuously growing.

Municipal waste

Due to the introduction of waste charges according to the polluter-pays-principle in 1994, the amount of waste generated has only developed in parallel with population growth. At the same time, the amount of waste recycled has increased.

Use of freshwater resources

In 2004, 8.99 million m³ freshwater was used (54 % groundwater, 46 % spring water). This corresponds to a per capita water consumption of 860 litres per day. 37 % freshwater consumption is caused by industry. 1.9 million m³ freshwater is used every year for the production of thermal energy.

For more information please contact the relevant national focal point. Contact details can be found on:
http://org.eea.eu.int/organisation/nfp-Eionet_group.html

Lithuania

Lithuania is performing well across most of the selected indicators. Emissions to air have decreased drastically over the last decade in line with all international obligations. Current efforts to reduce relatively high energy intensity should improve the situation and keep emissions at relatively low levels. Efforts to increase the production of electricity from renewables are still necessary. Share of organic farming is rapidly increasing indicating positive response to environmental actions.

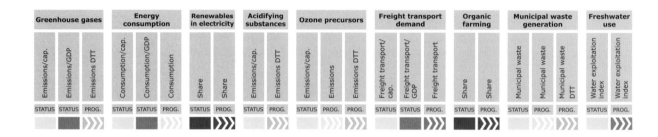

Greenhouse gas emissions

An almost threefold decrease of GHG emissions since 1990 was mainly driven by economic decline. The re-structuring of industry and more efficient use of energy resources allowed a high increase of GDP per capita without significant increase of GHG emissions. It is expected that closing the Ignalina nuclear power plant will affect GHG emissions, but that Lithuania will comply with its commitment to the Kyoto Protocol.

Energy consumption

Energy intensity in Lithuania is comparatively high. Sectoral analysis showed that households generate the highest share of total energy consumption. Therefore, it is expected that the new building renovation programme will increase energy efficiency significantly. The country manages to maintain a low GDP per capita compared with the other European countries and high annual GDP increase, by linking intensification of energy consumption with increased efficient energy use.

Renewable electricity

Much effort has been made to increase the share of renewable energy resources. Although renewables contribute nearly 10 % of total energy generation, Lithuania is still below EU targets. Intensification of hydro-energy generation is limited due to relatively low potential capacities of Lithuania's lowland rivers. However, Lithuania is committed to reach 7 % share of renewable electricity by 2010.

Emissions of acidifying substances

Emissions of acidifying substances have decreased significantly since 1990. The most significant decrease has been observed for SO_2 emissions. This was driven by re-structuring of big economic entities, introduction of more stringent requirements for SO_2 content in fuel and a shift to more environment-friendly fuels. Current emissions of acidifying substances are a few times below national limits.

Emissions of ozone precursors

Emissions of ozone precursors reduced drastically during the last decade. This was mainly due to changing fuel types and introduction of new techniques in the transport sector and fuel distribution systems. A decrease of livestock reduced CH_4 emissions from the agricultural sector. Since the measurement of tropospheric ozone in urban areas only began in 2003 it is not possible to evaluate changes of concentrations over the years. Observations in recent years showed that the eight hour maximum concentration limit might be exceeded occasionally in some of the measurement stations, although concentrations are below the risk threshold.

Lithuania

Country perspective

Population: 3 454 000
Size: 65 200 km²
GDP: 7 473 million euro

Freight transport demand

Freight transport is continuously increasing. Road freight transport is increasing most rapidly, followed by rail transport. Marine freight remains stable or is even decreasing. Rail freight is mainly determined by transit transport, while domestic freight is mostly transported by road.

Share of organic farming

The area of organic farming is continuously increasing in Lithuania. During 2004 the area of certified agricultural farming land increased by 20 000 ha and covered the total area of more than 40 000 ha — approximately 1.5 % of all farming land of the country. The number of organic farms increased rapidly in 2003–2004, growing by 60 % every year. In 2004 more than one thousand entities producing organic produce had been certified in Lithuania.

Municipal waste

Waste management is under reorganisation in Lithuania. The accounting system has also changed so that it is difficult to compare waste generation before and after the introduction of new waste accounting principles. It is estimated that volumes of collected municipal waste have been increasing continuously over the past ten years. Accounts over the last few years show that yearly amounts of collected waste are becoming more stable. The amount of total collected waste has increased following the improvement of the waste collection system. However, the broader introducing of waste recycling processes should to some extent compensate this increase. Comparison with other countries with significantly higher GDP per capita leads to the conclusion that municipal waste will increase in future.

Use of freshwater resources

Close to 4.5 thousand million m³ of water was abstracted from freshwater sources in 2004; 10 % less than in 2003. Use of freshwater resources in Lithuania is mostly dependent on energy production (95.8 % of the total water consumption in 2004). The total amount of water abstraction decreased in 2004. This could be generally explained by the lower activity of the Ignalina nuclear power plant and the Kruonis pumped storage plant. Higher water prices and the installation of water meters, encourage water savings. Metered water made up 69 % of total amount of withdrawn water in 2003. Losses constitute approximately 9–13 % of total water taken and can include water which a consumer has received but not paid for. A decrease of losses has been encouraged by modernization and reconstruction of the water supply systems and improving freshwater resources management in small towns supported by EU funds.

For more information please contact the relevant national focal point. Contact details can be found on:
http://org.eea.eu.int/organisation/nfp-Eionet_group.html

Country analysis | Country perspective

Luxembourg

Luxembourg performs less well than other countries on a per capita basis because it is a small country with high economic activity attracting people, especially workers, from abroad, thus increasing its population by a quarter. Although some decoupling has occurred, the characteristics of the country (size, commuters from abroad, rapidly increasing population) require specific policies for transport, land use and households.

Greenhouse gas emissions

Luxembourg has the most ambitious reduction target among EU countries (– 28 % by 2008–2012 relative to 1990). Due to important decreases in industrial emissions (steel industry moving from blast furnaces to electrical steelworks), this objective was already reached by 1995, but since 1999, total GHG emissions have been rising. From 2002, they were above the agreed reduction target due to rising transport emissions and the starting up of the country's first gas-heat cogeneration power plant (generating around 10 % of Luxembourg's greenhouse gases). However, despite an increasing population, domestic emissions remain stable. This is the result of a policy promoting renewable energy use by agriculture and the private and household sectors (through subsidies and fiscal reductions). Nevertheless, Luxembourg still has the highest emissions of GHG per capita in Europe. One significant reason is 'fuel tourism' which represents almost 40 % of the country's GHG emissions.

Energy consumption

Energy consumption changed dramatically since the early 1990s with coal tending towards zero and other fuels rising by more than 50 %. Between 1990 and 2002 industrial energy consumption decreased from 55 to 26 %, and transport consumption increased from 29 to 55 %. However total energy consumption is now only slightly above what it was in the early 1970s, hence a clear decrease in energy intensity (relative decoupling). Domestic final energy consumption only increased by 30 % since 1990 (whereas for transport the increase is 110 %), an encouraging figure compared with Luxembourg's population growth (+ 17 % since 1990). Consequently, as for GHG emissions, transport, and particularly 'fuel tourism', is the key sector penalising Luxembourg on a per capita basis.

Renewable electricity

Luxembourg imports most of its electricity needs from Germany and Belgium. To gain more control of the production and delivery of energy, it has been decided to develop national production of electricity. Renewable energies and new forms of energy production (cogeneration, such as the gas-heat cogeneration power plant) will be encouraged. Luxembourg aims at 5.7 % of final electricity production to be covered by nationally produced renewable sources. However, since the potential for water and wind renewable energies is almost exhausted, the 5.7 % objective will only be reached by reducing electricity consumption and promoting other renewable sources, such as biomass.

Emissions of acidifying substances

Emissions of the three main gases have all decreased since the early 90s: NO_X and NH_3 moderately; SO_2 significantly. In 2002, SO_2 emissions were only a sixth of 1990 levels. As for GHG emissions, the reasons are a decrease in industrial emissions (move from blast furnaces to electrical steelworks) and a strict policy for new industrial establishments. Luxembourg is well on track to meet its commitments for both SO_2 and NH_3. This is not the case for NO_X emissions which have risen

Luxembourg

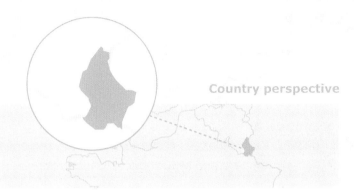

Country perspective

Population: 448 000
Size: 2 597 km²
GDP: 20 823 million euro

with the increasing population and national road traffic with a high share of diesel vehicles and 'fuel tourism'.

Emissions of ozone precursors

Luxembourg is not on track to meet targets for ozone precursors: in 2002, non-methane volatile organic compound (NMVOC) and NO_x levels were still around 30 points above the objectives set for 2010. However, absolute emissions of these two gases decreased since 1990. CO has had the most dramatic reductions: 2002 emissions are around 27 % of 1990 levels due to reductions in emissions from the steel industry.

Freight transport demand

Both passenger and freight road transport are key environmental problems for Luxembourg. Collaboration with neighbouring countries aims to tackle heavy road traffic from the foreign daily workforce. High investments are already foreseen to reinforce public transport. Government has set itself an ambitious objective of a 25–75 % modal split for passenger transport in Luxembourg by 2020. Reducing the gap in fuel prices between Luxembourg and border countries is also a goal for tackling 'fuel tourism'. Freight transport can only be addressed within an EU framework since Luxembourg is a land-locked country and located on one of the major north-south roads.

Share of organic farming

The share of organic farming in the total utilised agricultural area is clearly increasing: from 0.8 % in 2000 to 2.6 % in 2004. Organic farming concentrates mainly on dairy products, eggs, some cereals and meat (beef, poultry).

Municipal waste

Two factors explain the high per capita municipal waste generation figures. First, Luxembourg has a comprehensive waste collection system with 100 % of the population covered. Secondly, the 110 000 plus daily workforce from abroad generates waste which is counted against residents when per capita values are calculated. Nevertheless, Luxembourg residents generate too much waste. Fortunately, the share of municipal waste to be eliminated decreased between 1995 and 2002: – 20 % for landfill and – 15 % for incineration. At the same time, the recovery rate doubled to reach about 50 % of generated municipal waste. This encouraging rate is the result of a policy promoting voluntary waste separation and recovery.

Use of freshwater resources

Water abstraction per capita in Luxembourg is relatively moderate and remains stable. Since abstraction quantities are not a problem compared to the available surface and groundwater resources, water policy in Luxembourg focuses more on water quality and treatment of waste water.

For more information please contact the relevant national focal point. Contact details can be found on:
http://org.eea.eu.int/organisation/nfp-Eionet_group.html

Malta

Priorities for Malta include encouraging and developing markets for organic farming and renewables, as well as improving its data reporting in key areas such as waste, transport and air quality. Key challenges relate to improving the environmental performance of the transport and energy sectors and reducing waste generation.

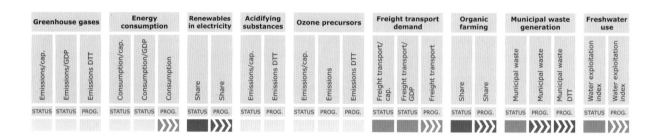

Greenhouse gas emissions

Malta's greenhouse gas emissions are relatively low when compared to EU averages, however they rose 44 % between 1990 and 2003.

Energy consumption

Although energy consumption may be relatively low at the European level, it is rising. Between 1990 and 2002 Malta's gross energy consumption rose by 61 %. A sharp rise in consumption in 2003 is likely to be due to increased use of air conditioning following the 2003 heat wave. The islands' energy intensity rose during the early 2000s after falling in the late 1990s.

Renewable electricity

The share of energy from renewable sources remains negligible in Malta's context. However a national renewable energy strategy is soon to be published for public consultation, providing indicative national targets in an EU context and outlining a way forward for Malta to increase the share of renewable energy sources in energy generation.

Emissions of acidifying substances and ozone precursors

While national data on quantities of acidifying emissions and ozone are not available, data on concentrations in 2004 indicate that the main issues of concern are particulates and sulphur dioxide, and nitrogen oxides in certain urban areas. Transboundary importation of ozone and sulphur dioxide is also of concern. There were significant decreases in sulphur dioxide and benzene concentrations during 2004 due to use of a cleaner fuel mix.

Freight transport demand

While national figures for freight transport km are not available, the number of vehicles on the road continues to rise, doubling between 1986 and 2004. The number of vehicles per capita was 0.7 in 2004. High rates of freight transport calculated on a European scale may be related to the small size and peripheral location of the island. The country imports many of its goods, which often need to travel long distances from continental centres of production.

Malta | PART C

Country perspective

Population: 399 000
Size: 320 km²
GDP: 3 095 million euro

Share of organic farming

Malta's share of land under organic farming remains relatively low, but there has been improvement over the last few years and 0.09 % of land is now under organic cultivation, with 80 % of this certified.

Municipal waste

The amount of municipal waste generated per capita in Malta is relatively high and rising. The quantity of municipal waste generated increased 53 % between 1996 and 2004. However, management systems are being put in place to encourage waste reduction, reuse and recycling.

Use of freshwater resources

Since it is based on abstraction of water, the performance regarding water in Malta appears favourable. However this is due to the fact that groundwater abstraction only accounts for 56 % of production — the remainder comes from desalination. This indicator does not reveal the fact that most of Malta's aquifers are currently over-exploited and at risk.

For more information please contact the relevant national focal point. Contact details can be found on:
http://org.eea.eu.int/organisation/nfp-Eionet_group.html

The Netherlands

European policies are leading to considerable emission reductions in the Netherlands but this is not always enough to meet EU environmental quality standards. Compared to other EU Member States, emissions per square kilometre are higher because the Netherlands is more densely built up and populated. The Netherlands is showing good progress in meeting its renewable energy target and more recently in the reduction of municipal waste.

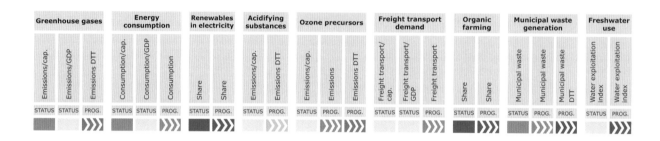

Greenhouse gas emissions

Emissions of greenhouse gases were 1 % lower in 2003 than in the base-year (1990/1995). Emissions of CO_2 increased considerably, but this was compensated by the reduction of non-CO_2 greenhouse gases. Compared to other EU Member States, the Netherlands has been quick to purchase emission reductions in other countries via the Kyoto mechanisms. By purchasing these foreign emission reductions, the Netherlands has a 50 % chance of achieving the 2008–2012 Kyoto commitment if domestic emissions remain roughly stable.

Energy consumption

Because it is energy intensive, the Dutch economy uses more energy than the European average. The yearly rate of energy saving amounts to 1 % (10-yearly average). In the period 1999–2002 the energy efficiency of industry improved by 0.7 %.

Renewable electricity

The share of nationally produced renewable energy was 1.8 % in 2004. The target for the share of renewable energy in the energy supply (5 % in 2010) will not be met. The target for renewable electricity will be met.

Emissions of acidifying substances

Although acidifying emissions are falling, the Netherlands will probably not comply with the EU national emission ceilings (NEC) directive for SO_2 and NO_x. Nevertheless, Dutch eco-efficiencies for SO_2 and NO_x are among the best in Europe. For NH_3 the Netherlands has a 50 % chance of achieving the national emission ceiling.

Emissions of ozone precursors

As in many parts of Europe, concentrations of particulate matter exceed European air quality standards to a considerable extent across wide areas of the Netherlands. Regional concentrations of particulate matter in the Netherlands as well as in Belgium, the German Ruhr region and Italy are relatively high. The air quality limits for particulate matter will still be exceeded on a large scale in 2010.

The Netherlands

Country perspective

Population: 16 215 000
Size: 41 530 km²
GDP: 385 436 million euro

Freight transport demand

Due to EU emission requirements for cars, transport-related NO_x-emissions are decreasing, despite the increase in traffic volume and the fuel shift from petrol to diesel. However, the 2010 emission target for NO_x for the transport sector will be exceeded. Improved environmental performance of trucks will have a major effect on future NO_x emissions. The Dutch air quality decree represents a strict implementation of the European air quality directive, particularly because of its integration with the spatial planning legislation. The standards for particulate matter and ozone are a major step in the protection of public health but are no guarantee. Long standing exposure to high particulate matter concentrations possibly has serious health effects.

Share of organic farming

Nowadays about 2 % of the Dutch agriculture consist of organic farming. If the future growth rate of organic farming remains the same as the current rate, the target (10 % in 2010) will not be met. High consumer prices interfere with a further development of Dutch organic farming.

Municipal waste

In the period 1990–2002 the amount of household waste increased by over 40 %. The amount of landfilled waste is strongly decreasing while recycling and incineration is increasing. Since 2000 the total amount of waste has been decreasing.

Use of freshwater resources

In the past 30 years the abstraction and use of freshwater in the Netherlands has been more or less stable. However, the use of groundwater has decreased significantly, while the use of surface water has increased. Surface water is the main resource for industry and the energy sector, while water companies use mainly groundwater.

The use of drinking water by households has been decreasing since the mid 1990s due to technological changes and changes in consumer behaviour.

For more information please contact the relevant national focal point. Contact details can be found on:
http://org.eea.eu.int/organisation/nfp-Eionet_group.html

Norway

Stretching far into the Arctic, Norway is sparsely populated with extensive wilderness. Due mainly to offshore petroleum extraction, shipping and some industries (metals and chemicals), Norway has some challenges meeting international air emissions commitments. Scoring relatively well by GDP (eco-efficient economy), its performance on municipal waste is average and below average for growth in the use of renewable electricity, organic farming and reducing ozone precursor emissions.

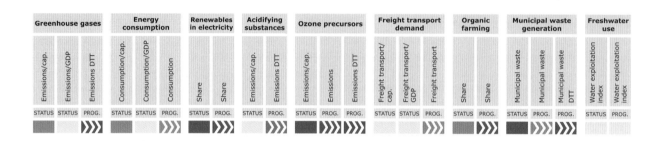

Greenhouse gas emissions

Norway has slightly higher GHG emissions per capita than the European average. Emissions are relatively low when measured in relation to GDP. While Norway is allowed to increase its emissions by 1 % by 2008–2012 relative to the base-year 1990, emissions were 5 % above the 1990 levels in 2002 (9 % in 2003). The achievement of Norway's Kyoto commitments will depend on contributions to emission reductions abroad (by making use of the mechanisms for joint implementation, clean development and emissions trading) in addition to preventing further growth in emissions at home.

Energy consumption

Norwegians have an energy consumption per capita well above the European average. This is explained by different factors: the country's large renewable energy resources, the cold climate, the wide geographical distribution of the population and the substantial factor of an energy-intensive industry. Measured against GDP, consumption is relatively low. Norway's growth in total energy consumption is close to the European average

Renewable electricity

Renewable energy produced by hydroelectric powerplants already accounts for as much as 99.5 % of total Norwegian electricity production. The share of renewable energy sources excluding large hydroelectric powerplants is very close to the average of the countries for which data are available. The growth in the share of such renewables in Norwegian electricity production has been lower than the average for EU-25.

Emissions of acidifying substances

Norwegian emissions of acidifying substances are somewhat lower in per capita terms than the European average. Norway is expected to meet its targets for 2010 under the Gothenburg protocol of the Convention on Long-range Transboundary Air Pollution for all components except for NO_X for which the gap against the target is still large. Shipping and stationary combustion in the oil and gas industry makes up the major part of Norwegian NO_X emissions.

Emissions of ozone precursors

Norway ranks as the second largest polluter in Europe when it comes to emissions of ozone precursors per capita. A major source is volatile organic compound (VOC) emissions during loading and storing of crude oil offshore. These emissions have decreased substantially in the last two to three years due to new technology. However, NO_X emissions still remain a considerable problem.

Norway

Country perspective

Population: 4 560 000
Size: 323 880 km²
GDP: 141 203 million euro

Freight transport demand

Considering the geographical distribution of its population, Norway's freight transport volume as measured per capita and in relation to GDP is relatively low. The growth in freight transport volume is close to the European average.

Share of organic farming

The share of organic farming in Norwegian agriculture is just about the European average. The growth in the share of such farming is also very close to the European average.

Municipal waste

The amount of municipal waste collected per capita in Norway is substantially higher than the EU-25 average. On the other hand the recycling rate is very high. The growth in the generation of municipal waste is slightly lower than EU-25 average.

Use of freshwater resources

Only 0.7 % of the water resources available each year in Norway is utilised before draining to the coast (97 %) or via rivers to neighbouring countries (3 %).

For more information please contact the relevant national focal point. Contact details can be found on:
http://org.eea.eu.int/organisation/nfp-Eionet_group.html

Poland

With the exception of organic farming, Poland is performing well across the scorecard and compared to the EU-25 average. In common with many of the EU-10 Member States eco-efficiency improvements in energy, greenhouse gas emissions and transport are priorities. Market development for organic farming and renewables is also needed. In the short term, Poland is not at risk of run-away trends worsening environmental conditions.

Greenhouse gas emissions

After 1990, total GHG emissions declined substantially, mainly due to restructuring or closure of heavily polluting and energy intensive industries. Under the Kyoto Protocol Poland has a reduction target of 6 % from the base-year and is 29 % below its linear target path. Emission trends until 2001 and projections for 2010 show that Poland is on track to meet its Kyoto target.

Energy consumption

Energy use has decoupled from economic growth in Poland, which means that the modernization of existing power generation facilities and the implementation of eco-efficient technologies, driven by, among other things, legal and economic instruments, is resulting in a successful decrease in energy consumption. Together with low per capita use this places Poland among the best performing countries in terms of energy consumption per capita.

Renewable electricity

The share of renewable energy accounts for about 2 % of total electricity consumption in Poland. Renewable energy sources mostly include biomass with a small but constantly increasing number of hydro and wind power plants.

Emissions of acidifying substances

Emissions of acidifying gases have been substantially reduced since 1990. Poland has already reached its reduction targets for NO_2 set for 2010 in the Gothenburg protocol under CLRTAP, is well below the reduction target set for NH_3 and is very close to reaching its SO_2 reduction target for 2010.

Emissions of ozone precursors

Compared to most of the EU countries, emissions of ozone precursors are low in Poland with a continual decrease of CH_4 and CO since 1990. For non-methane volatile organic compounds (NMVOCs), Poland is currently well below its emission reduction target set for 2010 in the Gothenburg protocol.

Freight transport demand

Transport volume, although increasing, is still decoupled from economic growth. However, since emissions from transport contribute significantly to air pollution and noise, more effort needs to be put into additional abatement measures to make the transport sector more environment-friendly.

Poland

Country perspective

Population: 38 195 000
Size: 312 690 km²
GDP: 141 807 million euro

Share of organic farming

Compared with most of the EU countries the area of organic farming in Poland is low but is still increasing systematically. It should be emphasised that in the period 1990–2002 the area of organic farming has expanded over two hundred times. In 2002 it covered about 0.3 % of total utilised agricultural area. It is projected that the interest in organic farming will increase in the next years. This will result from a major demand for natural food among Polish consumers and expected higher subsidies from EU funds.

Municipal waste

Household waste production in Poland is on the decrease. The figures show that Poland belongs to the group of countries which are leading Europe by producing less than 300 kg municipal waste per person. At the same time it must be stressed that further efforts towards the intensification of reuse, recovery and recycling are needed.

Use of freshwater resources

In Poland the level of water abstraction is on the decrease, with industry still remaining the main user. Traditionally, Poland represents a low level of water abstraction for agriculture in comparison with the EU average.

For more information please contact the relevant national focal point. Contact details can be found on:
http://org.eea.eu.int/organisation/nfp-Eionet_group.html

Portugal

Portugal is performing relatively well in many of the status indicators but is showing some poor developments in progress which may need to be reversed or slowed to avoid a worsening of the environmental situation. Areas for attention include air emissions and especially emissions of ozone precursors which risk causing substantial impacts on human health and ecosystems as a result of its southern location.

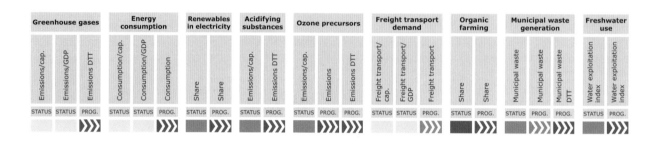

Greenhouse gas emissions

Portugal was the EU country after Lithuania which had the biggest GHG emissions reduction between 2002 and 2003 (6 %). This was mainly due to increased electricity production from hydroplants in 2003. This indicator strongly depends on the annual variations and availabilities affecting hydro-power. In 2003, GHG emissions were 37 % above 1990 levels, exceeding the Kyoto Protocol target by 10 % (taking into account only the main gasses responsible — CO_2, CH_4, N_2O — excluding land-use change and forestry, and including fires which were very severe in 2003).

Energy consumption

Energy intensity in Portugal has remained more or less steady as the energy efficiency of the economy has been maintained. The gradual and general introduction of natural gas (which, in 2003, made up 10.3 % of total primary energy consumption) and renewable energies, the improvement of the fossil fuels quality (such as the gradual reduction of sulphur levels in gasoline and fuel for diesel engines) and the promotion of energy and technological efficiency of some industrial processes, are the main drivers shaping the future environmental profile of this sector.

Renewable electricity

Portugal has a strong potential for renewable energies, especially solar, wind, hydro and biomass. Although an effort had been made over the last 15 years to introduce renewable energies, there is still potential for further growth. The absolute growth rate, however, shows an active uptake of renewable energies: between 1990 and 2003 the contribution of renewable energies to the energy balance increased from 3.5 to 4.2 Mtoe. The percentage of renewable energies in electricity consumption was 36 % in 2003, coming close to the target established by the EU for Portugal of 39 % in 2010. The annual contribution of renewable energies to the energy balance reflects the importance of hydro-power and its changeable character.

Emissions of acidifying substances

Portugal has been making significant efforts to reduce emissions and meet commitments, making significant progress in 2003. The emissions of acidifying substances decreased by 16 % between 1990 and 2003, mainly due to the reduction of SO_2 emissions by 37 %. This can be explained by changes in the energy production sector and significant improvements during this period in the quality of the fuel used. Also, 2003 was a year with high production from hydro-electric power plants.

Emissions of ozone precursors

The overall analysis of ozone precursor emissions must be made in connection with a local analysis of air quality, and especially the exceedances of targets established for each objective and time period. The complex morphology of the landscape means that under certain meteorological conditions, atmospheric

Portugal

Country perspective

Population: 10 191 000
Size: 91 980 km²
GDP: 100 758 million euro

pollutants tend to remain in the lower atmosphere, become recycled, resulting in the formation of secondary pollutants, such as ozone. Taking all these factors into account, even if agreed emission reductions are achieved, ozone episodes will occur in certain locations, requiring active information to the public.

Freight transport demand

Road transport is the largest consumer of energy in Portugal, as well as the main source of pollutant emissions among economic activities. Road transport accounts for 90 % (by volume) of overland freight transport. With current policies, a reverse in this trend is not expected in the coming years. Recent investments in some structural projects for passenger transportation should have positive impacts on the overall efficiency of this sector.

Share of organic farming

Portugal has good conditions for organic farming and the traditional agricultural practices are very close to organic production. However, the proportion of area occupied with organic farming is relatively small compared with the total agricultural area and with the EU average. There is a growing interest from consumers about these types of products. At the same time national production is not sufficient and producers have some difficulty putting their products on the market. Most organic products consumed in the country are imported. Over the past years the area and number of farms devoted to organic farming has grown progressively, but there still remains a huge potential.

Municipal waste

The generation of municipal waste in Portugal has remained steady over the last years, compatible with the 2005 national goal of 4.5 million tonnes/year. The generation of municipal waste per capita in Portugal is one of the lowest in the EU. In 2004, the final disposal of this type of waste is divided as follows: 66 % landfill, 20 % incineration, 7 % composting and 7 % recycling. Despite progress, these numbers are still far from of the established goals of 26 % composting and 26 % recycling in 2005.

Use of freshwater resources

Agriculture is the main water consumer in Portugal and the economic sector that exerts the greatest pressures on water resources. It is also the sector that uses water most inefficiently, with overall losses in the order of 30 %. The national program for the efficient use of water will allow the negative impacts on freshwater resources from agricultural, industrial and urban uses to be reduced.

For more information please contact the relevant national focal point. Contact details can be found on:
http://org.eea.eu.int/organisation/nfp-Eionet_group.html

Romania

Integration of environmental policy in the further development and implementation of sectoral and regional policies represents one of the main priorities of the Romanian government for 2005–2008. The national development plan for 2007–2013, the reference document for accession, is currently being prepared and this document will also serve as a basis for application to future cohesion and structural funding.

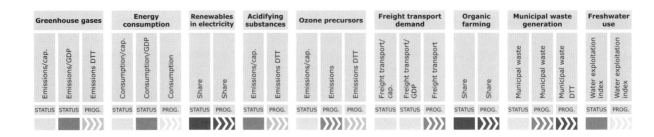

Greenhouse gas emissions

The trend in GHG emissions between 1989 and 2001 shows a decrease of 47 %. In 2001, GHG emissions represented 53 % (139 171 Gg CO_2-equavalents) of the 1989 base-year total, (261 355 Gg CO_2-equivalents). The decrease in GHG emissions over this period mainly resulted from a strong economic decline, associated with the transition to a market economy. To this came the added effect of the start-up and operation of the first reactor at the Cernavoda nuclear power plant in 1996. An unusual increase in the total annual emissions was recorded in 1995, when both the consumption in the energy sector and the production in various industrial branches increased significantly.

Based on these observations, it is very clear that Romania will meet its commitments to reduce GHG emissions in the Kyoto Protocol's first commitment period 2008–2012. Important changes regarding the GHG emissions are not expected to occur in the first commitment period, but an increase is likely. Some model assessments suggest that Romania's net GHG emissions in the 2008–2012 period could be between 175 000–200 000 Gg CO_2-equivalents., if the pace of economic growth increases.

In 1989, 83 % of GHG emissions came from the energy sector. Although this decreased to 79 % in 2001, the energy sector is still the main polluting sector in the Romanian economy. In the industrial sector, the largest CO_2 emissions came from mineral products. The industrial sector as a whole suffers from some data gaps and limited access to information.

Energy consumption

In 2005, a law was approved by governmental decision setting up the 'Emissions reducing national programme for large combustion plant pollutants'. The integrated pollution control licences contain details of the pollutant emissions monitoring.

Renewable electricity

In 2004, the Romanian government adopted a law (Governmental Decision no. 1429/2004) on certifying the origin of produced electrical energy from renewable sources, and another law (Governmental Decision no. 1892/2004) establishing a system to promote electrical energy from renewable energy resources.

Emissions of acidifying substances

The evaluation of acidifying substances emissions is a useful instrument for decision-makers to appreciate Romania's situation concerning compliance with its obligations towards its accession to the European Union. In 1999 Romania signed the Convention on Long-range Transboundary Atmospheric Pollution. This convention has been ratified by the Romanian Parliament through Governmental Decision 271/2003 and aims firstly to reduce acidification, eutrophication and tropospheric ozone pollution. Romania has pledged to comply with the levels of emission as established under the Gothenburg Protocol by 2010.

Romania

Country perspective

Population: 22 200 000
Size: 238 390 km²
GDP: 29 598 million euro

Emissions of ozone precursors

Tropospheric ozone monitoring is carried out in the Timişoara and Reşiţa stations that are part of the EuroAirnet network. The analysis of the data obtained from these two stations show that during 2003 the highest values were recorded between May and August and coincided with the atmospheric temperature increases and the diurnal period. At the Timişoara station, the daily maximum values (8-hour means) calculated from hourly data representing the long-term goal for human health protection, were within 6.37 µg/m³–90.78 µg/m³ and at Reşiţa station within 0.874 µg/m³–158.41 µg/m³. At the Reşiţa station 12 exceedances of the long-term goal for human health protection were recorded during the month of August. The maximum value was recorded on 13.08.2003 i.e. 158.41 µg/m³. At the Timişoara station no exceedances have been recorded (a long-term objective for human health protection).

Freight transport demand

Between 2000 and 2003 the amount of goods transported by road increased from 263 million tonnes to 276 million tonnes.

Share of organic farming

Between 2000 and 2003, the amount of chemical fertilisers used in agriculture was smaller than the amount of natural fertilizers. In 2000, the amount of chemical fertilisers used was 0.34 million tonnes and in 2003 the amount was 0.36 million tonnes. Use of natural fertilisers was 15.8 million tonnes in 2000 and 17.3 million tonnes in 2003.

Municipal waste

The quantity collected in 2003 was 0.292 t/capita and production of household waste is stable. Almost all of the collected municipal waste is landfilled. There are only a few pilot projects for separate collection and recovery of municipal waste. The national waste management plan establishes development of an integrated management system for municipal waste, comprising separate collection, treatment, recovery and disposal to licensed landfill sites.

Use of freshwater resources

Romanian water resources consist of surface (rivers and lakes) and underground waters. In 2004 the total water abstraction was 5 850 million m³ out of which: 12 % was for agriculture, 21 % was for human needs and 67 % was for industry.

The water abstraction decreased more than three times during the period 1990–2004 due to a combination of economic slow-down, reduction of the water used for technological processes, reduction of water losses, and implementation of the economic instruments and mechanisms in water management.

For more information please contact the relevant national focal point. Contact details can be found on:
http://org.eea.eu.int/organisation/nfp-Eionet_group.html

Slovak Republic

The Slovak Republic showed good environmental performance for the status indictors in 2002 but more average progress over the past 10 years. While these trends over time are not a cause for concern at present, the Slovak Republic needs to be vigilant that such trends do not worsen in the coming years which could give rise to unfavourable consequences for the environment.

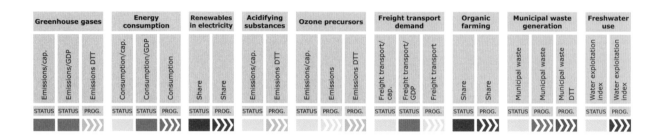

Greenhouse gas emissions

In accordance with the generally expected results, the aggregated emissions of GHGs in 2002 in the Slovak Republic significantly decreased against the base-year (1990) by approximately 21 Tg, equivalent to a decrease of about 29 %. This means that the 2008–2012 Kyoto target of 8 % reduction compared to 1990 levels will be successfully met. However, the longer-term global emissions of GHGs will need to be reduced by approximately 70 % compared to 1990 levels for which additional policy measures will be necessary.

Energy consumption

Although the final national energy consumption per capita in the Slovak Republic is rather low compared with other EU countries, it is expected to increase along with a growth in GDP and increased standards of living. The number of home appliances used by Slovak households is currently only about 50 % of the average in the EU Member States. The number of electrical appliances used in the Slovak service sector is only about 30 % of the EU average. Given the predicted increase in the number of appliances and the related increase in energy consumption it is essential for the country to become more active in the field of energy efficiency. Moreover, one of the main characteristics of the Slovak energy sector is the high level of energy intensity in comparison with the rest of the EU and some of its neighbouring countries. This is mainly due to a high level of energy demand from heavy industry (metallurgy, steel and machine works, chemicals). The objective to enhance energy efficiency has been declared in the energy policy of the Slovak Republic. The study also includes an action plan on energy efficiency for 2002–2012.

Renewable electricity

Utilisation of renewable energy sources will have a positive impact on the Slovak economy. Biomass has the largest potential for utilisation in the Slovak Republic (44 % of all renewable energy sources), followed by large water power stations (17.5 %), geothermal energy (16.6 %), solar energy (13.7 %), waste management (9.3 %), bio-fuels (6.6 %), small hydropower stations (2.7 %) and wind energy (1.6 %).

Results of the process are expected as follows: enhanced utilisation of domestic energy sources, reduced dependency on imported energy, enhanced foreign trade balance, enhanced safety and reliability of energy supplies, reduced greenhouse gas emissions and enhanced economic activities with respect to new production programmes and new jobs. The right location of renewable energy sources can become the key element of regional development and can contribute to better social and economic cohesion in the country. The conception of renewable energy sources utilisation in the Slovak Republic approved by the Slovak government in February 2002 is the key strategic document.

Population: 5 381 000
Size: 49 010 km²
GDP: 20 066 million euro

Emissions of acidifying substances and ozone precursors

Emissions of SO_2, NO_x, NH_3 and non-methane volatile organic compound (NMVOC) have declined compared with 1990 and the Slovak Republic is well on track to fulfil specified targets by 2010 in accordance with Directive 2001/81/EC on national emission ceilings for certain atmospheric pollutants and in accordance with the Gothenburg Protocol. According to the scorecard on ozone precursors, the Slovak Republic is one of the countries with the best progress.

Freight transport demand

There has been significant progress in freight transport demand in the Slovak Republic compared to 1995. Volume by rail has decreased and there is now no real alternative to freight transport by road. Nevertheless, the Slovak Republic is one of the best performing countries in terms of progress (1995–2003).

Share of organic farming

The share of organic farming area is 2.18 % of the total agricultural land area. A new action plan for organic farming in the Slovak Republic up to 2010 has been adopted, with the target to reach 5 % of organic farming area of the total agricultural area by 2010.

Municipal waste

The production of municipal waste is stable at the level of approximately 300 kg per person per year (1.6 million tonnes/year for Slovakia). The main part of this amount is disposed in landfills (78.2 % in 2002) and a small part (4.3 % in 2002) is incinerated. The amount of separately collected waste from households is increasing because the municipalities utilise the financial support of the Recycling Fund. The recycling infrastructure is growing fast. According to the development in municipal waste management during the monitored period and the latest data from 2004, the targets set in the waste management plan until 2005 (35 % material recovery, 15 % energy recovery and 50 % disposal in landfills) can in all probability be achieved.

Use of freshwater resources

Since 1990, the total water abstraction has decreased in the Slovak Republic. Surface water abstraction represents 60 % of all abstractions. The biggest consumer of surface water is the industry sector with 78 %. Surface water abstraction for water supplies represents approximately 10% of all abstractions; surface water abstraction for irrigation represents approximately 10.5 %. Groundwater abstraction represents 40 % of all abstractions in the Slovak Republic. The major part (approximately 75 %) is represented by public water supplies.

For more information please contact the relevant national focal point. Contact details can be found on:
http://org.eea.eu.int/organisation/nfp-Eionet_group.html

Slovenia

Slovenia is showing steady average performance and progress, with positive progress in per capita waste collected and in current water use per capita. Priorities for Slovenia include developing its eco-efficiency as shown in particular in the low rankings achieved for air emissions and energy intensity.

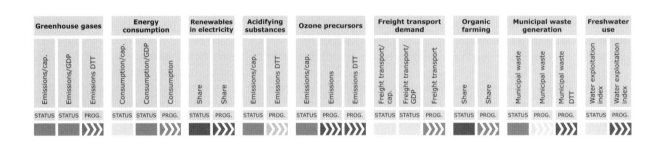

Greenhouse gas emissions

Although total GHG emissions have not changed significantly compared with the base-year, their distribution by sector changed considerably. Traffic emissions almost doubled, indicating the need to develop an integrated action programme. Emissions from fuel consumption in residential and commercial sectors, as well as emissions from waste, increased. With the loss of the Yugoslav markets, abandonment of non-profitable production and increase in productivity, emissions from manufacturing industries decreased. For the purposes of maintaining competitiveness, emissions trading and compliance with the Integrated Pollution Prevention and Control (IPPC) directive, the industrial sector is being encouraged to make use of existing best available technologies. The agricultural sector is also showing lower emissions, mostly as a result of a reduction in the number of livestock. Projections indicate that the number of cattle will rise again due to quotas determined for Slovenia. On the other hand, agricultural policy is expected to reduce agricultural emissions by introducing good agricultural practice in fertilising and establishing biogas installations for electricity and heating production. Forests cover more than half of Slovenia's land surface and constitute an important sink of greenhouse gases.

Energy consumption

Primary energy consumption in Slovenia has been growing since 1992 by 2.7 % per year (1992–2002). More than two-thirds comes from fossil fuels, with a growing share of natural gas. Slovenian energy intensity dropped considerably between 1995 and 1999, although at a lower rate in recent years. Slovenia's relatively high energy intensity is due to the current economic structure, where manufacturing industries contribute 27 % to the total value added. Of this, a large part is taken up by energy intensive industries (metal, paper and chemical).

Renewable electricity

Slovenia is one of those countries with a relatively large share (11 %) of renewable energy sources in its total primary energy consumption. In accordance with its natural characteristics, the major renewable sources are biomass (61 %) and hydropower (38 %). The exploitation of renewable energy sources in Slovenia is promoted through investment subsidies, CO_2 emission tax and the priority dispatching of electricity produced by qualified producers (mainly small hydro-electric power plants), including price incentives. There is still considerable potential to develop traditional renewable sources. There are plans to increase the exploitation of technically exploitable hydro potential from the current 43 % to 52 % by 2013 by building a chain of hydro-electric power plants along the Sava river. As annual increment well exceeds the level of tree-felling in Slovenia's forests, growing emphasis has recently been placed on the exploitation of wood biomass.

Slovenia

PART C

Country perspective

Population: 1 964 000
Size: 20 250 km²
GDP: 20 548 million euro

Emissions of acidifying substances

With the decrease, since 1990, of traditional livestock-oriented agriculture, by 2002 ammonia emissions had fallen by 20 %, close to the EU national emission ceilings (NEC) directive target. Volatile organic compound (VOC) emissions, on the other hand, show an upward trend, caused by growing road traffic and use of biomass for energy production. Due to a natural prevalence of carbonate rocks, Slovenia has never faced major acidification problems.

Emissions of ozone precursors

Ground-level ozone concentrations in summer occasionally exceed limit values, especially in the cities and western parts of Slovenia. The major part of this pollution comes from other countries. The apparent increase in VOC emissions is the result of changes, since 2000, in the method by which emissions are calculated rather than by a real increase. However, without additional measures, such as a reduction of the content of organic solvents in products, improvements to combustion plants and, most importantly, a renewed car fleet and limited road traffic growth, the NEC targets will not be easily met.

Freight transport demand

Compared to other countries, Slovenia has had a relatively low share of road freight transport (about two-thirds). The remaining third is shipped by rail, a fact that can be explained by the relatively well developed railway system, partly a legacy of the Austro-Hungarian Empire. As is typical for a small inland state, freight transport in Slovenia consists predominantly of international transport; with major improvements of the national motorway network and EU accession, transit across Slovenia has increased significantly, causing adverse environmental and other side-effects.

Share of organic farming

Small farms, the high nature value of agricultural land and low agricultural intensity, stimulate the development of organic farming in Slovenia. Measures under the agri-environmental programme, adopted in 1998, and activities of non-governmental associations encourage farmers to report and seek certifications for their organic products.

Municipal waste

Although, the apparent decrease in the amount of municipal waste collected over the past years can be the result of changes in methodology behind the statistics, a small but real decrease in the amount of waste collected has been noted over recent years. Measures taken to implement EU waste management directives, including management of packaging waste and the establishment of 'sort-by-source' systems, are expected to reduce generated waste further. Improvements are needed in waste management since a large proportion of generated waste currently goes to landfill.

Use of freshwater resources

The water use index in Slovenia is low (about 2.5 %) since, compared with consumption, available water is plentiful. The largest consumer of water is the energy sector with 70.9 %. Industry and public water supply sectors consumed 28.4 % of the total amount of water used, while the smallest share, 0.7 %, was consumed by agriculture.

For more information please contact the relevant national focal point. Contact details can be found on:
http://org.eea.eu.int/organisation/nfp-Eionet_group.html

Spain

Fast economic development has come with increasing urbanisation and intensive use of resources. At the same time environmental management has been strengthened, leading to improvements in protection of natural areas, waste management, water treatment and use of renewable energy. Spain also faces specific problems linked to its climatic and geographic characteristics: fires, droughts, erosion and flooding.

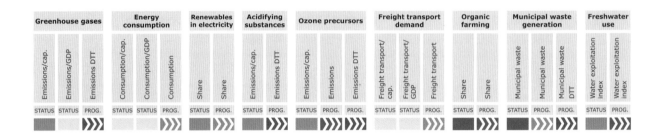

Greenhouse gas emissions

During 1990–2003, total greenhouse gas emissions increased by almost 40.6 %. This is 25.6 % above the Kyoto Protocol agreements for Spain, which allows for a 15 % increase above 1990 levels by 2008–2012. In 2003, total greenhouse gas emissions reached 402 million tonnes of CO_2-equivalent. Greenhouse gas emissions in Spain are similar to other large European countries. CO_2 emissions per capita (8 tonnes/inhabitant) are below the EU average.

Energy consumption

Per capita energy consumption in Spain is similar to other southern European countries. Spain ranks seventh in the intensity of final energy consumption. The 2004 figure of 171.5 toe/million euro, represented an increase of 3 % compared to the previous year. Final energy consumption in Spain reached 97.2 million toe in 2003, a 60 % increase since 1990.

Renewable electricity

The share of renewable electricity in Spain was above the EU average in 2004 (19.8 %) and very close to the EU-15 2010 target of 22.1 % (specific target of 24.9 % in 2010 for Spain), which puts Spain in fourth place among 30 European countries. The share of renewable energy in primary energy consumption was 6.3 % in 2004, compared to the EU-15 2010 target of 12 %. This share has not increased further because of the strong surge in primary energy consumption in the last years. Excluding large hydropower, the shares of renewables in electricity consumption in 2004 were 5.5 % for wind, 0.8 % for biomass and 0.72 % for other sources of energy (including solar photovoltaic). Spain is the world's second largest producer of electricity from wind. A reward system exists to increase energy production from renewable sources with a regulatory framework and a new plan for renewable energy 2005–2010 approved in August 2005.

Emissions of acidifying substances

Spain has relatively high emissions of acidifying substances per capita. In relation to its distance to target Spain ranks bottom: based on current trends it only appears feasible to meet the 2010 objective for SO_2. Total SO_2 emissions fell by over 38 % between 1990 and 2003. The combustion and transformation of energy is responsible for 72.5 % of these emissions. Total NO_X emissions increased 21 % during this period. The transport sector is responsible for 52.8 % of these emissions (35 % accounted for by road transport alone). Total NH_3 emissions increased by 21.1 %. Just over 87 % of total NH_3 emissions are accounted for by the agricultural sector.

Emissions of ozone precursors

Regarding emissions of tropospheric ozone precursors, Spain is in a very unfavourable position. Only significant reductions in CO emissions (– 32 % between 1990 and 2003) were made. Total NO_X emissions increased by almost 21 % in the same period. The

Spain

Country perspective

Population: 41 101 000
Size: 505 990 km²
GDP: 582 408 million euro

EU national emission ceilings (NEC) directive set for 2010 the target of not exceeding 847 ktonnes. In 2003, NO_x emissions were 1 411 ktonnes–574 ktonnes above the target value. Total non-methane Volatile organic compounds (NMVOC) emissions remained stable. In 2003 emissions were, for the first time, 1 % above 1990 levels. The NEC directive set for 2010 the target of not exceeding 662 ktonnes. In 2003, NMVOC emissions were 1 100 ktonnes, 438 ktonnes above the target value. Total methane emissions grew by 29 %. 63 % of the emissions are accounted for by agriculture and 27 % by waste treatment and elimination.

Freight transport demand

The Spanish economy is transport intensive. Both freight and passenger transport has grown above the EU average. Between 1990–2003 passenger transport demand increased by almost 84 % and freight demand by almost 100 %. Large infrastructure investments, particularly in high-capacity roads, the urbanisation of rural areas and the growth in the vehicle fleet have led to an increase in emissions (transport represents 24.4 % of total CO_2 emissions in Spain). During the last years, large investments have been made to improve urban and metropolitan transport. The government investment to public transport alone was EUR 650 million per year. The modernisation of the vehicle fleet and fuel improvements have slowed down the increase in transport emissions, but have not managed to reduce them. There is a clear decoupling between higher numbers of vehicles and the number of accidental deaths. The strategic plan on infrastructure and transport 2005–2020 envisages changes to the existing transport network, with more weight being given to public transport, an increase of rail and maritime transport and higher intermodality.

Share of organic farming

In 2004 organic farming covered 322 000 ha cultivated land and 412 000 ha used for the breeding of organic stock. The organic area represents 2.9 % of the total agricultural area, close to the EU-15 average. Between 1994 and 2004 organic farming increased about 61 000 ha per year, and between 2002 and 2004 it increased 10.7 %. Domestic demand for organic products is low and the vast majority of Spanish products are exported.

Municipal waste

Urban waste generation has been increasing throughout the last decade similar to other neighbouring European countries. In 2003, 500 kg per inhabitant per year was exceeded (501.88). Reaching the old 300 kg per capita benchmark seems quite difficult, particularly if one takes into consideration the large number of tourists visiting Spain, as well as current consumption levels and economic growth. However, the management of urban waste is improving with more composting and recycling. In recent years there has been a very important increase in recycling of packaging waste and in paper and cardboard collecting rates.

Use of freshwater resources

Total water abstraction in Spain is very high given available resources, with a growth of 3.4 % between 1997 and 2002. Economic development and urban expansion, together with the use of water for agricultural purposes (about 76 % of total water abstraction) relating to the Mediterranean climate of a large part of the country, exert strong pressures on water resources. Spain is seeking to abandon current management practices based on supply measures (such as transferring water resources between hydrological basins), towards demand management particularly by the construction of seawater desalination plants in tourist areas.

For more information please contact the relevant national focal point. Contact details can be found on:
http://org.eea.eu.int/organisation/nfp-Eionet_group.html

Sweden

Sweden shows very good progress and performance towards maintaining and improving its environment. Its 15 environmental quality objectives continue to help improve and strengthen cooperation between different sectors and stakeholders in society and to promote environmental integration in agriculture, forestry, energy and transport. However, additional effort is required to meet the ambitious environmental objectives and goals that have been set.

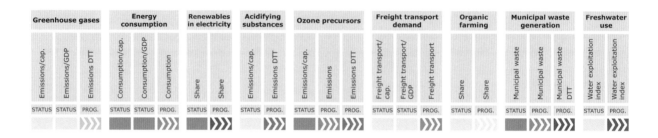

Greenhouse gas emissions

In Sweden emissions fell by 2.3 %, or approximately 1.7 million tonnes, between 1990 and 2003. Before the so called 'special checkpoint' in 2004 a forecast was made and according to that scenario, Sweden is expected to meet its Kyoto commitment by some margin. After 2010 however, a slight increase in carbon dioxide emissions is expected.

Energy consumption

Energy intensity decreased from the mid-1990s until 2003. There are government programmes which aim at increased energy efficiency and in recent years voluntary agreements with industrial companies have also been introduced.

Renewable electricity

A large share of Sweden's electricity is produced from hydropower. Production of renewable energy in Sweden increased during the 1990s, especially in district heating, but production from wind energy and combined heat and power (CHP) based on bioenergy are also increasing. The main policy instruments are energy and carbon taxes and the electricity certificate scheme

Emissions of acidifying substances

Swedish emissions of sulphur dioxide and ammonia are already below the national ceilings within the EU national emission ceilings (NEC) directive . However, further actions need to be taken to reduce the emissions of nitrogen oxides, mainly from heavy vehicles and mobile machinery. Sulphur emissions from international shipping have doubled since 1990 and today are greater in volume than total national emissions.

Emissions of ozone precursors

Swedish emissions of non-methane volatile organic compounds (NMVOC) decreased by 40 % between 1990 and 2003. However, during the past five years the decrease has slowed mainly because of an increase in household stationary combustion. Emissions of methane follow the same pattern as NMVOC. NO_x emissions decreased by about 35 % between 1990 and 2003 and continue to decrease.

Freight transport demand

The actual freight transport volume has increased by 5 % since 1997. But compared with a higher growth in GDP there is some progress towards decoupling

Sweden

Country perspective

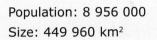

Population: 8 956 000
Size: 449 960 km²
GDP: 232 716 million euro

growth from pressure on the environment within the transport sector. Technical advances are resulting in more fuel-efficient engines, but heavier and faster vehicles and growth in traffic volume are offsetting this positive trend.

Share of organic farming

Sweden has one the highest shares of organic farming at 13 % (3.8 million ha) among European countries. The share of arable soils under organic farming has doubled since 1998 and is approaching the government target of 20 %. So far the increase has been most extensive in the dairy sector, and less in cereal production. The distribution of organic farming within the country is uneven with a higher share in the more extensive central and northern provinces.

Municipal waste

Generation of household waste has increased by about 27 % during the last ten years. However, recycling as well as incineration for energy generation has increased even faster, resulting in a low amount of disposal to landfill (9 % of total amount of household waste in 2004). Three instruments have contributed to this reduced use of landfill: one is a 2002 ban on sending separated combustible waste to landfill, extended to apply to all organic waste since January 2005; another has been successful packaging and waste paper recovery; a third has been a tax on disposal to landfill which has been in force since 2000.

Use of freshwater resources

The availability of water resources in Sweden is good and the rate of abstraction of water is not considered a problem. However, warm and dry summers may result in temporary shortages of water in the southern parts of Sweden. The total freshwater use in Sweden amounted to 2 695 million m³ in 2000, a slight decrease compared to 1995. Industrial use accounted for 60 % of the total withdrawal, households for 23 % and agricultural irrigation for 3.5 %.

One of the interim targets within the national environmental objective 'flourishing lakes and streams' is to adopt water protection plans, including water protection areas and protection regulations, for all public and large private surface water sources. If the objective is to be attained, regional work on water supply plans and the adoption by local authorities of protection areas for water sources must be stepped up.

For more information please contact the relevant national focal point. Contact details can be found on:
http://org.eea.eu.int/organisation/nfp-Eionet_group.html

Switzerland

Switzerland has remarkable natural scenery and wilderness. Despite high investment in environmental protection and some successes (e.g. stabilisation of forest area, improvement of river and lake water quality, reduction of air pollutants), much remains to be done to counteract pressures from economic activities, an increase in built-up areas, high population density and developed tourism industry, e.g. meeting air targets, abating noise, and protecting nature, landscapes and biodiversity.

Greenhouse gas emissions

Approximately 34 % of carbon dioxide (CO_2) emissions in Switzerland come from transport (excl. international air traffic). Swiss implementation of the Kyoto Protocol is based on the Swiss federal law on CO_2 emissions, which requires CO_2 emissions to be reduced by 10 % in total by 2010, using 1990 levels of emissions as a benchmark, with combustible fuels to be reduced by 15 % and moter fuels by 8 %. By 2003 there has been a 0.2 % increase, a 4.6 % reduction and an 8.1 % increase respectively.

Energy consumption

Yearly final energy consumption per capita has been fluctuating between 31 000 and 33 500 kilowatt hours (kWh) since 1990. In 2003 final energy consumption per capita stood at 32 750 kWh, a third of which was used for transport. At the same time, both the population and final energy consumption continue to increase in absolute terms. This is despite improvements made to installations and processes and resulting efficiency gains.

Renewable electricity

Consumption of renewable electricity has decreased from a high in 2001. The share of electricity in total consumption of renewable energy was 73 % in 2003. The biggest sources of renewables are hydro-electric power, biomass and waste incineration.

Emissions of acidifying substances and ozone precursors

Emissions of most air pollutants have decreased over the past few years. Nevertheless, high concentrations of low-lying ozone, particulate matter (PM_{10}) and nitrogen dioxide are still reported on a regular basis. Measures are still needed to improve air quality. It is still necessary to reduce emissions of ozone precursors (in particular NO_X and volatile organic compounds (VOCs) from transport and industry) in order to meet national legislation and international obligations.

Freight transport demand

There has been an increase in the volume of all types of transport in Switzerland, but the number of heavy goods vehicles has declined following a high in 1990–1991 and has almost returned to 1990 levels. However, there has been a dramatic doubling in the number of light goods vehicles (less than 3.5 tonnes) levels since 1990.

Share of organic farming

Organic farms are becoming more prevalent in Switzerland with over 6 100 organic farms in 2003. Environmental concerns in agriculture are addressed through certification of environmental management systems (EMS), which are intended to protect natural biodiversity, reduce nitrate pollution in soils and spring water, reduce phosphorous pollution in surface water, and ensure that farmers treat animals as humanely as

Population: 7 344 000
Size: 41 290 km²
GDP: 275 660 million euro

possible. In order to receive farm subsidies, farmers must obtain EMS certification by demonstrating that they: make a balanced use of fertilisers; use at least 7 % of their farmland as ecological compensation areas; regularly rotate crops; adopt appropriate measures to protect animals and soil; and make limited and targeted use of pesticides.

Municipal waste

There has been a rise in municipal waste generation, but the total amount of municipal waste going to landfill or incineration has declined since 1988, with a corresponding increase in recycling.

Use of freshwater resources

Water is Switzerland's major resource. 4 000–5 000 km of watercourses (roughly 10 % of the entire Swiss water distribution network) have been diverted to generate hydroelectric power. In addition, settlement areas, agriculture, businesses and industry all tap into the natural water cycle (mainly for drinking water consumption), placing addtional pressure on this valuable resource in the process. 80 % of all drinking water comes from groundwater sources. These sources are polluted by nitrates, pesticide residues and hydrocarbons. Nutrients and pesticides in water generally result from intensive farming (mainly from manure, tilling of the ground and pest management) as well as from settlement and urban areas (use of pesticides). Hydrocarbons in water mainly come from transport, businesses and industry.

For more information please contact the relevant national focal point. Contact details can be found on:
http://org.eea.eu.int/organisation/nfp-Eionet_group.html

Turkey

Although Turkey still has a great amount of preserved natural habitats and ecosystems, migration from countryside towards big cities, a high level of economic growth and high rates of population growth produce environmental pressures on these areas. Turkey has a large potential in terms of organic farming and renewable energy, particularly in wind, biomass and solar systems. Turkey has ratified the UN Framework Convention on Climate Change but not the Kyoto Protocol.

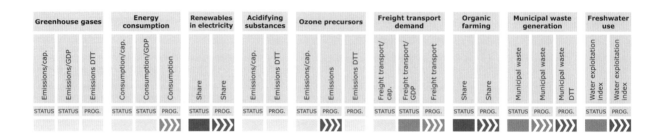

Greenhouse gas emissions

The shares of direct greenhouse gases (CO_2, CH_4 and N_2O) from fuel consumption in total man-made emissions were 90.4 %, 11.5 % and 14.5 % respectively in 1995 and 92.6 %, 7.7 % and 18.5 % in 2000. The shares of sectors in fuel consumption have also changed: while in 1995 28 % of CO_2 emissions were generated by electricity production, 29 % by manufacturing industry, and 21 % by other (residential, agriculture, etc.), in 2000 34 % were generated by electricity production, 32 % by manufacturing industries, 17 % by transportation and 16 % by other sectors.

Energy consumption

While total energy production is increasing, there are increases and decreases in energy consumption depending on the demand. Total energy consumption increased by almost 12.4 % between 1998 and 2003. Based on 2003 figures, 57.4 % of electricity is consumed in the industry sector, 22.5 % in households, and 11.5 % in trading establishments.

Renewable electricity

Turkey is located in a relatively advantageous geographical position as far as the use of solar energy in energy production is considered. The wind data measurements in Turkey show that some locations have high wind energy potential. These results have encouraged private firms to attempt installing wind power plants. Turkey also has a high potential to generate biogas by using organic waste from extensive stockbreeding and farming activities within the country.

Emissions of acidifying substances

Emissions of acidifying substances arise from fuel combustion, industrial processes, traffic and burning of agricultural residues. The power sector contributes 58 % of total SO_2 emissions, industrial fuel combustion 27 %, and industrial processes 2.5 %. Because of the high sulphur content of diesel fuel, household heating contributes to 10 % of total SO_2 emissions and mobile sources 2.5 %. For nitrogen oxides, mobile sources contribute to 32.5 % of total emissions, while the shares from other sources are: 20.1 % from household heating; 22.1 % from the power sector; 21.7 % from industrial fuel combustion; 2.4 % from industrial processes and 1.2 % from agricultural sources

Emissions of ozone precursors

Turkey is subject to relatively high solar radiation input, leading to ozone formation according to the abundance of the precursors. Non-methane volatile organic compound (NMVOC) emissions are distributed among sectors as follows: 30.4 % from mobile sources, 22.5 % from total fuel combustion and 47.1 % from industrial processes

Turkey

Population: 70 712 000
Size: 774 820 km²
GDP: 166 092 million euro

Freight transport demand

Based on the total freight transport figures between 1997–2003, it can be concluded that there is an increasing trend in total transport. From 1997 to 2003 total freight transport increased by 25.1 %, but for the last five years the increase was 3.4 %. Taking the high increase rates in production, import and export volumes from year 2004 into consideration, it is expected that freight transport demand will increase in the coming years.

Share of organic farming

Organic agriculture has started to develop recently in Turkey. There has been a 60 % increase in total production quantities between 2000 and 2003.

Municipal waste

As a result of improvements in waste management systems, municipal waste collected by or on behalf of the municipalities has increased significantly. The amount of municipal solid waste collected increased 47.09 % between 1994 and 2003. The amount of municipal waste composted, incinerated or disposed on controlled landfill sites was 5.6 % in 1994 but increased to 29.7 % in 2003. In 1994, 71 % of the total population received solid waste collection services. This increased to 77 % in 2003. Municipal waste per capita increased from 1.10 to 1.38 between 1994 and 2003 in Turkey.

Use of freshwater resources

Considering the net total water potential, the fact that it is not evenly distributed, and the population increase, which have all led to a decrease in the annually allocated available amount of water per person, would indicate that Turkey is water-stressed. The use of surface water resources is much higher in comparison to the share of groundwater abstractions in total. Total freshwater (surface and groundwater) abstraction increased by 32.9 % between 1995 and 2001. Around 84 % of total freshwater abstraction comes from surface water resources. When surface water and groundwater abstractions are considered separately; total abstraction of surface water increased by 35.0 % whereas groundwater abstraction increased by only 22.6 % in the same period. The share of surface water in total freshwater abstraction increased from 83.1 % to 84.4 % between 1995 and 2001. However, the share of groundwater abstraction in total decreased from 16.9 % to 15.5 % in the same period.

For more information please contact the relevant national focal point. Contact details can be found on:
http://org.eea.eu.int/organisation/nfp-Eionet_group.html

Country analysis | Country perspective

United Kingdom

The United Kingdom is fortunate to have implemented a series of structural economic changes in the recent past that have brought environmental improvements. The country shows relatively good environmental progress and status, and relatively high levels of eco-efficiency. However it has a more intermediate performance per capita. The United Kingdom is on track to meet its formal targets, except for the generation of municipal waste which is steadily increasing.

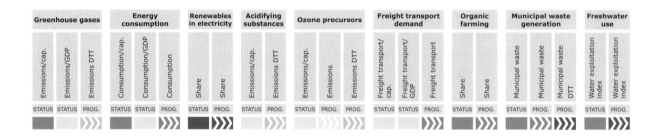

Greenhouse gas emissions

Estimates for 2003 show decreases between 1990 and 2003 of about 7 % for CO_2 and about 14 % for the basket of greenhouse gases. The estimates for 2003 show an increase since 2002, largely because of greater use of coal in electricity generation.

Energy consumption

Between 1980 and 2003 energy use for transport increased by 58 %, mainly as a result of an 80 % increase in road traffic over the same period and a levelling-off of domestic consumption.

Renewable electricity

One of the United Kingdom's renewable targets is that 10 % of electricity generated should be from renewable sources by 2010. Between 1990 and 2004 the percentage of electricity generated from renewable sources increased from 1.8 % to 3.6 %. The largest increase in electricity generated over the last six years was from landfill gas, from 1 185 GWh in 1998 to 4 004 GWh in 2004. Generation from wind power more than doubled.

Emissions of acidifying substances

In England 62 % of rivers were of good quality in 2004 compared with 43 % in 1990. In Northern Ireland quality fell in the mid-1990s and then recovered. In all years since 1993 over 90 % of rivers in Wales have been of good chemical quality. In Scotland, 87 % of rivers were of good quality in 2004, the same as in 2000, based on a combined chemical, biological and aesthetic assessment.

Emissions of ozone precursors

There is evidence that background levels of near ground level ozone have doubled over the past 100 years. Production of ozone is affected by the weather and by air pollutants blown over from mainland Europe. Ozone concentrations tend to be lower in urban areas where it is converted to NO_2 through chemical reaction with NO_x. The average number of days with exceedences per site varies greatly from one year to the next and there is no clear trend.

Freight transport demand

There was no significant change in the UK level of car traffic in 2003 but light van freight traffic increased by 5 %. Overall, motor vehicle traffic levels within the United Kingdom rose by almost 20 % between 1990 and 2003; car and taxi traffic was up by 17 % whilst other vehicle traffic rose by 30 %.

Share of organic farming

The UK area of land that was organically managed peaked at 741 000 ha in March 2003. Significant factors

United Kingdom

Country perspective

Population: 59 280 000
Size: 242 910 km²
GDP: 1 078 600 million euro

in this increase were: farmers seeking alternatives to conventional farming in response to falling income; the scope of organic farming being extended by the EU in July 1999 to include livestock production; and payment rates under organic farming support schemes being substantially increased.

Municipal waste

In England, the amount of household waste increased by around 15 % in total, and by 12 % per person, between 1996–1997 and 2002–2003. During 2002–2003 almost 26 million tonnes was collected by local authorities. Around 14.5 % of this waste was recycled or composted in 2002–2003. This compares with a UK target to recycle/compost 25 % of household waste by 2005–2006.

Use of freshwater resources

In England and Wales water is abstracted under licences, granted on the basis of the reasonable needs of the public, industry and agriculture, and availability of supplies. The amount abstracted has been generally rising since the mid-1990s. In 2002, 83 % of water abstracted was for the public water supply and electricity supply industry.

For more information please contact the relevant national focal point. Contact details can be found on:
http://org.eea.eu.int/organisation/nfp-Eionet_group.html

Country analysis

Methodology and main decision points

Since this is the first time that a scorecard assessment has been presented by the EEA some consideration of methodology is required. The methodology has been developed in conjunction with EEA countries and experts through a multi-year consultation process. This methodology section is designed for the lay reader, giving sufficient overview of some of the key decision points and their implications for the subsequent assessment and interpretation of the scorecard. A box at the end of this section describes some frequently asked questions, which includes technical explanations.

The scorecard is intended to provide a quick overview of country performance against multiple variables. It is primarily a communication tool which will be at its most powerful in electronic format where the user will be able to navigate easily between indicators, data graphs and scale (e.g. EU, country to regional) at the click of a button. On paper the scorecard is slightly more cumbersome but still an effective communication tool, allowing a quick overview of the performance of countries across the different indicators.

A number of key methodological and conceptual choices were made when creating the scorecard assessment presented in this report. Some of these choices are related to:

- the underpinning rationale;

- the issue of what is measured (e.g. how the indicators are selected and where the data come from);

- the process by which the methodology was developed and tested;

- the communicability of the assessment (e.g. matching different design options with the target audiences); and

- testing of the accuracy, adequacy and relevance of the results against existing knowledge for the different users.

Each of these points is briefly discussed in turn below. More detailed methodological information, which can be used for example by practitioners who wish to undertake their own scorecard assessment, will be published in a separate technical document.

Underpinning rationale

The scorecard allows policy-makers to understand country contributions to European environmental performance and distance-to-target results. A breakdown at country level enables decisions-makers at national level to assess how their country performs in a specific environmental area compared with Europe and with other countries, something that many countries already do in their own national reports. By examining performance differences the expectation is to learn lessons and, by sharing experiences, begin to improve overall European environmental performance.

The management of the environment benefits from an increased understanding of the interactions and links between the different parts. Assessment tools such as the scorecard, which gather together diverse information on different countries and environmental issues, allow the reader to begin to identify and tease out such linkages.

It is becoming increasingly difficult to interpret the large amount of information that is now available on the environment. The EEA's core set of indicators (CSI) aims to simplify and focus attention on key information, helping to assimilate and improve the quality of existing information. Joining together this information and using it, such as in the form of a scorecard, helps to develop new insights into the data from which new discoveries can be made.

While there is a need and desire to combine information and indicators, even to the extent of creating a single composite indicator or index, there are obvious pitfalls in doing this, which include oversimplification and lack of credibility or legitimacy with the countries. The scorecard is therefore a cautious step in the direction of a more integrated approach which allows comparative environmental performance of countries to be assessed while maintaining transparency by not combining the separate indicators into a single composite indicator.

However, there are limits to the usefulness of such quantitative information alone since it is unable to capture the richness of information and knowledge about environmental management and performance. Adding qualitative information at country level to the set of scorecard indicators thus allows a more complete understanding to emerge than through a purely quantitative indicator based assessment. From this broader information base, lessons can be learned that lead to improved European environmental policy-making and performance.

Also by involving all EEA member countries in the process contributing their different country views and understandings about the meaning of the data, an EU and a wider pan-European perspective on environmental performance emerges.

What variables to measure

Over the past ten years the EEA has developed and published over 450 different environmental indicators for its environmental assessments. In 2004 a small core set consisting of 37 of these indicators was published.

The scorecard methodology is applied to a sub-set of nine indicators in the core set. It is expected that at some future stage more of the 37 core set indicators may be used in the scorecard. The nine indicators are listed in Table 2.

Since the scorecard is intended to give insights into progress with environmental performance the indicators in the scorecard need to relate to policy levers, i.e. points that policy can affect and on which policy is targeted [1]. Therefore, suitable scorecard indicators measure pressures, drivers or responses which represent policy leverage points rather than 'state' or 'impact' indicators. Although some 'state' and

Table 2   Summary table of the composition of the EEA scorecard

EEA core set of indicators code	Indicator name	Environmental issue	Position in DPSIR (main effect)
CSI 10	Greenhouse gas emissions and removals	Climate change	Pressure
CSI 28	Total energy consumption (energy intensity)	Energy	Response
CSI 31	Renewable electricity	Energy	Response
CSI 01	Emissions of acidifying substances	Air pollution	Pressure
CSI 02	Emissions of ozone precursors	Air pollution	Pressure
CSI 36	Freight transport demand	Transport	Driver
CSI 26	Area under (and share of) organic farming	Agriculture	Response
CSI 16	Municipal waste generation	Waste	Pressure
CSI 18	Use of freshwater resources (water exploitation index)	Water	Pressure

[1] In this section 'targets' are used generically to refer to all legal and policy aims, objectives and goals which range from soft qualitative objectives to hard quantitative targets.

Country analysis

'impact' indicators are highly politically relevant they do not represent areas that can readily be affected by policy action and their importance is often related to raising awareness of an issue and monitoring longer-term trends. The selection of indicator for the scorecard is primarily a case of ensuring that indicators are policy levers, i.e. that they represent and measure environmental pressures, policy responses or driving forces.

In addition the scorecard indicators should ideally cover hard quantifiable or time-bound policy targets since this reflects the agreements made by the EU and the countries on how to manage the environment and measure progress. If the targets are softer the data need to be highly comparable and suitable for country benchmarking.

With these considerations, Table 3 summarises the entire EEA CSI, focusing on their suitability for the EEA scorecard. In particular, emphasis is given to the type of targets for which the indicator is relevant and also to the country coverage.

As has been well-documented in existing management literature on performance metrics: 'what is measured is managed'. Thus, most indicators also tend to show trends of improvement over time. The presence of **hard targets** (²) however puts such improvements into perspective and illustrates the need for timely action in dealing with environmental problems in order to stabilise trends before they are amplified through runaway or rebound effects. It is this time perspective that is lost from **softer targets** where it would be easy to be lulled into a false sense of security based on the direction of change but which would not take into account the rate (or the ultimate size) of the change that would be needed.

Through the scorecard development process countries expressed a preference, wherever possible, for selecting scorecard indicators for which hard political targets exist. However, in the scorecard several indicators are included where the targets are old or soft. Some specific cases that need mentioning are:

- Municipal waste generation (CSI 16): This had a hard target under the fifth environment action programme (5th EAP) but this target was not included in the sixth environment action programme (6th EAP). However, in its analyses the EEA has often referred to this older target and its inclusion enables municipal waste rather than the more specific packaging waste to be included. Municipal waste has the advantage of directly covering the household sector. As such it is an indicator of sustainable consumption that is directly visible to the citizens of Europe, and as a consequence highly politically relevant. There are some definitional issues surrounding this and other waste indicators, but the data are the best available and normalised as far as possible by Eurostat. The indicator, Generation and recycling of packaging waste (CSI 17) suffers from similar definition problems and is a less visible issue for the general public. However this indicator could be included in future versions of the scorecard if so desired.

- Area under organic farming (CSI 26) is a very politically visible indicator for agriculture, with robust data and one that allows the issue of sustainable consumption to be further addressed. Although organic farming has an extremely soft policy target, the alternative indicator, Gross nutrient balance (CSI 25) has no policy target and data availability is currently only EU-25 or less.

(²) Hard targets refer to those for which a specific quantitative value has been agreed through a process of political negotiation, and are contrasted with soft targets where qualitative objectives or aims have been agreed through a process of political negotiation. An example of a soft target would be to decouple freight transport demand from economic growth, or to increase the share of organic farming; while hard targets would be e.g. from Kyoto Protocol or NEC directive.

- Use of freshwater resources (CSI 18): No targets exist for the use of freshwater resources but the water exploitation index is a robust and well established indicator measure that allows country benchmarking to be undertaken on the basis of an ecological stress factor. The indicator is included in the scorecard in order to capture the important environmental issue of water.

Another consideration when building the scorecard was to present a balanced set of issues. As discussed above, this was one of the key reasons for including indicators on water and organic farming to allow some coverage of water resources and the agricultural sector, even though there are no hard targets in these areas. However, the indicator and issue selection for the scorecard was however ultimately bound by the coverage of environmental issues in the CSI which covers six environmental themes (air pollution and ozone depletion, climate change, waste, water, biodiversity and terrestrial environment) and four sectors (agriculture, energy, transport and fisheries).

Some key environmental issues are not covered in the scorecard for a combination of reasons to do with data quality and target availability. Some of these issues are outlined below:

- Only one biodiversity indicator with policy leverage is included in the CSI. This indicator, Designated areas (CSI 08), has a soft qualitative target but only allows regional comparisons or comparisons between a subset of countries and is therefore not suitable for country benchmarking.

- For the fisheries sector the two policy leverage indicators, Aquaculture production (CSI 33) and Fishing fleet capacity (CSI 34), have no targets and soft qualitative targets respectively and neither is suitable for country benchmarking.

- The one terrestrial indicator that is a policy leverage point, Progress in management of contaminated sites (CSI 15), has a soft target but is not suitable for country benchmarking.

In conclusion the EEA scorecard is composed of nine indicators, all of which represent points of policy leverage, have robust data with comparable country coverage and mostly well defined policy targets.

What process to use to develop and test the scorecard

The process by which an assessment is developed is at least equally important to the final results, and it is through the process that the assessment is given legitimacy and credibility with the intended users. The EEA therefore developed the scorecard through process of extensive consultation with its member countries and in particular with the experts on environmental reporting within those countries.

The idea for the country scorecard initially grew out of a demand from the countries themselves. In particular they wanted to know how to interpret European level results, including distance to targets, and felt that a breakdown to country level would be more meaningful and would allow decision-makers at national level to assess how their particular country was performing in a specific environmental area. If benchmarks of some sort were additionally provided, a country could also begin to benchmark its own performance against that of other European countries.

In response to this request from the countries, work was begun to create a summary scorecard for the member countries of the EEA which in 1998/99 numbered 18 (the EU-15 plus Norway, Liechtenstein and Iceland). This process of development took five years, during which the number of EEA member countries grew to 31, and in 2005 the prototype scorecard was fully coupled to the CSI.

How to communicate it

Once the legitimacy and credibility of the scorecard is ensured through a transparent and participatory process, the key factor determining the scorecard's fitness for purpose is its communicability.

Country analysis

The scorecard is predominantly a visual communication tool which is also ideally suited for web-based publishing. The scorecard has visual elements in the form of the full scorecard but also the graphs that systematically present the lower layers of information.

Additionally, the text-based assessments allow space for the information to be enriched by country-specific and thematic information, some of which is qualitative.

Some of the key decision points in the scorecard development related to:

- the choice of visually distinct symbols for the progress and status indicators;

- the colour scheme that differentiates between the hard quantitative distance-to-target assessments (red/green) and the softer assessments that measure relative progress (shades of blue) — overall the lighter the colour or tone the better the performance;

- the presentation and design of the underlying information graphs;

- the visual communication of issues relating to uncertainty and differential quality of the data and indicators; and

- the language to use in the assessments.

5. Testing: does it work?

The adequacy of a small number of indicators such as those included in the scorecard needs to be constantly monitored by testing against the result from using more indicators. A scorecard can never provide a definitive assessment of environmental performance, but is rather a tool for allowing patterns in relative environmental performance to emerge between countries and between groups of countries. As discussed above the development of the scorecard included decision points on issues such as balance between themes and between the number of indicators in each theme. Testing is continuously required to understand the effect that adding another theme or sector (or of adding another indicator for e.g. transport or waste) would have on the pattern of results. Linked to this issue is the need to understand any possible cross linkages or co-dependencies within any set of scorecard indicators.

The results of the scorecard also need to be compared against other global measures of environmental performance to look for wider patterns and trends. The EEA plans to investigate the results from the scorecard against other indicator initiatives that have included EEA member countries. These include some of the composite environmental indicators that have been published (e.g. the environmental sustainability index) and this exercise will allow the further development of some key questions that have emerged: in particular the extent of overlap between models of environmental sustainability at national and at European level.

The fitness of the scorecard for the purpose of underpinning policy-making will also be continuously tested, especially with regard to the need for early warning, while avoiding bias in the form of false negative or false positive results. This is linked to a deeper understanding both of the responsiveness of the indicator to policy action and to the certainty of the wider cause-and-effect science framework in which the indicator sits.

Methodology and main decision points

PART C

Frequently asked questions concerning the scorecard methodology

I. What is the scorecard?

The scorecard refers to a communication tool used to facilitate the presentation of environmental country performance which allows the user to 'drill down' to different levels of aggregation in environmental information and to make relevant and informative comparisons against performance benchmarks.

The scorecard is made up of the overall scorecard summary table comprising: the individual scorecards for each of the thematic indicators, supported by graphs showing the country rankings for progress and status; and the country scorecards supported by graphs showing country ranks across the range of scorecard indicators.

In developing the scorecard a level of statistical analysis is used to underpin the assessment.

II. What is the difference between status, progress and distance-to-target?

A. STATUS indicators compare environmental performance in the year for which we have latest available data (usually 2002). These are absolute data normalised by population in a country unless otherwise stated.

Status indicators are shown on the scorecard in three shades of blue, where the lightest refers to countries performing in the top 25 % of the range of values (best performers); mid-blue in the intermediate 50 % of values; and dark blue in the lowest 25 % of the range of values.

B. PROGRESS indicators are a measure of change over time (ten years, usually 1992–2002 depending on time series availability) compared to average EU-25 ten-year change.

Progress indicators are marked on the scorecard in three shades of blue with the lightest blue identifying the best performing countries that are in the top 25 % of the range of values (i.e. the difference between change over ten years in a country and the average change in the EU-25). The darkest blue identifies those countries in the lowest performing 25 % of the range of values and mid-blue refers to the intermediate 50 % of the range of values.

C. DISTANCE TO TARGET (DTT) indicators cover those indicators for which there are hard policy targets and compare the absolute values in the latest year available against the linear target path. They are identified by their red-green colouring on the scorecard.

Countries that are not on track to meet the target are shown on the scorecard as red; those on track are shown in two shades of green: countries close to the target line (+ or – 5 percentage points) are marked in dark green. Given the uncertainties and the closeness they are to the target line some caution should be exercised and for this reason such countries are differentiated from the best performers which are marked in light green showing that they are firmly on track, even taking into account the uncertainties.

III. How do I interpret the colour scoring for Status indicators?

The status indicators rank countries in terms of normalised absolute performance (latest available data) and simply labels the countries as those falling into the range of values in the top 25 %, the lowest 25 % and the middle 50 % of the range. So, although the ranking is based on absolute environmental performance, the countries' final score in terms of their colour in the scorecard, is relative to the performance of all other countries.

Year by year, this methodology means that countries must continue to change along with their peers in order to maintain their position in the ranking, as it is the distance between a country's performance and that of the top and lowest performers in the set that determine their colour on the scorecard.

IV. How do I interpret the colour scoring for DTT indicators?

Is it possible, for example, that all the countries are close to target but those further away (although still very close) come up red? The answer is no, since the colour scheme is calculated on an absolute basis — thus, if all countries were similarly on or off track to meeting the target they would all be the same colour.

Country analysis

V. How do I interpret performance and progress when they are very different?

A country that has already made good progress may have difficulty going further or may begin to focus attention elsewhere. Consequently such a country may show little further progress or may even decline a little. It would, however, be unfair to show them as only having poor progress, ignoring their current high status. Alternatively, countries that have traditionally shown very poor environmental performance, due for example to their particular geographic or climatic predispositions, or to an inherited socio-economic or environmental legacy, can be encouraged by showing how much they have progressed, even if the performance to date has brought them to a status which is still relatively poor compared to other countries.

VI. How do we interpret the different performance values for populations and GDP?

Comparing performance against GDP is a measure of eco-efficiency, while it also contains a subtle implicit assumption that richer countries (with high GDP) are allowed to pollute more. Similarly, comparing performance only against population size might give rise to the tacit implication that countries with large populations do not have to bother about eco-efficiency.

The ten new EU Member States have lower GDPs than the 15 older members and in general favour comparison by population (per capita) — in fact they strongly object to GDP being used for comparisons. Some of the older EU Member States who are performing well in terms of eco-efficiency, are consuming and polluting more than their per capita equitable share of global resources, and thus prefer comparison by GDP.

The choice made in the scorecard is to show the per capita indicator or the indicator as a share (for example of total utilised agricultural area or total electricity consumption). In addition the indicator is shown per GDP when this is relevant to the target (for freight transport for example where the target relates to decoupling demand from economic growth) or where this presentation is the one used in the structural indicators or other indicator initiatives.

VII. Why do we not make an overall ranking of countries?

A selection of environmental indicators is used to create this scorecard, but there might be others that would need to be included to get a balanced composite indicator; and it is likely that some differential weighting between them would be desirable. Because the scorecard is not a composite indicator no overall country ranking is presented.

However, key pointers for interpretation include being aware of the following general categories of country performance: (i) countries that are performing well but progressing poorly and are therefore likely to be heading towards environmental deterioration unless action is taken to remedy the situation; (ii) countries that are performing poorly and progressing poorly and which are therefore heading towards very rapid worsening of already poor environmental conditions; and (iii) countries that are performing poorly but progressing well which suggests that action has been taken to address some of the issues and that environmental condition is likely to improve in the future; as well as (iv) countries that are performing well and progressing well and are therefore well on the way to preserving their good environmental condition for the future.

VIII. How sensitive are the results to the uncertainties in the data?

The scorecard methodology addresses some of the issues concerning uncertainty but is to some extent dependent on the accuracy and reliability of the data submitted by the countries. The indicators and data used are generally of very good quality. For more details on the uncertainty related to individual indicators please refer to the EEA CSI and the indicator specifications (see www.eea.eu.int/coreset and http://themes.eea.eu.int/IMS/About/EEACSITopicsAndQuality2004.pdf).

Specific to the scorecard, however, is the detailed 'drilling down' into indicator data to country level or lower and the comparison of changes over time. Errors for a single indicator may differ between countries, with some countries providing less robust data in general for several indicators due for example to: differences in national or regional monitoring systems; federal state organisation compared to national state organisation; or the institutional organisation and environmental governance in a country. Errors can also differ between years in a time-series (as the number of monitoring stations increases over ten years for example), and such changes would be especially relevant in the progress and distance-to-target assessments where a base-year is used. Together such systematic patterns of errors could potentially have direct effects on the reliability and accuracy of the scorecard and DTT results, but all efforts are taken to minimise and foresee such problems in advance, for example by using an alternative base year or verifying a data point against external sources.

Distance-to-target assessments are based on an assumption of linear distance to target, which is a weak but standard methodology. The way countries move towards their targets is usually non-linear, and often targets are met only at the last moment. However the linear methodology is currently standard for determining distance to target.

Outliers can affect the categories that the countries fall into. Any big outliers are therefore removed before the categorisation is made. This is the case in particular for area under organic farming, municipal waste generation and ozone precursor emissions.

IX. Is the scorecard approach too brutal?

Clearly countries remain sensitive to the outcomes of the scorecard, particularly if the results are not favourable. Given the use of a limited number of indicators and the way in which relative performance is highlighted, the presentation of the scorecard can be considered to be relatively hard and frank.

The EEA response to this critique has been: to ensure that the scorecard has legitimacy with its member countries by using a consultative approach to develop the methodology; to use only Core set indicators with the highest quality, validated, data; and to use a clear and transparent colouring scheme in the scorecard which leaves the reader in no doubt when an indicator is assessed on the basis of an agreed target (red/green) and when it is assessed against some other benchmark such as an EU-25 average (blue).

The countries are additionally invited to contribute their own voice to the overall assessment and interpretation of the scorecard results as part of the country analysis section. This section provides an opportunity for countries to raise issues specific to their situation and to bring in relevant information that serve to balance the scorecard results.

X. How do we expect readers such as policy-makers to react to the scorecard?

It is expected that these results will stimulate discussion among both policy-makers and citizens in the EEA member countries, and lead to an examination of the factors that lie behind the occasionally large differences in environmental performance and progress seen in Europe.

XI. Is the scorecard a valid measure of European environmental performance?

Since the scorecard covers a number of essential environmental issues, uses high quality data and indicators and has legitimacy from country participation and involvement in the assessment and the methodology; the EEA scorecard is a valid tool to further the understanding of the underlying factors behind European environmental performance.

Country analysis

Table 3   Mapping the EEA core set of indicators in terms of suitability for the EEA scorecard

The table below summarises the entire EEA core set of indicators (CSI), focusing on their suitability for the EEA scorecard. In particular emphasis is given to the type of targets that the indicator is relevant for and also to the country coverage. For indicators to be suitable for the scorecard, they need to represent points of policy leverage, i.e. measure pressures, drivers or response. In addition, they ideally need to cover hard quantifiable or time-bound policy targets. If the targets are softer, the data need to be highly comparable and suitable for country benchmarking.

| No policy leverage points |
| Not suitable for country comparisons |
| No suitable targets |
| Potential indicator for the scorecard |
| Included in scorecard |

Scorecard indicator	CSI	Indicator name	Environmental issue	DPSIR (main part)	Targets ([3])	Comparability between countries ([4])
Yes	CSI 01	Emission of acidifying substances	Air pollution	P	4	4
Yes	CSI 02	Emission of ozone precursors	Air pollution	P	4	4
No (not suitable for country benchmarking, no suitable target)	CSI 03	Emission of primary particles and secondary particulate precursors	Air pollution	P	1	2
No (no policy leverage)	CSI 04	Exceedance of air quality limit values in urban areas	Air pollution	I	2	0
No (no policy leverage)	CSI 05	Exposure of ecosystems to acidification, eutrophication and ozone	Air pollution	I	4	2
No (not possible for country benchmarking)	CSI 06	Production and consumption of ozone depleting substances	Air pollution	P	4	0
No (no policy leverage)	CSI 07	Threatened and protected species	Biodiversity	I	2	1
No (no targets and not suitable for country benchmarking)	CSI 08	Designated areas	Biodiversity	R	2	2
No (no policy leverage)	CSI 09	Species diversity (bears, wolves, farmland birds woodland, park and garden birds, butterflies)	Biodiversity	S	2	2

([3]) Does the indicator monitor progress toward the quantified targets?
    0 = No targets
    1 = Targets but the indicator do not fully reflect these
    2 = Qualitative targets (generic)
    3 = Qualitative targets (specific) or Quantified targets not time bound
    4 = Quantified targets time bound (Ref: Topic descriptions and quality evaluations for indicators in the EEA core set, EEA, 2004.)

([4]) Space and temporal coverage and representativeness for countries (country comparison):
    0 = Country comparison relevant but not possible for the moment
    1 = Country comparison not relevant (e.g. temperature)
    2 = Regional comparison or between subset of countries
    3 = Absent from scoring (blank)
    4 = Possible to use indicator for country benchmarking (Ibid.).

Methodology and main decision points — PART C

Scorecard indicator	CSI	Indicator name	Environmental issue	DPSIR (main part)	Targets (4)	Comparability between countries (5)
Yes	CSI 10	Greenhouse gas emissions and removals	Climate change	P	4	4
No (CSI 10 chosen instead)	CSI 11	Projections of GHG emissions, removals	Climate change	P	4	4
No (state indicator — no policy leverage)	CSI 12	Global and European temperature	Climate change	S	4	1
No (state indicator — no policy leverage)	CSI 13	Atmospheric greenhouse gas concentrations	Climate change	S	3	1
No (state indicator — no policy leverage)	CSI 14	Land take	Terrestrial	S	2	2
No (not suitable for national benchmarking; no suitable target)	CSI 15	Progress in management of contaminated sites	Terrestrial	R	2	2
Yes (5th EAP target used)	CSI 16	Municipal waste generation	Waste	P	1	4
Under consideration but very small and specific waste stream only	CSI 17	Generation and recycling of packaging waste	Waste	P/R	4	4
Yes (although there are no targets water coverage is desirable in the scorecard and the indicator is suitable for benchmarking)	CSI 18	Use of freshwater resources (water exploitation index)	Water	P	2	4
No (state indicator — no policy leverage)	CSI 19	Oxygen consuming substances in rivers	Water	S	1	4
No (state indicator — no policy leverage)	CSI 20	Nutrients in freshwater	Water	S	2	4
No (state indicator — no policy leverage)	CSI 21	Nutrients in transitional, coastal and marine waters	Water	S	2	2
No (state indicator — no policy leverage)	CSI 22	Bathing water quality	Water	S	3	4
No (state indicator — no policy leverage)	CSI 23	Chlorophyll in transitional, coastal and marine waters	Water	S	1	2
No (no suitable target and not suitable for country benchmarking)	CSI 24	Urban wastewater treatment	Water	R	1	2
No (insufficient country coverage — and no target)	CSI 25	Gross nutrient balance	Agriculture	P	0	4
Yes (no target but good benchmarking data)	CSI 26	Area under organic farming	Agriculture	R	1	4
No (total energy consumption used instead as more environmentally sound than just focusing on the sectors)	CSI 27	Final energy consumption by sector	Energy	D	2	4
Yes	CSI 28	Total energy intensity	Energy	R	1	4
Yes (used in indicator above)	CSI 29	Total energy consumption by fuel	Energy	D	2	4
No (renewables in electricity chosen instead)	CSI 30	Renewable energy consumption	Energy	D	4	4
Yes	CSI 31	Renewable electricity	Energy	R	4	4
No (state indicator — no policy leverage)	CSI 32	Status of marine fish stocks	Fisheries	S	2	2
No (data quality and comparability issues)	CSI 33	Aquaculture production	Fisheries	D	0	2
No (only soft targets and not suitable for country benchmarking)	CSI 34	Fishing fleet capacity	Fisheries	D	2	2
No (Temporarily removed pending methodological revisions — will then be included to balance freight transport demand)	CSI 35	Passenger transport demand	Transport	D	3	4
Yes	CSI 36	Freight transport demand	Transport	D	3	4
No (insufficient countries covered as yet)	CSI 37	Use of cleaner and alternative fuels	Transport	R	4	4

Part C

Country analysis | Ranking graphs

Austria

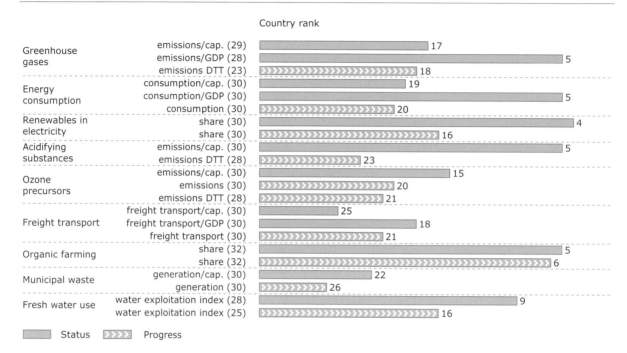

Note: Ranks: low rank indicates good performance (number in brackets indicates total number of countries ranked). DTT indicates a distance-to-target analysis.

Belgium

Country rank

Greenhouse gases	emissions/cap. (29)	26
	emissions/GDP (28)	16
	emissions DTT (23)	14
Energy consumption	consumption/cap. (30)	25
	consumption/GDP (30)	18
	consumption (30)	10
Renewables in electricity	share (30)	15
	share (30)	11
Acidifying substances	emissions/cap. (30)	12
	emissions DTT (28)	21
Ozone precursors	emissions/cap. (30)	20
	emissions (30)	22
	emissions DTT (28)	24
Freight transport	freight transport/cap. (30)	22
	freight transport/GDP (30)	15
	freight transport (30)	11
Organic farming	share (32)	19
	share (32)	17
Municipal waste	generation/cap. (30)	9
	generation (30)	9
Fresh water use	water exploitation index (28)	
	water exploitation index (25)	

Status — Progress

Note: Ranks: low rank indicates good performance (number in brackets indicates total number of countries ranked). DTT indicates a distance-to-target analysis.

Ranking graphs

Bulgaria

Country rank

Category	Indicator	Rank
Greenhouse gases	emissions/cap. (29)	8
	emissions/GDP (28)	27
	emissions DTT (23)	
Energy consumption	consumption/cap. (30)	6
	consumption/GDP (30)	29
	consumption (30)	5
Renewables in electricity	share (30)	28
	share (30)	26
Acidifying substances	emissions/cap. (30)	29
	emissions DTT (28)	8
Ozone precursors	emissions/cap. (30)	9
	emissions (30)	8
	emissions DTT (28)	4
Freight transport	freight transport/cap. (30)	2
	freight transport/GDP (30)	19
	freight transport (30)	1
Organic farming	share (32)	31
	share (32)	32
Municipal waste	generation/cap. (30)	16
	generation (30)	2
Fresh water use	water exploitation index (28)	27
	water exploitation index (25)	2

■ Status ▶▶▶▶ Progress

Note: Ranks: low rank indicates good performance (number in brackets indicates total number of countries ranked). DTT indicates a distance-to-target analysis.

Cyprus

Country rank

Category	Indicator	Rank
Greenhouse gases	emissions/cap. (29)	
	emissions/GDP (28)	
	emissions DTT (23)	
Energy consumption	consumption/cap. (30)	12
	consumption/GDP (30)	15
	consumption (30)	25
Renewables in electricity	share (30)	29
	share (30)	28
Acidifying substances	emissions/cap. (30)	27
	emissions DTT (28)	26
Ozone precursors	emissions/cap. (30)	17
	emissions (30)	29
	emissions DTT (28)	22
Freight transport	freight transport/cap. (30)	1
	freight transport/GDP (30)	2
	freight transport (30)	18
Organic farming	share (32)	30
	share (32)	29
Municipal waste	generation/cap. (30)	26
	generation (30)	24
Fresh water use	water exploitation index (28)	28
	water exploitation index (25)	

■ Status ▶▶▶▶ Progress

Note: Ranks: low rank indicates good performance (number in brackets indicates total number of countries ranked). DTT indicates a distance-to-target analysis.

Country analysis | Ranking graphs

Czech Republic

Country rank

Category	Indicator	Rank
Greenhouse gases	emissions/cap. (29)	24
	emissions/GDP (28)	24
	emissions DTT (23)	7
Energy consumption	consumption/cap. (30)	21
	consumption/GDP (30)	26
	consumption (30)	7
Renewables in electricity	share (30)	17
	share (30)	14
Acidifying substances	emissions/cap. (30)	16
	emissions DTT (28)	6
Ozone precursors	emissions/cap. (30)	13
	emissions (30)	5
	emissions DTT (28)	8
Freight transport	freight transport/cap. (30)	20
	freight transport/GDP (30)	24
	freight transport (30)	12
Organic farming	share (32)	8
	share (32)	8
Municipal waste	generation/cap. (30)	3
	generation (30)	5
Fresh water use	water exploitation index (28)	14
	water exploitation index (25)	3

Status   Progress

Note: Ranks: low rank indicates good performance (number in brackets indicates total number of countries ranked). DTT indicates a distance-to-target analysis.

Denmark

Country rank

Category	Indicator	Rank
Greenhouse gases	emissions/cap. (29)	22
	emissions/GDP (28)	11
	emissions DTT (23)	17
Energy consumption	consumption/cap. (30)	17
	consumption/GDP (30)	4
	consumption (30)	11
Renewables in electricity	share (30)	2
	share (30)	1
Acidifying substances	emissions/cap. (30)	20
	emissions DTT (28)	14
Ozone precursors	emissions/cap. (30)	25
	emissions (30)	19
	emissions DTT (28)	18
Freight transport	freight transport/cap. (30)	16
	freight transport/GDP (30)	9
	freight transport (30)	3
Organic farming	share (32)	7
	share (32)	7
Municipal waste	generation/cap. (30)	27
	generation (30)	20
Fresh water use	water exploitation index (28)	8
	water exploitation index (25)	9

Status   Progress

Note: Ranks: low rank indicates good performance (number in brackets indicates total number of countries ranked). DTT indicates a distance-to-target analysis.

Ranking graphs

Estonia

Country rank

Greenhouse gases	emissions/cap. (29)	25
	emissions/GDP (28)	28
	emissions DTT (23)	3
Energy consumption	consumption/cap. (30)	16
	consumption/GDP (30)	28
	consumption (30)	1
Renewables in electricity	share (30)	24
	share (30)	19
Acidifying substances	emissions/cap. (30)	26
	emissions DTT (28)	3
Ozone precursors	emissions/cap. (30)	24
	emissions (30)	4
	emissions DTT (28)	5
Freight transport	freight transport/cap. (30)	29
	freight transport/GDP (30)	29
	freight transport (30)	30
Organic farming	share (32)	13
	share (32)	12
Municipal waste	generation/cap. (30)	7
	generation (30)	10
Fresh water use	water exploitation index (28)	12
	water exploitation index (25)	4

Status   Progress

Note: Ranks: low rank indicates good performance (number in brackets indicates total number of countries ranked). DTT indicates a distance-to-target analysis.

Finland

Country rank

Greenhouse gases	emissions/cap. (29)	27
	emissions/GDP (28)	21
	emissions DTT (23)	16
Energy consumption	consumption/cap. (30)	28
	consumption/GDP (30)	25
	consumption (30)	21
Renewables in electricity	share (30)	3
	share (30)	3
Acidifying substances	emissions/cap. (30)	15
	emissions DTT (28)	10
Ozone precursors	emissions/cap. (30)	27
	emissions (30)	17
	emissions DTT (28)	13
Freight transport	freight transport/cap. (30)	26
	freight transport/GDP (30)	20
	freight transport (30)	16
Organic farming	share (32)	6
	share (32)	5
Municipal waste	generation/cap. (30)	10
	generation (30)	11
Fresh water use	water exploitation index (28)	6
	water exploitation index (25)	19

Status   Progress

Note: Ranks: low rank indicates good performance (number in brackets indicates total number of countries ranked). DTT indicates a distance-to-target analysis.

Country analysis | Ranking graphs

France

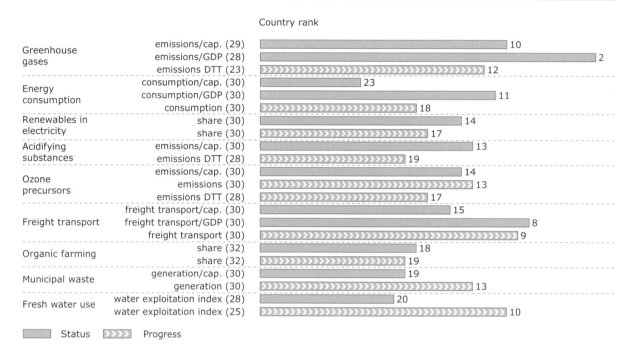

Note: Ranks: low rank indicates good performance (number in brackets indicates total number of countries ranked). DTT indicates a distance-to-target analysis.

Germany

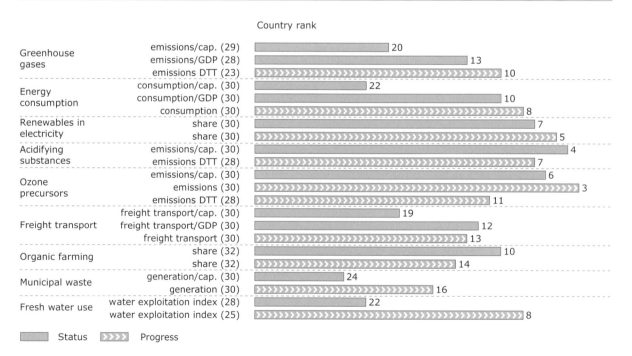

Note: Ranks: low rank indicates good performance (number in brackets indicates total number of countries ranked). DTT indicates a distance-to-target analysis.

Greece

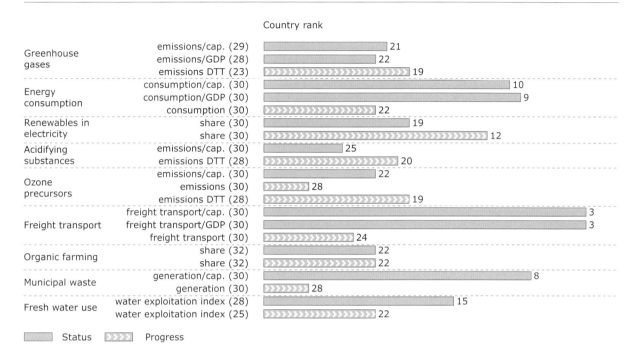

Note: Ranks: low rank indicates good performance (number in brackets indicates total number of countries ranked).
DTT indicates a distance-to-target analysis.

Hungary

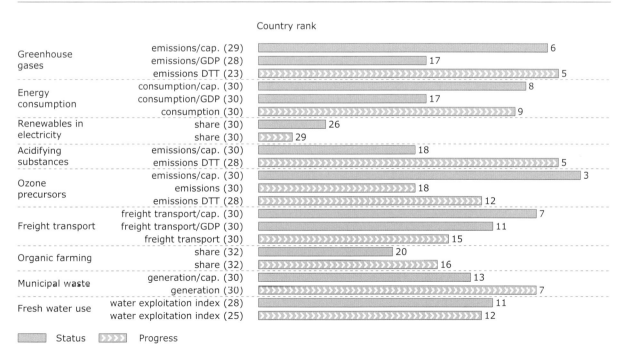

Note: Ranks: low rank indicates good performance (number in brackets indicates total number of countries ranked).
DTT indicates a distance-to-target analysis.

Country analysis | Ranking graphs

Iceland

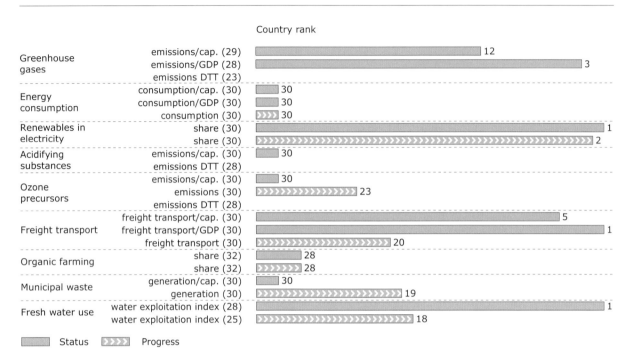

Note: Ranks: low rank indicates good performance (number in brackets indicates total number of countries ranked). DTT indicates a distance-to-target analysis.

Ireland

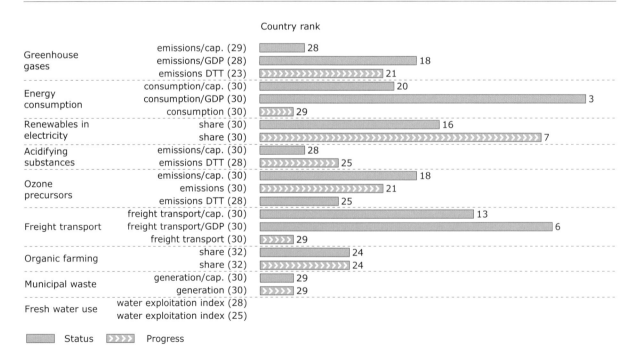

Note: Ranks: low rank indicates good performance (number in brackets indicates total number of countries ranked). DTT indicates a distance-to-target analysis.

Italy

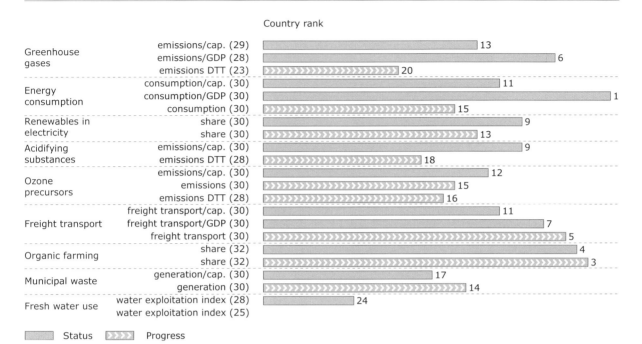

Note: Ranks: low rank indicates good performance (number in brackets indicates total number of countries ranked). DTT indicates a distance-to-target analysis.

Latvia

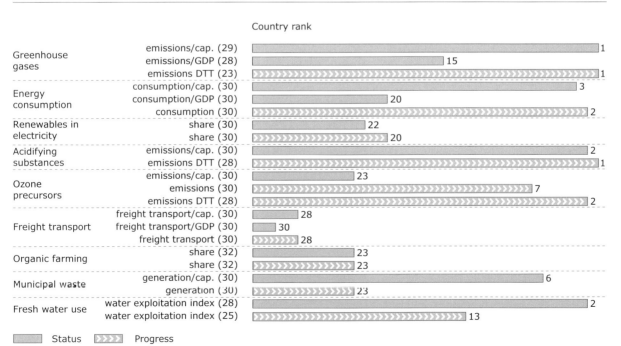

Note: Ranks: low rank indicates good performance (number in brackets indicates total number of countries ranked). DTT indicates a distance-to-target analysis.

Country analysis | Ranking graphs

Liechtenstein

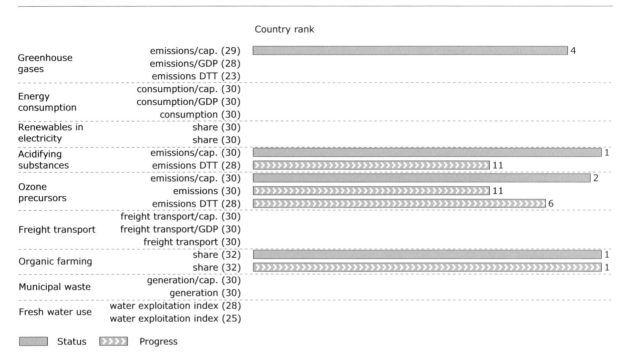

Note: Ranks: low rank indicates good performance (number in brackets indicates total number of countries ranked). DTT indicates a distance-to-target analysis.

Lithuania

Note: Ranks: low rank indicates good performance (number in brackets indicates total number of countries ranked). DTT indicates a distance-to-target analysis.

Ranking graphs

Luxembourg

Category	Indicator	Country rank
Greenhouse gases	emissions/cap. (29)	29
	emissions/GDP (28)	14
	emissions DTT (23)	
Energy consumption	consumption/cap. (30)	8 (progress)
	consumption/GDP (30)	29
	consumption (30)	16
Renewables in electricity	share (30)	12 (progress)
	share (30)	13
Acidifying substances	emissions/cap. (30)	18 (progress)
	emissions DTT (28)	19
Ozone precursors	emissions/cap. (30)	15 (progress)
	emissions (30)	28
	emissions DTT (28)	12 (progress)
Freight transport	freight transport/cap. (30)	20 (progress)
	freight transport/GDP (30)	30
	freight transport (30)	26
Organic farming	share (32)	25 (progress)
	share (32)	21
Municipal waste	generation/cap. (30)	21 (progress)
	generation (30)	25
Fresh water use	water exploitation index (28)	21 (progress)
	water exploitation index (25)	7

Status ▶▶▶▶ Progress

Note: Ranks: low rank indicates good performance (number in brackets indicates total number of countries ranked). DTT indicates a distance-to-target analysis.

Malta

Category	Indicator	Country rank
Greenhouse gases	emissions/cap. (29)	5
	emissions/GDP (28)	8
	emissions DTT (23)	
Energy consumption	consumption/cap. (30)	4
	consumption/GDP (30)	2
	consumption (30)	24 (progress)
Renewables in electricity	share (30)	29
	share (30)	27
Acidifying substances	emissions/cap. (30)	
	emissions DTT (28)	
Ozone precursors	emissions/cap. (30)	
	emissions (30)	
	emissions DTT (28)	
Freight transport	freight transport/cap. (30)	27
	freight transport/GDP (30)	27
	freight transport (30)	14 (progress)
Organic farming	share (32)	
	share (32)	31 (progress)
Municipal waste	generation/cap. (30)	18
	generation (30)	30 (progress)
Fresh water use	water exploitation index (28)	25
	water exploitation index (25)	5 (progress)

Status ▶▶▶▶ Progress

Note: Ranks: low rank indicates good performance (number in brackets indicates total number of countries ranked). DTT indicates a distance-to-target analysis.

Country analysis | Ranking graphs

The Netherlands

Country rank

Category	Indicator	Status	Progress
Greenhouse gases	emissions/cap. (29)	23	
	emissions/GDP (28)	12	
	emissions DTT (23)		13
Energy consumption	consumption/cap. (30)	24	
	consumption/GDP (30)	13	
	consumption (30)		17
Renewables in electricity	share (30)	12	
	share (30)		6
Acidifying substances	emissions/cap. (30)	7	
	emissions DTT (28)		16
Ozone precursors	emissions/cap. (30)	7	
	emissions (30)		10
	emissions DTT (28)		14
Freight transport	freight transport/cap. (30)	24	
	freight transport/GDP (30)	17	
	freight transport (30)		8
Organic farming	share (32)	16	
	share (32)		18
Municipal waste	generation/cap. (30)	20	
	generation (30)		12
Fresh water use	water exploitation index (28)	13	
	water exploitation index (25)		21

Note: Ranks: low rank indicates good performance (number in brackets indicates total number of countries ranked). DTT indicates a distance-to-target analysis.

Norway

Country rank

Category	Indicator	Status	Progress
Greenhouse gases	emissions/cap. (29)	19	
	emissions/GDP (28)	4	
	emissions DTT (23)		
Energy consumption	consumption/cap. (30)	27	
	consumption/GDP (30)	12	
	consumption (30)		19
Renewables in electricity	share (30)	10	
	share (30)		21
Acidifying substances	emissions/cap. (30)	11	
	emissions DTT (28)		24
Ozone precursors	emissions/cap. (30)	29	
	emissions (30)		25
	emissions DTT (28)		27
Freight transport	freight transport/cap. (30)	14	
	freight transport/GDP (30)	5	
	freight transport (30)		23
Organic farming	share (32)	11	
	share (32)		11
Municipal waste	generation/cap. (30)	28	
	generation (30)		15
Fresh water use	water exploitation index (28)		
	water exploitation index (25)		

Note: Ranks: low rank indicates good performance (number in brackets indicates total number of countries ranked). DTT indicates a distance-to-target analysis.

Poland

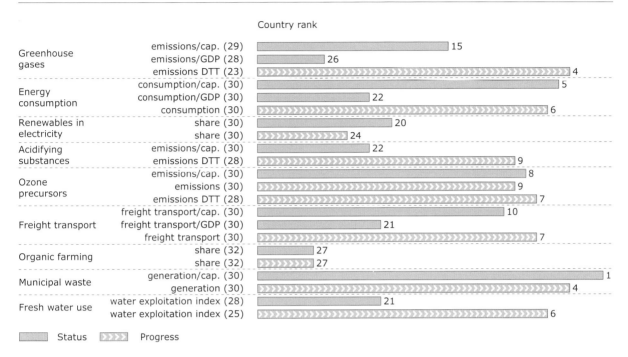

Note: Ranks: low rank indicates good performance (number in brackets indicates total number of countries ranked). DTT indicates a distance-to-target analysis.

Portugal

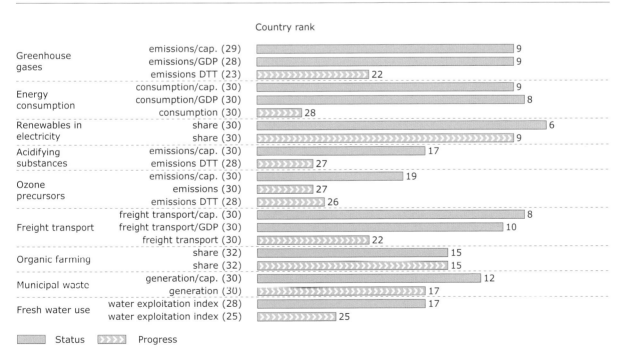

Note: Ranks: low rank indicates good performance (number in brackets indicates total number of countries ranked). DTT indicates a distance-to-target analysis.

Romania

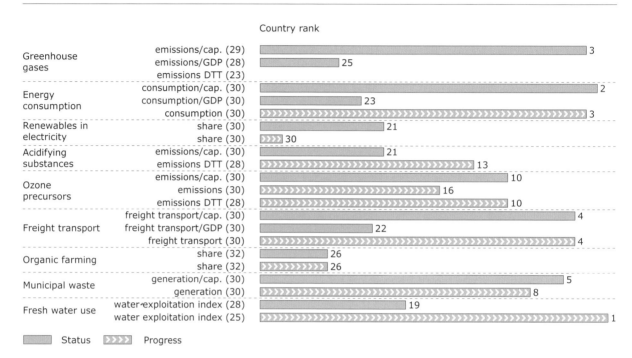

Note: Ranks: low rank indicates good performance (number in brackets indicates total number of countries ranked). DTT indicates a distance-to-target analysis.

Slovak Republic

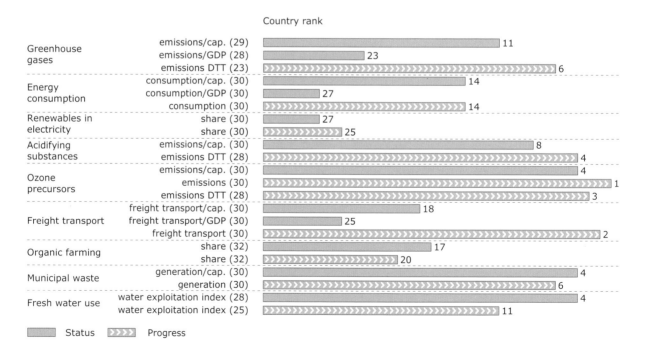

Note: Ranks: low rank indicates good performance (number in brackets indicates total number of countries ranked). DTT indicates a distance-to-target analysis.

Ranking graphs

Slovenia

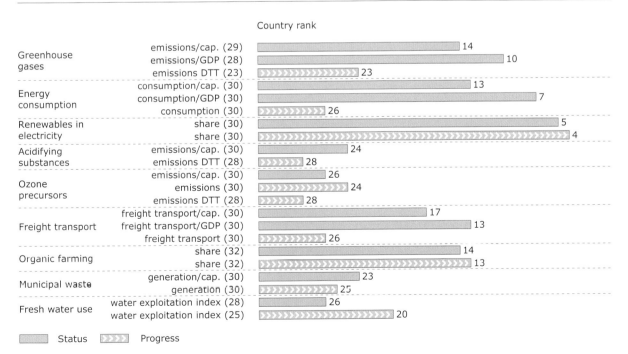

		Country rank
Greenhouse gases	emissions/cap. (29)	16
	emissions/GDP (28)	19
	emissions DTT (23)	15
Energy consumption	consumption/cap. (30)	15
	consumption/GDP (30)	19
	consumption (30)	23
Renewables in electricity	share (30)	11
	share (30)	15
Acidifying substances	emissions/cap. (30)	23
	emissions DTT (28)	17
Ozone precursors	emissions/cap. (30)	16
	emissions (30)	26
	emissions DTT (28)	23
Freight transport	freight transport/cap. (30)	12
	freight transport/GDP (30)	14
	freight transport (30)	17
Organic farming	share (32)	12
	share (32)	10
Municipal waste	generation/cap. (30)	11
	generation (30)	3
Fresh water use	water exploitation index (28)	3
	water exploitation index (25)	14

Status / Progress

Note: Ranks: low rank indicates good performance (number in brackets indicates total number of countries ranked). DTT indicates a distance-to-target analysis.

Spain

		Country rank
Greenhouse gases	emissions/cap. (29)	14
	emissions/GDP (28)	10
	emissions DTT (23)	23
Energy consumption	consumption/cap. (30)	13
	consumption/GDP (30)	7
	consumption (30)	26
Renewables in electricity	share (30)	5
	share (30)	4
Acidifying substances	emissions/cap. (30)	24
	emissions DTT (28)	28
Ozone precursors	emissions/cap. (30)	26
	emissions (30)	24
	emissions DTT (28)	28
Freight transport	freight transport/cap. (30)	17
	freight transport/GDP (30)	13
	freight transport (30)	26
Organic farming	share (32)	14
	share (32)	13
Municipal waste	generation/cap. (30)	23
	generation (30)	25
Fresh water use	water exploitation index (28)	26
	water exploitation index (25)	20

Status / Progress

Note: Ranks: low rank indicates good performance (number in brackets indicates total number of countries ranked). DTT indicates a distance-to-target analysis.

Country analysis | Ranking graphs

Sweden

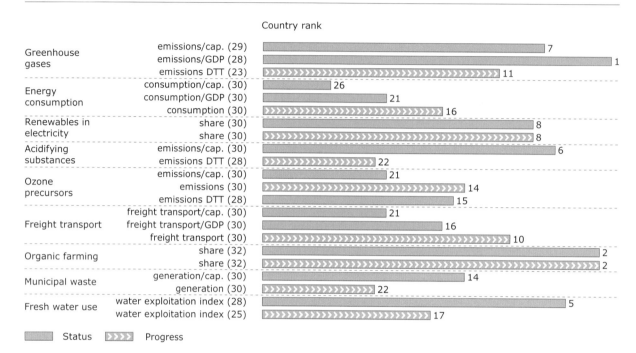

Note: Ranks: low rank indicates good performance (number in brackets indicates total number of countries ranked). DTT indicates a distance-to-target analysis.

Switzerland

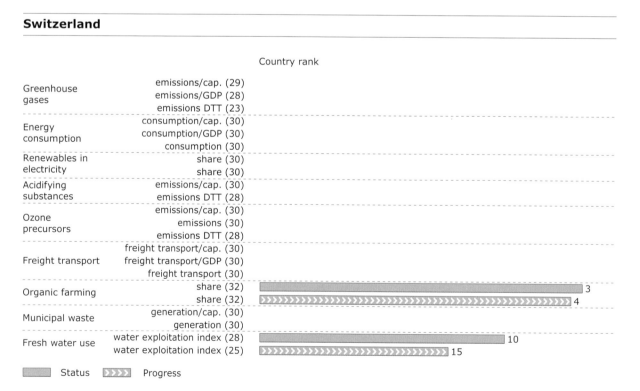

Note: Ranks: low rank indicates good performance (number in brackets indicates total number of countries ranked). DTT indicates a distance-to-target analysis.

Ranking graphs

Turkey

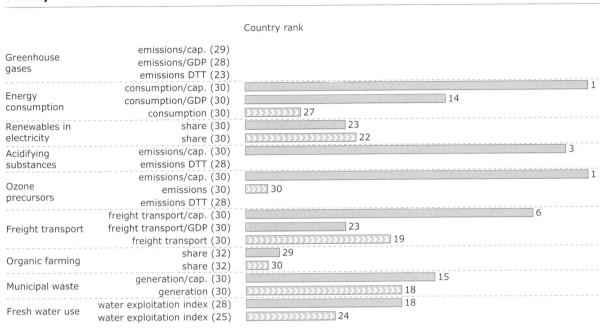

Note: Ranks: low rank indicates good performance (number in brackets indicates total number of countries ranked). DTT indicates a distance-to-target analysis.

United Kingdom

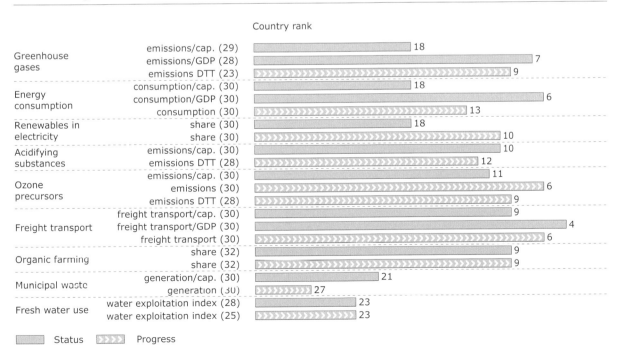

Note: Ranks: low rank indicates good performance (number in brackets indicates total number of countries ranked). DTT indicates a distance-to-target analysis.

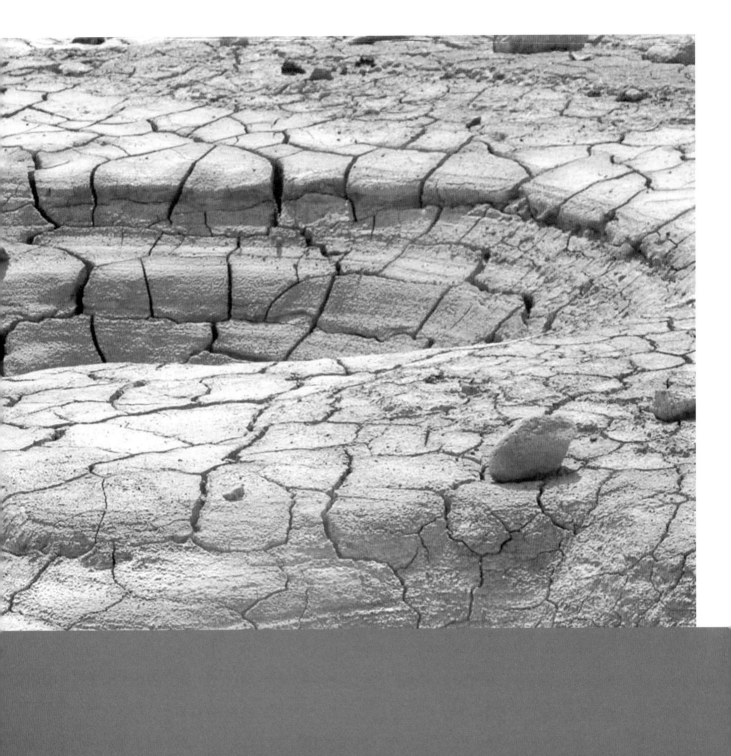

D Bibliography

D

Bibliography

Introduction 529
Previous state-of-the-environment reports 530
Signals reports 531

Reports 2000–2005, by theme:
    Air pollution 532
    Climate change 534
    Terrestrial environment and biodiversity 536
    Waste and material flows 537
    Water 539
    Agriculture 542
    Energy 543
    Transport 544
    Other issues 546
    Policy measures and instruments 549
    Eionet development and information systems, by theme 552

Condensed list of reports 2000–2005, by series:
    Environmental assessment reports (2000–2003) 566
    Environmental issue reports (2000–2004) 566
    EEA reports (2004–2005) 567
    EEA briefings (2003–2005) 567
    Topic reports (2000–2003) 567
    Technical reports (2000–2003) 568
    EEA technical reports (2004–2005) 570

PART D

PART D Bibliography

# Introduction

The EEA founding regulation mandates the Agency 'to publish a report on the state of, trends in and prospects for the environment every five years, supplemented by indicator reports focusing upon specific issues'.

Part D provides an overview of the reports EEA has published since the previous five-year report, *Environment in the European Union at the turn of the century*. It covers all environmental reports issued in the period January 2000 to November 2005 and the electronic version also includes hyperlinks to the products. Corporate reports, such as annual reports and yearly work programmes, as well as promotional brochures, are excluded from this overview.

Until 2003, EEA reports were published in four series: Environmental assessment reports, Environmental issue reports, Topic reports and Technical reports. In 2004, these were replaced by two new series: EEA reports and EEA technical reports. In addition, EEA briefings were introduced in 2003.

Although the EEA continues its publication of indicator reports, the regular indicator reporting is increasingly undertaken via the Internet only. The core set of indicators presented in parts B and C of this report will be regularly updated on the EEA web site.

Most of the reports are presented with either the full or slightly shortened version of the summary text as written at the time of their launch on the web site. Reports that have been published by other organisations under licence agreements with the EEA, and are not available on the EEA web site, are not included in this overview.

The reports are arranged as follows. After presenting all previous state-of-the-environment reports and Signals reports, the reports are listed by theme in line with the groups in the Core set of indicators. Titles that are of a more technical nature and contribute to the development of Eionet and information systems are listed separately at the end. They are also grouped by theme.

Within each theme, the reports are listed chronologically and even though a report cuts across two or more themes it is listed only once. This is contrary to the Report section of the EEA web site where reports are assigned to all the themes they cover. When a report exists in more than one language version, a link is provided to each. Finally, all the reports are listed again in a more condensed list, grouped by the series they belong to.

These reports, and the previous state-of-the-environment reports, establish the knowledge base upon which the present report relies. In addition, the 2005 report is built on nine sub-reports which are in the process of being published. The first three reports are included in this overview, marked as 'Sub-report to the present report'.

The EEA web site (www.eea.eu.int) has a comprehensive list of all the EEA products and services and includes reports issued prior to the period covered by this overview, as well as future productions.

Previous state-of-the-environment reports

Europe's environment: the third assessment

This is the third Pan-European state-of-the-environment report produced by the EEA. It was prepared for the 'Environment for Europe' Ministerial Conference being held under the auspices of the UN Economic Commission for Europe in Kiev, Ukraine on 21–23 May 2003. This assessment is the most comprehensive up-to-date overview currently available of the state of the environment on this continent.

Published on: 12 May 2003
Environmental assessment report No 10

Languages available: EN RU

Summary available in: BG CS DA DE EL EN ES ET FI FR HU IS IT LT LV NL NO PL PT RO RU SK SL SV TR

Environment in the European Union at the turn of the century

The EEA founding regulation mandates EEA to publish a report on the state of, trends in and prospects for the environment every five years. This report is published when EEA has operated in Copenhagen for close to five years. Earlier (November 1995), EEA has published *Environment in the European Union 1995* (not available on the web).

Published on: 1 June 1999
Environmental assessment report No 2

Languages available: EN FR IT

Summary available in: DA DE EL EN ES FI FR IS IT NL NO PT SV

Europe's environment: The second assessment

A report on the changes in the Pan-European environment as a follow-up to *Europe's environment — The Dobris assessment (1995)* requested by the environment ministers for the whole of Europe to prepare for the fourth ministerial conference in Aarhus, Denmark, June 1998.

Published on: 23 June 1998

Languages available: DA DE EL EN ES FI FR IS IT NL NO PT RU SV

Summary available in: DA DE EL EN ES FI FR IS IT NL NO PT RU SV

Europe's environment — The Dobris assessment

This report is a result of more than three years' work following the request of the first Pan-European conference of environment ministers, which took place at Dobris Castle, near Prague, from 21 to 23 June 1991. This conference called for the preparation of a state-of-the-environment report for Europe and invited the European Commission to take responsibility for the work.

Published on: 12 September 1995

Languages available: EN

Summary available in: DA DE EN ES FI FR IS IT NT NO PT SV

Bibliography PART D

Signals reports

EEA Signals 2004

The 2004 edition of the EEA's regular survey of environmental trends in its member countries covers aspects of agriculture, water pollution, nature protection, packaging waste, energy, transport, air pollution and climate change. It also provides an environmental perspective on the economic and social situation in Europe, including trends in demography and resource use, in the context of progress towards sustainability.

Published on: 1 June 2004

Languages available: BG CS DA DE EL EN ES ET FI FR HU IS IT LT LV MT NL NO PL PT RO SK SL SV TR

Environmental signals 2002 — Benchmarking the millennium

This, the third of the EEA's Environmental signals reports, provides an insight into the state of Europe's environment and is targeted at high-level policy-makers in EEA member countries and the EU, as well as the wider public. The publication of this report demonstrates that the annual routine of reporting on the state of the environment, and above all the progress that has been made, is now well established.

Published on: 14 May 2002
Environmental assessment report No 9

Languages available: EN

Summary available in: BG CS DA DE EL EN ES ET FI FR HU IS IT LT LV NL NO PL PT RO SK SL SV

Environmental signals 2001

Environmental signals 2001 is the second issue in the series of regular indicator-based reports produced by the EEA for high-level policy-makers in EEA member countries and the EU. The main aim is to present key environmental indicators in order to report, on a regular and consistent basis, on progress in a number of policy areas at the European level. The report also presents a benchmark of countries' performance ('name and shame, name and fame').

Published on: 29 May 2001
Environmental assessment report No 8

Languages available: EN

Summary available in: DA DE EL EN ES FI FR IS IT NL NO PT SV

Environmental signals 2000

This first edition of Environmental signals marks the beginning of a new period in reporting on the environment for policy-makers and the public by the EEA. Previous reports such as *Europe's environment — The second assessment* and *Environment in the European Union at the turn of the century* are comprehensive documents containing detailed information to support the development of strategic, long-term environmental policies and to provide general public information.

Published on: 2 May 2000
Environmental assessment report No 6

Languages available: DE EL EN ES FR IT NL

Bibliography | Reports 2000–2005, by theme

Air pollution

Air pollution by ozone in Europe in summer 2004

In summer 2004, the levels of ground-level ozone were high in southern Europe with widespread exceedances of the information threshold value (180 µg/m3), based on data submitted to the European Commission under the EU ozone directive.

Published on: 8 June 2005
Technical report No 3/2005

Languages available: EN

Previous 'Air pollution by ozone in Europe' reports:
Topic report 3/2003, 28 October 2003 (summer 2003)
Topic report 6/2002, 16 October 2002 (summer 2002)
Topic report 13/2001, 26 October 2001 (summer 2001)
Topic report 1/2001, 28 May 2001 (1999 and summer 2000)
Topic report 10/2000, 20 November 2000 (1998 and summer 1999)

Annual European Community CLRTAP emission inventory 1990–2002

This report is the annual European Community CLRTAP emission inventory presenting the European Community air pollution data from the years 1990 to 2002.

Published on: 5 January 2005
Technical report No 6/2004

Languages available: EN

Previous 'EC CLRTAP emission inventory' reports:
Technical report 91, 13 December 2002 (1990–2000)
Technical report 73, 10 June 2002 (1990–1999)
Technical report 52, 13 December 2000 (1980–1998)

Air pollution and climate change policies in Europe: exploring linkages and the added value of an integrated approach

There is an increasing awareness in both the science and policy communities of the importance of addressing the linkages between the traditional air pollutants and greenhouse gases. Many of the traditional air pollutants and greenhouse gases have common sources, their emissions interact in the atmosphere, and separately or jointly they cause a variety of environmental impacts on the local, regional and global scales.

Published on: 16 November 2004
Technical report No 5/2004

Languages available: EN

Exploring the ancillary benefits of the Kyoto Protocol for air pollution in Europe

This report explores the potential ancillary benefits for air pollution — in terms of reductions in pollutant emissions and changes in control costs and environmental impacts — resulting from different ways of implementing the Kyoto Protocol on climate change in Europe. It is based on a comparison of three climate policy scenarios, which differ in their use of the Kyoto mechanisms, with a baseline scenario for 2010.

Published on: 13 April 2004
Technical report No 93

Languages available: EN

Bibliography **PART D**

Air pollution in Europe 1990–2000

This report provides an overview and analysis of the air pollution situation in Europe in the year 2000 and the preceding decade. It is based on indicators for underlying driving forces, emissions, air quality, deposition of pollutants and the effectiveness of policies and measures. The report covers the 31 EEA member countries and Switzerland.

Published on: 31 March 2004
Topic report No 4/2003

Languages available:  EN

Previous related report:
Topic report 4/2002, 20 February 2003 (Air quality in Europe: state and trends 1990–1999)

Emissions of atmospheric pollutants in Europe, 1990–1999

An analysis of trends in emissions of the main air pollutants in EU Member States and candidate counties between 1990 and 1999 is presented in this report. Emissions data are shown and analysed by country and economic sector. They are also compared with agreed international emission ceilings and reduction targets.

Published on: 30 January 2003
Topic report No 5/2002

Languages available: EN

Previous related report:
Topic report 9/2000, 19 October (Emissions of atmospheric pollutants in Europe, 1980–1996)

Air quality in larger cities in the European Union

In this report, the air quality in about 200 urban agglomerations within the EU is calculated for a reference year (1995 or 1990) and for the year 2010, assuming the 'Auto-Oil II' programme base case scenarios. The parameter calculated is the urban background air pollution concentration, which is representative of the concentration in most of the urban area, with the exception of places under direct influence of emission sources, such as street traffic.

Published on: 20 June 2001
Topic report No 3/2001

Languages available: EN

Bibliography | Reports 2000–2005, by theme

Climate change

How much biomass can Europe use without harming the environment?

Extending biomass use to produce energy (bioenergy) will both help reduce greenhouse gas emissions and meet the European renewable energy targets. However, biomass production may create additional environmental pressures, such as on biodiversity, soil and water resources.

Published on: 18 October 2005
Briefing No 2/2005

Languages available: EN

Climate change and a European low-carbon energy system

This report presents an assessment of possible greenhouse gas emission reduction pathways made feasible by global action and a transition to a low-carbon energy system in Europe by 2030. It analyses trends and projections for emissions of greenhouse gases and the development of underlying trends in the energy sector. It also describes the actions that could bring about the transition to a low-carbon energy system in the most cost-effective way.

Sub-report to the present report

Published on: 29 June 2005
EEA Report No 1/2005

Languages available: EN

Annual European Community greenhouse gas inventory 1990–2003 and inventory report 2005

This report is the annual submission of the greenhouse gas inventory of the European Community to the United Nations Framework Convention on climate change. It presents greenhouse gas emissions between 1990 and 2003 by individual member state and by economic sector.

Published on: 21 June 2005
Technical report No 4/2005

Languages available: EN

Previous 'EC GHG inventory' reports:
Technical report 2/2004, 14 July 2004 (1990–2002, rep. 2004)
Technical report 95, 5 May 2003 (1990–2001, rep. 2003)
Technical report 75, 29 April 2002 (1990–2000, rep. 2002)
Technical report 60, 23 April 2001 (1990–1999)
Technical report 41, 23 May 2000 (1990–1998)

Climate change and river flooding in Europe

Extreme floods are the most common type of natural disaster in Europe. Climate change, including the increasing intensity of heavy rainfall, is projected to make extreme river floods even more frequent in some areas, especially in central, northern and northeastern Europe.

Published on: 6 April 2005
Briefing No 1/2005

Languages available: BG CS DA DE EL EN ES ET FI FR HU IS IT LT LV MT NL NO PL PT RO SK SL SV TR

Bibliography   PART D

 **Analysis of greenhouse gas emission trends and projections in Europe 2004**

This report provides a detailed analysis and background information for the EEA report *Greenhouse gas emission trends and projections in Europe 2004.*

Published on: 21 December 2004
Technical report No 7/2004

Languages available:

Previous related reports:
Technical report No 4/2004, 26 October 2004 (2003)
Technical report No 77, 1 July 2003 (Projections)
Topic report No 7/2002, 10 February 2003 (Trends)
Topic report 1/2002, 23 August 2002 (Comparison of national and EU-wide projections of GHG emissions)

 **Greenhouse gas emission trends and projections in Europe 2004**

Projections show that the pre-2004 EU Member States (EU-15) could cut their total emissions to 7.7 % below 1990 levels by 2010 with existing domestic policies and measures already being implemented and additional policies and measures currently planned.

Published on: 21 December 2004
EEA Report No 5/2004

Languages available:

Previous 'GHG emission trends and projections' reports:
Environmental issue report 36, 5 December 2003 (2003)
Environmental issue report 33, 5 December 2002 (2002)
Topic report 10/2001, 24 October 2001 (trends 1990–1999)
Topic report 6/2000, 24 October 2001 (trends 1990–1998)

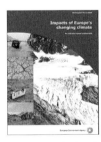 **Impacts of Europe's changing climate**

The impacts of climate change on Europe's environment and society are shown in this report. Past trends in the climate, its current state and possible future changes are presented using 22 selected indicators. For almost all of these a clear trend exists and impacts are already being observed. The report highlights the need to develop strategies at European, national, regional and local level for adapting to climate change.

Published on: 18 August 2004
EEA Report No 2/2004

Languages available:

Terrestrial environment and biodiversity

Halting the loss of biodiversity in Europe

This briefing focuses on the urgent need for information based on monitoring and indicators to support the achievement of a significant reduction in the rate of biodiversity loss by 2010 and the EU's more ambitious goal of halting the loss of biodiversity by 2010.

Published on: 9 February 2004
Briefing No 1/2004

Languages available: BG DE EL EN ES ET FI FR IS LV NL NO PL PT SL SV TR

Towards an urban atlas: Assessment of spatial data on 25 European cities and urban areas

The goal of this report is to disseminate the first results of a research project dedicated to measuring and assessing urban dynamics through the creation of land use databases for 25 cities and urban areas within Europe. The report, published together with the Joint Research Centre, describes the work developed by the Murbandy/Moland project (Monitoring urban dynamics/Monitoring land use changes), with the aim of proposing a methodology for strategic monitoring of the impact of urban development on the environment.

Published on: 29 July 2002
Environmental issue report No 30

Languages available: EN

Europe's biodiversity — biogeographical regions and seas

This is the first report to describe the condition and main influences on main habitat types and species groups in the eleven biogeographical regions and the seven seas around Europe. The report covers Europe from the Arctic to the Mediterranean and from the Atlantic to the Ural Mountains. It is based on expert contributions and national comments.

Published on: 31 May 2002
(10 chapters available on the web in November 2005. Publication of remaining chapters ongoing)

Languages available: EN

Down to earth: Soil degradation and sustainable development in Europe

This report is a broad assessment of soil conditions in the Pan-European region. It describes the degradation of soil, its relevance in the European and global contexts, and the links between soil and sustainable development. The report is mainly addressed to policy makers and the general public.

Published on: 15 December 2000
Environmental issue report No 16

Languages available: DE EN ES FR

Bibliography PART D

Waste and material flows

Case studies on waste minimisation practices in Europe

Ten successful case studies are presented in this report, which seeks to support and encourage waste minimisation in EEA countries. Focusing on waste prevention and recycling, the case studies cover five themes: producer responsibility, voluntary agreements, legislative requirements, information programmes and waste taxes.

Published on: 29 January 2003
Topic report No 2/2002

Languages available: EN

Biodegradable municipal waste management in Europe

This report is intended as guidance to help decision-makers in their efforts to comply with the EU Landfill Directive's targets for reducing the landfilling of biodegradable municipal waste. The first two parts of the report provide information on the strategies and instruments available to achieve this goal while the third part focuses on technology and market issues.

Published on: 22 March 2002
Topic report No 15/2001

Languages available: EN

Review of selected waste streams

Five waste streams are reviewed in this technical report: sewage sludge, construction and demolition waste, waste oils, waste from coal-fired power plants and biodegradable municipal waste. The review focuses on the quantities of waste produced and on systems used to collect and treat them in several European countries and regions.

Published on: 25 January 2002
Technical report No 69

Languages available: EN

Material flow-based indicators in environmental reporting

This report explores in detail material flows in the economy which have a possible influence on the environment. It attempts to define the parameters that need to be taken into account to steer human development towards ecological sustainability. The report proposes a measure for resource productivity in the economy and shows how quantitative targets for sustainable resource consumption can be derived and used to define performance indicators.

Published on: 15 October 2001
Environmental issue report No 14

Languages available: EN

Bibliography | Reports 2000–2005, by theme

 **Waste management facilities — Electronic catalogue**

This report provides information about EEA member countries; facilities for the treatment of hazardous and non-hazardous wastes, including their type, location and capacity. It is intended to support policy making since information about available treatment facilities is a prerequisite for setting realistic and achievable targets for the various waste management options. The report contains proposals for further data collection to expand current information.

Published on: 3 October 2001
Technical report No 65

Languages available:

 **Total material requirement of the European Union**

Contains detailed outcomes of the total material requirement (TMR) study reported on in *Environmental signals 2000*. It describes the methodology to calculate aggregated material flow indicators and presents a first application on the EU level. The study contains more complete time series than the *Environmental signals 2000* chapter. Technical details are in Technical report no 56.

Published on: 12 March 2001
Technical report No 55

Languages available: EN

Related report:
Technical report No 56, 20 February 2001 (Total material requirement of the European Union — Technical part)

Bibliography PART D

Water

Priority issues in the Mediterranean environment

In 1999, recognising the lack of timely and targeted information for action, EEA and UNEP/MAP collaborated on a joint publication: *The State and pressures of the marine and coastal Mediterranean environment*. Further cooperation to provide more in-depth assessments, which form the basis for global action to reverse the present trends in the region, has continued and is exemplified by this joint report.

Published on: 10 November 2005
EEA report 5/2005

Languages available: EN

Europe's water: An indicator-based assessment

This report assesses the quality and quantity of Europe's water. Its geographical scope is the EU, EFTA and EU accession and candidate countries. Four water issues are assessed: ecological quality, eutrophication and organic pollution of water, hazardous substances and water quantity. This is done on the basis of 57 indicators selected for their representativeness and relevance.

Published on: 27 November 2003
Topic report No 1/2003

Languages available: EN

Summary available in: BG CS DA DE EL EN ES ET FI FR HU IS IT LV NL NO PL PT RO SK SL SV TR

Status of Europe's water

This paper gives a brief overview of the state of Europe's water, highlighting the issues on which progress towards environmental improvement is and is not being made. The briefing is based on the report *Europe's water: an indicator-based assessment*.

Published on: 27 November 2003
Briefing No 1/2003

Languages available: BG CS DA DE EL EN ES ET FI FR HU IS IT LT LV NL NO PL PT RO SK SL SV TR

Hazardous substances in the European marine environment — Trends in metals and persistent organic pollutants

This report provides an assessment of trends in concentrations of six hazardous substances in mussels and fish in Europe's seas since 1985. The trends are linked to reduction measures, foodstuff limit values and trends in inputs to coastal and marine waters. The results are presented in a series of maps.

Published on: 27 October 2003
Topic report No 2/2003

Languages available: EN

The rivers of the Black Sea

This report presents the results of a study of the discharge of river water and river sediment load into the Black Sea. Intended for Black Sea researchers, it is published in English and Russian as a contribution to cooperation between EEA and the Black Sea Commission. The volume of river water and load is investigated by individual river, by region and in total over the whole area.

Published on: 7 August 2002
Technical report No 71

Languages available: EN RU

Eutrophication in Europe's coastal waters

Eutrophication is the enhanced primary production of marine algae due to excessive supply of nutrients from human activities, independent of the natural productivity level of the sea area in question. The main objectives of this report are to evaluate the causes, state and development of eutrophication in European coastal waters and identify areas where more monitoring data are needed to improve the assessment.

Published on: 18 October 2001
Topic report No 7/2001

Languages available: EN

Sustainable water use in Europe — Part 3: Extreme hydrological events: floods and droughts

This report gives an overview of the main natural and artificial causes and impacts of floods and droughts in western and central European countries, as well as policy responses. It aims to help policy- and decision-makers in their work on preventing and managing such extreme events, as well as the European Commission's work in the field of civil protection.

Published on: 10 August 2001
Environmental issue report No 21

Languages available: EN

Sustainable water use in Europe — Part 2: Demand management

The EEA and its European Topic Centre on Inland Waters is undertaking an assessment of the sustainable use of water in Europe. This report describes the second part of that assessment and looks in particular at demand-side management of water across Europe. It is intended that the report will serve as a source of comparative data to support the assessment of policies in place and a source of information for the development of new demand management policies.

Published on: 5 April 2001
Environmental issue report No 19

Languages available: EN

Bibliography   PART D

Sustainable use of Europe's water? State, prospects and issues

This report is intended to give ministers, senior civil servants, other policy-makers and others with an interest to protect our waters a broad overview of the major water issues in Europe. It represents a distillation of the work undertaken by the EEA and its European Topic Centre for Inland Waters (ETC/IW). The report provides, for each issue, a summary of our scientific and technical knowledge of the problem, an analysis of its causes, an indication of the actions taken and their effects, and an assessment of what further needs to be done.

Published on: 13 October 2000
Environmental assessment report No 7

Languages available: DA DE EN ES FI FR IS IT NL NO PT SV

State and pressure of the marine and coastal Mediterranean environment

Human activity, including tourism, puts increasing pressure on a Mediterranean Sea that is still remarkably alive in spite of many problems, particularly in coastal areas. The report, prepared by the EEA and its European Topic Centre on the Marine and Coastal Environment in cooperation with the Mediterranean action plan, presents an overview of the Mediterranean marine and coastal environment.

Published on: 28 February 2000
Environmental assessment report No 5

Languages available: EN ES
Summary available in: DA DE EL EN ES FI FR IS IT NL NO PT SV

Nutrients in European ecosystems

This report presents a Pan-European overview of the geographical distribution, and severity of adverse effects, of excessive anthropogenic inputs of nutrients in European ecosystems.

Published on: 11 January 2000
Environmental assessment report No 4

Languages available: EN

Groundwater quality and quantity in Europe

This report presents the first Pan-European overview of groundwater quality and quantity based on measured data. Data and information provided from 37 countries have been used in this assessment report. Different indicators have been used to assess the pressures on groundwater quality and quantity related to, in particular, nitrate, pesticides and groundwater abstraction. The report also identifies a need for improved information on groundwater across Europe.

Published on: 7 January 2000
Environmental assessment report No 3

Languages available: EN

Agriculture

High nature value farmland — Characteristics, trends and policy challenges

Farmland supports many habitats and species of European conservation concern. In 2003, Europe's environment ministers agreed to identify all farmland areas with high nature value and take conservation measures. This report shows that these areas cover roughly 15–25 % of the European countryside and suffer from land abandonment and intensification. Current policy measures appear insufficient to prevent further biodiversity decline.

Published on: 29 April 2004
EEA Report No 1/2004

Languages available: EN TR

Agriculture and the environment in the EU accession countries — Implications of applying the EU common agricultural policy

Agriculture is a key influence on the environment in the acceding countries. This report reviews the pressures from agriculture on water, soil and air. It also shows the importance of agricultural management for biodiversity and landscapes. Introduction of the CAP will bring positive and negative changes. Which tendency will predominate depends largely on policy implementation by the new EU Member States.

Published on: 28 April 2004
Environmental issue report No 37

Languages available: EN

Genetically modified organisms (GMOs): The significance of gene flow through pollen transfer

This report, written for the EEA by experts from the European Science Foundation, considers the significance of the transfer by pollen of genes from six major genetically modified (GM) crop types that are close to commercial release in the EU. Oilseed rape, sugar beet, potatoes, maize, wheat and barley are reviewed in detail using recent and current research findings to assess their potential environmental and agronomic impacts. The report also includes a short review of the current status of GM fruit crops in Europe.

Published on: 21 March 2002
Environmental issue report No 28

Languages available: EN

Bibliography **PART D**

Energy

 **Energy subsidies and renewables**

This briefing presents the main findings from the EEA technical report *Energy subsidies in the European Union: A brief overview*. While solid fuels received the largest share of subsidies, renewables received significantly higher support per unit of energy produced. Governments seem therefore to recognise that renewable energy is a much less mature industry with a greater need for technological and market support to enable full commercial development.

Published on: 4 June 2004
Briefing No 2/2004

Languages available: BG CS DA DE EL EN ES ET FI FR HU IS IT LT LV NL NO PL PT RO SK SL SV

 **Energy subsidies in the European Union: A brief overview**

This report synthesises data from a range of sources to estimate the size of subsidies to the energy sector in the 15 pre-2004 Member States of the EU. It finds that the level of subsidies to fossil fuels remains high despite their environmental impacts, while support for renewable energy sources is increasing steadily.

Published on: 3 June 2004
Technical report No 1/2004

Languages available:

 **Energy and environment in the European Union**

This report provides policy-makers with the information needed to assess how effectively environmental policies and concerns are being integrated into energy policies. This is achieved with the help of selected indicators measuring progress. The report concludes that, while there have been some successes, overall progress has so far been insufficient.

Published on: 30 May 2002
Environmental issue report No 31

Languages available: EN

Summary available in: BG CS DA DE EL EN ES ET FI FR HU IS IT LT LV NL NO PL PT RO SK SL SV

 **Renewable energies: success stories**

Based on a series of case studies, this study identifies the factors that can influence successful implementation of specific renewable energy technologies in EU Member States. It aims to facilitate greater use of renewable energy sources and contribute to efforts by the EU and its Member States to meet targets for increasing power from renewables by 2010.

Published on: 6 December 2001
Environmental issue report No 27

Languages available:

Transport

Transport biofuels: exploring links with the energy and agriculture sectors

Transport biofuels are being promoted as a useful means of greening the transport sector. However, impacts on the development of renewable energy and the intensity of agricultural land use need to be taken into account when assessing the overall environmental benefits.

Published on: 10 November 2004
Briefing No 4/2004

Languages available: BG CS DA DE EL EN ES ET FI FR HU IS IT LT LV MT NL NO PL PT RO SK SL SV TR

Ten key transport and environment issues for policy-makers

Transport volumes are growing at a rate where many of the improvements brought about by new technology are being partly or fully negated. Emissions of harmful pollutants are decreasing steadily, but there are indications that the test cycles, on which the emissions are calculated, fail to represent real world driving conditions. Therefore air quality is not improving as would be expected from the emission figures. The TERM 2004 report looks at trends in transport emissions and other transport impacts over the decade from the early 1990s to the early 2000s.

Published on: 19 October 2004
EEA Report No 3/2004

Languages available: EN TR

Transport and environment in Europe

Growing transport volumes are leading to increased pressure on the environment especially in relation to climate change and biodiversity loss. Present efforts to counteract these trends are at best only slowing down the rate of increase. On the positive side, technological improvements are delivering reductions in air pollution from road transport despite the growth in traffic volumes. Even so, more is needed to solve the problem of urban air pollution. This briefing looks at developments from the early 1990s to the early 2000s.

Published on: 19 October 2004
Briefing No 3/2004

Languages available: BG CS DA DE EL EN ES ET FI FR HU IS IT LT LV MT NL NO PL PT RO SK SL SV TR

Transport price signals

The EEA, jointly with the Commission's Directorate-General for the environment, Directorate-General for transport and energy and the statistical office Eurostat, developed the transport and environment reporting mechanism (TERM). TERM aims to monitor progress in integrating environmental concerns into transport policy throughout Europe and comprises 40-odd indicators, which cover all relevant aspects of the transport and environment system.

Published on: 20 September 2004
Technical report No 3/2004

Languages available: EN

Bibliography PART D

Paving the way for EU enlargement — Indicators of transport and environment integration

This third report developed under the EU's transport and environment reporting mechanism (TERM) is the first to include the 13 accession countries (ACs). As in previous reports, the TERM indicators are used to answer a set of policy questions related to the integration of environmental concerns into transport policies.

Published on: 28 November 2002
Environmental issue report No 32

Languages available: EN

Summary available in: BG CS DA DE EL EN ES ET FI FR HU IS IT LT LV NL NO PL PT RO SK SL SV

Road freight transport and the environment in mountainous areas

This report was developed to support an analysis by the European Commission of the effectiveness of the 'ecopoints' system for limiting the environmental impact of heavy-duty vehicles transiting through Austria. The analysis culminated in a communication issued by the Commission in February 2001. The report takes stock of the environmental pressures that growing freight transport by road is exerting on the environment in the EU and investigates the specific problems of traffic impacts in sensitive mountain areas.

Published on: 7 December 2001
Technical report No 68

Languages available: EN

TERM 2001 — Indicators tracking transport and environment integration in the European Union

Main report, presentation by Domingo Jiménez-Beltrán, Executive Director at the Joint transport and environment Informal Council, 14–16 September 2001, Leuven, Ottignies- Louvain-la-Neuve and presentation of key findings in English and summaries in all official languages.

Published on: 11 September 2001
Environmental issue report No 23

Languages available: EN

Summary available in: DA DE EL EN ES FI FR IS IT NL NO PT SV

Are we moving in the right direction? Indicators on transport and environmental integration in the EU

The transport and environment reporting mechanism (TERM) was set up to enable policy-makers to gauge how this integration is progressing in the transport sector. At the core of this first TERM report is a set of 31 indicators that are benchmarked against a number of policy objectives and targets.

Published on: 20 July 2000
Environmental issue report No 12

Languages available: EN

Summary available in: DA DE EL EN ES FI FR IS IT NL NO PT SV

Other issues

European environment outlook report

Protecting our environment is a key element in ensuring sustainable livelihoods for today's and future generations. Indeed, the most recent Eurobarometer surveys show that as Europeans we regard the protection of our environment to be one of the six key priorities for the EU. Issues of particular concern are water and air pollution, man-made disasters, and climate change. In addition, new challenges arising from diffuse sources of pollution, changing consumption patterns, and the possibility of sudden extreme environmental changes all need to be addressed.

Sub-report to the present report

Published on: 11 September 2005
EEA Report No 4/2005

Languages available: EN

Mapping the impacts of recent natural disasters and technological accidents in Europe

This is the first EEA publication to address the impacts of natural disasters and technological accidents across Europe. Focusing on major events between 1998 and 2002, the report adds value to existing studies by bringing together available information on their human and economic costs and adding the environmental perspective. The impacts of these events are documented through a large number of maps, illustrations and case studies.

Published on: 23 March 2004
Environmental issue report No 35

Languages available: EN TR

Arctic environment: European perspectives

The Arctic is one of the planet's last pristine areas where indigenous peoples pursue their traditional lifestyles. However, as Europe's dependence on the Arctic's resources grows, the region is coming under increasing pressure from unsustainable development, land fragmentation, climate change and pollution. EEA and UNEP have jointly prepared this report to raise awareness among policy makers and the public about the issues involved and why Europe should care.

Published on: 15 March 2004
Environmental issue report No 38/2004

Languages available: EN

Children's health and environment: A review of evidence.

This publication provides an overview of the available evidence of the relationship between the physical environment and children's health. It identifies both research needs and policy priorities to protect children's health from environmental hazards. The report was prepared by the WHO European Centre for Environment and Health, Rome Operational Division, with support from the EEA. It is based on background papers prepared for the Third Ministerial Conference on Environment and Health, held in London in 1999.

Published on: 12 April 2002
Environmental issue report No 29

Languages available: EN

Bibliography PART D

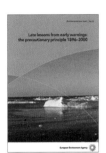

Late lessons from early warnings: the precautionary principle 1896–2000

This report is about gathering information on the hazards raised by human economic activities and its use in taking action to protect better the environment and the health of the species and ecosystems that are dependent on it. The study aims to contribute to better and more accessible science-based information and more effective stakeholder participation in the governance of economic activity so as to help minimise environmental and health costs and maximise innovation.

Published on: 10 January 2002
Environmental issue report No 22

Languages available: DE EN ES FR

Designing effective assessments: The role of participation, science and governance, and focus

This publication reports on a workshop held in March 2001 to discuss how environmental assessments can be conducted more effectively. The Copenhagen workshop focused on the issues of participation, science and governance, and scope. It was co-organised by the Global environmental assessment project and the EEA, with contributions from the European forum on integrated environmental assessment (EFIEA).

Published on: 3 December 2001
Environmental issue report No 26

Languages available: EN

Scenarios as tools for international environmental assessments

Scenarios can be useful tools in international environmental assessments for evaluating future environmental problems, and this report is intended to contribute to improving scenario-building and analysis for decision-makers. It proposes a step-wise approach to scenario development and draws on some relevant scenario-building exercises, including recent activities of the Intergovernmental Panel on climate change and World Water Commission.

Published on: 6 November 2001
Environmental issue report No 24

Languages available: EN

Environmental benchmarking for local authorities: From concept to practice

This report addresses the use of benchmarking as a management tool in the context of local authority actions towards sustainability management. The growing interest at the European policy-making level in stimulating the use of sustainability management tools and exchange of best practices among local authorities has motivated its publication. It is also a recognition that local authority's actions play a vital role in responding to the challenges of enhancing the state of the environment not only through policy-making, but also in the provision of better services and in a better informed planning process.

Published on: 30 July 2001
Environmental issue report No 20

Languages available: EN

The European environment | State and outlook 2005 547

Bibliography | Reports 2000–2005, by theme

 **Cloudy crystal balls: An assessment of recent European and global scenario studies and models**

This report presents an inventory of existing scenarios studies relevant for Europe in the context of sustainable development, including European and global models. It also provides a useful review of a selected number of scenarios and pinpoints their strengths and weaknesses.

Published on: 15 December 2000
Environmental issue report No 17

Languages available: EN

 **The dissemination of the results of environmental research**

This report reviews relevant knowledge and identifies issues for consideration in implementing the EEA's mission to facilitate the dissemination of policy-relevant environmental scientific research. It begins by examining and analysing literature relating to research-policy interactions, primarily from the social sciences; it then relates experiences from disseminating the research of the UK Global environmental change programme. After summarising existing European environmental research and dissemination efforts, it concludes by offering recommendations to the EEA on how to proceed with its mandate.

Published on: 22 November 2000
Environmental issue report No 15

Languages available: EN

 **Information tools for environmental policy under conditions of complexity**

This report contains a description of environmental policy tools, gives an overview of post-Normal science (bridge between complex systems and environmental policy), ecological economics, the incommensurability principle, the multicriteria evaluation and the NAIDE method (fuzzy information).

Published on: 10 March 2000
Environmental issue report No 9

Languages available: EN

Policy measures and instruments

Effectiveness of packaging waste management systems in selected countries: an EEA pilot study

Packaging waste is an important and growing waste stream. The amended packaging and packaging waste directive has recently been adopted, and work is underway to develop EU thematic strategies on waste prevention and recycling and on the sustainable use and management of natural resources. The packaging and packaging waste directive is one of the few environmentally-related directives to contain directly measurable, quantitative targets.

Published on: 7 October 2005
EEA Report No 3/2005

Languages available:

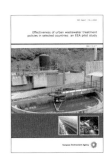

Effectiveness of urban wastewater treatment policies in selected countries: an EEA pilot study

Water pollution caused by wastewater persists despite three decades of effort to clean up European surface waters and despite the requirements of the urban waste water treatment directive (UWWTD). Several EU Member States have yet to satisfy the requirements of the directive. A European Commission report released in 2004 noted that several countries had failed to designate sensitive areas and were behind schedule in establishing the capacity of sewage treatment as required by the directive deadlines in 1998 and 2000.

Published on: 7 October 2005
EEA Report No 2/2005

Languages available:

Environmental policy integration in Europe — Administrative culture and practices

This paper presents an overview of administrative culture and practices for EPI in Europe, and investigates some of the main management styles used. It reviews institutional structures and practices in the EU-25, the candidate and applicant countries, the countries of the European free trade area (EFTA) and the countries of eastern Europe, Caucasus and central Asia (EECCA).

Sub-report to the present report

Published on: 6 July 2005
Technical report No 5/2005

Languages available:

Environmental policy integration in Europe — State of play and an evaluation framework

Article 6 of the European Community treaty states that 'environmental protection requirements must be integrated into the definition and implementation of the Community policies and activities (...) in particular with a view to promoting sustainable development'. Environmental policy integration (EPI) involves a continual process to ensure environmental issues are reflected in all policy-making.

Published on: 27 May 2005
Technical report No 2/2005

Languages available:

Bibliography | Reports 2000–2005, by theme

Reporting on environmental measures: Are we being effective?

Assessing the effects of EU environmental legislation, and whether specific measures have actually been effective in delivering expected results, is an essential task, yet one that is not always carried out. The REM report examines this issue and tries to respond to a set of questions on why, what and how effects and effectiveness should be assessed. The report aims to contribute to the development of a more effective and streamlined future regime for reporting on environmental measures.

Published on: 29 November 2001
Environmental issue report No 25

Languages available:

Business and the environment: current trends and developments in corporate reporting and ranking

This report is aimed at the business and financial sectors, which are beginning to develop strategies on integrating the environment into their activities, following the current legal and political moves to promote the 'integration' of environmental issues into economic activities. It provides an overview of these and other developments in corporate environmental reporting (CER), in corporate environmental performance indicators (EPIs), and in corporate environmental performance ranking tools (EPRTs).

Published on: 19 February 2001
Technical report No 54

Languages available:

Environmental taxes — Recent developments in tools for integration

The use of environmental taxes in Europe is widening and more tax bases are being used. The EEA reports on developments in the use and impact of environmental taxes and on progress made with ecological tax reform.

Published on: 13 December 2000
Environmental issue report No 18

Languages available:

Bibliography **PART D**

Eionet development and information systems, by theme

Air quality

EMEP/Corinair emission inventory guidebook — Third edition
This third edition of the emission inventory guidebook prepared by the UNECE/EMEP task force on emissions inventories and projections provides a comprehensive guide to state-of-the-art atmospheric emissions inventory methodology. Its intention is to support reporting under the UNECE Convention on Long-range Transboundary Air Pollution and the EU directive on national emission ceilings.
Published on: 19 January 2004
Technical report No 30

Euroairnet — Status report 2000
This is the latest status report on Euroairnet, the Europe-wide air quality monitoring network coordinated by the EEA. Euroairnet's main goal is to improve significantly the extent and quality of reporting of air quality in Europe, and thus to improve the basis for representative assessments of European air quality. The report describes the network's status as of November 2000.
Published on: 20 February 2003
Technical report No 90

Air quality in the Phare countries 1997
This report provides an overview of the state of air quality in 13 central and eastern European countries involved in the European Union's Phare programme. Based on Euroairnet monitoring network data for 1997 as reported by these countries, it also provides information on emission sources and economic sectors contributing to air pollution, impacts on health and ecosystems as well as measures taken to reduce the air pollution and its impacts.
Published on: 8 February 2002
Topic report No 16/2001

Air quality — Annual topic update 2000
This report provides an overview of work in the field of air quality conducted in 2000 by the EEA's European Topic Centre on Air Quality in cooperation with the Phare Topic Link on Air Quality.
Published on: 3 October 2001
Topic report No 9/2001

Air emissions — Annual topic update 2000
This report provides a summary of the activities and products of the EEA's European Topic Centre on Air Emissions (ETC/AE) during 2000. These include assistance to countries through provision of software tools for compiling and reporting emission estimates, analysis of air emission trends from sectors such as energy and transport, and analysis of progress towards achieving emission reduction targets.
Published on: 13 August 2001
Topic report No 5/2001

Eionet workshop on air quality monitoring and assessment
The fourth Eionet workshop on air quality management and assessment was held in Santorini 23–24 September 1999. This report deals with the discussions on the current European air quality issues and the work of the EEA's European Topic Centre on Air Quality (ETC/AQ) and its partners in the Phare Topic Link on Air Quality (PTL/AQ) with institutions and experts from European countries, the European Commission, and with collaborating international institutions.
Published on: 31 May 2001
Technical report No 57

Air quality — Annual topic update 1999
This report provides an overview of work in the field of air quality conducted in 1999 by the EEA's European Topic Centre on Air Quality in cooperation with the Phare Topic Link on Air Quality as part of the EEA work programme. Products are listed and plans for 2000 are described.
Published on: 10 October 2000
Topic report No 7/2000

Evaluation of the air quality model documentation system
The Internet air quality model documentation system (MDS) has been developed by the ETC-AQ to provide information and guidance to users in selection of air pollution models for a specified application. After three years of operation, its functionality and utility is evaluated through a survey of its users. The report summarises answers to a questionnaire from 41 respondents.
Published on: 4 October 2000
Technical report No 48

Air emissions — Annual topic update 1999
This report presents a summary of the activities and products of the EEA's European Topic Centre on Air Emissions in 1999. The EEA's European Topic Centre (ETC/AE) was appointed in December 1994 by EEA to act as a centre of expertise on air emissions for use by the Agency in support of its mission.
Published on: 13 September 2000
Topic report No 8/2000

Biodiversity

An inventory of biodiversity indicators in Europe, 2002
This review of biodiversity-related indicators was undertaken to support the development of a core set of environmental indicators by the EEA. Among its main conclusions is that a great variety of indicators has been developed to assess aspects of biodiversity at the national, international or global scale, but only a limited number of them are actually in use on a regular basis. The main content of the report is a summary of 655 such indicators.
Published on: 12 May 2004
Technical report No 92

Nature conservation annual topic update 2000

The work of the EEA's European Topic Centre for Nature Conservation between 1995 and 2000 is presented in this report. The EEA's European Topic Centre's main activities were support to the European Commission and Member States on the EU Birds and Habitats Directives, contributions to EEA reporting and development of EUNIS (European nature information system database on selected species, habitat types and sites).

Published on: 13 December 2002

Topic report No 3/2002

Towards a global biological information infrastructure

This report describes the state of the art in biodiversity informatics in the field of entomology. The seven papers in this volume are the results of a one-day symposium that was held during the XXI International Congress of Entomology in Iguassu Falls, Brazil, on 24 August 2000.

Published on: 18 February 2002

Technical report No 70

The global biodiversity information facility — Architectural and implementation issues

This paper supports the OECD Megascience forum for biological informatics on the technical aspects of its plan for the Global biological information facility. The paper identifies the capacity to produce homepages for all species of organisms as the main goal of GBIF, but in such a way that the homepages are dynamically derived from online databases.

Published on: 5 October 2000

Technical report No 34

Nature conservation — Annual topic update 1999

Over the year, ETC on Nature conservation has continued to develop the European nature information system UNIS. The EEA's European Topic Centre has also provided support to the NATURA 2000 network and to the EEA's reporting activities.

Published on: 13 August 2000

Topic report No 5/2000

Climate change

The EU reporting obligations under the United Nations framework convention on climate change (UNFCCC) and the monitoring mechanism

This report is part of the project undertaken with the participation of the firm Garrigues and Andersen to carry out for the EEA the study of the reporting obligations in the framework of environmental conventions and provide support to international activities, also to support the European Commission (Environment DG) as representative of the European Community as party to the conventions.

Published on: 29 January 2001

Technical report No 46

Terrestrial environment

Assessment and reporting on soil erosion
This report presents the results of a peer review of work on soil erosion carried out by the EEA up until 2001. It also summarises the conclusions of an expert workshop on indicators for soil erosion, held in Copenhagen in March 2001.
Published on: 27 May 2003
Technical report No 94

Corine land cover update 2000: Technical guidelines
This report has been prepared for those interested in updating the Corine land cover database. The document describes the methodological steps to be followed in developing an update in a consistent and harmonised way to ensure comparability among results across Europe.
Published on: 18 December 2002
Technical report No 89

Assessment of data needs and data availability for the development of indicators on soil contamination
Information on the national relevance and data availability of a proposed list of policy-relevant indicators on soil contamination from local and diffuse sources is provided in this report. It constitutes the final output of a working group on soil contamination indicators and analyses the results of a consultation of partners within the European environment information and observation network (Eionet). The outcomes of this consultation are feeding into the establishment of a regular process of data collection and assessment.
Published on: 13 November 2002
Technical report No 81

Proceedings of the technical workshop on indicators for soil sealing
This report contains the proceedings of a technical workshop on soil sealing indicators held in Copenhagen in March 2001. The report also presents the results of a European information and observation network consultation on a series of 'hot-spot' maps for soil degradation and provides recommendations to the EEA for further activities. The outcomes feed into the work of the EEA's European Topic Centre on Terrestrial Environment.
Published on: 18 September 2002
Technical report No 80

Eionet technical workshop on indicators for soil contamination
This report contains the proceedings of a European environmental information and observation network (Eionet) workshop on indicators for soil contamination held in Vienna in January 2001. The main objectives of the workshop were to: 1) discuss the development of a selection of soil contamination indicators; 2) get information on national data availability. The outcome of the workshop feeds into the work of the EEA's European Topic Centre on Terrestrial Environment.
Published on: 2 September 2002
Technical report No 78

Bibliography | Reports 2000–2005, by theme

Second technical workshop on contaminated sites
This report contains the proceedings and follow-up to date of the second workshop on contaminated sites held in Dublin in November 1999. The report analyses the results of a test regional data collection and presents a proposal for regional indicators on contaminated sites.
Published on: 29 August 2002
Technical report No 76

European soil monitoring and assessment framework
This report contains a summary of EEA work on the development of a European soil monitoring and assessment framework for soil since 1999. The framework is being designed as a basis for the work needed to support policy developments within the European Community as well as the EEA reporting activities on soil. In particular the report contains the proceedings and summarises the results of an Eionet workshop organised by the EEA's European Topic Centre on Soil in Vienna in November 1999, which started off the development of the framework.
Published on: 6 December 2001
Technical report No 67

Towards agri-environmental indicators: integrating statistical and administrative data with land cover information
This publication is the result of a close collaboration between four services of the European Commission – Directorates-General for agriculture, environment, Eurostat and the Joint Research Centre (Ispra) – and the EEA. It presents work carried out by the teams for integrating statistical and administrative data with land cover information. The different articles are a contribution to the development of agri-environment indicators using geo-referenced information.
Published on: 26 October 2001
Topic report No 6/2001

Land cover — Annual topic update 2000
This report summarises the work and products of the EEA's former European Topic Centre on Land Cover (ETC/LC) in 2000 and its achievements since 1995. The report also looks briefly at the main objectives of the new European Topic Centre on Terrestrial Environment (ETC/TE), which has now replaced the ETC/LC.
Published on: 9 August 2001
Topic report No 4/2001

Towards spatial and territorial indicators using land cover data
The present report focuses on the state of development of spatial and territorial indicators based on land cover as contribution to the environmental indicator based reporting. The report summarises the joint efforts during 1999–2000 of the EEA's European Topic Centre on Land Cover (ETC/LC) and the Phare Topic Link for Land Cover (PTL/LC) in this area.
Published on: 6 July 2001
Technical report No 59

Proposal for a European soil monitoring and assessment framework
This report has been prepared in the framework of the EEA annual work programme, as part of the work plan of the former EEA's European Topic Centre on Soil. It contains a proposal for the establishment of a comprehensive framework for monitoring, assessing and reporting on soil conditions in Europe.
Published on: 6 July 2001
Technical report No 61

Land cover — Annual topic update 1999
The *Land cover — annual topic update 1999* provides an overview of progress made in 1999 on monitoring, data collection, information, assessment and reporting on land cover related environmental issues. Special attention is given to the plans for updating the Corine land cover (CLC) database and providing information on land cover and use changes for different policy areas. The report includes a list of available products prepared by ETC/LS during 1999 and plants for further development during 2000.
Published on: 26 October 2000
Topic report No 4/2000

Corine land cover technical guide — Addendum 2000
This technical guide is an addendum to the Corine land cover technical guide, published by the European Commission within the frame of the Corine programme (CEC, 1994). This addendum 2000 was prepared by the EEA's European Topic Centre on Land Cover and Phare Topic Link on Land Cover to support the land cover monitoring activities of the EEA (EEA multi-annual work programme 1998–2002, project 1.3.5 Land cover).
Published on: 29 May 2000
Technical report No 40

Waste and material flows

Assessment of information related to waste and material flows — a catalogue of methods and tools
A brief overview of available tools for assessing information relevant to waste and material flows is given in this technical report. Its objective is to help practitioners select the appropriate methodological tools for addressing the main policy questions related to waste and material flows.
Published on: 17 June 2003
Technical report No 96

Eionet workshop on waste and material flows
This report reflects the content of the 4th annual Eionet waste workshop held in Bratislava, Slovak Republic in October 2001. It covers the main presentations and the outcomes of the discussions held during the various sessions of the workshop, which was attended for the first time by representatives of the new EEA member countries.
Published on: 18 April 2002
Technical report No 63

Bibliography | Reports 2000–2005, by theme

Hazardous waste generation in EEA member countries
The purpose of this report is to review existing data on hazardous waste in Europe and assess their comparability. Drawing on data from 15 EEA countries and two regions, it uses the European Union's hazardous waste list as a common classification basis for comparing them.
Published on: 15 February 2002
Topic report No 14/2001

Waste — Annual topic update 2000
This report outlines the activities and achievements of the EEA's European Topic Centre on Waste (ETC/W) in 2000, its last year of operation before its mandate was widened to include material flows.
Published on: 3 October 2001
Topic report No 8/2001

Information about waste management facilities in EEA member countries
This study describes the present situation on available data and databases for waste management facilities in EEA countries based on a questionnaire inquiry. Strong and weak points are discussed and possible requirements for improvements are evaluated.
Published on: 26 October 2000
Technical report No 43

Household and municipal waste: Comparability of data in EEA member countries
The report investigates comparability between existing household and municipal waste data for the years 1993–1996 in EEA member countries except Liechtenstein.
Published on: 10 October 2000
Topic report No 3/2000

Waste — Annual topic update 1999
This report outlines the objectives and results of the EEA's European Topic Centre on Waste during 1999.
Published on: 28 August 2000
Topic report No 2/2000

Development and application of waste factors — an overview
The immediate objective of the report is to provide an overview on waste factors, their derivation and application and the experiences made, based on reports and literature available. This report has been prepared by the EEA's European Topic Centre on Waste (ETC/W) as part of the work programme of the EEA.
Published on: 28 August 2000
Technical report No 37

Dangerous substances in waste
This report focuses specifically on environmental risks associated with the main final disposal technologies, landfilling and incineration.
Published on: 14 February 2000
Technical report No 38

Bibliography

Water

Eurowaternet quantity — Technical guidelines for implementation
It is difficult to obtain a comparable overview of how much water is available in Europe due to huge differences in the way countries estimate their water resources. These guidelines propose a new approach to assessing water quantity, which will make national comparisons easier. They form part of Eurowaternet, the EEA's water monitoring and information network.
Published on: 8 September 2003
Technical report No 99

Eurowaternet: towards an index of quality of the national data in Waterbase
This report provides guidance to member countries on quality control procedures for Eurowaternet data which need to be taken into account at national and regional levels. The guidance focuses on quality control of sampling and analysis. It describes tests on the integrity of the data carried out by the EEA's European Topic Centre on Water before the data are uploaded to the Waterbase database. An attempt to develop a semi-quantifiable index of the quality of data held in Waterbase is also described.
Published on: 9 July 2003
Technical report No 98

Eurowaternet: Technical guidelines for implementation in transitional, coastal and marine waters
These technical guidelines detail the content and format of data required by the EEA from the marine conventions and national sources. The information collected on Europe's transitional, coastal and marine waters is used for the production of indicators in the EEA's core indicator set.
Published on: 7 July 2003
Technical report No 97

Testing of indicators for the marine and coastal environment in Europe — Part 3: Present state and development of indicators for eutrophication, hazardous substances, oil and ecological quality
This report describes the present state and development of policy-relevant indicators for eutrophication, hazardous substances, oil and ecological quality in the marine and coastal environment in Europe. The third of three volumes on development of coastal indicators, it focuses on pressure, state and impact indicators.
Published on: 4 July 2003
Technical report No 86

Testing of indicators for the marine and coastal environment in Europe — Part 2: Hazardous substances
This report, the second in a three-volume series on development of coastal indicators by EEA, explains the process and results of a study of potential state and pressure indicators for hazardous substances in coastal waters. Due to a lack of reliable and comparable European data, the testing mainly concentrated on the method for aggregating data on six substances monitored in the North East Atlantic Ocean. The report's primary aim is to test indicators for EEA reporting.
Published on: 5 February 2003
Technical report No 85

Bibliography | Reports 2000–2005, by theme

Testing of indicators for the marine and coastal environment in Europe. Part 1: Eutrophication and integrated coastal zone management

The development of coastal indicators by EEA in the period 1998–2001 is described in this report, the first of three volumes. It explains the process and results of a study of potential state and pressure indicators for eutrophication. The report also describes the development of, and information gathering for, a response indicator measuring progress in integrated coastal zone management at subnational (regional) level. The report's primary aim is to test indicators for EEA reporting and not to undertake assessments.

Published on: 19 December 2002

Technical report No 84

Remote sensing's contribution to evaluating eutrophication in marine and coastal waters

The aim of this report is to examine whether estimates of the plant pigment chlorophyll-a from satellite data can support the evaluation of eutrophication in European marine and coastal waters. Satellite images are a good indicator of the occurrence of algal blooms in relation to eutrophication but overestimate true chlorophyll-a concentrations in open sea areas by 60–70 % and even more in coastal waters. The report recommends using chlorophyll-a images from March to April/May as an indicator of eutrophication in the North Sea, Skagerrak and Kattegat since they reflect the magnitude of the spring bloom in these areas.

Published on: 13 December 2002

Technical report No 79

Marine and coastal environment — Annual topic update 2000

This report summarises the main activities carried out by the EEA's European Topic Centre on Marine and Coastal Environment in 2000, its final year of operation.

Published on: 14 November 2001

Topic report No 11/2001

Marinebase — Database on aggregated data for the coastline of the Mediterranean, Atlantic, North Sea, Skagerrak, Kattegat and Baltic

As a part of the work programme of the EEA, the Norwegian institute for water research (NIVA) has compiled a first version of a European marine database 'Marinebase'. This database on water quality parameters covers aggregated data related to eutrophication (nutrients, chlorophyll and oxygen in bottom water) and harmful substances (mainly metals in sediments and biota).

Published on: 6 August 2001

Technical report No 58

Inland waters — Annual topic update 2000

This report describes the work carried out by the EEA's European Topic Centre on Inland Waters in 2000 and highlights the progress made.

Published on: 29 June 2001

Topic report No 2/2001

Guidelines of the EC reporting obligations under the Barcelona convention and its protocols in force

The following report, carried out with the support of Garrigues & Andersen, is an in-depth study of the reporting obligations outlined in the framework of the Barcelona Convention on the protection of the Mediterranean Sea against pollution (Barcelona Convention). The report is one of three that the EEA has produced on reporting obligations of environmental conventions, as agreed with Directorate general environment of the European Commission (DG ENV).

Published on: 6 April 2001

Technical report No 45

Marine and coastal environment — Annual topic update 1999

This report presents a summary of the main activities carried out by the EEA's European Topic Centre on Marine and Coastal Environment (ETC/MCE) in 1999 including the development of the database Marinebase and thematic maps and indicators on eutrophication and hazardous substances in European coastal waters.

Published on: 22 January 2001

Topic report No 11/2000

Inland waters — Annual topic update 1999

This report describes the work carried out by the EEA's European Topic Centre on Inland Waters in 1999 and highlights the progress made.

Published on: 28 March 2000

Topic report No 1/2000

Pilot implementation Eurowaternet — Groundwater

Data and information for this report were provided by twelve countries to enable the Agency to successfully test the proposed Eurowaternet methodology for demonstrating the state and trends of groundwater in selected groundwater bodies with reference to nitrate, nitrite, ammonium and dissolved oxygen.

Published on: 28 March 2000

Technical report No 39

Agriculture

Towards agri-environmental indicators: integrating statistical and administrative data with land cover information.

This publication is the result of a close collaboration between four services of the European Commission — Directorates-General for agriculture, environment, Eurostat and the Joint research centre (Ispra) — and the EEA. It presents work carried out by the teams for integrating statistical and administrative data with land cover information. The different articles are a contribution to the development of agri-environment indicators using geo-referenced information.

Published on: 26 October 2001

Topic report No 6/2001

Calculation of nutrient surpluses from agricultural sources — Statistics spatialisation by means of Corine land cover

The assessment of potential losses of agricultural fertilisers to waters require prior calculation of nutrient balance (surplus). This study shows how the use of Corine land cover allows calculating comparable surplus values at the river catchment level, despite heterogeneity and high degree of aggregation of the available statistics for the two test basins (Loire, (F) and Elbe (D, CZ)).

Published on: 29 January 2001

Technical report No 51

Fisheries

An indicator-based approach to assessing the environmental performance of European marine fisheries and aquaculture

This scoping study forms the basis for the EEA's present and future development of indicators on fisheries and aquaculture. Based on a review of fisheries and aquaculture indicators developed by international and regional fisheries and environmental organisations, it presents a list of 52 potential candidate indicators. Twenty-nine of these are being recommended for inclusion in a core set of indicators on fisheries and aquaculture for coastal and marine as well as inland waters. A storyline links these indicators to each other and relates them to policy objectives.

Published on: 31 March 2003

Technical report No 87

Transport

National and central estimates for air emissions from road transport

This report evaluates emissions of greenhouse gases and air pollutants (CO_2, NO_x, VOCs) from road transport in the 15 EU Member States between 1981 and 1998 as well as emission projections to 2010/2020, based on information available at the end of 2000. It also compares different methods for preparing emission projections and proposes several actions to improve their quality. The report provides background information for transport emission indicators included in the EEA's publications *Environmental signals 2001* and *TERM 2001*.

Published on: 30 May 2002

Technical report No 74

Copert III Computer programme to calculate emissions from road transport — User manual

This report is designed in order to help Copert III users to produce in a short time a complete annual national emission data set from road transport (and off road machinery). The manual is divided in several chapters. The different chapters include all information needed to build a complete data set, assuming that the user has no former experience in using Copert III but he is quite familiar with the methodology and the terminology used.

Published on: 27 November 2000

Technical report No 50

Bibliography   PART D

Copert III Computer programme to calculate emissions from road transport — Methodology and emission factors
This report is the third update of the initial version for the Corinair 1985 emissions inventory (1989) and firstly updated in 1991 for the Corinair 1990 inventory (1993) and included in the *Atmospheric emission inventory guidebook (1996)*. The second update of the methodology (1997) was introduced in the software tool Copert II (1997) and a further update of the Guidebook was prepared.
Published on: 22 November 2000
Technical report No 49

Information systems

EEA core set of indicators — Guide
This guide provides information on the quality of the 37 indicators in the EEA core set. Its primary role is to support improved implementation of the core set in the EEA, EEA's European Topic Centres and the European environment information and observation network (Eionet). In parallel, it is aimed at helping users outside the EEA/Eionet system make best use of the indicators in their own work. It is hoped that the guide will promote cooperation on improving indicator methodologies and data quality as part of the wider process to streamline and improve environmental reporting in the EU and beyond.
Published on: 6 April 2005
Technical report No 1/2005

Development of common tools and an information infrastructure for the shared European environment information system
The elements of a new data collection network called Reportnet are described in this report, which reflects a shared concern among the major stakeholders that European environmental reporting needs to be revised. The elements described consist of developing of an environmental information infrastructure and a set of new services based on it, aimed at streamlining reporting as well as making it less burdensome.
Published on: 13 December 2002
Technical report No 83

Land and commerce registry as an instrument for sustainability
This 'experts corner' report examines how land and mercantile registries can be used as legal instruments for environmental protection. It comprises a set of coordinated papers analysing different aspects of the issue.
Published on: 2 October 2002
Technical report No 88

Implications of EEA/EU enlargement for state-of-the-environment reporting in the EU and EEA Member States
This report examines the implications for national and European state-of-the-environment reports of the forthcoming enlargement of the EU from 15 Member States to as many as 28 over the next decade. It results from discussions among state-of-the-environment reporting practitioners from over 30 European countries gathered in the EEA expert group on guidelines and state-of-the-environment reporting.
Published on: 27 September 2002
Technical report No 82

DAFIA II — further development of data-flow analysis for integrated assessments

This report describes the further development of data-flow analysis for integrated assessment (DAFIA) as a tool for supporting data delivery and data manipulation. DAFIA focuses on visualisation of data flows for structuring the coherence of core indicators. It will be applicable as a planning tool in integrated assessments and for documenting data used and results achieved.

Published on: 24 May 2002

Technical report No 72

The ShAIR scenario

Recent efforts to improve the integrated assessment approach to air pollution and greenhouse gases are documented in this report. Among other things, the study evaluates the methodologies used in the baseline scenario developed for the EEA's report *Environment in the European Union at the Turn of the Century* and presents revised projections for air pollution and CO_2 emissions.

Published on: 8 March 2002

Topic report No 12/2001

Data collection guidelines for the Kiev report

This technical report aims to provide guidance to all involved in data collection for the indicator-based state of the environment report that the EEA has been asked to provide for the next conference of environment ministers of all European countries, to be held in Kiev in May 2003. The report summarises current knowledge of the contents of international databases to be used for the Kiev report.

Published on: 13 November 2001

Technical report No 66

Participatory integrated assessment methods

This report gives a critical appraisal of participatory methods of Integrated environmental assessment (IEA) and evaluates the applicability of these methods to the EEA's work on IEA's. The report is a background document on using participatory integrated assessments in the context of preparations for the EEA's next report on the state of and outlook for the Pan-European environment, scheduled for publication in 2004/2005.

Published on: 13 September 2001

Technical report No 64

EEA support to the European Community in reporting obligations within the framework of international environmental conventions

This report provides the first compendium of the 64 international environmental conventions signed and ratified by the European Community as well as a 'road map' of the main reporting obligations arising from them. The report is intended to clarify the obligations and to contribute to efforts initiated by UNEP to streamline them and thus prevent 'reporting fatigue' among public authorities.

Published on: 9 July 2001

Technical report No 62

Guide to tools
This technical report summarises the software developments performed by the ETC/CDS from the year 1996 to the year 2000. The available instruments form a well tested toolbox for anybody who wants to register environmental information for the purpose of facilitating and improving the access to this information.
Published on: 13 February 2001
Technical report No 44

Establishment of a European green (and sustainable) chemistry award
Green and sustainable chemistry is a new concept and research area. The EEA has taken the initiative to investigate the possibilities to establish a European green and sustainable chemistry award. This report illustrates the contest and the steps to establish such award.
Published on: 5 February 2001
Technical report No 53

Questions to be answered by a state-of-the-environment report — The first list
This document is the final report of a study undertaken by the EEA together with its expert group on guidelines and the state-of-the-environment reporting in 1999–2000. It lists questions of policy makers and the general public that could be used to set up a state of the environment report or an environmental indicator report. It should be seen as a 'living' document that can be amended in the future through new versions, and certainly not as a definitive and universal one.
Published on: 11 October 2000
Technical report No 47

Condensed list of reports 2000–2005, by series

Environmental assessment reports (2000–2003)

3 (2000): Groundwater quality and quantity in Europe
4 (2000): Nutrients in European ecosystems
5 (2000): State and pressure of the marine and coastal Mediterranean environment
6 (2000): Environmental signals 2000
7 (2000): Sustainable use of Europe's water? State, prospects and issues
8 (2001): Environmental signals 2001
9 (2002): Environmental signals 2002 — Benchmarking the millennium
10 (2003): Europe's environment: the third assessment

Environmental issue reports (2000–2004)

9 (2000): Information tools for environmental policy under conditions of complexity
12 (2000): Are we moving in the right direction? Indicators on transport and environmental integration in the EU
14 (2001): Material flow-based indicators in environmental reporting
15 (2000): The dissemination of the results of environmental research
16 (2000): Down to earth: Soil degradation and sustainable development in Europe
17 (2000): Cloudy crystal balls: An assessment of recent European and global scenario studies and models
18 (2000): Environmental taxes — Recent developments in tools for integration
19 (2001): Sustainable water use in Europe — Part 2: Demand management
20 (2001): Environmental benchmarking for local authorities: From concept to practice
21 (2001): Sustainable water use in Europe — Part 3: Extreme hydrological events: floods and droughts
22 (2002): Late lessons from early warnings: the precautionary principle 1896–2000
23 (2001): TERM 2001 — Indicators tracking transport and environment integration in the European Union
24 (2001): Scenarios as tools for international environmental assessments
25 (2001): Reporting on environmental measures: Are we effective?
26 (2001): Designing effective assessments: The role of participation, science and governance, and focus
27 (2001): Renewable energies: success stories
28 (2002): Genetically modified organisms (GMOs): The significance of gene flow through pollen transfer
29 (2002): Children's health and environment: A review of evidence
30 (2002): Towards an urban atlas: Assessment of spatial data on 25 European cities and urban areas
31 (2002): Energy and environment in the European Union
32 (2002): Paving the way for EU enlargement — Indicators of transport and environment integration
33 (2002): Greenhouse gas emission trends and projections in Europe, 2002
35 (2004): Mapping the impacts of recent natural disasters and technological accidents in Europe
36 (2003): Greenhouse gas emission trends and projections in Europe, 2003
37 (2004): Agriculture and the environment in the EU accession countries
38 (2004): Arctic environment: European perspectives

Bibliography

EEA reports (2004–2005)

1/2004:	High nature value farmland — Characteristics, trends and policy challenges
2/2004:	Impacts of Europe's changing climate
3/2004:	Ten key transport and environment issues for policy-makers
5/2004:	Greenhouse gas emission trends and projections in Europe 2004
1/2005:	Climate change and a European low-carbon energy system
2/2005:	Effectiveness of urban wastewater treatment policies in selected countries: an EEA pilot study
3/2005:	Effectiveness of packaging waste management systems in selected countries: an EEA pilot study
4/2005	European Environment Outlook
5/2005:	Priority issues in the Mediterranean environment

EEA briefings (2003–2005)

1/2003:	Status of Europe's water
1/2004:	Halting the loss of biodiversity in Europe
2/2004:	Energy subsidies and renewables
3/2004:	Transport and environment in Europe
4/2004:	Transport biofuels: exploring links with the energy and agriculture sectors
1/2005:	Climate change and river flooding in Europe
2/2005:	How much biomass can Europe use without harming the environment?

Topic reports (2000–2003)

1/2000:	Inland Waters — Annual topic update 1999
2/2000:	Waste — Annual topic update 1999
3/2000:	Household and municipal waste: Comparability of data in EEA member countries
4/2000:	Land Cover — Annual topic update 1999
5/2000:	Nature Conservation — Annual topic update 1999
6/2000:	European Community and Member States greenhouse gas emission trends 1990–1998
7/2000:	Air Quality — Annual topic update 1999
8/2000:	Air Emissions — Annual topic update 1999
9/2000:	Emissions of atmospheric pollutants in Europe, 1980–1996
10/2000:	Air pollution by ozone in Europe in 1998 and summer 1999
11/2000:	Marine and Coastal Environment — Annual topic update 1999
1/2001:	Air pollution by ozone in Europe in 1999 and the summer of 2000
2/2001:	Inland Waters — Annual topic update 2000
3/2001:	Air quality in larger cities in the European Union
4/2001:	Land cover — Annual topic update 2000
5/2001:	Air Emissions — Annual topic update 2000
6/2001:	Towards agri-environmental indicators: integrating statistical and administrative data with land cover information
7/2001:	Eutrophication in Europe's coastal waters
8/2001:	Waste — Annual topic update 2000

PART D Bibliography | Condensed list of reports 2000 – 2005, by series

9/2001: Air Quality — Annual topic update 2000
10/2001: European Community and Member States greenhouse gas emission trends 1990–1999
11/2001: Marine and coastal environment — Annual topic update 2000
12/2001: The ShAIR scenario
13/2001: Air pollution by ozone in Europe in summer 2001
14/2001: Hazardous waste generation in EEA member countries
15/2001: Biodegradable municipal waste management in Europe
16/2001: Air quality in the Phare countries 1997
1/2002: Analysis and comparison of national and EU-wide projections of greenhouse gas emissions
2/2002: Case studies on waste minimisation practices in Europe
3/2002: Nature conservation Annual topic update 2000
4/2002: Air quality in Europe: state and trends 1990–1999
5/2002: Emissions of atmospheric pollutants in Europe, 1990–1999
6/2002: Air pollution by ozone in Europe in summer 2002
7/2002: Greenhouse gas emission trends in Europe, 1990–2000
1/2003: Europe's water: An indicator-based assessment
2/2003: Hazardous substances in the European marine environment — Trends in metals and persistent organic pollutants
3/2003: Air pollution by ozone in Europe in summer 2003
4/2003: Air pollution in Europe 1990–2000

Technical reports (2000–2003)

30 (2004): EMEP/Corinair Emission Inventory Guidebook. Third edition
34 (2000): The Global Biodiversity Information Facility — Architectural and Implementation Issues
37 (2000): Development and application of waste factors — an overview
38 (2000): Dangerous substances in waste
39 (2000): Pilot implementation Eurowaternet — Groundwater
40 (2000): Corine land cover technical guide — Addendum 2000
41 (2000): Annual European Community Greenhouse Gas Inventory 1990–1998
43 (2000): Information about waste management facilities in EEA member countries
44 (2001): Guide to Tools
45 (2001): Guidelines of the EC reporting obligations under the Barcelona convention and its protocols in force
46 (2001): The EU reporting obligations under the United Nations Framework Convention on Climate Change (UNFCCC) and the monitoring mechanism
47 (2000): Questions to be answered by a state-of-the-environment report — The first list
48 (2000): Evaluation of the Air Quality Model Documentation System
49 (2000): Copert III Computer programme to calculate emissions from road transport — Methodology and emission factors
50 (2000): Copert III Computer programme to calculate emissions from road transport — User manual
51 (2001): Calculation of nutrient surpluses from agricultural sources — Statistics spatialisation by means of Corine land cover
52 (2000): Annual European Community CLRTAP emission inventory 1980–1998
53 (2001): Establishment of a European Green (and Sustainable) Chemistry Award
54 (2001): Business and the environment: current trends and developments in corporate reporting and ranking
55 (2001): Total material requirement of the European Union
56 (2001): Total material requirement of the European Union — Technical part

Bibliography

57 (2001): Eionet workshop on air quality monitoring and assessment
58 (2001): Marinebase — Database on aggregated data for the coastline of the Mediterranean, Atlantic, North Sea, Skagerrak, Kattegat and Baltic
59 (2001): Towards spatial and territorial indicators using land cover data
60 (2001): Annual European Community Greenhouse Gas Inventory 1990–1999
61 (2001): Proposal for a European soil monitoring and assessment framework
62 (2001): EEA support to the European Community in reporting obligations within the framework of international environmental conventions
63 (2002): Eionet workshop on waste and material flows
64 (2001): Participatory integrated assessment methods
65 (2001): Waste management facilities — Electronic catalogue
66 (2001): Data collection guidelines for the Kiev report
67 (2001): European soil monitoring and assessment framework
68 (2001): Road freight transport and the environment in mountainous areas
69 (2002): Review of selected waste streams
70 (2002): Towards a global biological information infrastructure
71 (2002): The rivers of the Black Sea
72 (2002): DAFIA II — further development of data-flow analysis for integrated assessments
73 (2002): Annual European Community CLRTAP emission inventory 1990–1999
74 (2002): National and central estimates for air emissions from road transport
75 (2002): Annual European Community Greenhouse Gas Inventory 1990–2000 and Inventory Report 2002
76 (2002): Second technical workshop on contaminated sites
77 (2003): Greenhouse gas emission — projections for Europe
78 (2002): Eionet technical workshop on indicators for soil contamination
79 (2002): Remote sensing's contribution to evaluating eutrophication in marine and coastal waters
80 (2002): Proceedings of the Technical Workshop on Indicators for Soil Sealing
81 (2002): Assessment of data needs and data availability for the development of indicators on soil contamination
82 (2002): Implications of EEA/EU enlargement for state-of-the-environment reporting in the EU and EEA Member States
83 (2002): Development of common tools and an information infrastructure for the shared European environment information system
84 (2002): Testing of indicators for the marine and coastal environment in Europe. Part 1: Eutrophication and integrated coastal zone management
85 (2003): Testing of indicators for the marine and coastal environment in Europe — Part 2: Hazardous substances
86 (2003): Testing of indicators for the marine and coastal environment in Europe — Part 3: Present state and development of indicators for eutrophication, hazardous substances, oil and ecological quality
87 (2003): An indicator-based approach to assessing the environmental performance of European marine fisheries and aquaculture
88 (2003): Land and commerce registry as an instrument for sustainability
89 (2003): Corine land cover update 2000: Technical guidelines
90 (2003): Euroairnet — Status report 2000
91 (2001): Annual European Community CLRTAP emission inventory 1990–2000
92 (2004): An inventory of biodiversity indicators in Europe, 2002
93 (2004): Exploring the ancillary benefits of the Kyoto Protocol for air pollution in Europe
94 (2003): Assessment and reporting on soil erosion
95 (2003): Annual European Community greenhouse gas inventory 1990–2001 and inventory report 2003
96 (2003): Assessment of information related to waste and material flows — a catalogue of methods and tools

97 (2003):   *Eurowaternet: Technical guidelines for implementation in transitional, coastal and marine waters*
98 (2003):   *Eurowaternet: towards an index of quality of the national data in Waterbase*
99 (2003):   *Eurowaternet Quantity — Technical guidelines for implementation*

EEA technical reports (2004–2005)

1/2004:   *Energy subsidies in the European Union: A brief overview*
2/2004:   *Annual European Community greenhouse gas inventory 1990–2002 and inventory report 2004*
3/2004:   *Transport price signals*
4/2004:   *Analysis of greenhouse gas emission trends and projections in Europe 2003*
5/2004:   *Air pollution and climate change policies in Europe: exploring linkages and the added value of an integrated approach*
6/2004:   *Annual European Community CLRTAP emission inventory 1990–2002*
7/2004:   *Analysis of Greenhouse gas emission trends and projections in Europe 2004*
1/2005:   *EEA core set of indicators — Guide*
2/2005:   *Environmental policy integration in Europe — State of play and an evaluation framework*
3/2005:   *Air pollution by ozone in Europe in summer 2004*
4/2005:   *Annual European Community greenhouse gas inventory 1990–2003 and inventory report 2005*
5/2005:   *Environmental policy integration in Europe — Administrative culture and practices*

European Environment Agency

The European environment — State and outlook 2005

2005 — 576 pp. — 21 x 29.7 cm
ISBN 92-9167-776-0

SALES AND SUBSCRIPTIONS

Publications for sale produced by the Office for Official Publications of the European Communities are available from our sales agents throughout the world.

How do I set about obtaining a publication?
Once you have obtained the list of sales agents, contact the sales agent of your choice and place your order.

How do I obtain the list of sales agents?
- Go to the Publications Office website http://publications.eu.int/
- Or apply for a paper copy by fax +352 2929 42758